TAILORED ORGANIC–INORGANIC MATERIALS

TAILORED ORGANIC–INORGANIC MATERIALS

Edited by

**ERNESTO BRUNET, JORGE L. COLÓN
AND ABRAHAM CLEARFIELD**

Published by John Wiley & Sons, Inc., Hoboken, New Jersey

Published simultaneously in Canada

For general information on our other products and services or for technical support, please contact our Customer Care Department within the United States at (800) 762-2974, outside the United States at (317) 572-3993 or fax (317) 572-4002.

Wiley also publishes its books in a variety of electronic formats. Some content that appears in print may not be available in electronic formats. For more information about Wiley products, visit our web site at www.wiley.com.

Library of Congress Cataloging-in-Publication Data:

Tailored organic-inorganic materials / edited by Ernesto Brunet, Jorge L. Colón, Abraham Clearfield.
    pages   cm
  Includes bibliographical references and index.
  ISBN 978-1-118-77346-8 (hardback)
1. Laminated materials.   2. Nanocomposites (Materials)   3. Membranes (Technology)
4. Chemistry, Inorganic.   5. Organic compounds–Synthesis.   6. Surface chemistry.
I. Brunet, Ernesto.   II. Colón, Jorge L.   III. Clearfield, Abraham.
  TA418.9.L3T345 2014
  620.1'18–dc23

                                                                        2014032137

Printed in the United States of America

10   9   8   7   6   5   4   3   2   1

CONTENTS

LIST OF CONTRIBUTORS

Pilar Aranda Instituto de Ciencia de Materiales de Madrid, ICMM-CSIC, Cantoblanco, Madrid, Spain

Ernesto Brunet Departamento de Química Orgánica, Facultad de Ciencias, Universidad Autónoma de Madrid, Madrid, Spain

Bruno Bujoli CNRS, CEISAM UMR 6230, University of Nantes, Nantes Cedex, France

Aurelio Cabeza Departamento de Química Inorgánica, Cristalografía y Mineralogía, Universidad de Málaga, Málaga, Spain

Barbara Casañas Department of Chemistry, University of Puerto Rico-Río Piedras Campus, San Juan, Puerto Rico

Abraham Clearfield Department of Chemistry, Texas A&M University, College Station, TX, USA

Rosario M. P. Colodrero Departamento de Química Inorgánica, Cristalografía y Mineralogía, Universidad de Málaga, Málaga, Spain

Jorge L. Colón Department of Chemistry, University of Puerto Rico-Río Piedras Campus, San Juan, Puerto Rico

Ferdinando Costantino Dipartimento di Chimica, Biologia e Biotecnologie, University of Perugia, Perugia, Italy and CNR – ICCOM, Sesto Fiorentino, Firenze, Italy

Margarita Darder Instituto de Ciencia de Materiales de Madrid, ICMM-CSIC, Cantoblanco, Madrid, Spain

Konstantinos D. Demadis Crystal Engineering, Growth and Design Laboratory, Department of Chemistry, University of Crete, Crete, Greece

María de Victoria-Rodríguez Departamento de Química Orgánica, Facultad de Ciencias, Universidad Autónoma de Madrid, Madrid, Spain

Agustin Diaz Department of Chemistry, Texas A&M University, College Station, TX, USA

Laura Jiménez García-Patrón Departamento de Química Orgánica, Facultad de Ciencias, Universidad Autónoma de Madrid, Madrid, Spain

Hussein Hindawi Departamento de Química Orgánica, Facultad de Ciencias, Universidad Autónoma de Madrid, Madrid, Spain

Gary B. Hix School of Science and Technology, Nottingham Trent University, Nottingham, United Kingdom

Andrea Ienco CNR – ICCOM, Sesto Fiorentino, Firenze, Italy

Paul-Alain Jaffrès CEMCA, CNRS UMR 6521, Université Européenne de Bretagne, Université de Brest, Brest, France

Olga Juanes Departamento de Química Orgánica, Facultad de Ciencias, Universidad Autónoma de Madrid, Madrid, Spain

Giuseppe Leone CNR-ISMAC, Istituto per lo studio delle Macromolecole, Milano, Italy

Pascual Olivera-Pastor Departamento de Química Inorgánica, Cristalografía y Mineralogía, Universidad de Málaga, Málaga, Spain

Konstantinos E. Papathanasiou Crystal Engineering, Growth and Design Laboratory, Department of Chemistry, University of Crete, Crete, Greece

Clémence Queffelec CNRS, CEISAM UMR 6230, University of Nantes, Nantes Cedex, France

Giovanni Ricci CNR-ISMAC, Istituto per lo studio delle Macromolecole, Milano, Italy

Elena Rodríguez-Payán Departamento de Química Orgánica, Facultad de Ciencias, Universidad Autónoma de Madrid, Madrid, Spain

Juan Carlos Rodríguez-Ubis Departamento de Química Orgánica, Facultad de Ciencias, Universidad Autónoma de Madrid, Madrid, Spain

Jean-Michel Rueff ENSICAEN, CNRS UMR 6508, Laboratoire CRISMAT, Caen, France

Eduardo Ruiz-Hitzky Instituto de Ciencia de Materiales de Madrid, ICMM-CSIC, Cantoblanco, Madrid, Spain

Marco Taddei Dipartimento di Chimica, Biologia e Biotecnologie, University of Perugia, Perugia, Italy

PREFACE

The idea for this book was formulated at the International Union of Pure and Applied Chemistry (IUPAC) conference held in Puerto Rico, more precisely at a symposium on *Layered Materials* that was a significant part of it. The symposium was the outcome of Co-editor – and Professor at the University of Puerto Rico – Jorge Colon's persuasive talks with his former mentor, Prof. Abraham Clearfield, in order to convince the latter to organize a scientific meeting on the *Layered Materials* topic. The symposium was a complete success and featured an array of excellent presentations and debates by an outstanding number of exceptional speakers. Yet, unbeknownst to the organizer and participants, a representative from Wiley was in the audience. At a certain point of the conference, she made contact with the organizer and Profs. Ernesto Brunet (*Autonomous University of Madrid*, Spain) and Jorge Colón as well, expressing that 'Wiley would be highly interested in publishing a book along the lines presented in the symposium.' She also pointed out that our possible co-edition from three different parts of the globe (Texas, the Caribbean and Spain) would be a plus in the achievement of the book. The three of us finally agreed and put the show on the road under the ambitious, broad title of 'Tailored Organic–Inorganic Materials'. This is the result of it.

The honours of Chapter 1 titled 'Zirconium Phosphate Nanoparticles and Their Extraordinary Properties' unanimously corresponded to Prof. Abraham Clearfield (with A. Diaz of the Chemistry Department at Texas A&M University) due to his gigantic contribution to the field of *Layered Materials* and many others. He humorously complains that despite the intended broad book scope, 'I can't seem to escape from metal Phosphonate Chemistry'. Nevertheless, his early discovery of crystalline alpha-zirconium phosphate (α-ZrP) and of its structure and his development of multitude of applications make him, no doubt, the most prominent author to explain the foundations and the most recent discoveries of this area. The latest functionalization of the surfaces of α-ZrP, as well as novel drug delivery processes, is described. The potential of the use of α-ZrP as nanoparticles with functionalized surfaces opens up the possibility of many more applications in polymer composites, gels and surfactants.

Chapter 2, 'Tales from the Unexpected: Chemistry at the Surface and Interlayer Space of Layered Organic–Inorganic Hybrid Materials Based on γ-Zirconium Phosphate', is by Co-editor Prof. Ernesto Brunet from the Autonomous University of Madrid and some of his colleagues, from which Prof. Hussein Hindawi from Al-Azhar University of Gaza stands out. As the authors point out, the confinement of organic molecules within layered inorganic salts by either topotactic exchange or intercalation allows their structure to be modified in many ways and within a reasonable range of predictability. This confinement leads to the production of multifunctional composites with unusual properties that are difficult to achieve otherwise. Reading this chapter will indeed reveal to you the unexpected.

Chapter 3, authored by K. E. Papathanasiou and K. D. Demadis from the University of Crete, is titled 'Phosphonates in Matrices' and treats the acid–base chemistry of phosphonic acids and the very many applications based on this knowledge. This is a much needed topic because of the greater complexity of phosphonic acid groups and the effects of the ligands to which they are bonded.

Chapter 4, authored by A. Cabeza, P. Olvera-Pastor and R. M. P. Colodrero, University of Malaga, Spain, is titled 'Hybrid Materials Based on Multifunctional Phosphonic Acids'. It is an extremely thorough treatment of the subject with 274 references. Almost any moiety can affix phosphonic acid groups; and furthermore, additional functional groups such as amine, hydroxyl, ether, carboxylate, and so on may form part of the attachment. Also, the ligand may have di-, tri-, tetra- or more phosphonic acid groups. All of these factors influence the acid–base character of the final product and their behaviour in synthetic procedures. This chapter goes a long way in understanding this bewildering array of phosphonic acids and their applications.

Chapter 5, authored by Prof. Ferdinando Costantino *et al.*, *University of Perugia and CNR*, Italy, is titled 'Hybrid Multifunctional Materials Based on Phosphonates, Phosphinates and Auxiliary Ligands'. Incidentally, two of us (Profs. Clearfield and Brunet) had had a fruitful association with Profs. Julio Alberti and Umberto Costantino of the *University of Perugia*. It is refreshing to see how the younger generation has embraced the succession of their illustrious forbearers. This chapter illustrates the difficulties and rewards of using phosphonic acids to prepare coordination polymers or metal–organic frameworks (MOFs). They also describe the differences in using phosphonate ligands as well as a second ligand such as aza-heterocycles and carboxylates of which very little is known. In addition, they point out that phosphinates display high thermal stability and, depending on the type of coordinating metal, have interesting magnetic and optical properties.

Chapter 6, authored by E. Ruiz-Hitzky, P. Aranda and M. Darder from the *Institute of Materials Science of Madrid*, Spain, presents 'Hybrid and Biohybrid Materials Based on Layered Clays'. This chapter is a very thorough treatment of the topic with 334 references. We were fascinated with the huge array of compounds that could be intercalated within the clay mineral structures and the enormous range of their applications. Many clay minerals are abundant, can be exfoliated and present active surfaces. Let your imagination visualize what can be done with such materials.

Chapter 7, 'Fine-Tuning the Functionality of Inorganic Surfaces Using Phosphonate Chemistry', authored by B. Bujoli and C. Queffelec from the University of Nantes, follows a very fine review by the same authors and their associates in *Chemical Reviews*. The authors address the surface modification of zirconium phosphate (ZrP) without specifying whether alpha, gamma, theta and so on are used. The authors refer to the major challenges of designing biomaterial surfaces on ZrP for many applications such as bio-sensors, biomedical devices, catalysts, biomedical implants and many other possibilities.

Chapter 8, titled 'Photofunctional Polymer/Layered Silicate Hybrids by Intercalation and Polymerization Chemistry', was chosen as the subject to explore by G. Leone and G. Ricci of the Institute of Macromolecular Studies of Milan, Italy. Polymer composites based on micrometre-sized particles resulted in an enormous transformation in the chemical design, engineering and performance of structured materials. More recently, the use of nanometre-sized particles such as platelets, fibres and tubes further advanced this field of endeavour. This chapter deals with the preparation of photofunctional layered silicate-based hybrids. The interweaving of the advancements in nanometre particle developments to this particular specialty is fascinating, specifically π-conjugated molecules and polymers for application in π-conjugated LEDs and polymer light-emitting devices (PLEDs).

The last chapter dealing with phosphonic acids is Chapter 9, 'Rigid Phosphonic Acids as Building Blocks for Crystalline Hybrid Materials', authored by J-M. Rueff *et al.*, *University of Caen*, France. In this chapter, the authors discuss hybrid structures in which the functional groups (largely $-PO_3H_2$) are bonded directly to heteroaromatic rings. A significant effort is devoted to methods of synthesis of these materials including hybrids obtained from polyphosphonic acids and heterofunctional precursors. The difficulty in predicting the outcome of the synthetic reactions of phosphonic acids relative to carboxylic acids is treated in some detail.

Last but not least, Chapter 10 is presented by Co-editor Jorge L. Colón and B. Casañas (Department of Chemistry, University of Puerto Rico, San Juan, PR) and is titled 'Drug Carriers Based on Zirconium Phosphate Nanoparticles'. This chapter describes the intercalation of a number of anticancer drugs and insulin between the layers of θ-ZrP. The resultant structures and release of the drugs in cancer cells are thoroughly treated. Significant progress has now been made in trials with cancer cells and in the diversity of therapeutic agents that can be intercalated.

Some final words just to say that the authors of this volume have long laboured in the topics described herein. It is therefore pleasing to see how the pursuit of metal phosphonate chemistry has become embraced worldwide and at an advanced level of comprehension. This is particularly important because of the difficulty of predictions of synthetic outcomes because of the complexity of the phosphonic acids. The chapters in this book provide details on the effect of functionalization and form of the phosphonic acids in the final outcome of the hybrid. Additional systematic studies should finally lead to modicum of predictability in metal phosphonate chemistry.

Of no less importance are the many variations in behaviour of *layered materials* as to intercalation, particle size, surface functionalization and layer composition. At present, the field is in flux. An enormous effort has been expended on silica with great rewards. However, each layered material has its own properties and behaviour. No doubt, further studies, aside from graphene, will also bring about rich rewards.

ERNESTO BRUNET, JORGE L. COLÓN AND ABRAHAM CLEARFIELD
Co-editors

1

ZIRCONIUM PHOSPHATE NANOPARTICLES AND THEIR EXTRAORDINARY PROPERTIES

ABRAHAM CLEARFIELD AND AGUSTIN DIAZ

Department of Chemistry, Texas A&M University, College Station, TX, USA

1.1 INTRODUCTION

The first report of a crystalline form of zirconium phosphate was in 1964. Up to that time, only an amorphous white fine powder was known. The transformation from the amorphous to crystalline is a slow process. It is therefore possible to control the size of the particles from very small, approximately 50 nm, to micro size and to large crystals. These particles are layered and exhibit the ability to exchange positively charged species for protons, to undergo intercalation behaviour and exfoliation of the layers. In addition, it has been shown that the surface of the particles may be functionalized by bonding to silanes, isocyanates and epoxides. By first replacing the surface protons by Zr^{4+} or Sn^{4+}, bonding may be extended to include phosphates and phosphonic acids. Attachment of a functional group to the surface bonding ligands including the phosphates or phosphonic acids allows this large class of functionalized molecules to be utilized for a variety of applications. Because of the extraordinary properties of this compound, a great variety of potential and realized uses have been invoked. As a result, from 1964 to the present, more than 10,000 scientific papers have been published describing the chemistry and applications of this remarkable compound. This phenomenon continues as every year a few hundred new papers appear in the chemical literature. Among the many uses in addition to ion exchange are catalysis, polymer composites, proton conduction, drug delivery and many more, as will be described in this chapter.

Tailored Organic–Inorganic Materials, First Edition. Edited by Ernesto Brunet, Jorge L. Colón and Abraham Clearfield.
© 2015 John Wiley & Sons, Inc. Published 2015 by John Wiley & Sons, Inc.

1.2 SYNTHESIS AND CRYSTAL STRUCTURE OF α-ZIRCONIUM PHOSPHATE

We shall begin by describing the synthesis and structure of α-zirconium phosphate (α-ZrP), $Zr(O_3POH)_2 \cdot H_2O$. The addition of phosphoric acid to a soluble zirconium salt results in the precipitation of an amorphous white solid. This solid was observed to incorporate ions that may be in the solution. Interest in zirconium phosphate was keyed by the advent of nuclear energy. In swimming pool reactors, ionic species formed in the wastewater. These ions needed to be removed before the water could be reused. Because of the high temperature of the water and the radioactivity of the ions, organic resins were unsuitable for this purpose. It was felt that inorganic materials would serve the purpose and much work was concentrated on the amorphous zirconium phosphate [1, 2]. Unfortunately, the hot water hydrolysed the phosphate to hydroxide.

Our initial effort involved crystallizing the amorphous powder by adding excess phosphoric acid and heating the mix under reflux [3]. The single crystals for structure determination were prepared in 9 M H_3PO_4 in sealed tubes at 180°C. The availability of the single crystals resulted in the determination of their structure [4, 5].

α-ZrP is a layered compound as shown in Figure 1.1 and has the composition $Zr(O_3POH)_2 \cdot H_2O$ and an interlayer distance of 7.56 Å. The zirconium ions are arranged at the corners of a parallelogram with alternate Zr ions above and below the mean plane of the layer. The phosphate groups sit alternately above and below the mean plane of the layer with three oxygen atoms of the phosphate group bonding to three of the Zr^{4+} ions forming a triangle in half of the parallelogram. The Zr^{4+} are six coordinate with oxygen contributions from six phosphate groups in adjacent

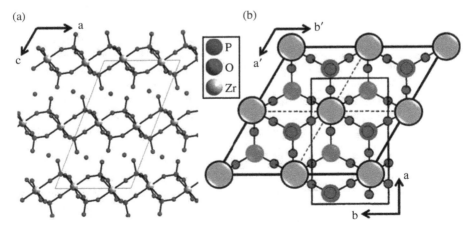

FIGURE 1.1 Schematic representation of α-ZrP viewed along the *b*-axis showing its unit cell and the formed layered structure (a) and along the *c*-axis (b) showing the surface of the layers. The view along the *c*-axis (b) is showing the relationship of the pseudo-hexagonal cell (black rectangle) to the true, monoclinic unit cell. The hydrogen atoms were omitted for clarity (Taken from: *Chem. Mater.* 2013, *25*, 723–728; DOI: 10.1021/cm303610v).

parallelograms. The P–OH groups form a double layer in the interlayer space. We shall refer to this compound as α-ZrP.

Before discussing any applications, some additional information on crystal growth and formation of the crystals is required. The solubility of the amorphous powder increases with an increase in concentration of phosphoric acid [6]. Thus, it is possible to control the growth of particles of α-ZrP by varying the concentration of the acid. In this way, we are able to grow nanoparticles of less than 100 nm through up to 4 µm-sized particles and single crystals [7]. In fact, at present, particles of about 50 nm in size and only a few layers thick have been obtained [8]. Table 1.1 shows the various ways that were used to control particle size of the zirconium phosphate by varying the concentration of phosphoric acid, use of a hydrothermal technique at temperatures from 120 to 200°C or addition of HF as a solubilizing agent [7].

Of course, there is also time and temperature control to consider. The HF forms a complex ZrF_6^{2-} that releases Zr^{4+} slowly at temperatures above 60°C. The decreased yield in the HF method results from increased solubility of the ZrF_6^{2-} ions. An example of crystal growth is shown by the electron micrograph patterns in Figure 1.2. The numbers indicate the concentration of phosphoric acid and the reflux time.

TABLE 1.1 The yield and typical particle length of α-zrp samples prepared

Sample	Yield (%)	Typical particle length (nm)
ZrP(3 M)[a]	96.4	50–100
ZrP(6 M)[a]	87.3	100–200
ZrP(9 M)[a]	85.8	100–200
ZrP(12 M)[a]	86.6	150–300
ZrP(HT3 M-200-5)[b]	85.6	100–200
ZrP(HT6 M-200-5)[b]	90.7	150–250
ZrP(HT9 M-200-5)[b]	98.0	150–250
ZrP(HT12 M-200-5)[b]	97.3	200–400
ZrP(HT3 M-200-24)[c]	89.3	300–500
ZrP(HT6 M-200-24)[c]	97.1	600–800
ZrP(HT9 M-200-24)[c]	96.0	800–1000
ZrP(HT12 M-200-24)[c]	93.7	1000–1200
ZrP(HF1)[d]	83.5	1000–2000
ZrP(HF2)[d]	72.0	1000–3000
ZrP(HF3)[d]	53.5	1500–3500
ZrP(HF4)[d]	41.8	2000–4000

Taken from *New J. Chem.*, 2007, 31, 39–43; DOI: 10.1039/B604054C.
[a]Refluxing method: 3.0/6.0/9.0/12.0 M H_3PO_4 at 100°C for 24 h. ZrP([H_3PO_4]).
[b]Hydrothermal method: 3.0/6.0/9.0/12.0 M H_3PO_4 sealed into a Teflon®-lined pressure vessel and heated at 200°C for 5 h. ZrP(HT[H_3PO_4]-temperature-time(h)).
[c]Hydrothermal method: 3.0/6.0/9.0/12.0 M H_3PO_4 sealed into a Teflon-lined pressure vessel and heated at 200°C for 24 h. ZrP(HT[H_3PO_4]-temperature-time(h)).
[d]Hydrofluoric acid method: HF solution (5.0 M) at molar ratio of F^-/Zr^{4+} = 1, 2, 3 and 4 refluxed at 100°C for 24 h. ZrP(HF molar ratio of F^-/Zr^{4+}).

Reflux method

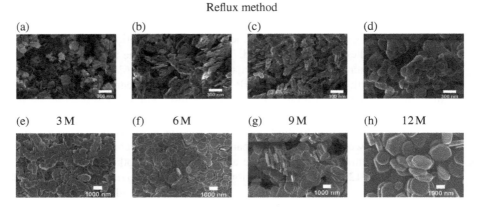

Hydrothermal method

FIGURE 1.2 SEM images of α-ZrP particles prepared by refluxing the amorphous precipitated particles (top) and hydrothermally (bottom) for 24 h in increasing concentrations of H_3PO_4 (a and e) 3 M, (b and f) 6 M, (c and g) 9 M and (d and h) 12 M (Taken from: *New J. Chem.* 2007, *31*, 39–43; DOI: 10.1039/B604054C).

Camino Trobajo et al. observed a new phenomenon dealing with the synthesis of α-ZrP [9]. In their preparation of amorphous zirconium phosphate, they added a solution of zirconyl chloride in 2 M HCl to a solution of 1.25 M H_3PO_4 with constant stirring. The white solid that was obtained was washed with dilute phosphoric acid and dried in air. To their surprise, the X-ray pattern was that of crystalline α-ZrP. However, when they followed the method described by Clearfield and Stynes [3], they did obtain the amorphous solid. What Trobajo et al. found was that in their preparations there was always a significant amount of phosphoric acid present in the solid and this may have keyed the conversion to crystals. We shall return to this point again.

It is now necessary to describe several other features of the α-ZrP crystals that relate to their usefulness [10–12]:

1. The particles are ion exchangers in which cations readily replace the protons

$$Na^+ + Zr(HPO_4)_2 \cdot H_2O \leftrightarrows Zr(NaPO_4)(HPO_4) \cdot xH_2O + H^+$$

2. The layers can intercalate molecules by means of acid–base reactions

$$RNH_2 + Zr(HPO_4)_2 \cdot H_2O \leftrightarrows Zr(RNH_3PO_4)(HPO_4) \cdot xH_2O$$

3. Solid–solid exchange [13–15]

4. $Zr(HPO_4)_2 \cdot H_2O + 2/z\ MCl_z \xrightarrow[1\,h]{100°C} Zr(M_{2/x})(PO_4)_2 + 2HCl \uparrow + H_2O$

Layered compound   Colloidal suspension

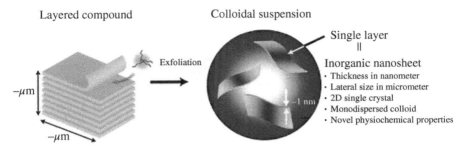

Exfoliation

$-\mu m$

$-\mu m$

Single layer
‖
Inorganic nanosheet
· Thickness in nanometer
· Lateral size in micrometer
· 2D single crystal
· Monodispersed colloid
· Novel physiochemical properties

$-1\ nm$

FIGURE 1.3 Schematic model illustrating the exfoliation of a layered compound into colloidal nano-sheets (Taken from: *Adv. Mater.* 2010, *22*, 5082–5104; DOI: 10.1002/adma.201001722).

5. The driving force here is the removal of HCl at elevated temperature. For example,

$$2LiCl + Zr(HPO_4)\cdot H_2O \xrightarrow[1\,h]{100^\circ C} Zr(LiPO_4) + 2HCl\uparrow + H_2O$$

The reverse reaction is also easily done, for example, treatment of the metal zirconium phosphate with gaseous HCl at 120°C. Similar reactions have been carried out with zeolites [16]. A summary of intercalation reactions of α-ZrP has been published [17]. Shortly after this publication, Mallouk et al. [11] examined the intercalation/exfoliation reactions of α-ZrP by use of atomic force microscopy (AFM) and transition electron microscopy (TEM). They utilized tetra(*n*-butyl)ammonium hydroxide (TBA+OH−) as the exfoliant (Figure 1.3). The rate-determining step was found to be the opening of the interlamellar space increasing this space. Then rapid diffusion of TBA+ ions into the galleries occurs. A hydrolysis reaction occurs around the edges forming 4 nm hydrated zirconia particles around the layer edges. This reaction introduces phosphate ion into the solution and creates an equilibrium state. The hydrolysis may be suppressed by operating at 0°C.

Another way to exfoliate the layers is to titrate the crystals with propylamine [18, 19], using only one mole of amine per mole of α-ZrP. The process is slow enough that the amines arrange themselves at every other P–OH group. This places the propylammonium ions 10.6 Å apart and moves the layers so that the interlayer distance is more than 10 Å. Water can now flood the interlayer space and exfoliate the layers. Smaller amines also exfoliate the layers, but butylamine does not. Rather, it forms a series of phases as the amount intercalated increases. That is the case for amines with larger alkyl chains [20].

1.3 ZIRCONIUM PHOSPHATE-BASED DIALYSIS PROCESS

Given all these attributes of the crystalline particles, it is interesting that the first commercial application was for the amorphous ZrP. A group at the NIH wished to design a portable kidney dialysis system. People who suffer from renal problems

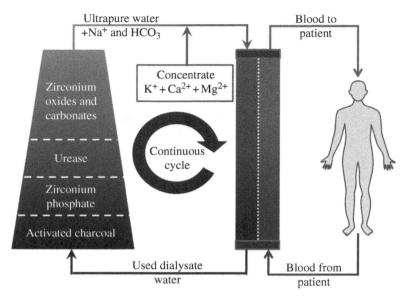

FIGURE 1.4 Schematic representation of the dialysis process.

need to undergo dialysis to remove all the toxins that accumulate in their system. This requires the use of an artificial kidney membrane with a flow of water across the membrane to wash the toxins down the drain. Being hooked up to this apparatus for several hours with a loud pump powering 100–250 gallons of water across the membrane was not an event to look forward to, as well as waiting for a room to become available and turning yellow while sitting in the hospital waiting room.

In the portable unit, the wash water would have to be recycled so sorbents were required. It turned out that activated charcoal could remove all of the toxins except urea. To remove the urea, after testing many sorbents, they fixed on putting the enzyme urease on amorphous zirconium phosphate. The urea was converted to NH_4CO_3 that then reacted with ZrP as in

$$\left(NH_4\right)_2 CO_3 + Zr\left(HPO_4\right)_2 \cdot nH_2O \rightarrow Zr\left(NH_4PO_4\right)_2 \cdot nH_2O + CO_2 + H_2O$$

In this process, it is necessary that the dialysate be slightly basic. This results in some hydrolysis of the zirconium phosphate accompanied by release of a small amount phosphate that is prevented from entering the dialysate by sorption with a layer of hydrous zirconia placed below the layer of ZrP (Figure 1.4). With purchase of this unit, the process of dialysis could be carried out in your own dwelling administered by a trained family member. All the systems were miniaturized so that the unit weighed about 60 lbs and operated quietly and with temperature control for the patients' comfort. Many hospitals use these units rather than the older method [21].

1.4   ZrP TITRATION CURVES

You might wonder why the amorphous ZrP rather than the crystalline form was used and that gives me a chance to describe this ZrP system in more detail. If we examine the titration curves (NaOH/NaCl) of the amorphous and crystalline samples shown in Figure 1.5, the curves are entirely different [22]. Note that with each increase in the addition of sodium ion, the pH increases for the gel. This is similar to polymer type ion exchangers where the Na^+ is distributed equally throughout the solid. In contrast, it turns out that with the fully crystalline solid, the sodium ion enters from the edges [10] and immediately forms a new phase, $Zr(O_3PONa)(O_3POH)\cdot5H_2O$. This means that there are two solid phases within the same crystal. The phase rule can explain this behaviour:

$$f = C - P + 2$$

where f = the degrees of freedom, C = number of components and P = the number of phases. In ion exchange with the crystalline phase, there are two solid phases and the solution phase, the components are also three, one choice being the sodium ion concentration of the solid phase and of the solution phase and the solution hydrogen ion concentration. Since the pressure and temperature are constant, f = O. As a result, the exchange, in contrast to the gel, takes place at constant pH and constant sodium ion concentration (Figure 1.5) [22]. Thus, all sodium ion added at constant pH is taken up by the solid particles, and the reaction is

$$Zr(O_3POH)_2 \cdot H_2O + Na^+ + 4H_2O \rightarrow Zr(O_3PONa)(O_3POH)\cdot5H_2O$$

When all the α-ZrP is converted to the half sodium ion phase, the pH rises to that of the new phase and a second reaction at a new constant pH takes place:

$$Zr(O_3PONa)(O_3POH)\cdot5H_2O + Na^+ \rightarrow Zr(O_3PONa)_2 \cdot 3H_2O + H^+ + 2H_2O$$

The liberated H^+ is neutralized by the base from NaOH addition. The interlayer spacing for the half Na^+ ion phase is 11.8 Å and for the fully exchanged phase 9.8 Å.

An interesting fact is that as the gel is treated to produce crystals, the crystallinity can be changed very slowly. This is illustrated in Figure 1.6. It is seen that the gel has no discernible X-ray pattern [23]. However, upon refluxing at 48 h in 0.5 M H_3PO_4 (0.5 : 48), several very broad peaks appear. This broadness of peaks is the result of the nanoscale of the particles and their disorder. Interestingly, the first peak at $2\theta \equiv 8°$ gives a d-spacing of 11.0 Å. This first peak is the 002 or interlayer spacing value. Notice that at a slightly higher concentration of acid (0.8 : 48), the 2θ gives the d-spacing value that is very close to 7.6 Å. The other unit cell dimensions also change very slowly until a pattern like that of the 12 : 48 (refluxing for 48 h in 12 M H_3PO_4) sample is attained.

As the ZrP particles increase in crystallinity, the shape of the titration curves changes as shown in Figure 1.7. What transpires is that the sodium ion no longer

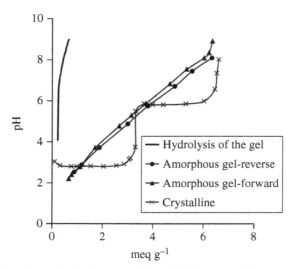

FIGURE 1.5 Potentiometric titration curve for α-ZrP. Titrant: 0.100 M (NaCl + NaOH); ×, highly crystalline sample; ▲, amorphous gel forward direction; ●, reverse direction. Solid line at far left is the extent of hydrolysis for the gel (Taken from: *Ind. Eng. Chem. Res.* 1995, *34*(8), 2865–2872; DOI: 10.1021/ie00047a040).

spreads throughout the particles but forms solid solution phases of different compositions of increasing Na+ content [24]. Only when full crystallinity is achieved is the state of no degrees of freedom observed. The approximate sizes of the particles as obtained from X-ray peak broadening observed in the patterns in Figure 1.6 are given in Table 1.2 [23, 24]. Also, the ratio of phosphorus to zirconium for 0.5 : 48 was 1.934 and for 2.5 : 48, 1.959, as determined by chemical analysis. Only when the crystallinity is more than that of sample 12 : 48 is the P:Zr ratio 2, and even here, one needs to be careful that some hydrolysis has not occurred.

The very small particle size that we described in Table 1.2 was known to us early on as shown in Figure 1.2. The thickness of the 0.5 : 48 particles is 70 Å or 6 layers. Presumably, the non-refluxed particles are even smaller, especially in length. Under the conditions of rapid precipitation on mixing a soluble zirconium compound with H_3PO_4, the probability of achieving a perfectly regular layer and stacking these layers parallel to each other is small. The gel particles are of the order of 0.1 μm or even smaller [24]. One can imagine that the phosphate groups are tilted randomly away from their equilibrium positions in the crystals so that some P–OH groups point towards and others away from each other. This creates high electrostatic stresses near the layers but weak forces between the layers. Thus, water is sorbed as a means of reducing the coulombic forces through solvation. This is similar to the swelling exhibited by organic cation-exchange resins.

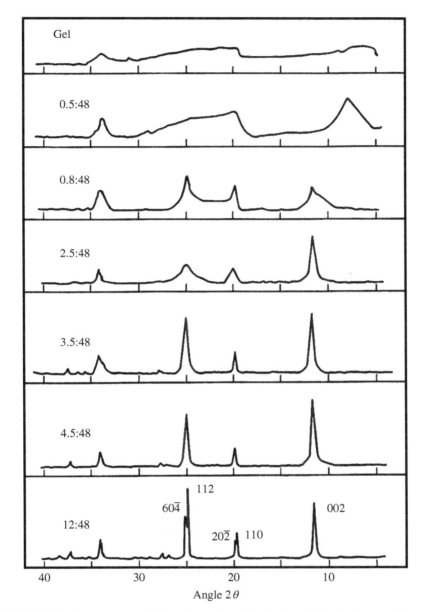

FIGURE 1.6 X-ray patterns of zirconium phosphates having different degrees of crystallinity. The numbers indicate the concentration of H_3PO_4 in which the gel was refluxed and the time of reflux in hours (Taken from: *Annu. Rev. Mater. Sci.* 1984, *14*, 205–229; DOI: 10.1146/annurev.ms.14.080184.001225).

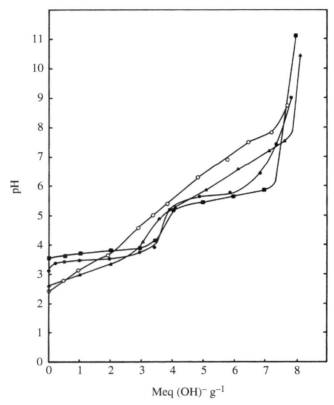

FIGURE 1.7 Titration curves for α-ZrP of low and intermediate crystallinities: 0.8:48 (○), 4.5:48 (●), 2.5:48 (▲), 12:48 (■). The numbers indicate the concentration of H_3PO_4 in which the gel was refluxed and the time of reflux in hours (Taken from: *Annu. Rev. Mater. Sci.* 1984, *14*, 205–229; DOI: 10.1146/annurev.ms.14.080184.001225).

TABLE 1.2 Crystallite sizes of zirconium phosphate gels

Sample[a]	d_{002}		Uncorrected d_{hkl}		Corrected d_{hkl}	
	Size (Å)	No. of layers	$d_{112, 202}$	$d_{112, 20\bar{4}}$	$d_{112, 202}$	$d_{112, 20\bar{4}}$
0.5 : 48	70	6				
0.8 : 58	100	13	210 (39)	260	300 (56)	360
2.5 : 48	330	44	250 (46)	100	350 (65)	140
3.5 : 48	330	44	550 (100)	360	770 (144)	510
4.5 : 48	400	53	770 (144)	550	1090 (200)	770

Taken from: *Ion Exchange and Membranes*, 1972, Vol. 1, 91–107.
[a] The numbers indicate the concentration of H_3PO_4 in which the gel was refluxed and the time of reflux in hours (i.e. 0.5 : 48 = 05 M H_3PO_4 refluxed for 48 h).

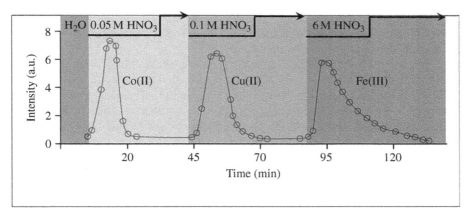

FIGURE 1.8 Chromatographic separation of three transition metal ions using the semi-amorphous ZrP 0.5 : 48 (Figure 1.6). The higher charge ion Fe (III) is more strongly held on the exchanger (Taken from: Clearfield, A., Jahangir, L.M. in Recent Developments in Separation Science, Navratil J. D. and Li, N. N. Eds, CRC Press, Boca Raton, FL. 1984 Vol. VIII. Ch. 4).

1.5 APPLICATIONS OF ION-EXCHANGE PROCESSES

In 1984, Clearfield and postdoc Jahangir published a paper titled 'New Tools for Separations' that listed 147 references [25]. It covered a wide range of topics including the differences in behaviour of the amorphous, semi-crystalline and crystalline forms of ZrP. Topics included separations of alkali metal ions, divalent and polyvalent cations, separation of lanthanides and actinides, nuclear waste processing and water purification. This paper also introduces θ-ZrP, a form of α-ZrP but containing 6 mol of water and an interlayer spacing of 10.4 Å. The amorphous gel phase was shown to behave as a weak field ion exchanger in the sense of Eisenman's theory [26]. The selectivity for the alkali cations is $Cs^+ > Rb^+ > K^+ > Na^+ > Li^+$ [27]. As the cation content increases, the selectivities change as predicted by Eisenman's theory to the strong field order that is just the reverse of the above order.

Selectivities of cations with higher charge are preferred by the amorphous (gel) ZrP, and some interesting separations of radioisotopes have been carried out using chromatographic separations. An example is given in Figure 1.8.

1.6 NUCLEAR ION SEPARATIONS

The amorphous zirconium phosphate was found to be active in the separation of radioactive ions because of its resistance to ionizing radiation, oxidizing media and strong acid solutions [28]. In general, it was found that separations for ions with the same charge were low, whereas those with different oxidation states were high [29, 30]. Fletcher Moore described a separation of Cm from Am by oxidizing americium to the plus five oxidation state where Cm does not oxidize. The Cm was then preferentially solved by the amorphous ZrP, while Am(V) was only weakly sorbed [30]. Additional separations of actinides from lanthanides and actinides from each other were affected by the redox method [31, 32].

Subsequently, we developed a number of monophenyldiphosphonic acid phosphates of Zr^{4+} and Sn^{4+}. These compounds were found to be highly selective for ions of 3+ or 4+ charge but not for those of lower charge [33]. These ion exchangers are currently being utilized to develop separations of lanthanides from actinides and actinides from each other [34–36]. Such separations are required for the nuclear fuel cycle intended to recover fuel values from the reactor spent rods.

Interest in zirconium phosphate as an ion exchanger and sorbent in its many forms has existed to this day with about 30 papers a year on this subject. We provide only few examples.

Because of the versatility of the zirconium and titanium phosphates, new uses and novel forms are prepared to meet current needs. A composite titanium and zirconium phosphate cation exchanger was prepared by sol–gel mixing of polyaniline into precipitated ZrTi phosphate [37]. The best sample had an ion-exchange capacity of 4.52 meq g^{-1} with excellent chemical and thermal stability. It was found to remove heavy toxic metals, especially Pb(II) and Hg(II), from waste solutions.

Pb^{2+}, Zn^{2+}, Cd^{2+} and Ca^{2+} were exchanged by a nanoparticle, nearly amorphous sample, of α-ZrP [38]. The selectivity sequence is as in the order listed above. In a separate study, a similar ZrTi phosphate modified with Al^{3+} or Fe^{3+} was found to sorb uranium [39].

Studies have shown that hydrogen uranyl phosphate on *Serratia* sp. as a film showed 100% removal of ^{90}Sr, ^{137}Cs and ^{60}Co. However, the authors wished to use a non-toxic metal to replace the toxic U. Zr was found to fit the bill [40]. Zirconium in the form of glycerol 2-phosphate was immobilized as a biofilm onto polyurethane foam.

1.7   MAJOR USES OF α-ZrP

The list of uses and potential uses of α-ZrP is quite extensive, but among the most promising ones are ion exchanger, sensors, drug delivery, polymer composites, antimicrobials, fire retardants, non-soluble surfactants and catalysts. Some of them have been already mentioned, but not all of them will be described. For those described applications will be interspersed with additional properties of the α-ZrP particles both amorphous and crystalline.

1.8   POLYMER NANOCOMPOSITES

Polymer nanocomposites exhibit significantly enhanced physical and mechanical properties as opposed to conventional micrometre-scale inorganic filler-reinforced polymer composites [41]. Nanofillers have been based on TiO_2, $CaCO_3$, SiO_2 and clays, among others [42]. One difficulty encountered is incomplete dispersion of the nanofillers within the polymer. By exfoliating a montmorillonite clay and polymerizing nylon in the exfoliated media, a fairly uniform composite with the clay was obtained [43]. In general, clay-based composites exhibit enhanced modulus and gas barrier properties of the polymer but also significant reduction in ductility and toughness [44]. Also, it is difficult to achieve very high purity, narrow particle size distribution and controlled aspect ratio of a clay. Many of the clay-based nanocomposites exhibit incomplete exfoliation of the clay, leading to inconsistent results.

Our original concern was to produce pure nanofillers of complete exfoliation so as to gain fundamental structure–property relationships. Our first effort involved the use of nanoparticles of α-ZrP with an epoxy polymer (diglycidyl ether of bisphenol A) [45]. A gel was formed in methyl ethyl ketone (MEK) by intercalating Jeffamine M 715 $[CH_3(CH_2CH_2O)_{14}CH_2CH_2NH_2]$ between the α-ZrP layers. The X-ray powder pattern, shown in Figure 1.9 (top left), consists of a successive order of 001 peaks. The MEK was removed and the gel (5.2 wt%) was combined with the epoxy and a curing agent. Polymerization was carried out at 130°C. The resultant composite in Figure 1.9 indicates a uniform distribution of the ZrP throughout the polymer. The tensile module increased by 50%, and the yield strength improved by 10%. However, the ductility (elongation at break) was drastically reduced. The effect of incorporating the Jeffamines on the physical properties was not determined.

In a subsequent study, a 1 and 2% (vol%) α-ZrP dispersion in the epoxide was examined, and detailed facture toughness values are found to be similar to that of the neat epoxy [46]. There was no sign of crack deflection. Rather, the nano-platelets are broken into two halves as the crack propagated through them, a sign of strong bonding to the epoxy.

The effect of nano-platelets on the rheological behaviour of epoxy monomers with variations in nano-platelet exfoliation level of aspect ratio was investigated [47]. The results show that the presence of exfoliated nano-platelets in epoxy can significantly influence viscosity and lead to shear-thinning phenomena, especially when the aspect ratio of the nano-platelets is high. The Krieger–Dougherty model was employed to describe quantitatively the effectiveness of the nano-platelets upon the resultant rheological behaviour of the composites [48].

Studies on the effect of aspect ratios [49] and surface functionalization [50] by employing a combination of a long-chain alkylamine and a short-chain bulky amine were investigated. Gels of α-ZrP were prepared by treatment of α-ZrP particles

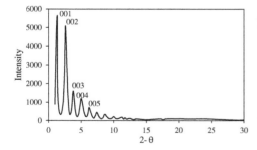

FIGURE 1.9 X-ray powder pattern of α-ZrP intercalated with Jeffamine M715 (left). The layered character of the compound is observed. The interlayer spacing is 73.27 Å. TEM image of M715-α-ZrP/epoxy showing high magnification of uniform dispersion and exfoliation of α-ZrP layers (right) (Taken from: *Chem. Mater.* 2004, *16*, 242–249; DOI: 10.1021/cm030441s).

(25–80 nm) with propylamine and mixed with polystyrene in a water-soluble organic solvent [51]. Hung et al. used melt compounding to incorporate ZrP into a styrene–butadiene polymer [52]. It was shown how the physics and chemistry incompatibility of the filler and polymer can be overcome.

Nanocomposites of poly(ethylene terephthalate) and α-ZrP or zirconium phenylphosphonate (ZrPPh) were prepared by melt extrusion [53]. Two and five weight percent composites were prepared in a twin-screw extruder and samples obtained by injection moulding. Many polymer composites have been prepared using ZrP or its derivatives aimed towards better fire retardancy or as proton conductors for fuel cells. These composites will be described later in connection with fuel cells. For our final effort, we attempted a nanocomposite for polypropylene (PP).

Preparation of PP nanocomposites appears to be most challenging. The low compatibility between PP and inorganic nano-platelets leads to poor dispersion and adhesion of nano-platelets in the PP mix, resulting in the low performance of the nanocomposites. However, because of the significant commercial importance of PP, improvement in the thermal stability and mechanical strength would dramatically enhance the utilization of PP in several engineering applications. The problem is the poor compatibility between the nanoparticle and the PP. Many efforts have been attempted to alter the nanoparticle, the PP or both to achieve compatibility [54–60].

Our own effort at compatibility was to prepare a mixed derivative, $Zr(O_3POH)_{2-x}(O_3P-CH_3)_x$ as shown in Figure 1.10, with $x = 0.66$, 1.0, 1.33 [54]. The X-ray patterns showed interlayer spacings of 8.4, 8.6 and 8.8 Å (7.6 Å for α-ZrP), respectively, for the three samples. The methyl groups impart a measure of hydrophobicity to the particles, and when sonicated in toluene, the 8.4 Å sample swelled to 9.4 Å. The reactions were carried out in cold toluene. Sonication was continued to the point (6 h) where very little of the nanoparticles precipitated. Gaseous propylene was dissolved in the toluene and an Figure 1.10 Schematic of reaction for the synthesis of α- ZrP and its methyl/

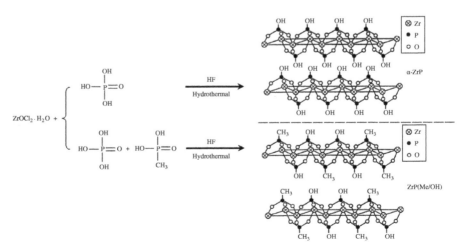

FIGURE 1.10 (Continued)

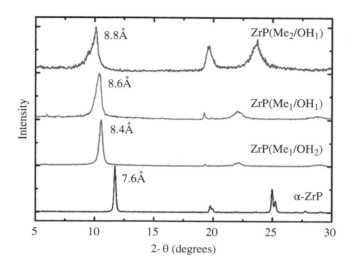

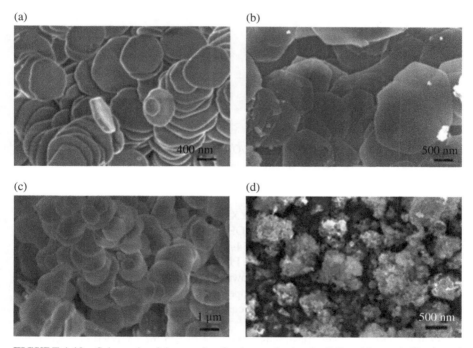

FIGURE 1.10 Schematic of the reaction for the synthesis of α-ZrP and its methyl/hydroxyl mixed derivative. At the bottom is the XRD (left) and SEM image (right) of (a) α-ZrP and its methyl/hydroxyl mixed derivatives: (b) ZrP(Me$_1$/OH$_2$), (c) ZrP(Me$_1$/OH$_1$) and (d) ZrP(Me$_2$/OH$_1$) (Taken from: *Chem. Mater.* 2009, 21, 1154–1161; DOI: 10.1021/cm803024e).

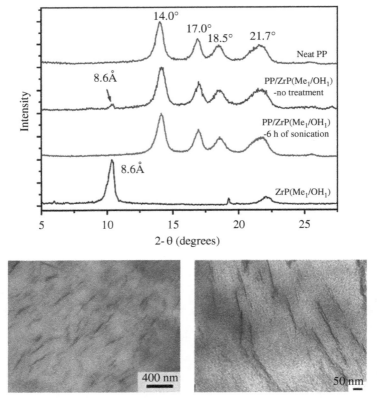

FIGURE 1.11 XRD pattern of PP/ZrP(Me$_1$/OH$_1$) nanocomposites (top). TEM images (bottom) of PP/ZrP(Me$_1$/OH$_2$) nanocomposite prepared with ultrasonicated ZrP(Me$_1$/OH$_2$) (Taken from: *Chem. Mater.* 2009, *21*, 1154–1161; DOI: 10.1021/cm803024e).

hydroxyl mixed derivatives, above, and the XRD and SEM images of the mixed derivatives (page 15) as (a) α-ZrP, (b) ZrP (Me$_{1.5}$/OH$_2$); (c) ZrP(Me$_1$/OH$_1$); (d) ZrP (Me$_2$/OH$_1$). (Taken from Chem. Mater. 2009, 21, 1154–1161). organometallic catalyst added. To stop the reaction, 10% HCl was added. The PP was of the isotactic structure. The dispersion of the particles is fairly uniform as shown in Figure 1.11. However, complete exfoliation did not occur. The layers are grouped in stacks of two to five layers. It was reasoned that better compatibility would require a somewhat longer carbon chain balanced with better exfoliation character.

What the industry really desires is to be able to add the nanoparticles when the polypropylene is fluid at elevated temperature. In this way, they would not have to change their synthesis procedures. However, that is a daunting task. More recent work has concentrated on different approaches to achieve good composites.

Hung et al. claim that they were able to add α-ZrP nanoparticles to a styrene–butadiene copolymer during melt compounding [52]. Composites of polystyrene and polyethylene-vinyl acetate were prepared with organically modified α-ZrP, that is, alkyl chains attached to the phosphonate head [61]. They used melt blending as a means of dispersing the particles into the polymer. A styrene–butadiene rubber was combined with α-ZrP that was modified with intercalated alkylamines of different chain lengths used to move the layers apart. The intercalation mechanism is described in some detail [62]. Casciola et al. [63] used dodecyl groups bonded to α-ZrP to prepare polymer composites with molten polyethylene. A number of starch polymer–α-ZrP composites were also prepared [64]. Many polymer composites are either prepared for flame retardation [61] or proton conduction that will be discussed in another section.

There is a need by the packaging industry to have wrapping materials and plastic bottles that prevent diffusion of O_2 through the package. The use of clays and other inorganic materials has been somewhat successful but not adopted.

1.9   MORE DETAILS ON α-ZrP: SURFACE FUNCTIONALIZATION

At this juncture, it is necessary to describe additional properties of the ZrP particles. It is now well known that silanes will bond to silica and to silicon to form self-assembled monolayers [65]. Since then, a veritable cornucopia of SAMs have been produced on these surfaces [66–73]. In addition, the SAMs may be functionalized by surface reactions or by prefunctionalization prior to preparing the SAM [71, 72]. As a result, a wide range of applications for SAMs have emerged. The functionalization of surfaces allows the worker to change the surface properties such as friction, wettability, adhesion, and so on. Furthermore, the end groups of the silanol may be changed in an almost unlimited number of ways for applications as chemical sensors and biosensors; in microelectronics, thin film technology and cell adhesion photolithography; and in a variety of important protective coatings, composites and catalytic materials [67, 72].

A schematic representation of the surface of α-ZrP is shown in Figure 1.1(b). The surface is just a slice of one of the layers and therefore should react with a number of ligands. We have recently shown that the surface indeed does bond with silanes, epoxides, isocyanates and acrylates directly. The silanes are suspended in hot toluene and allowed to react with dewatered α-ZrP [74]. In addition to these ligands, we have been able to bond polyethylene glycols (PEGs) to the surface in two ways. In the first, we used a carbodiimide [75], N,N'-diisopropylcarbodiimide, to activate the surface as depicted in Figure 1.12. The other technique is to add Zr^{4+} or Sn^{4+} to the α-ZrP in water. These ions replace the protons on the surface, producing an arrangement similar to the arrangement of the metal ions within the layers

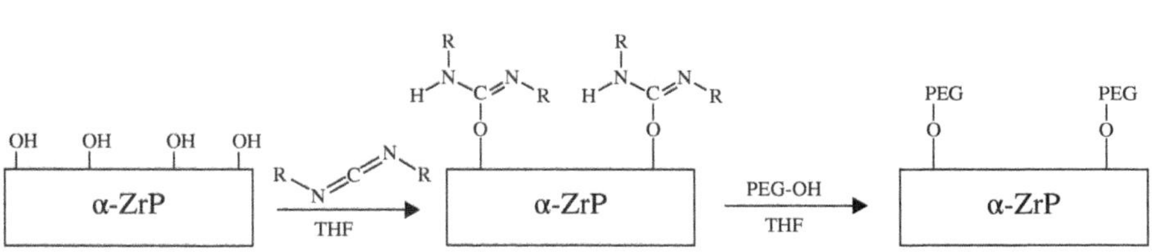

FIGURE 1.12   Reaction scheme for the surface modification via phosphate activation by a carbodiimide.

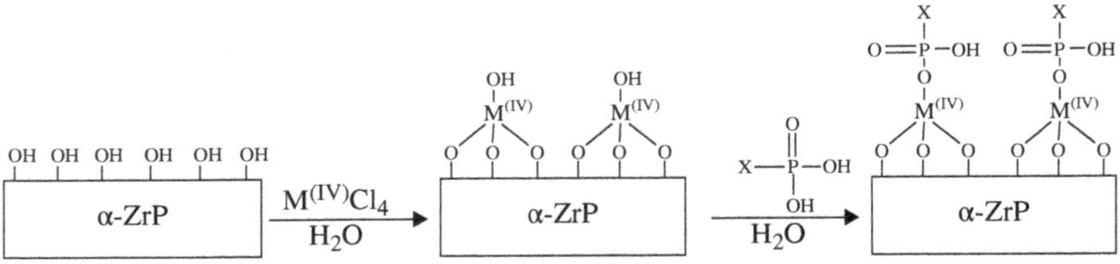

FIGURE 1.13 Reaction scheme for the surface modification using tetravalent metals to coordinate to the phosphate groups on the surface of ZrP, followed by the addition of a phosphonic acid to complete the coordination.

(Figure 1.13). As a result, any phosphate or phosphonic acid can be affixed to the metal ion surface. PEGs can readily be converted to a phosphate by oxidation of the alcohol group with $POCl_3$. In fact, a whole variety of alcohols may be oxidized and bonded to the surface metal ions. A pictorial summary of the surface functionalized in this manner is provided in Figure 1.13 and the totality of reactions in Figure 1.14. The ability to carry out these functionalizations opens up vast new possibilities that will be described in what follows.

1.10 JANUS PARTICLES

Janus particles were named after the Roman God of doors by Nobel Laureate Pierre-Gilles de Gennes [76]. These compounds have two halves that differ in chemical properties. An example is a layered compound that has one ligand on the topside and a different ligand on the underside. There are many ways of preparing Janus particles such as layer-by-layer self-assembly or in general shielding part of the particle while coating the unshielded part (Figure 1.15) [77–81]. While many uses are proposed for these materials, the problem is the lack of methods to prepare them in quantity [80, 81]. Amphiphilic Janus particles absorb to interfaces and foam surfaces [82, 83]. Janus particles can play a role in catalysis [84] and applications in display technology, switching between dark and light sides using magnetic or electric fields [85].

Our synthesis of Janus particles involved reaction of the α-ZrP nanoparticles with octadecylisocyanate to cover both outer surfaces [86]. The particles are then exfoliated and separated using foams or water–oil mixtures (Figure 1.16).

Janus particles are strongly adsorbed onto interfaces, where they act as surfactants. Our amphiphilic α-ZrP nano-sheets have a large aspect ratio due to their very small layer thickness as shown earlier by Alberti et al. [19]. Their large lateral surface area offers strong adsorption energy at the oil–water interface. An oil-in-water emulsion stabilized by the α-ZrP Janus particles as surfactant was prepared at room temperature by sonication. This emulsion was stable for months, whereas an emulsion prepared from only α-ZrP particles was stable for only a few hours [86].

These initial experiments are very promising. Given the fact that the size of the ZrP particles can be controlled and there is a large choice of what is put on the surface, the

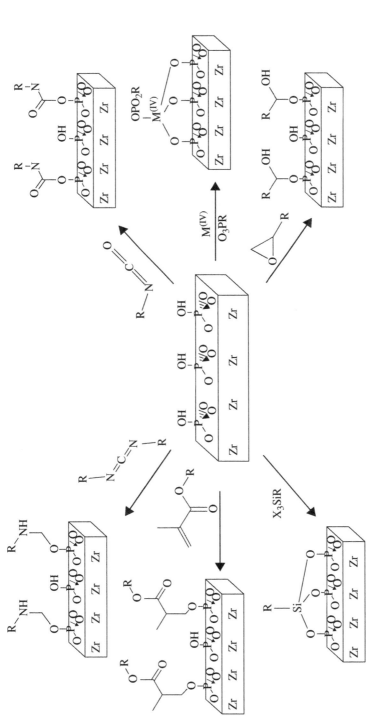

FIGURE 1.14 The many ways that the surface of α-ZrP may be functionalized.

(a)   (b)   (c)   (d)   (e)

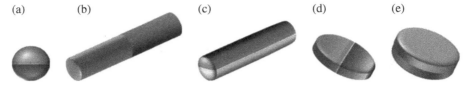

FIGURE 1.15 Schematic representation of the possible Janus particle architectures: (a) sphere, (b and c) cylinders and (d and e) discs. The light and dark portions differ in hydrophobic/hydrophilic character (Taken from: *Soft Matter* 2008, *4*, 663–668.

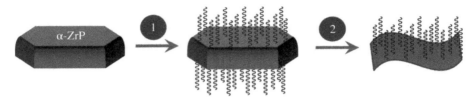

FIGURE 1.16 Schematic representation of the fabrication of surface-modified amphiphilic nano-sheets. The initial step (1) consists of grafting a coupling agent over the surface of α-ZrP. Subsequently, the exfoliation of the crystals is carried out (2) to obtain the surface-modified amphiphilic nano-sheets.

Janus particles can be tailored for many applications. To increase the percentage of Janus particles in each preparation, it is necessary to decrease the number of layers per particle, and we are now studying how this may be accomplished.

1.11 CATALYSIS

α-ZrP has a long history of use as a catalyst. It is mildly acidic and can be altered by surface area expansion, by intercalating metal ions, by formation of porous types and by intercalation of catalytically active molecules. Presumably, the first catalytic reaction using α-ZrP was between Cu^{2+}-exchanged α-ZrP and carbon monoxide [87]. High levels of oxidation were obtained with O_2 at 250–300°C. Subsequently, it was determined that the Cu^{2+} could easily be reduced to Cu metal by H_2 at 150°C [88]. The copper metal was deposited around the sides and top and bottom of the ZrP crystals, while H^+ replaced the Cu^{2+}. Interestingly, on standing in air at room temperature, the copper metal oxidized back to Cu^{2+} and diffused back into the ZrP layers. Presumably, the reaction is as shown in the equation, but no attempt to isolate H_2O_2 was made:

$$Cu^\circ + Zr\left(O_3POH\right)_2 \xrightarrow[RT]{O_2} ZrCu^{2+}\left(PO_4\right)_2 + H_2O_2$$

By 1980, reactions in which ZrP was used as a catalyst included dehydrogenation, isomerization, polymerization and alkylation [89]. The oxidative dehydrogenation of cyclohexene to benzene was effected with $ZrCu(PO_4)_2$. Complete oxidation to CO_2

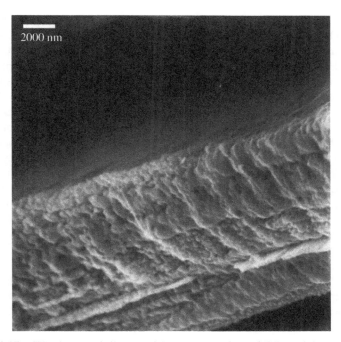

FIGURE 1.17   SEM image of silver particles on the surface of ZrP particles, grown by the reduction of silver ions previously loaded into the layers of α-ZrP (Taken from: *J. Chem. Soc. Faraday Trans.* 1984, *80*, 1579).

and H_2O could also be achieved [90, 91]. We also were able to prepare nanometre-sized clusters of Cu by limiting the amount of copper ion exchanged into the α-ZrP layers. Ag^+ was also reduced to $Ag°$ with H_2 at temperatures as low as 60°C and also showed reversible behaviour (Figure 1.17) [92]. A good summary of our early work is available [93]. This paper discusses the acidity of ZrP surfaces and reactions such as dehydration of alcohols to olefins, polymerization of olefins, oxidation reactions with transition elements between the layers, hydrogenation reactions, oxidative dehydrogenation and other reactions.

Costantino et al. carried out base catalysis using the Michael reaction as an example [94]. To increase the surface area of the catalyst, they exfoliated the layers with propyl-amine titration and then removed the amine with HCl under sonication. This procedure increased the surface area from 0.5 to 17 m^2 g^{-1}. A number of reactions with 85–98% yields using the Na^+ form of α-ZrP were recorded, for example, as given below:

Almost all catalytic reactions with α-ZrP are now carried out with high-surface-area materials. One method involves separate nucleation and ageing steps (SNAS) [95]. The ZrP particles were prepared in a colloid mill, refluxed in 15 M H_3PO_4 and then spray-dried to form microspheres with a diameter of 5–45 μm. These spheres were then used to convert fatty acid methyl esters to monoethanolamines. The fatty acids are high cost and need to produce high-cost products [96]. Fatty acid derivatives combined with amino alcohols, through acylations, can produce high-cost amides that are used in pharmaceutical and surfactant products. These microspheres were shown to effect the reactions in high yield under mild conditions. Other catalytic reactions to note are the intercalation of Cu(Salen) into α-ZrP to oxidize cyclohexene with dry tert-butyl hydroperoxides [97]. Olefin oxidation with dioxygen was catalysed by porphyrins and phthalocyanines intercalated into α-ZrP [98]. Alvaro and Johnstone prepared large-surface-area ZrP that could be highly loaded with Pd, Pt and Ni ions [99]. The ions were readily reduced to the metal state with H_2 at 400°C or with sodium tetrahydroborate at room temperature. These materials were very effective in the hydrogenation of alkenes.

The reader should be satisfied with these few examples to recognize the widespread use of α-ZrP as a catalyst and host for a variety of molecules, ions and metals to develop a useful search of the literature. A search of the literature from 2007 to 2013 yield 131 papers showing that the search for new catalysts based upon ZrP is an ongoing process.

1.12 CATALYSTS BASED ON SULPHONATED ZIRCONIUM PHENYLPHOSPHONATES

Another facet of the catalysis story is to recognize that there is a large literature on the preparation of zirconium and tin (IV) phenylphosphonates and their sulphonated forms. Therefore, it is necessary to begin with some facts about zirconium phenylphosphonate, $Zr(O_3PC_6H_5)_2$. This compound has a layered structure similar to that of α-ZrP. Heating at 140°C for 4 days produced the nanoparticles shown in Figure 1.18. Heating at 200°C hydrothermally for 30 days was necessary to obtain well enough formed crystals to determine the structure from X-ray powder data [100]. A view of the structure is shown in Figure 1.19. Because the phenyl rings are only 5.3 Å apart, the π–π overlap is strong and the layers resist interlayer expansion. However, by preparing mixed derivatives with H_3PO_4 to include phosphate groups (Figure 1.20), they can function as does α-ZrP. It was also advantageous to sulphonate the phenyl ring (Figure 1.21). The early history of preparing sulphonates of phenylphosphonate and phosphate derivatives is detailed in a recent book [101]. However, the catalytic behaviour of this molecule is of interest here.

Sulphonated or zirconium sulphophenyl phosphonate (ZrSPP) is a strong Brönsted acid and may be used in organic solvents permitting many catalytic reactions [102]. A series of papers using ZrSPP as catalyst were carried out by Curini et al. at the University of Perugia [103, 104]. The reactions they reported on include the preparation of cyclic and acyclic β-amino alcohols by addition of amines to epoxides [103], cyclic ketals and thioketals [104]. Early papers by

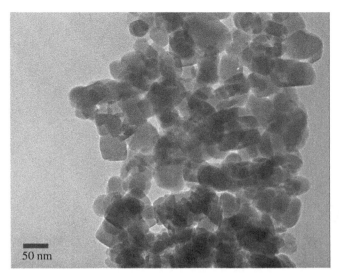

FIGURE 1.18   SEM image of zirconium phenylphosphonate particles grown hydrothermally at 140°C, 3 days, HF/Zr = 3 (Taken from: Clearfield, A. In *Environmental Applications of nanomaterials: Synthesis, Sorbents and Sensors*; Fryxell, G. E., Guozhong, C., Eds.; Imperial College Press, London, UK, 2007, p. 89).

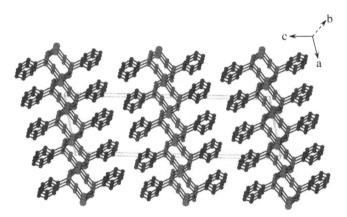

FIGURE 1.19   Schematic structure of zirconium phenylphosphonate as viewed down the *b*-axis direction (Taken from: Clearfield, A. In *Environmental Applications of nanomaterials: Synthesis, Sorbents, and Sensors*; Fryxell, G. E., Guozhong, C., Eds.; Imperial College Press: London, UK, 2007, p. 89).

the Curini group used a mixed ligand catalyst, $Zr(O_3PCH_3)_{1.2}(O_3PC_6H_4SO_3H)_{0.8}$ [105, 106]. It was found to be an efficient catalyst for the tetrahydropyranylation of alcohols and phenols. Additional related papers in use of such catalysts appeared regularly [107]. It was also shown that the mixed methane-sulphophenyl

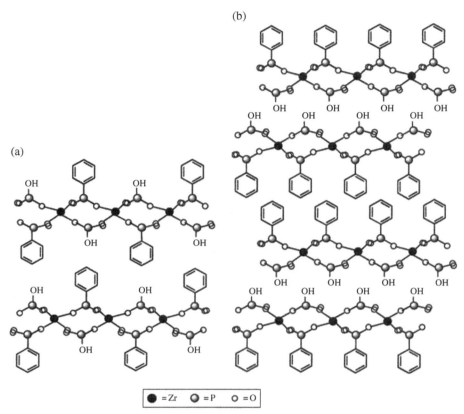

FIGURE 1.20 Schematic representation of two mixing patterns of zirconium phenylphosphonate phosphates. (A) Layers of mixed phenyl and phosphate groups (B) Staged structure with alternating layers of phenyl and phosphate groups. (Taken from: *Appl. Catal. A: Gen.* 2009, *353*, 236–242.

phosphonate mediated a regioselective synthesis of 2,3-disubstituted tetrahydro-2H-indazols. *In vivo* evaluation of these compounds proved the presence of anti-inflammatory activity without any gastric injury [108]. More recent work in China showed that the ZrSPP was an efficient and stable solid acid catalyst for the carbonylation of formaldehyde [109].

The next episode in catalysis results from our ability to functionalize the surface of the ZrP layers. In a separate set of α-ZrP nanoparticles, tris(2,2′bipyridyl) ruthenium(II), $Ru(bpy)_3^{2+}$, was intercalated into α-ZrP [74]. These intercalated particles were also surface functionalized with octadecyltrichlorosilane (OTS). The particles were highly hydrophobic as both surfaces contained the C_{18} silane. Two bottles were filled with a mixture of hexanes and water. To one is added the $Ru(bpy)_3^{2+}$/ZrP and to the other the surface-functionalized nanoparticles with $Ru(bpy)_3^{2+}$ between the layers. The hydrophilic particles reside solely in the water,

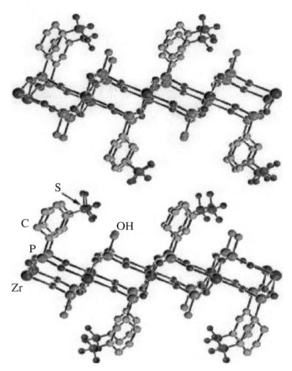

FIGURE 1.21 Schematic representation of zirconium phosphate sulphophenylen-phosphonate (Taken from: *Solid State Ion.* 2005, *176*, 2893–2898; DOI: 10.1016/j.ssi.2005.09.042).

and the hydrophobic C_{18}-functionalized particles are all in the organic layer (Figure 1.22) [74].

The fact that the α-ZrP particles can be functionalized to the point where they can be dispersed in non-polar liquids indicates that they have a role to play in catalysis. An organometallic catalyst may be tethered to the support layer via flexible linkers. The dispersion of these nanoparticles in the fluid is as close as possible to being a homogeneous catalyst but with the advantage that the catalyst can be easily recovered and reused. The products would remain in the solvent and are recovered by conventional methods. Our first experiment involved the addition of Wilkinson's catalyst to propyl (diphenylphosphine) triethoxysilyl linker bonded to the surface of α-ZrP (Figure 1.23). 1-Dodecene was hydrogenated in toluene in very high yield. The reaction was repeated 15 times with no decrease in yield and no detectable catalyst in the toluene solvent.

In addition to its catalytic usage, ZrSPP is an excellent proton conductor. This fact requires that we pursue the applications to fuel cell technology.

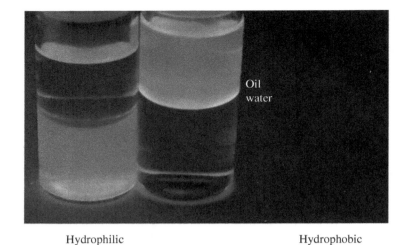

FIGURE 1.22 Zirconium phosphate, previously loaded with tris-(2,2′-bipyridine) ruthenium(II), surface modified with octadecyltrichlorosilane (OTS), making the nanoparticles compatible with non-polar solvent (Taken from: *Chem. Mater.* 2013, *25*, 723–728; DOI: 10.1021/cm303610v).

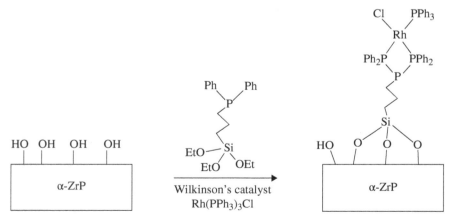

FIGURE 1.23 Scheme showing the addition of the silyl linker and catalyst on the surface of α-ZrP particles.

1.13 PROTON CONDUCTIVITY AND FUEL CELLS

α-ZrP is inherently a proton conductor. Early studies showed that the proton conductivity depends upon its crystallinity and the relative humidity [110, 111]. Table 1.3 summarizes the conductivity of α-ZrP of different crystallinites. The specific conductance decreases as the crystallinity increases. The explanation is that the surface area decreases in the same order and proton conduction is higher on the surface because of the higher surface water content and a lower order of confinement [112]. The protons in the bulk of the particles move slowly parallel to the layers to access the solution.

Subsequently, we prepared $Zr(O_3POH)_{1.1}(O_3PC_6H_4SO_3H)_{0.9}$, from the corresponding phenylphosphonate phosphate, in fuming sulphuric acid at 60°C [113]. Alberti et al. prepared a similar compound of composition $Zr(O_3C_6H_4SO_3H)_{0.73}$ $(O_3PCH_2OH)_{1.27}$ [114]. This compound was shown to be a pure proton conductor. Arrhenius plots at several RH are shown in Figure 1.24. The highest conductivity at 295 K was 1.65×10^{-2} S cm^{-1}. Somewhat higher conductivities were obtained at 278 K for more highly sulphonated Zr and Ti compounds as shown in Table 1.4. The zirconium and titanium phenylphosphonates were sulphonated in fuming sulphuric acid and recovered by addition of methanol to the diluted acid followed by centrifugation [115]. More recently, chlorosulphonic acid was used in the preparation for better control of the reaction.

The titanium compound experiences a certain amount of cleavage of the P–C bond as the formula derived from elemental analysis and thermogravimetric analysis was $Ti(O_3POH)_{0.25}(O_3PC_6H_5)_{0.12}(O_3PC_6H_4SO_3H)_{1.63} \cdot 3.64H_2O$ as opposed to the fully sulphonated zirconium phosphonate, $Zr(O_3PC_6H_4SO_3)_2 \cdot 3.6H_2O$. These sulphonated derivatives are among the best known proton conductors [115].

Before proceeding to discuss fuel cell electrodes, forgive me for pointing out a little investigated fact about the ZrSPP materials. They are excellent ion-exchange materials that arise from the ready displacement of the sulphonic acid protons. This is illustrated in Table 1.5 [116]. The increase in K_d values as a function of ion exchange and size allows for potential easy separations of these ions. We also note that the compound with the greater amount of sulphonic acid groups gave much

TABLE 1.3 Specific conductance of α-ZrP samples obtained by different methods of preparation and ordered according to their degree of crystallinity ($T = 25$°C)

Sample	Preparation method	Specific conductance (Ω^{-1} cm^{-1})
1	Precipitation at room temperature, amorphous	8.4×10^{-3}
2	Precipitation at room temperature, amorphous	3.5×10^{-3}
3	Refluxing method $(7 : 48)^a$, semi-crystalline	6.6×10^{-4}
4	Refluxing method $(10 : 100)$, crystalline	9.4×10^{-5}
5	Refluxing method $(12 : 500)$, crystalline	3.7×10^{-5}
6	Slow precipitation from HF solutions, crystalline	3.0×10^{-5}

^{a}Numbers in parentheses indicate the concentration of H_3PO_4 in molarity and the number of hours refluxed, respectively.

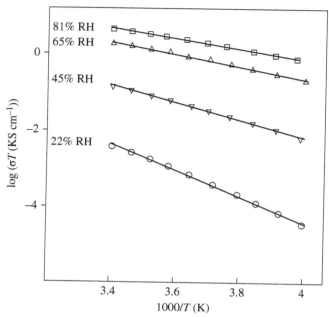

FIGURE 1.24 Arrhenius plots of log (σT) as a function of $1000/T$ for $Zr(O_3PC_6H_4SO_3H)_{0.73}$ $(O_3PCH_2OH)_{1.27}$ at different relative humidities (Taken from: *Solid State Ionics* 1992, *50*, 315–322; DOI: 10.1016/0167-2738(92)90235-H).

TABLE 1.4 **Conductivity in reciprocal $\Omega \cdot cm$ at 5°C as a function of relative humidity for zirconium and titanium sulphophenylphosphonates**

Sample	Relative humidity (%)				
	20 (3)[a]	30 (3)	50 (4)	65 (4)	85 (4)
EWS-3-89 (Zr)	5.0×10^{-6}	2.0×10^{-6}	1.1×10^{-3}	7.8×10^{-3}	2.1×10^{-2}
EWS-4-1 (Ti)	4.0×10^{-5}	3.0×10^{-3}	1.2×10^{-2}	7.2×10^{-2}	1.3×10^{-1}

[a]Number in parenthesis is the estimated error in the humidity measurement.

higher K_d values. A thermodynamic treatment of this ion-exchange behaviour is available [27, 117].

There is currently a great effort directed towards development of workable fuel cells because of their energy efficiency. Efficiencies as high as 70–80% are possible for fixed site units and 40–50% for transportation applications versus the current 20–35% with internal combustion engines [118]. For transportation purposes, proton-exchange membrane fuel cells (PEMFC) using H_2 as the fuel are preferred. A schematic drawing of such a fuel cell is given in Figure 1.25. The key ingredient in the PEMFC is the solid polymer membrane that conducts protons from the anode to the cathode. The most preferred membranes are fluorocarbon in nature to which are

TABLE 1.5 Distribution coefficient for alkali and alkaline earth metal ions on exchanger MY-IV-95, MY-VI-2 and amorphous zirconium phosphate for comparison

Ion	K_d (ml g^{-1})		
	MY-IV-95^{a,b}	MY-VI-2^{a,c}	Amorphous ZrP^{a,d}
Li$^+$	110	—	7
Na$^+$	205	—	11
K$^+$	1,500	650	120
Cs$^+$	6,500	—	1600
Mg^{2+}	21,000	9,800	—
Ca^{2+}	8,9000	37,000	—
Ba^{2+}	400,000	190,000	—

$^a K_d$ at pH = 2.00 and a metal loading of 0.1 meq g^{-1}.
^{b}Zr(O$_3$PC$_6$H$_4$SO$_3$H)$_{0.767}$(O$_3$POH)$_{1.23}$.
^{c}Zr(O$_3$PC$_6$H$_4$SO$_3$H)$_{0.43}$(O$_3$POH)$_{1.57}$.
^{d}Calculated from selectivity coefficient given in Ref. [27].

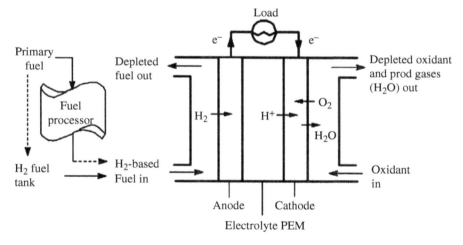

Anode (fuel) reaction: $H_2 = 2 H^+ + 2 e^-$
Cathode (oxidant) reaction: $1/2 O_2 + 2 H^+ + 2 e^- = H_2O$
Total reaction: $H_2 + 1/2 O_2 = H_2O$

FIGURE 1.25 Schematic representation of a proton-exchange membrane fuel cell (PEMFC) system using on-board or on-site fuel processor or on-board H$_2$ fuel tank (Taken from: *Catal. Today*, 2002, 77, 17–49; DOI: 10.1016/S0920-5861(02)00231-6).

affixed sulphonic acid groups (Nasicon). These membranes are excellent electronic insulators with high proton conductivities of the order of 10^{-2} S cm^{-1} [or (Ω cm)$^{-1}$]. The membranes require water as the main proton conduction medium and so work best at temperatures below 100°C. The problem arises from the fact that the H$_2$ must be generated by an on-site fuel processor from methane or methanol. In the process, CO is also generated, which poisons the anode catalyst, platinum [119]. However, operation of the fuel cell at 120–130°C would eliminate this problem, but the membrane requires a considerable external pressure to avoid water loss [118, 119]. At temperatures of 140–160°C, direct methanol fuel cells (DMFC), where methanol is used directly as the fuel, are possible, but suitable membranes are not yet available. In the most recent work, polymer composites containing zirconium phosphates or sulphonated zirconium phenylphosphonate show considerable promise.

1.14 GEL SYNTHESIS AND FUEL CELL MEMBRANES

In one of the most active groups synthesizing fuel cell membranes, Alberti et al. have demonstrated that with the addition of a soluble metal salt, such as alkoxide or carboxylate, in a polar organic solvent to which phosphoric acid and a sulphophenylphosphonic acid have been added, a clear solution is initially obtained. Heating to temperatures above 40°C results in the formation of a mixed derivative such as $Zr(HPO_4)_x(O_3PC_6H_4SO_3H)_{2-x}$ [120, 121]. Many other compositions of the gel may be envisioned. However, the clear solutions have been added to solutions of polymers used as conducting membranes such as Nafion®, polyether ketones and polyvinyl fluoride [120–122]. Thin films of the composites may be cast out and the solvent evaporated at elevated temperatures. In this process, it would seem that many of the arylphosphonates we have described here may also be utilized in this way with or without the addition of phosphoric acid. The process may also be utilized with porous membranes where the pores are filled with the clear solution and the solvent evaporated.

This field continues to be highly active. One of the major problems faced is the high cost of Nafion, and recently, a major increase in its price was installed. Thus, much activity is based upon membrane materials other than Nafion. Alberti et al. found that a sulphonated poly(ether etherketone) (SPEEK) composite membrane exhibited excellent medium-temperature performance at high relative humidity (75%) and 160°C at 4×10^{-2} S cm^{-1}. Subsequently, ZrP or a combination of Zr and Ti phosphates was added [123, 124]. The search for new materials and compositions is an active one as attested to the many recent publications.

We will provide two recent examples. Lee et al. developed a new process for the preparation of nanoparticles of α-ZrP, ZrSPP [$Zr(O_3PC_6H_4SO_3H)_2$], and zirconium sulphate, $Zr(SO_4)_2 \cdot 4H_2O$ [125]. A zirconium oxide powder was prepared by hydrolysis and condensation of zirconium butoxide in isopropanol in the presence of acetylacetone. HNO_3 was added, and after stirring to homogenize the mixture, it was left standing for 6 h. The powder was recovered and dried at 80°. The oxide was then converted to ZrP in 2 M H_3PO_4 at 80°C and to sulphophosphonate by treatment with sulphuric acid. The oxide was also treated separately to obtain the sulphate

Objective in this research

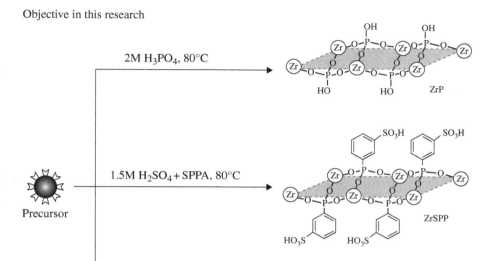

FIGURE 1.26   The novel mild synthesis routes to ZrP, ZrSPP and ZrS from the same precursor at 80°C under mild acidic conditions (Taken from: *J. Mater. Chem.* 2010, *20*, 6239–6244; DOI: 10.1039/C0JM00130A).

(Figure 1.26). The authors claim that this mild method of preparation allows the oxide products to be dispersed in polymer electrolytes at a very high loading with subsequent conversion to the Zr compound conductors. This paper also contains an excellent list of references.

It is pointed out by Pica et al. [126] that Nafion membranes filled with ZrP and other inorganic fillers 'represent the best compromise between chemical stability and proton conductivity but suffer from a relatively low mechanical stability'. However, it is pointed out that a new perfluorosulphonic acid compound that has shorter side chains exhibits a higher crystallinity at a given equivalent weight. This means that for a lesser mass, it may be possible to have high stability for the membrane. Therefore, this new shorter-chain perfluorosulphonate was filled with different amounts of ZrP nanoparticles by solution casting. The membrane exhibited somewhat lower conductivity but an improvement in mechanical properties up to 10 wt% of ZrP [126].

In spite of all the fine research that has been carried out, an ideal PEMFC membrane has not yet been forthcoming. An ideal situation would be to replace platinum with a less expensive catalyst but also more importantly one that would be uninfluenced by carbon monoxide. Then research could concentrate on membranes that have the necessary proton conductivity together with the required mechanical properties.

1.15   ELECTRON TRANSFER REACTIONS

In our initial attempts at carrying out electron transfer reactions, we found that the intercalation of electron donors between the layers of α-ZrP was a difficult task. Therefore, we utilized zirconium phosphate sulphophenyl phosphonate, (ZrPS) $Zr(HPO_4)(O_3PC_6H_4SO_3H)$. This compound has an interlayer spacing of 16.1 Å and readily takes up large molecules [113]. Our interest was to quantitatively evaluate the nature of the chemical environment within ZrPS and the rate of diffusion within the interlayer space. In this regard, luminescence quenching studies with $Ru(bpy)_3^{2+}$ intercalated into ZrPS were initiated [127, 128]. It was found that the $\pi-\pi*$ bond at 285 nm in aqueous solution of $Ru(bpy)_3^{2+}$ is red shifted to 317–320 nm. The metal-to-ligand charge transfer (MLCT) bond is also red shifted, the degree of shift depending upon the amount of $Ru(bpy)_3^{2+}$ intercalated. These several interactions are enhanced by increased packing of $Ru(bpy)_3^{2+}$ within the layers. Subsequently, the delay kinetics of the self-quenching of $Ru(bpy)_3^{2+}$ was determined. The luminescence decay data indicates that a multiplicity of binding sites is present in ZrPS that can explain the self-quenching.

Methyl viologen (MV^{2+}) was also used as a quenching agent. The MV^{2+} was found to lie flat in the α-ZrP layers. These quenching curves were fit to two models of which Albery's dispersed kinetics model gave very excellent duplication of the decay curves [129].

The importance of photoinduced electron transfer reactions is their interest in photosynthesis and is of practical importance for light energy conversion [130]. A major problem is to achieve long-lived charge separation by inhibiting charge recombination reactions. The idea is to create a system to improve the efficiency of energy storage by preventing rapid back electron transfer reactions [131].

Krishna and Kevan, in trying to place N,N,N',N'-tetramethylbenzidine (TMB) in the α-ZrP layers, only succeeded to cover the surface. Nevertheless, after being photoirradiated by >370 nm visible light at room temperature, the colourless sample turned green and became ESR active, confirming the photoionization of the ZrP–TMB samples. The longer the irradiation, up to one hour, of the samples, the stronger the ESR original. The half-life of the photoinduced TMB$^+$ is about 10 h at room temperature [131]. This is longer than the half-life in micelles and in amorphous silica.

The problem of solving the intercalation of large molecules into ZrP was subsequently solved by Marti and Colon [132]. They synthesized the θ-ZrP phase as described by Kijima [133]. The theta phase is α-ZrP with 5–6 mol of water between the layers creating a 10.3 Å interlayer spacing [134, 135]. Intercalation of $Ru(bpy)_3^{2+}$ was carried out at room temperature with θ-ZrP in water. After

(a)   (b)   (c)

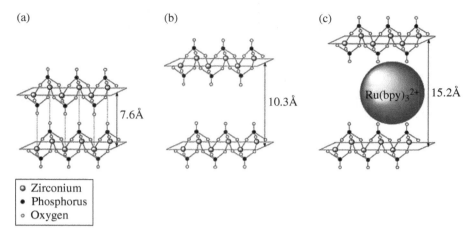

7.6Å

10.3Å

Ru(bpy)$_3^{2+}$ 15.2Å

○ Zirconium
● Phosphorus
○ Oxygen

FIGURE 1.27   Idealized representation of three different zirconium phosphate phases: (a) α-ZrP, (b) expanded 10.3 Å phase and (c) Ru(bpy)$_3^{2+}$-exchanged ZrP. Hydrogen atoms and water molecules are not shown for clarity (Taken from: *Inorg. Chem.* 2003, *42*, 2830–2832; DOI: 10.1021/ic025548g).

several days, all the interior sites of the host were occupied by the ruthenium cation as shown figuratively (Figure 1.27).

Subsequently, bis(n^5-cyclopentadienyl) iron(II), ferrocene (Fc), was intercalated into the θ-ZrP layers where it becomes the ferrocenium ion Fc$^+$. It was found that the intercalated Fc$^+$ is stabilized within the ZrP layers and remains electroactive. It is capable of oxidizing cytochrome C and quenching the excited state of Ru(bpy)$_3^{2+}$ in aqueous solution [136].

The ability to functionalize the surface of α-ZrP opens up more possibilities for electron transfer reactions. For example, the surface coverage can be changed in an almost unlimited way and electron acceptors bonded to each new surface to assess the electron pathways. Also, it is easy to design electron-conducting polymer composites by adding acceptor groups in the polymer and making the donor in ZrP have functional groups compatible with the polymer.

Our first step in this effort was to intercalate Ru(bpy)$_3^{2+}$ within the layers of θ-ZrP. The platelets were then functionalized with OTS in toluene. There was no significant leaching of the Ru(bpy)$_3^{2+}$ as no change in colour of the solvent was observed. The particles were extremely hydrophobic in character so C$_{60}$ fullerene, also hydrophobic, was used as the acceptor. The luminescence spectra of this system suspended in 1,2-dichlorobenzene with different concentrations of C$_{60}$ from 0 to 300 μm were determined. The results are shown in Figure 1.28.

This result shows that the microenvironment of the intercalated Ru(bpy)$_3^{2+}$ is intact in the interlayer region and was not affected by the surface modification reaction. The inset in Figure 1.28 also shows the Stern–Volmer plot for the quenching of the surface-modified Ru(bpy)$_3^{2+}$/ZrP by C$_{60}$ (inset), indicating an entirely static

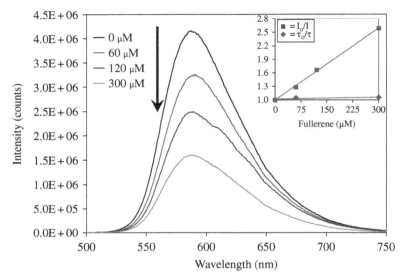

FIGURE 1.28 Luminescent spectra of OTS surface-modified Ru(bpy)$_3^{2+}$/ZrP with different concentrations of C$_{60}$ in 1,2-dichlorobenzene. Inset: Stern–Volmer plot for the quenching with C$_{60}$ using steady-state fluorescence intensity (squares) and fluorescence lifetime (diamonds). λ_{ex} = 440 nm.

quenching mechanism due to the lack of overlapping Stern–Volmer plots for the steady-state fluorescence intensity and time-resolved lifetime measurements. Based on the Stern–Volmer plots, we obtained a static quenching constant (K_s) value of 9.9×10^2 M^{-1}. These reactions now open a wide gamut of new applications for the well-known and versatile zirconium phosphate family. More experiments on surface modification on ZrP are in progress, especially in pursuit of electron-conducting polymer composites.

It should also be mentioned that Brunet et al. have been studying electron transfer reactions using γ-ZrP as the host material for the electron donors [137].

1.16 DRUG DELIVERY

Inorganic layered nanomaterials are receiving attention for biomedical applications because of their size, structure and shape. Kumar and co-workers reported the intercalation of several proteins and enzymes into ZrP, mostly the γ-phase [138]. They generally used the exfoliation method of encapsulation and reported improvement in stabilization, thermal exposure and activity of the intercalates [139, 140].

Our initial effort in drug delivery was to encapsulate insulin into θ-ZrP [141]. This was done using water as solvent. The pH was kept at three, which imparted a positive charge onto the insulin molecules and a measure of solubility. A slow ion-exchange process occurs in which a phase with an interlayer spacing of 26.2–29.3 Å is obtained. However, not all of the particles were intercalated, which may have resulted in some of the θ-ZrP reverting to α-ZrP as peaks due to this latter phase appear in the X-ray powder pattern. The insulin-containing particles were roughly 150 nm in length. Release of the insulin takes place in about 30 min at pH 7.4. The main reason for this study was to determine whether an alternative delivery system, oral rather than injection, would serve diabetic patients. The nanoparticles may need some surface coverage to clear the acidity of the stomach but would release slowly once in the bloodstream.

Our current research is centred on the delivery of anticancer drugs directly into the cancerous tumours, avoiding interaction with healthy tissue. To affect this salutary result, we make use of α-ZrP nanoparticles as host for the anticancer agents. This compound can be prepared in a form that readily intercalates between their layers high amounts of the anticancer drugs (ca. 30% w/w), such as cisplatin and doxorubicin, among others, and later releases the drugs under pH stimulus. Under standard biological conditions, ZrP is stable and releases its drug in response to the lower pH of the cancer cells and endosomes. Moreover, the extreme conditions of the lysosome and the peroxisomes that are triggered during apoptosis may dissociate the ZrP to form phosphate ion and hydrous zirconia. These products have been found to be biologically benign [142].

It is important to remark that the drug would be biologically inactive while intercalated inside the layers of ZrP and therefore excluded from the external medium and then released and become active once it has reached its target.

The potential for using inorganic layered nanoparticles (ILN) as non-viral vectors and drug carriers has been explored by many investigators [141, 143–146]. However, not many materials have the ability to effectively deliver these anticancer agents in high dosage with minimum damage to healthy tissue. Perhaps the best example is the use of layered double hydroxides [143, 144, 147]. These compounds sorb anions, whereas ZrP is a cation exchanger but also intercalates neutral molecules. There are several advantages to using ZrP over other materials. ZrP nanoparticles have a platelet shape and as we will show can be made small enough to penetrate into cancer cells [146]. ZrP can be compared to highly studied, and widely used in biomedicine, spherical-shaped porous silica nanomaterials [148]. Ferrari et al. have recently referred to spherical nanoparticles as having the worst shape in terms of their margination properties, penetration through vascular fenestrations and adherence to endothelial walls [148–150]. On the contrary, the novel platelet shape of ZrP is very promising to overcome all these disadvantages present in spherical-shaped nanoparticles.

Cisplatin [cis-diamminedichloroplatinum(II)] $(NH_3)_2PtCl_2$ and doxorubicin (Adriamycin) are among the most potent anticancer agents approved for use in humans. However, their clinical effectiveness is limited by significant side effects due to lack of specificity and emergence of drug resistance. We believe that this side effect can be significantly avoided if we encapsulate the drugs into the layer of ZrP,

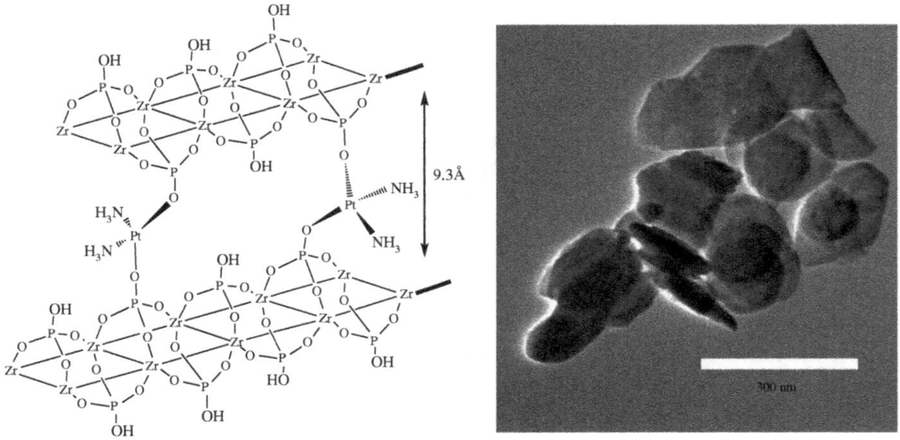

FIGURE 1.29 Idealized representation of cisplatin in the galleries of zirconium phosphate, bonded to the phosphates of the layers in a cross-linked fashion. TEM images of cisplatin–ZrP 1 : 1 intercalation product (Taken from: Diaz, A. Ph. D Thesis, University of Puerto Rico, San Juan 2010).

excluding them from the media until they reach the acidic environment of the tumour. We will describe our preliminary results of the intercalation of both cisplatin and doxorubicin, respectively, into ZrP. The samples were examined by X-ray powder diffraction (XRPD), scanning and transmission electron microscopy, spectroscopic methods and modelling as shown on the left half of Figure 1.29).

In the case of cisplatin, elemental analysis showed that most of the Cl was displaced by the bonding of cisplatin to the layers, producing a new phase with an interlayer distance of 9.3 Å. TEM images show an average particle size of ca. 130 nm [151]. The release profile of the cisplatin–ZrP intercalation product at different pH values in a phosphate buffer shows to be 2.5 times faster at pH 5.4 than at physiological pH. The latter pH approaches the typical pH of the acidic environment of tumour cells, endosomes and lysosomes. The potential inhibition of cell growth by ZrP nanoparticles was evaluated on human breast adenocarcinoma (MCF-7) and T cell lymphoblast-like cell (CEM) lines after a 24 h treatment using the MTT assay procedure. ZrP alone did not affect the viability of MDF-7 and slightly reduced viability of CEM cells. Treatment of CEM cell line with cisplatin reduced the viable cells with an IC_{50} value of 1 μm. However, ZrP–cisplatin (1 : 1) at the concentration of 100 μm reduced 30% of the cell growth.

On the other hand, the intercalation of doxorubicin into ZrP produces a new phase with an interlayer distance of 20.3 Å, A representative model of the ZrP: Cis-platinum phase and its X-ray powder pattern are shown in Figure 1.30. TEM images show an average particle size of ca. 150 nm. The potential inhibition of cell growth by ZrP nanoparticles was evaluated on human breast adenocarcinoma (MCF-7) cell lines after a 24 h treatment using the MTT assay procedure. The cell

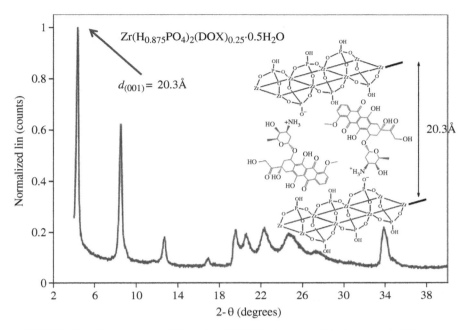

FIGURE 1.30 X-ray powder diffraction (XRPD) of doxorubicin-intercalated ZrP. *Inset:* schematic representation of the intercalation of doxorubicin into α-ZrP by direct ion exchange; the image shows the intercalation of doxorubicin between the layers of ZrP causing the swelling of the particles, forming a new phase with an interlayer distance of 20.3 Å (Modified from: *Chem. Commun.*, 2012, *48*, 1754–1756; DOI: 10.1039/C2CC16218K).

TABLE 1.6   Cell viability assay for doxorubicin in solution and doxorubicin ion exchanged into ZrP

Cell line	Time (h)	IC$_{50}$		
		Doxorubicin (nM)	ZrP–doxorubicin (nM)	Improvement
MCF-7	24	91.6	31.4	~3×
	48	7.9	2.2	~4×
MDA-MB-231	24	1740	830	~2×
	48	523	199	~3×
	96	2.2	0.8	~3×

viability reveals that the doxorubicin–ZrP intercalation product exhibited a significant higher cytotoxicity than the free doxorubicin. The IC$_{50}$ value for the doxorubicin–Zr nanoparticles (31.41 nM) was found to be ca. 3 times lower than free doxorubicin (91.64 nM). Confocal images show the uptake of the ZrP loaded with doxorubicin into the cell within the first 60 min of incubation (Table 1.6). The

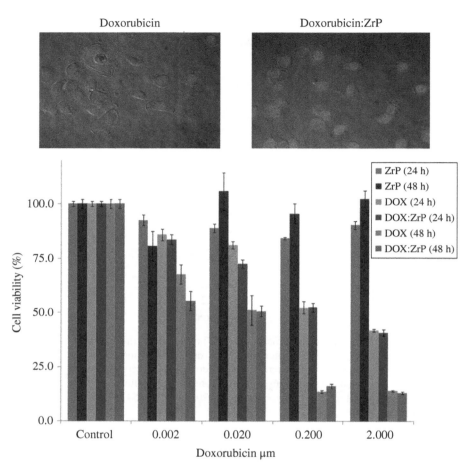

FIGURE 1.31 CLSM images and MTT results in breast cancer cells; top panels show CLSM images of MCF-7 cells treated with DOX (left) and DOX:ZrP (right) nanoplatelets showing the higher uptake of the DOX:ZrP nanoplatelets into the cells. The bottom panel shows MTT assay results in MCF-7 cell lines at 24 h and 48 h of exposure of DOX:ZrP, with their respective controls ($n = 3$, $p < 0.05$ for both 24 h and 48 h of exposure). (Modified from: *Chem. Commun.*, 2012, *48*, 1754–1756; DOI: 10.1039/C2CC16218K, Top panels and Nanoscale, 2013, *5*, 2328–2336. Ref. 151 for bottom graph).

drug release curves are shown in Figure 1.31, and the cellular uptake as revealed by confocal microscope is shown in Figure 1.32.

These results are sufficiently optimistic to consider carrying out *in vivo* studies. In addition, the surfaces require functionalization in order to protect the

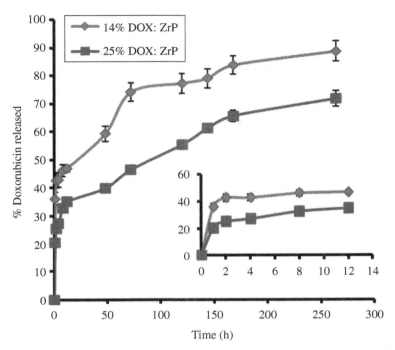

FIGURE 1.32 *In vitro* release profile of doxorubicin (DOX) from different loading level of DOX:ZrP NPs. Drug release study was performed at 37 °C in *simulated body fluid (SBF) pH 7.4*, under shaking (100 rpm) Mean ± SD, *n*=3. (Taken from: *Nanoscale*, 2013, *5*, 2328–2336; DOI: REF. 151, 10.1039/C3NR34242E).

drug from premature release. This we are doing given the fact that the surface layers of ZrP are amenable to forming bonds with such a great array of compounds.

1.17 CONCLUSIONS

We have now completed our excursion through α-ZrP chemistry from its beginnings to the present. But this excursion did not carry us completely through the length and breadth of the subject. We did not include sensors and biosensors, flame retardants, antibiomaterials, and so on. In addition, highly porous forms have been prepared that improve catalytic activity, sorption phenomena and other useful applications. Even after almost 50 years of study, the story is not complete. If we add to this story a similar one for γ-ZrP, also first prepared by the senior author, an equivalently varied and important narrative would obtain. As a start, the reader may wish to consult the excellent review article by Bruno Bujoli et al. [152].

REFERENCES

1. Kraus, K. A., Phillips, H. O. *J. Am. Chem. Soc.* **1956**, *78*, 694.
2. Kraus, K. A., Phillips, H. D., Carlson, T. A., Johnson, J. S. In *Proc. Second Int. Conf.*, United Nations, **1958**, p. 1832.
3. Clearfield, A., Stynes, J. A. *J. Inorg. Nucl. Chem.* **1964**, *26*, 117.
4. Clearfield, A., Smith, G. D. *Inorg. Chem.* **1969**, *8*, 431.
5. Troup, J. M., Clearfield, A. *Inorg. Chem.* **1977**, *16*, 3311.
6. Clearfield, A., Thomas, J. R. *Inorg. Nucl. Chem. Lett.* **1969**, *5*, 775.
7. Sun, L., Boo, W. J., Sue, H.-J., Clearfield, A. New *J. Chem.* **2007**, *31*, 39.
8. Pica, M., Donnadio, A., Capitani, D., Vivani, R., Troni, E., Casciola, M. *Inorg. Chem.* **2011**, *50*, 11623.
9. Trobajo, C., Khainakov, S. A., Espina, A., Garcia, J. R. *Chem. Mater.* **2000**, *12*, 1787.
10. Alberti, G. *Acc. Chem. Res.* **1978**, *11*, 163.
11. Kaschak, D. M., Johnson, S. A., Hooks, D. E., Kim, H.-N., Ward, M. D., Mallouk, T. E. *J. Am. Chem. Soc.* **1998**, *120*, 10887.
12. Clearfield, A., Ed. *Inorganic Ion Exchange Materials*; CRC Press, Boca Raton, FL, **1982**.
13. Clearfield, A., Troup, J. M. *J. Phys. Chem.* **1970**, *74*, 2578.
14. Clearfield, A., Jirustithipong, P. Fast Ion Transp. Solids electrodes electrolytes, *Proc. Int. Conf.*, **1979**, p. 153.
15. Jerus, P., Clearfield, A. *J. Inorg. Nucl. Chem.* **1981**, *43*, 2117.
16. Clearfield, A., Saldarriaga-Molina, C. H., Buckley, R. H., Solid-Solid Ion Exchange, II. Zeolites in Proc. Int. Conf. Mol. Sieves, Uytterhoeven, J. B., Ed. Univ. Leuven Press, Leuven, Belgium, **1973**, 241.
17. Clearfield, A. *Prog. Intercalation Res.* **1994**, *17*, 223.
18. Clearfield, A., Tindwa, R. M. *J. Inorg. Nucl. Chem.* **1979**, *41*, 871.
19. Alberti, G., Casciola, M., Costantino, U. *J. Colloid Interface Sci.* **1985**, *107*, 256.
20. Sun, L., O'Reilly, J. Y., Kong, D., Su, J. Y., Boo, W. J., Sue, H. J., Clearfield, A. *J. Colloid Interface Sci.* **2009**, *333*, 503.
21. Gordon, A., Better, O. S., Greenbaum, M. A., Marantz, L. B., Gral, T., Maxwell, M. H. *Trans. Am. Soc. Artif. Intern. Organs.* **1971**, *17*, 253.
22. Clearfield, A. *Ind. Eng. Chem. Res.* **1995**, *34*, 2865.
23. Clearfield, A. *Annu. Rev. Mater. Sci.* **1984**, *14*, 205.
24. Clearfield, A., Oskarsson, A., Oskarsson, C. *Ion Exch. Membr.* **1972**, *1*, 91.
25. Clearfield, A., Jahangir, L. M. In *Recent Developments in Separation Science*; Navratil, J. D., Ed., CRC Press, Boca Raton, FL, **1984**, Vol. VIII, Ch.4.
26. Eisenman, G. *Biophys. J.* **1962**, *2*, 259.
27. Kullberg, L. H., Clearfield, A. *J. Phys. Chem.* **1981**, *85*, 1578.
28. Gal, I., Ruvarac, A. *J. Chromatogr. A* **1964**, *13*, 549.
29. Horwitz, E. P. *J. Inorg. Nucl. Chem.* **1966**, *28*, 1469.
30. Moore, F. L. *Anal. Chem.* **1971**, *43*, 487.
31. Shafiev, A. I., Efremov, Y. V., Andreev, V. P. *Radiokhimiya* **1973**, *15*, 265.
32. Shafiev, A. I., Efremov, Y. V., Nikolaev, V. M., Yakovlev, G. N. *Radiokhimiya* **1971**, *13*, 129.

33. Cahill, R., Shpeizer, B., Peng, G. Z., Bortun, L., Clearfield, A. In *Separation of F Elements*; Nash, K. L., Choppin, G. R., Eds., Plenum Press Div Plenum Publishing Corporation, New York, **1995**, p. 165.

34. Burns, J. D., Clearfield, A., Borkowski, M., Reed, D. T. *Radiochim. Acta* **2012**, *100*, 381.

35. Burns, J. D., Morkowski, M., Clearfield, A., Reed, D. T. *Radiochim. Acta* **2012**, *100*, 901.

36. Burns, J. D., Shehee, T. C., Clearfield, A., Hobbs, D. T. *Anal. Chem.* **2012**, *84*, 6930.

37. Khan, A. A., Paquiza, L. *Desalination* **2011**, *265*, 242.

38. Pan, B., Zhang, Q., Du, W., Zhang, W., Pan, B., Zhang, Q., Xu, Z., Zhang, Q. *Water Res.* **2007**, *41*, 3103.

39. Zhuravlev, I., Zakutevsky, O., Psareva, T., Kanibolotsky, V., Strelko, V., Taffet, M., Gallios, G. *J. Radioanal. Nucl. Chem.* **2002**, *254*, 85.

40. Mennan, C., Paterson-Beedle, M., Macaskie, L. *Biotechnol. Lett.* **2010**, *32*, 1419.

41. Wang, Z., Pinnavaia, T. *J. Chem. Mater.* **1998**, *10*, 1820.

42. Chan, C.-M., Wu, J., Li, J.-X., Cheung, Y.-K. *Polymer* **2002**, *43*, 2981.

43. Usuki, A., Kojima, Y., Kawasumi, M., Okada, A., Fukushima, Y., Kurauchi, T., Kamigaito, O. *J. Mater. Res.* **1993**, *8*, 1179.

44. Wang, H., Zeng, C., Elkovitch, M., Lee, L. J., Koelling, K. W. *Polym. Eng. Sci.* **2001**, *41*, 2036.

45. Sue, H. J., Gam, K. T., Bestaoui, N., Spurr, N., Clearfield, A. *Chem. Mater.* **2003**, *16*, 242.

46. Boo, W. J., Sun, L. Y., Liu, J., Clearfield, A., Sue, H. J., Mullins, M. J., Pham, H. *Compos. Sci. Technol.* **2007**, *67*, 262.

47. Sun, L., Boo, W.-J., Liu, J., Clearfield, A., Sue, H.-J., Verghese, N. E., Pham, H. Q., Bicerano, J. *Macromol. Mater. Eng.* **2009**, *294*, 103.

48. Krieger, I. M., Dougherty, T. *J. Trans. Soc. Rheol.* **1959**, *3*, 137.

49. Boo, W. J., Sun, L., Warren, G. L., Moghbelli, E., Pham, H., Clearfield, A., Sue, H. *J. Polymer* **2007**, *48*, 1075.

50. Boo, W. J., Sun, L., Liu, J., Clearfield, A., Sue, H.-J. *J. Phys. Chem. C* **2007**, *111*, 10377.

51. Casciola, M., Alberti, G., Donnadio, A., Pica, M., Marmottini, F., Bottino, A., Piaggio, P. *J. Mater. Chem.* **2005**, *15*, 4262.

52. Hung, Y., Carrot, C., Chalamet, Y., Dal Pont, K., Espuche, E. *Macromol. Mater. Eng.* **2012**, *297*, 768.

53. Brandão, L. S., Mendes, L. C., Medeiros, M. E., Sirelli, L., Dias, M. L. *J. Appl. Polym. Sci.* **2006**, *102*, 3868.

54. Sun, L., Liu, J., Kirumakki, S. R., Schwerdtfeger, E. D., Howell, R. J., Al-Bahily, K., Miller, S. A., Clearfield, A., Sue, H.-J. *Chem. Mater.* **2009**, *21*, 1154.

55. Wang, Z. M., Han, H., Chung, T. C. *Macromol. Symp.* **2005**, *225*, 113.

56. Manias, E., Touny, A., Wu, L., Strawhecker, K., Lu, B., Chung, T. C. *Chem. Mater.* **2001**, *13*, 3516.

57. Hasegawa, N., Okamoto, H., Kawasumi, M., Kato, M., Tsukigase, A., Usuki, A. *Macromol. Mater. Eng.* **2000**, *280–281*, 76.

58. Kim, D. H., Fasulo, P. D., Rodgers, W. R., Paul, D. R. *Polymer* **2007**, *48*, 5308.

59. Chung, T. C. *J. Organomet. Chem.* **2005**, *690*, 6292.

60. Alberti, G., Costantino, U., Környei, J., Giovagnotti, M. L. L. *React. Polym. Ion Exch. Sorb.* **1985**, *4*, 1.

61. Lu, H., Wilkie, C. A. *Polym. Adv. Technol.* **2011**, *22*, 1123.

62. Dal pont, K., Gérard, J. F., Espuche, E. *Eur. Polym. J.* **2012**, *48*, 217.

63. Casciola, M., Capitani, D., Donnadio, A., Munari, G., Pica, M. *Inorg. Chem.* **2010**, *49*, 3329.

64. Pica, M., Donnadio, A., Bianchi, V., Fop, S., Casciola, M. *Carbohydr. Polym.* **2013**, *97*, 210.

65. Sagiv, J. *J. Am. Chem. Soc.* **1980**, *102*, 92.

66. Wasserman, S. R., Tao, Y.-T., Whitesides, G. M. *Langmuir.* **1989**, *5*, 1074.

67. Ruckenstein, E., Li, Z. F. *Adv. Colloid Interface Sci.* **2005**, *113*, 43.

68. Sieval, A. B., Linke, R., Zuilhof, H., Sudholter, E. J. R. *Adv. Mater.* **2000**, *12*, 1457.

69. Haensch, C., Hoeppener, S., Schubert, U. S. *Chem. Soc. Rev.* **2010**, *39*, 2323.

70. Park, J.-W., Park, Y.-J., Jun, C.-H. *Chem. Commun.* **2011**, *47*, 4860.

71. Buriak, J. M. *Chem. Commun.* **1999**, 1051.

72. Ulman, A. *Chem. Rev.* **1996**, *96*, 1533.

73. Herzer, N., Hoeppener, S., Schubert, U. S. *Chem. Commun.* **2010**, *46*, 5634.

74. Díaz, A., Mosby, B. M., Bakhmutov, V. I., Martí, A. A., Batteas, J. D., Clearfield, A. *Chem. Mater.* **2013**, *25*, 723.

75. Ramachandran, R., Paul, W., Sharma, C. P. *J. Biomed. Mater. Res. Part B* **2009**, *88*, 41.

76. de Gennes, P. G. *Rev. Modern Phys.* **1992**, *64*, 645.

77. Zhang, C., Liu, B., Tang, C., Liu, J., Qu, X., Li, J., Yang, Z. *Chem. Commun.* **2010**, *46*, 4610.

78. Ling, X. Y., Phang, I. Y., Acikgoz, C., Yilmaz, M. D., Hempenius, M. A., Vancso, G. J., Huskens, J. *Angew. Chem. Int. Ed. Engl.* **2009**, *48*, 7677.

79. Walther, A., Miller, A. H. E. *Soft Matter.* **2008**, *4*, 663.

80. Du, J., O'Reilly, R. K. *Chem. Soc. Rev.* **2011**, *40*, 2402.

81. Perro, A., Reculusa, S., Ravaine, S., Bourget-Lami, E., Dugutt, E. *J. Mater. Chem.* **2005**, *15*, 3745.

82. Jiang, S., Chen, Q., Tripathy, M., Luijten, E., Schweizer, K. S., Granick, S. *Adv. Mater.* **2010**, *22*, 1060.

83. Binks, B. P., Fletcher, P. D. I. *Langmuir* **2001**, *17*, 4708.

84. Faria, J., Ruiz, M. P., Resasco, D. E. *Adv. Synth. Catal.* **2010**, *352*, 2359.

85. Crowley, J. M., Sheridon, N. K., Romano, L. *J. Electrostat.* **2002**, *55*, 247.

86. Mejia, A. F., Diaz, A., Pullela, S., Chang, Y.-W., Simonetty, M., Carpenter, C., Batteas, J. D., Mannan, M. S., Clearfield, A., Cheng, Z. *Soft Matter.* **2012**, *8*, 10245.

87. Kalman, T. J., Dudukovic, M., Clearfield, A. *Adv. Chem. Ser.* **1974**, *133*, 654.

88. Clearfield, A., Pack, S. P. *J. Catal.* **1978**, *51*, 431.

89. Clearfield, A., Thakur, D. S. *J. Catal.* **1980**, *65*, 185.

90. Clearfield, A. *J. Mol. Catal.* **1984**, *27*, 251.

91. Cheung, H. C., Clearfield, A. *J. Catal.* **1986**, *98*, 335.

92. Cheng, S., Clearfield, A. *J. Catal.* **1985**, *94*, 455.

93. Clearfield, A., Thakur, D. S. *Appl. Catal.* **1986**, *26*, 1.

94. Costantino, U., Marmottini, F., Curini, M., Rosati, O. *Catal. Lett.* **1993**, *22*, 333.

95. Zhao, Y., Li, F., Zhang, R., Evans, D. G., Duan, X. *Chem. Mater.* **2002**, *14*, 4286.

96. Zhang, F., Xie, Y., Lu, W., Wang, X., Xu, S., Lei, X. *J. Colloid Interface Sci.* **2010**, *349*, 571.

97. Khare, S., Chokhare, R. *J. Mol. Catal. A Chem.* **2012**, *353*, 138.

98. Niño, M. E., Giraldo, S. A., Paez-Mozo, E. A. *J. Mol. Catal. A Chem.* **2001**, *175*, 139.

99. Alvaro, V. F. D., Johnstone, R. A. W. *J. Mol. Catal. A Chem.* **2008**, *280*, 131.

100. Poojary, M. D., Hu, H. L., Campbell, F. L., III, Clearfield, A. *Acta Crystallogr. Sect. B Struct. Sci.* **1993**, *B49*, 996.

101. Clearfield, A. in *Metal Phosphonate Chemistry: From Synthesis to Applications*, Clearfield, A.; Demadis, K. Eds. RSC Publ. Cambridge, U.K. **2013**, Ch. 1.

102. Wang, Z., Heising, J. M., Clearfield, A. *J. Am. Chem. Soc.* **2003**, *125*, 10375.

103. Curini, M., Epifano, F., Marcotullio, M. C., Rosati, O. *Eur. J. Org. Chem.* **2001**, 4149.

104. Curini, M., Epifano, F., Marcotullio, M. C., Rosati, O. *Synlett.* **2001**, 1182.

105. Curini, M., Epifano, F., Marcotullio, M. C., Rosati, O., Costantino, U. *Tetrahedron Lett.* **1998**, *39*, 8159.

106. Curini, M., Rosati, O., Pisani, E., Costantino, U. *Synlett* **1996**, 333.

107. Curini, M., Marcotullio, M. C., Pisani, E., Rosati, O., Costantino, U. *Synlett* **1997**, 769.

108. Rosati, O., Curini, M., Marcotullio, M. C., Macchiarulo, A., Perfumi, M., Mattioli, L., Rismondo, F., Cravotto, G. *Bioorg. Med. Chem.* **2007**, *15*, 3463.

109. Curini, M., Rosati, O., Costantino, U. *Curr. Org. Chem.* **2004**, *8*, 591.

110. Alberti, G., Casciola, M., Costantino, U., Levi, G., Ricciardi, G. *J. Inorg. Nucl. Chem.* **1978**, *40*, 533.

111. Casciola, M., Costantino, U. *Solid State Ionics* **1986**, *20*, 69.

112. Clearfield, A., Berman, J. R. *J. Inorg. Nucl. Chem.* **1981**, *43*, 2141.

113. Yang, C. Y., Clearfield, A. *React. Polym.* **1987**, *5*, 13.

114. Alberti, G., Casciola, M., Costantino, U., Peraio, A., Montoneri, E. *Solid State Ionics* **1992**, *50*, 315.

115. Stein, S. E. W., Clearfield, A., Subramanian, M. A. *Solid State Ionics* **1996**, *83*, 113.

116. Kullberg, L. H., Clearfield, A. *Solvent Extr. Ion Exch.* **1989**, *7*, 527.

117. Kullberg, L. H., Clearfield, A. *Solvent Extr. Ion Exch.* **1990**, *8*, 187.

118. Song, C. *Catal. Today.* **2002**, *77*, 17.

119. Hirschenhofer, J. H., Stauffer, D. B., Engleman, R. R., Klett, M. G. *Fuel Cell Handbook, Fourth Edition (DOE/FETC-99/1076)*; Department of Energy, Morgantown, WV, **1998**.

120. Alberti, G., Casciola, M., Massinelli, L., Bauer, B. *J. Membr. Sci.* **2001**, *185*, 73.

121. Alberti, G., Casciola, M. *Annu. Rev. Mater. Sci.* **2003**, *33*, 129.

122. Alberti, G., Casciola, M., Pica, M., Di Cesare, G. *Ann. N. Y. Acad. Sci.* **2003**, *984*, 208.

123. Casciola, M., Alberti, G., Ciarletta, A., Cruccolini, A., Piaggio, P., Pica, M. *Solid State Ionics* **2005**, *176*, 2985.

124. Alberti, G., Casciola, M., Pica, M., Tarpanelli, T., Sganappa, M. *Fuel Cells* **2005**, *5*, 366.

125. Lee, J.-M., Kikuchi, Y., Ohashi, H., Tamaki, T., Yamaguchi, T. *J. Mater. Chem.* **2010**, *20*, 6239.

126. Pica, M., Donnadio, A., Casciola, M., Cojocaru, P., Merlo, L. *J. Mater. Chem.* **2012**, *22*, 24902.

127. Colon, J. L., Yang, C. Y., Clearfield, A., Martin, C. R. *J. Phys. Chem.* **1988**, *92*, 5777.

128. Colon, J. L., Yang, C. Y., Clearfield, A., Martin, C. R. *J. Phys. Chem.* **1990**, *94*, 874.

129. Albery, W. J., Bartlett, P. N., Wilde, C. P., Darwent, J. R. *J. Am. Chem. Soc.* **1985**, *107*, 1854.

130. Kevin, L. *Photoinduced Electron Transfer Part B: Experimental Techniques and Medium Effects*; Elsevier Science Ltd, Amsterdam, **1988**.

131. Krishna, R. M., Kevan, L. *Microporous Mesoporous Mater.* **1999**, *32*, 169.

132. Marti, A. A., Colon, J. L. *Inorg. Chem.* **2003**, *42*, 2830.

133. Kijima, T. *Bull. Chem. Soc. Jpn.* **1982**, *55*, 3031.

134. Clearfield, A., Duax, W. L., Medina, A. S., Smith, G. D., Thomas, J. R. *J. Phys. Chem.* **1969**, *73*, 3424.

135. Alberti, G., Costantino, U., Gill, J. S. *J. Inorg. Nucl. Chem.* **1976**, *38*, 1733.

136. Santiago, M. B., Declet-Flores, C., Díaz, A., Vélez, M. M., Bosques, M. Z., Sanakis, Y., Colón, J. L. *Langmuir* **2007**, *23*, 7810.

137. Brunet, E., Alonso, M., Cerro, C., Juanes, O., Rodriguez-Ubis, J. C., Kaifer, A. E. *Adv. Funct. Mater.* **2007**, *17*, 1603.

138. Kumar, C. V., Chaudhari, A. *J. Am. Chem. Soc.* **2000**, *122*, 830.

139. Bhambhani, A., Kumar, C. V. *Microporous Mesoporous Mater.* **2008**, *110*, 517.

140. Jagannadham, V., Bhambhani, A., Kumar, C. V. *Microporous Mesoporous Mater.* **2006**, *88*, 275.

141. Díaz, A., David, A., Pérez, R., González, M. L., Báez, A., Wark, S. E., Zhang, P., Clearfield, A., Colón, J. L. *Biomacromolecules* **2010**, *11*, 2465.

142. Saxena, V., Diaz, A., Clearfield, A., Batteas, J. D., Hussain, M. D. *Nanoscale* **2013**, *5*, 2328.

143. Choy, J.-H., Kwak, S.-Y., Jeong, Y.-J., Park, J.-S. *Angew. Chem. Int. Ed.* **2000**, *39*, 4041.

144. Yang, J.-H., Han, Y.-S., Park, M., Park, T., Hwang, S.-J., Choy, J.-H. *Chem. Mater.* **2007**, *19*, 2679.

145. Oh, J.-M., Choi, S.-J., Kim, S.-T., Choy, J.-H. *Bioconjug. Chem.* **2006**, *17*, 1411.

146. Diaz, A., Saxena, V., Gonzalez, J., David, A., Casanas, B., Carpenter, C., Batteas, J. D., Colon, J. L., Clearfield, A., Delwar Hussain, M. *Chem. Commun.* **2012**, *48*, 1754.

147. Choi, S.-J., Choi, G. E., Oh, J.-M., Oh, Y.-J., Park, M.-C., Choy, J.-H. *J. Mater. Chem.* **2010**, *20*, 9463.

148. Gentile, F., Curcio, A., Indolfi, C., Ferrari, M., Decuzzi, P. *J. Nanobiotechnology* **2008**, *6*, 9.

149. Decuzzi, P., Lee, S., Bhushan, B., Ferrari, M. *Ann. Biomed. Eng.* **2005**, *33*, 179.

150. Ferrari, M. *Nat. Nanotechnol.* **2008**, *3*, 131.

151. Díaz, A., González, M. L., Pérez, R. J., David, A., Mukherjee, A., Báez, A., Clearfield, A., Colón, J. L. *Nanoscale* **2013**, *5*, 11456.

152. Queffélec, C., Petit, M., Janvier, P., Knight, D. A., Bujoli, B. *Chem. Rev.* **2012**, *112*, 3777.

2

TALES FROM THE UNEXPECTED: CHEMISTRY AT THE SURFACE AND INTERLAYER SPACE OF LAYERED ORGANIC–INORGANIC HYBRID MATERIALS BASED ON γ-ZIRCONIUM PHOSPHATE

ERNESTO BRUNET, MARÍA DE VICTORIA-RODRÍGUEZ, LAURA JIMÉNEZ GARCÍA-PATRÓN, HUSSEIN HINDAWI, ELENA RODRÍGUEZ-PAYÁN, JUAN CARLOS RODRÍGUEZ-UBIS AND OLGA JUANES

Departamento de Química Orgánica, Facultad de Ciencias, Universidad Autónoma de Madrid, Madrid, Spain

2.1 INTRODUCTION

The first part of this chapter's title is taken after that of Roald Dahl's book and TV series. It is not that the chemistry herein would have anything sinister or any wryly comedic undertone, but it happens that every time we have planned chemistry with laminar organic–inorganic materials, using zirconium phosphate as the inorganic scaffold, an unexpected, twisted ending always occurred. In short, this chapter will show several practical examples of *serendipity*, the finding of the unexpected, from the relatively simple process of intercalating organic species between the layers of zirconium phosphate.

Feynman's famous queries ('What would the properties of materials be if we could really arrange the atoms (molecules) the way we want them? [1] What could we do with layered structures with just the right layers?') struck us long ago when we started addressing the problem of building new porous solids with controlled geometry and

Tailored Organic–Inorganic Materials, First Edition. Edited by Ernesto Brunet, Jorge L. Colón and Abraham Clearfield.
© 2015 John Wiley & Sons, Inc. Published 2015 by John Wiley & Sons, Inc.

properties, a challenge of enormous technological and scientific importance. Along the past 20 years, a great deal of chemists, physicists, engineers and even biologists have worked intertwined in accomplishing Feynman's challenge. Yet, in the pursuit of this ambitious goal, science appears to still be far from achieving a once-and-for-all solution. Fortunately, again in Feynman's words, 'this leaves plenty of room at the bottom', a fact that has led to a restless change in chemical concepts. Old traditional chemical principles viewed through the optics of supramolecular chemistry [2] (chemistry beyond the molecule) and crystal engineering [3] (the understanding of molecular interactions under crystal packing) have enlightened the development of many different chemistries (sol–gel [4], self-assembly [5], film-making [6], zeolite based [7], etc.) and the production of novel materials with very interesting new-fangled properties. However, despite the myriad intelligent findings achieved in the past four decades reigned by this conceptual revolution, the science behind the discovery of new materials still allows for many serendipity-driven breakthroughs, the alma mater of basic research.

When the appropriate organic species are forced to stay between the layers of certain inorganic salts, new materials arise where the behaviour of the organic molecules is drastically modified. This is the consequence of two main factors. Firstly, the inorganic scaffold forces the organic molecules to stay in a different arrangement to that they attain when pure. But the *fact of confinement* is perhaps more determinant for the organic molecules to change their properties or even display new ones at the supramolecular level. To this effect, we and others have found that zirconium phosphate is a versatile layered inorganic salt, kind of 'carving board' where organic compounds can be attached, either covalently or by ionic forces, by mild hydrothermal reactions. This extremely simple chemical and conceptual approach produces complex 3D structures with remarkable properties and potential applications that are only limited by the imagination of the researcher. A representative coverage of our progress in this field will be presented in this chapter, embroidering past and ongoing results and ideas concerning the chemistry of metal phosphates/phosphonates essentially in relation with the following topics: (i) molecular recognition, (ii) chemically driven porosity changes, (iii) chiral memory and supramolecular chirality, (iv) luminescence signalling, (v) photoinduced electron transfer, (vi) gas storage and (vii) drug confinement.

2.2   THE INORGANIC SCAFFOLD: γ-ZIRCONIUM PHOSPHATE (MICROWAVE-ASSISTED SYNTHESIS)

The confinement of organic molecules within layered inorganic salts is profusely achieved by nature in the mineral world. Natural clays are an excellent example [8]. There are many other instances, either natural or artificial, like layered double hydroxides [9] (LDHs), layered perovskites, niobates [10], and so on. A remarkable feature of these different layered salts is their general ability to form intercalates by various chemical means, namely, ion-exchange processes. This outstanding property allows their structure to be modified in many ways to produce advanced multifunctional composites with unusual combinations of properties that are difficult to achieve otherwise.

Among all these natural and man-made layered structures, tetravalent metal phosphates stand out, especially those of Zr due to their high stability and ease of use.

Zirconium phosphate [11] can be synthesized in either of the three allotropic forms shown in Figure 2.1, namely, **α-** (a), **γ-** (b) and **λ-ZrP** (c). The composite structures based on zirconium phosphate found in the literature are derived from these three arrangements with slight variations if at all [12]. The three of them contain Zr in octahedral coordination. **α-ZrP** bears only one kind of phosphate (navy blue tetrahedrons), whereas two of them can be distinguished in **γ-ZrP**, one internal (green tetrahedrons), arranged within the layers in two slightly different levels, and another external (light blue tetrahedrons), exposed at the surface of both faces of every lamellae. **λ-ZrP** contains a single level of internal phosphates (green tetrahedrons), but the unique feature of this structure is the fact that every Zr atom completes

(a)

(b)

(c)

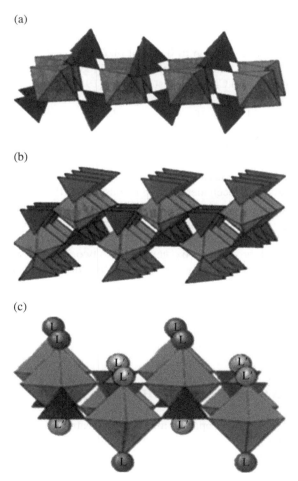

FIGURE 2.1 Most common structures of zirconium phosphates: **α-** (a), **γ-** (b) and **λ-ZrP** (c).

its octahedral coordination by bonding to one negatively charged ligand (blue L) and to another neutral one (orange L′) in order to achieve electroneutrality.

As simple as these arrangements may seem, their structure was difficult to determine because single crystals can seldom be obtained from them. One of the few examples is the case of **α-ZrP**, first synthesized in 1964 by Clearfield and Stynes [13], whose structure was accurately determined when [14] single crystals of enough quality could be collected. The synthesis [15] of **α-ZrP** in essence comprises a simple reflux (lower crystallinity) or hydrothermal treatment at 200 °C (higher crystallinity) of zirconium oxychloride in 3–6 M phosphoric acid. If the reaction medium contains phosphorus acid ($H-PO_3H_2$) or organic phosphonates ($R-PO_3H_2$), they can be incorporated into the **α-ZrP** structure where the phosphates are partially or totally replaced by phosphite or phosphonate tetrahedrons, respectively, with their correspondingly H or R groups pointing to the interlayer region. There are excellent reviews [16], profusely cited along the years [17],[1] covering this aspect that allows the direct synthesis of umpteen different organic–inorganic structures. **λ-ZrP** was discovered by Clearfield as well [18]. Its synthesis was accomplished by heating at 120 °C for several days a solution of 0.05 mol of $ZrOCl_2$ in 44% HF (Zr:F ratio 1 : 25), 0.1 mol of 85% H_3PO_4 and 0.8 mol of DMSO. Three years later, Alberti described the same structure but with Cl anions instead of F− [19].

The salts **α-** and **γ-ZrP** share the same empirical formula, this fact initially misleading to the consideration of **γ-ZrP** to be a different hydrate of **α-ZrP** [20]. The realization that **α-** and **γ-ZrP** were actually two different structures had to wait 19 years since its discovery, when Clayden [21] showed that the solid-state ^{31}P-NMR of **α-** and **γ-ZrP** displayed one and two signals, respectively. Clearfield's Rietveld refinement of the powder X-ray pattern of **γ-ZrP** put a definite end to previous speculations (Figure 2.1) [22]. As it has been mentioned previously, **γ-ZrP** comprises two planes of octahedral Zr atoms bridged by two series of mirror-image planes of tetrahedral phosphates bonded by four oxygen atoms to four different metals (green tetrahedrons). In the surface of the two faces of a layer, one may find a set of phosphates (blue tetrahedrons) bonded through two oxygen atoms to two different metals. This leads to the conclusion that **γ-ZrP** is better described by the formula $Zr(PO_4)(H_2PO_4)$, while $Zr(HPO_4)_2$ corresponds to **α-ZrP**, the number of atoms being in fact identical for the two phases. Electroneutrality requires that the external phosphates be bonded to the metals using neutral P=O and anionic P–O bonds. Therefore, every surface phosphate points two acidic P–OH bonds towards the interlayer region, whereas in **α-ZrP**, the surface phosphates only have one OH group left.

2.3   MICROWAVE-ASSISTED SYNTHESIS OF γ-ZrP

The classical procedure for preparing involves the heating of a mixture of $ZrOCl_2$, HF and $NH_4H_2PO_4$ (1 : 6 : 50 molar ratio) in water at 90–100 °C. After a minimum of 3 days, multigram quantities of **γ-ZrP** can be easily obtained after the necessary

---

[1] There is a number of recent review papers concerning the use of phosphonates in the formation of open frameworks (MOFs) which are out of the scope of this chapter [17].

filtration, centrifugation and NH_4^+/H^+ exchange with acid. This hydrothermal process seems to be especially appropriate to be tried under microwave irradiation. However, quite surprisingly, there are no precedents in the literature of such an attempt. Table 2.1 summarizes our results.

Figures 2.2 and 2.3 show the powder XRD patterns and the solid-state non-CP ^{31}P-NMR spectra, respectively, of two batches of γ-ZrP prepared by the classical method or under microwatt irradiation (procedure 4 of Table 2.1 as a representative example).

TABLE 2.1   Results of different reaction conditions for the preparation of γ-ZrP

Parameters and methods	Non-microwatt	Microwatt				
		1	2	3	4	5
$ZrOCl_2 \cdot 8H_2O$ (mg)[a]	21,660	217	217	433	433	433
Synthesis	72 h at 100 °C	40 min at 180 °C				20 min at 180 °C
NH_4^+/H^+ exchange	HCl 1 N 36 h 25 °C/HCl 0.01 N 2.5 h 25 °C				HCl 1 N 17 min 90 °C/HCl 0.01 N 3 min 70 °C	
Yield (mg)	9500 (45%)	170 (80%)	172 (80%)	381 (89%)	369 (86%)	395 (92%)
d_{001} acid form (nm)	1.22	1.22	1.23	1.23	1.22	1.22
% N acid form	0.02	0.09	0.06	0.10	0.03	0.02

[a]Starting amount of $ZrOCl_2 \cdot 8H_2O$ in the reaction mixture (total capacity 2 l without microwatt and 20 ml batches with microwatt).

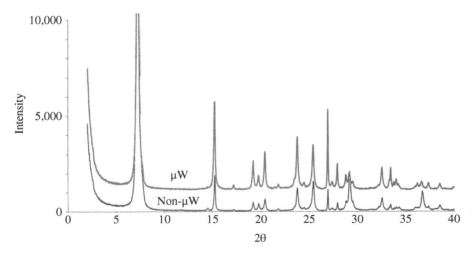

FIGURE 2.2 Comparison between the XRD powder patterns of γ-ZrP prepared with and without microwatt irradiation.

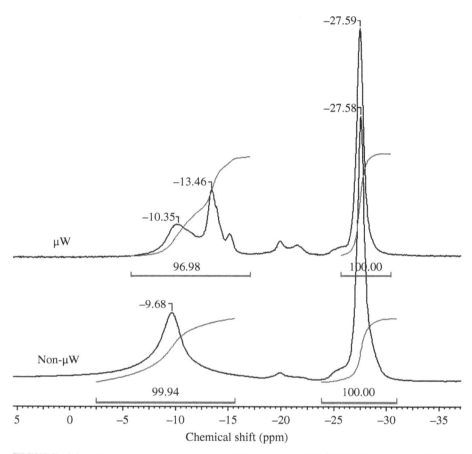

FIGURE 2.3 Comparison between the solid-state non-CP ^{31}P-NMR spectra of γ-ZrP prepared with and without microwatt irradiation.

It can be seen that the quality of the two materials is strikingly similar. The NMR traces show the resonances of the internal phosphates at −27.6 ppm. However, the signals of the surface phosphates are more complex in the case of the microwatt irradiation, suggesting an inhomogeneous degree of hydration of the interlayer space.

The overall result of the reaction under microwatt irradiation (including the NH_4^+/ H^+ exchange with acid) is the drastic increase and reduction, respectively, of yield and time. This is evidenced even with the use of a small microwatt oven. In order to reach the scale of 9.5 g achieved by the classical process after ca. 110 h (non-microwatt data in Table 2.1), the reaction had to be repeated ca. 24 times by means of a robotic vial changer, taking only 16–24 h and an important reduction of ca. 50% of the starting materials (e.g. 10.4 g of $ZrOCl_2 \cdot 8H_2O$ instead of 21.7 g needed in the classical reaction).

2.4 REACTIONS

2.4.1 Intercalation

The Brönsted acid OH groups of the surface phosphates pointing towards the inter-layer region of either α- or γ-phases are responsible of the intercalation processes in ZrP that have been explored for more than four decades. Not only amines can inter-calate within the layers of ZrP, but hydrogen-bonding donor–acceptor molecules of far lower basicity like alkanols, polyols, ketones, sulphoxides, and so on can be intercalated as well. Powder XRD readily shows (Figure 2.4) how the basal spacing of γ-ZrP increases with the length of the amine alkyl chain [23].

The experimental interlayer distance together with the elemental analyses (1 : 1 Zr/amine ratio) of the amine intercalates suggests the formation of a double layer of amines in a similar fashion as phospholipids arrange in a cellular membrane or in a liposome, pointing the polar head towards the ZrP polar surface and the non-polar alkyl chains to the interior. The slope of the straight line arising in the plot of inter-layer distance versus the amine length (0.19 nm) of Figure 2.4 is much lower than the basal spacing increase expected from the equivalent of the two C–C bonds added in the amine double layer (3.1 nm), suggesting a ca. 60° tilting, relative to the average plane of the layers, of the alkyl chains as Figure 2.4 sketches.

These intercalation processes turn out to be very important in the chemistry of ZrP because most of its chemical transformations had to be preceded by interlayer separation either by these intercalation reactions or other appropriate

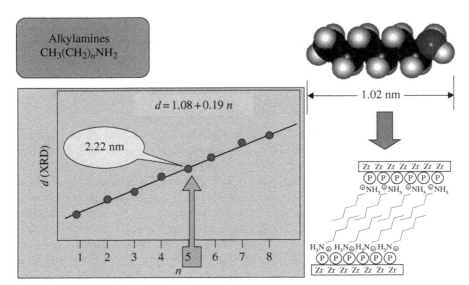

FIGURE 2.4 Double-layer formation in the intercalation of alkylamines within the layers of γ-ZrP.

exfoliation process. Table 2.2 summarizes some of the most outstanding materials mainly based on **α-ZrP** and their novel applications found in the most recent literature [24].

2.4.2   Microwave-Assisted Intercalation into γ-ZrP

One simple way of intercalating basic species into **γ-ZrP** involves the treatment for 48 h at 25 °C of the inorganic salt with a solution of the base in water/acetone. This process is amenable to microwatt irradiation, and Figure 2.5 shows the powder XRD patterns of the reaction of **γ-ZrP** with butyl-, hexyl-, octyl- and decylamine for 11 min at 105 °C in water/acetone.

It can be seen that despite the extremely short reaction time, the interlayer distance increased to 1.79, 2.12 and 2.42 nm for butyl-, hexyl- and octylamine that fairly compares to the distances (1.80, 2.22 and 2.67 nm) correspondingly obtained without microwatt irradiation. Elemental analyses of the three intercalates gave the same formulae [(**ZrPO$_4$**)(**H$_2$PO$_4$**)(**amine**)] as those obtained for the intercalation without microwatt irradiation, that is, one amine per surface phosphate. In contrast, the XRD pattern for the reaction with decylamine showed a much less intense peak at 2.54 nm and much farther from the expected distance of 3.13 nm observed for the intercalate prepared without microwatt irradiation. The much more intense peak at 6.9° suggests that a large amount of **γ-ZrP** remained unreacted. Nevertheless, the second trace in Figure 2.5 belonging to the decylamine intercalate, prepared under microwatt irradiation at the same temperature (105 °C) but for an extended period of 44 min, showed an intense peak corresponding to a distance of 3.02 nm. This fact suggests that the more hydrophobic the amine, the longer the time needed for it to get into the highly polar interlayer space of **γ-ZrP**.

In conclusion, the synthesis of **γ-ZrP** and the intercalation of basic species into it can be accomplished under microwatt irradiation very efficiently in terms of time, energy and reactants.

2.4.3   Phosphate/Phosphonate Topotactic Exchange

The external phosphates of **γ-ZrP** can be replaced by other phosphorus functions without altering the overall structure of the layers. This is the meaning of the term *topotactic*, a unique feature of the γ phase of ZrP product of the internal phosphates that maintain the layered integrity of the **γ-ZrP** structure. The lack of these two kinds of phosphates in **α-ZrP** prevents this phase to undergo this reaction. It is nevertheless true that **α-ZrP** may be made to contain phosphonates, but the materials must be obtained by direct reaction of all components. The same appears to be true for the λ phases. The topotactic exchange confers **γ-ZrP** with enormous versatility that turns this layered inorganic phase into a sort of carving board ready for the design and preparation of a wide variety of organic–inorganic materials [71]. Figure 2.6 summarizes some examples of the materials and applications that can be achieved with this synthetic rationale.

TABLE 2.2 Recent examples of intercalation into ZrP phases

Intercalating species	Uses [references]
None	Ra-223 radioisotope generator [25]
Octadecyltrichlorosilane	Photoinduced electron transfer [26]
Anticancer drugs	Biomedicine [27]
Insulin	Biomedicine [28]
$[Ru(phend)_2bpy]^{2+}$	Biosensing [29], electrochemical studies [30]
Co(II)- and Ni(II)-1,10-phenanthroline complexes	Thermal and kinetic studies [31], structural studies [32]
Diethylenetriamine	Absorption of carboxylic acids [33]
Melamine	Formaldehyde absorption [34]
$Ru(bpy)_3Cl_2$	Photophysical measurements [35]
Tris(2,2′-bipyridine)ruthenium(2+)	Structural studies [36, 37], luminescence studies [38]
Metalloporphyrin	Catalysis [39]
Butylamine	Sorption of metal ions [40]
Biogenic amines	Structural studies [41]
Proteins	Structural studies [42]
Haemoglobin	Electrochemical properties [43]
Enzymes	Structural studies [44]
Sulphanilic acid	Proton conductivity [45]
Pt(II) complexes	Luminescence studies [46]
Ionic liquids	Structural studies [47], catalysis [48]
Diaminonaphthalenes	Structural studies [49]
Ferrocene	Redox studies [50]
Cetyltrimethylammonium bromide	Structural studies [51]
1,2-Alkanediols	Structural studies [52]
Rhenium complexes	Structural studies [53]
Methyl viologens	Photochemical studies [54]
2,9-Dimethyl-1,10-phenanthroline	Complexation studies [55]
Zn and Cd complexes	Structural studies [56]
Amines	Adsorption of aldehyde fragrances [57]
DNA and proteins	Protein activity [58]
Glucosamine and chitosan	Structural studies [59]
Polymers	Structural studies [60]
Jeffamines	Structural studies [61]
Hexadecylamine	Structural studies [62]
1-Pyremethylamine	Photolysis studies [63]
Propylamine	Gel formation [64]
1,10-Phenanthroline	Complex formation [65]
Porphyrins	Structural studies [66]
Glucose oxidase	Glucose sensing [67]
Proteins	Structural studies [68]
Surfactants	Structural studies [69]
Acrylamide	Polymerization studies [70]

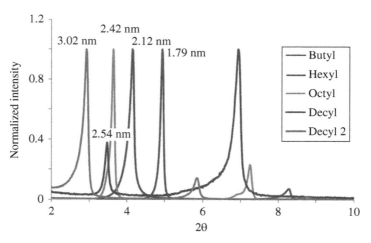

FIGURE 2.5 Powder XRD patters of γ-ZrP intercalates with the indicated alkylamines prepared under microwatt irradiation (see text).

Looking up Figure 2.6, one realizes that the topotactic exchange of the surface phosphates in **γ-ZrP** is a fascinating reaction. As simple as it appears to be, it is rather complex. Let us carefully analyse what it takes. In the first place, **γ-ZrP** is totally insoluble in any solvent exception made of HF where the layered structure is destroyed. Therefore, the reaction has to take place in the liquid–solid interface, and the crystalline compound has first to be transformed into a colloidal dispersion [72]. It is at that point when the phosphate/phosphonate can take place. Yet, after the chemical modifications have taken place by the fast topotactic reactions, the crystalline, layered structure has again to recompose. The colloidal dispersion is in fact reached by exfoliation of **γ-ZrP** in mixtures of water/acetone (from 0.7 : 0.3 to 0.2 : 0.8 in volume) at 60–80 °C [73]. After 30–60 min, the appropriate phosphonate is then added, and in a variable time, depending on the nature of the organic moiety, the layered solid flocculates back including the phosphonates between the layers and releases the exchange phosphate into the solution. As yet, no direct evidence of the chemical mechanism of this reaction in the solid–liquid interface has been published. Figure 2.7 outlines a plausible process.

The octahedra $Zr^{IV}O_6$ have to be formed by the contribution of four P–O$^-$ and two P=O groups in order to keep electroneutrality. One of the two Zr–O bonds coming from the P=O groups might thus be broken by the action of water (A) on the exfoliated lamellae, the attacked surface phosphate still remaining attached to another metal (B). The incoming phosphonate must approach the activated Zr bonded to water (C). The latter is then displaced and the phosphonate becomes bonded to Zr through its P=O bond (D). The surface phosphorus is finally detached by the action of one of the P–OH groups of the phosphonic acid (E). This process is repeated many times, until a critical amount of surface

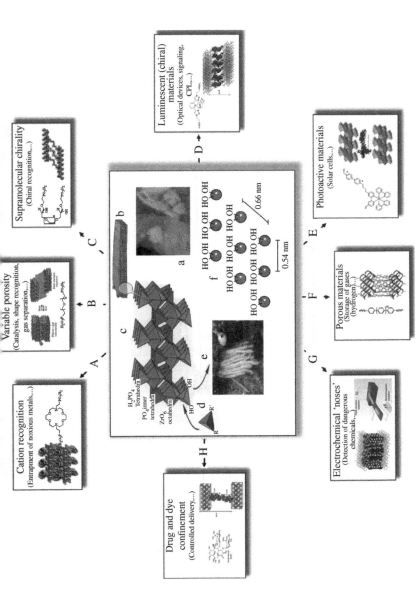

FIGURE 2.6 Central rectangle: (a) Scanning-electron microscopic view of the plates of a typical zirconium phosphate material formed by 50–100 layers; (b) sketch of the layers; (c) chemical structure of a portion of one layer showing the inner phosphates, metals, and the surface phosphates; (d) the surface phosphates can be replaced by phosphonates or other phosphorus functions that end up attached to the layer like the octopus eggs of the inset (e) that were laid at the ceiling of a small underwater cave; (f) horizontal and vertical separations of surface phosphates showing the available area around them. Periphery: A–H Possible applications of the organic–inorganic materials built from **γ-ZrP** (see text).

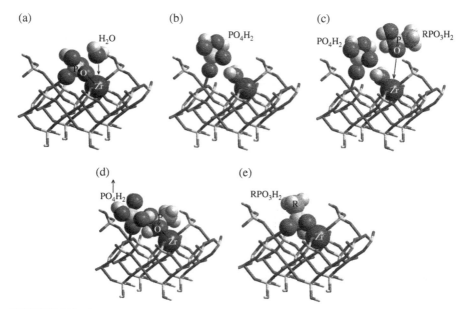

FIGURE 2.7 Steps (a–e) of a plausible mechanism for the topotactic exchange of phosphate by methylphosphonate in **γ-ZrP** (see text).

phosphates is replaced, and gradually stops right before the material flocculates. There are no studies in the literature concerning the thermodynamic parameters of this process, probably because the colloidal milieu makes the measurement difficult.

Flocculation takes place at a different pace depending primarily on the chemical structure of the attacking phosphorus functions. Besides, the extent of the topotactic exchange heavily relies on the bulkiness of the phosphonate and on the flocculation rate. Further stirring of the flocculated dispersion may allow very little if any additional exchange because the process is strongly slowed down due to the hampered diffusion of the attacking phosphonates along the already heavily filled interlayer space. Nonetheless, we have been able to perform sequential topotactic exchange with a second, much less bulkier phosphonate (*vide infra*). It can be estimated that the flocculation rate has to be various orders of magnitude slower than the single event of a phosphate/phosphonate replacement just considering that a typical sized particle of, say, $1\,\mu m^2$ bears ca. 28,000 phosphates on each of its two faces (see central rectangle in Figure 2.6).

This enthralling reaction can in principle be performed with any kind of phosphonate, rigid or flexible, polar or non-polar, or chiral or non-chiral, or with any conceivable chemical function, thus constituting a set of conceptual and practical tools to the smart design of organic–inorganic materials of any imaginable nanoscaled structure (Figure 2.6) whose arrangement can be predicted with quite reasonable accuracy.

It should be pointed out that diphosphonates render a 3D pillared arrangement of the inorganic lamellae that end up covalently bonded to one another by appropriate organic moieties. In this case, the mechanism of the topotactic exchange is even more intriguing. The comparison of the results obtained with four different diphosphonates, namely, alkyl, ethylenepolyoxa, polyphenyl and polyphenylethynyl, has allowed us to conclude that the outcome of the topotactic exchange reaction is determined by the following factors:

1. Initial stoichiometry of reactants
2. Cross-sectional area of the phosphonates
3. Conformational flexibility of phosphonates
4. Affinity of the organic chain for the lamellae surface
5. Mode of self-aggregation of phosphonates in the reaction media

2.5   LABYRINTH MATERIALS: APPLICATIONS

The layers of **γ-ZrP** containing organic phosphonates can be considered as a multiple-storey labyrinth where chemical processes of various kinds may take place. Our research group have picked on those indicated in Figure 2.6. The inclusion of ethylenepolyamine or polyoxa derivatives (paths A, B and C) or polyphenyl chains (paths F and G) rendered materials with binding and recognition properties towards quite a varied range of guests. If the organic moiety bears special functions, one may easily get solid materials with, for instance, luminescence or photoactive properties (paths D and E). Last but not least, **γ-ZrP** may be used as a reservoir for pharmacologically active compounds that could be released at the appropriate rate (path H). As we stated earlier, the number of structures and applications of these **γ-ZrP**-based materials is only limited by the imagination of the researcher.

2.5.1   Recognition Management

The aforementioned realization that the layers of **γ-ZrP** containing organic phosphonates can be considered as a multiple-storey labyrinth immediately hits the idea of molecular recognition. In principle, if one were able to control the level of incorporation of organic molecules within the layers, the resulting material would be porous enough to allow guest molecules to travel between the lamellae and differentially interact with them. One way of achieving this goal is to include crown ethers.

The first report considering the idea of forming supramolecular crown ethers inside the layers of ZrP was published almost 30 years ago [74]. **γ-ZrP** was reacted with $CH_3(OCH_2CH_2)_nOPO_3^{2-}$ ($n = 1$–3), and half of the phosphates were replaced by the phosphonates. The topotactically exchanged material displayed recognition capabilities towards alkali salts of soft anions not shown by the phosphonates in solution.

Real crown ethers were inserted into the layers of **γ-ZrP** either by intercalation or topotactic exchange for the first time 12 years later by us [75]. Our first attempts were made by intercalation of a series of cyclams as those shown in Figure 2.8 [76]. Unfortunately, the intercalation was very difficult to control and the resulting materials ended up with very low porosity and with no use in our hands in the field of cation recognition. Nevertheless, the structural studies led to some 'unexpected tales'.

The lactams (Figure 2.8, left) led to larger-than-expected interlayer distances (>2 nm) on the ground of the average diameter of the rings (0.8 nm). Elemental analyses indicated a high level of incorporation (66–94%), well above that likely (<50%) taking into account the cross section of the lactams. The experimental results had thus to be explained by considering a double layer of cyclams, slightly tilted with respect to the perfect perpendicular arrangement. We found an explanation to this behaviour in the formation of hydrogen-bonded dimers. The presence of alkyl groups vaguely affected the tilting and the level of incorporation as well. However, when the lactams were reduced to the corresponding amines (Figure 2.8; right), the intercalation behaviour was totally different. The parent compound (R=H) led to a material with an interlayer distance (1.7 nm) and incorporation level (46%) compatible with a monolayer of cyclam between the **γ-ZrP** lamellae. Surprisingly, the replacement of H by Me led to a shorter interlayer distance, suggesting that the Me-cyclam had to severely bow to adjust within the layers, thus showing how very small structural

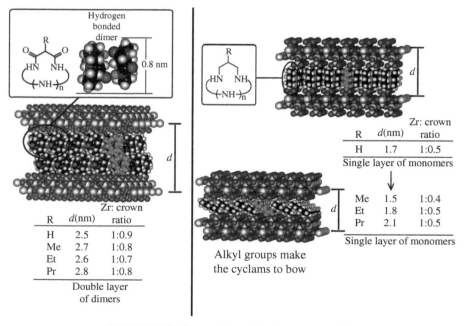

FIGURE 2.8   Intercalation of cyclams into **γ-ZrP**.

changes in the organic intercalate may strongly affect the structure of the whole scaffold. This unexpected notion could be used to finely tune the interlayer shape of similar materials.

The seminal paper on the topotactic exchange with various crown ethers [77] of different sizes, containing one or two phosphonate groups, allowed for a higher control of the number of incorporated species, giving rise to highly porous materials. The ensuing structures were not absent of surprises. For instance, it should be mentioned that the phosphate/crown ether phosphonate replacement occurred with an intriguing odd–even effect depending on the size of the crowns. Figure 2.9 summarizes the results. In simple water/acetone exfoliating conditions, the even-membered crowns pendent from opposing faces imbricated and the material ended up as a single layer of crowns, whereas the odd-membered counterpart did not imbricate and yielded a material with a double layer of crowns. Unexpectedly, the previous intercalation of butylamine produced slightly different results (Figure 2.9).

Yet, despite all these interesting structural findings, we believe the true importance of the work is the achievement for the first time ever of multiple, sequential exchange reactions, namely, the further replacement with methylphosphonate of the remaining phosphates from the initial crown ether phosphonate reaction (Figure 2.10). In this way, the polarity and acidity of the inorganic surface were drastically altered, and the effect on the recognition abilities of the materials towards Na^+ and K^+ ions was striking (Figure 2.10). Competitive experiments showed that the recognition abilities towards the alkaline ions, exerted in the solid–liquid interface by the confined crowns, were very different to those commonly exhibited in solution. Furthermore, the second phosphate/methylphosphonate exchange completely altered the crowns' response to the ions, thus showing that the nature of the whole matrix structure is crucial in the overall recognition process.

Another way of approaching the building of supramolecular crown ethers is the use of diphosphonates derived from linear polyethylene glycol chains, in order to

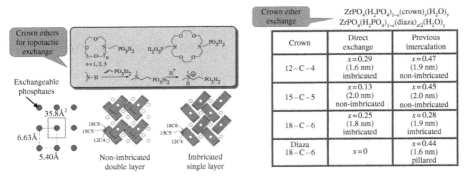

FIGURE 2.9 Results of the topotactic exchange on γ-ZrP of the indicated crown ethers by direct reaction of the exfoliated salt or with previous intercalation of butylamine.

Ion selectivity
in water Na^+ / K^+

Material	Singly exchanged	Doubly exchanged
ZrP	0.14/0.16	0.06/0.02
12C4	0.66/0.10	0.50/0.23
15C5	0.94/1.27	—
18C6	0.27/0.99	0.84/0.68
Diaza-18C6	0.42/0.86	0.72/0.22

FIGURE 2.10 Amount of Na^+ and K^+ ions relative to Zr found by atomic absorption measurements in the indicated materials (see Figure 2.23); in the doubly exchanged materials, the remaining phosphates of the first crown ether phosphonate reaction were replaced by methylphosphonate.

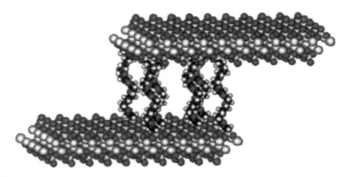

FIGURE 2.11 Idealized molecular models of **γ-ZrP** exchanged with pentaethyleneglycol diphosphonate as a representative example.

create pillared materials with polyethyleneoxa chains (Figure 2.11). By varying the chain length, one would have easy access to a set of different labyrinths with sticky columns of different height and different porosity.

We then prepared the appropriate polyethyleneoxa diphosphonates of different lengths (from di- to hexaethyleneglycol) and performed the topotactic exchange at 25% level into **γ-ZrP** [78]. The analytical data were compatible with the expected pillared materials of general molecular formula:

$$Zr(PO_4)(H_2PO_4)_{0.75}\left[HO_3P(CH_2)_2(OCH_2CH_2O)_n(CH_2)_2PO_3H\right]_{0.125} (n = 2-6).$$

In part of the prepared materials, the remaining surface phosphates [$(H_2PO_4)_{0.75}$] were quantitatively replaced by hypophosphite [$(H_2PO_2)_{0.75}$] as corroborated by the analytical data. In that way, we had at our disposal two sets of materials. We termed them *polar/polar* (polar columns and polar surface phosphates) and *polar/non-polar* (polar columns and non-polar surface hypophosphites).

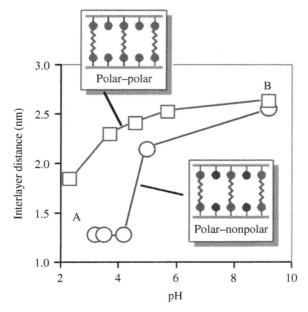

FIGURE 2.12   Molecular models of **γ-ZrP** exchanged with pentaethyleneglycol diphospho-nate with variable interlayer distance (see text); plot of interlayer distance variation with pH of the indicated materials (see text).

Figure 2.12 summarizes the corresponding *tale from the unexpected*. It shows the observed variation of interlayer distance (as measured by XRD) when the materials containing pentaethyleneglycol diphosphonate ($n = 5$), as a representative example, were treated with methylamine in aqueous dispersion. To our astonishment, the *polar/polar* material steadily augmented its interlayer distance with the increasing amount of the small intercalating amine. The starting interlayer distance at low pH is way shorter than that expected considering the length of the pentaethyleneoxa chains. Therefore, in the absence of methylamine, the chains have to arrange in a parallel fashion to the inorganic layers, leading to a scarcely porous material. Nonetheless, at high pH, the distance heavily increases, reaching a value compatible with somewhat extended, perpendicular pentaethyleneoxa columns. Yet methylamine is so small that the distance increment cannot be understood solely in terms of the intercalated amine (see Figure 2.4).

A reasonable explanation is as follows. When methylamine is not present, the oxygen atoms of the polar columns are very good receptors of hydrogen bonding from the surface phosphates, and the layers are compressed to one another (Figure 2.13a). Methylamine reacts with the surface phosphates and progressively disrupts the web of hydrogen-bonding interactions, thus giving the columns no choice but to stand up (Figure 2.13b). Unexpectedly, the overall porosity of the material is thus heavily increased by the mild acid–base reaction with the smallest alkylamine.

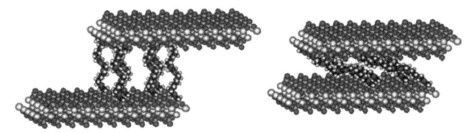

FIGURE 2.13 Molecular models of γ-ZrP exchanged with pentaethyleneglycol diphosphonate showing the experimental, variable interlayer distance as a consequence of the intercalation of methylamine.

In the case of the *polar/non-polar* material, the interlayer distance varies much more abruptly, within an extremely narrow pH range (see plot in Figure 2.12). It should be noted that the replacement of the surface phosphates by hypophosphite greatly diminishes the overall acidity of the material, the only acidic OH groups being those of the phosphonates. In the absence of amine, the oxygen atoms of the polar columns still can establish O···H–P hydrogen bonds with the surface hypophosphite groups, surely much weaker than those established with the phosphates (O···H–O–P). When sufficient amount of amine is present, the few remaining acidic OH groups exclusively belonging to the phosphonates are quickly neutralized, and the methylammonium ions act as wedges that pull up the columns to suddenly rise. Therefore, the interlayer distance is swiftly doubled, and the overall porosity of the material is drastically increased. To the best of our knowledge, this odd supramolecular behaviour in the solid state, where the porosity of a layered material is abruptly increased in response to a simple acid–base reaction in the solid–liquid interface, has never been observed before [79].

2.5.1.1 Chirality at Play Molecular modelling of the *polar/polar* and *polar/non-polar* materials in the extended arrangement, that is, when the amine is present, suggests that in order to fit the largest achieved interlayer distances, the polyethyleneoxa columns had to be helicoidally arranged (Figure 2.13). Polyethylene glycols are actually known to attain that conformation in aqueous solution, yet, what makes a notable difference is that in our materials, the polyethyleneoxa columns are covalently attached by both ends to the inorganic layers, thus conferring the whole scaffold a supramolecular dimension. It can be assumed that provided the necessary conditions are established, the random P/M helicity of the chains could in principle be directed towards homochirality, which could be visible by optical rotation measurements: another *tale from the unexpected*.

Dispersions of small amounts in water/acetone of either native γ-ZrP or exchanged at 25% with hexaethyleneglycol diphosphonate (*n*=6; *cf.* molecular formula in previous paragraphs) displayed no sizeable optical rotation as it could have easily been anticipated. However, the material pillared with hexaethyleneglycol diphosphonate intercalated with (+)-phenethylamine rendered a relatively large value of optical rotation. But to our surprise, when the enantiomerically pure amine was smoothly replaced with hexylamine, the optical rotation kept amazingly showing through. NMR experiments showed that the (+)-phenethylamine was no longer in the material. Hence, the only possibility left to explain the remaining optical activity must be laid upon the polyethyleneoxa columns that acquired a certain bias in their initially random P/M helicity due to the concourse of the chiral amine, the bias being maintained when the latter was replaced by the achiral one. This is an expression of supramolecular chiral memory; no similar cases can be found in the literature [80].

The tale from the unexpected went a step further. What would happen if the chiral influence were intrinsic to the polyethyleneoxa chains? To answer this question, enantiomerically pure diphosphonates were prepared by the addition of R-glycidol to diethyl vinylphosphonate and subsequent epoxide aperture with polyethylene glycols (scheme in Figure 2.14) [81]. Analogously to the symmetric diphosphonates, the enantiomerically pure counterparts were topotactically exchanged into γ-ZrP at the 25% level, and the materials were submitted to the usual characterization techniques whose overall result was compatible with the formation of the expected organic–inorganic materials.

In the case of the pillared materials **D25*** and **H25***, intercalation of amines of increasing length forced the intrinsically chiral polyethyleneoxa columns to acquire different conformations. The longer the amine, the more elongated the organic pillar. Figure 2.14 shows the result of these intercalation experiments. As anticipated, the basal spacing progressively increased with the length of the amine. On the contrary and to our delight, the optical rotation presented a maximum value when butylamine (**D25***) or hexylamine (**H25***) were intercalated. Molecular modelling of the different intercalates (Figure 2.15), performed by keeping the interlayer distance fixed at the corresponding experimental value and allowing the organic chain to freely reach the most stable conformation, showed that the maximum expression of helicity was precisely achieved at the spacing attained with hexylamine in **H25*** as a representative example. No other reasonable explanation could be found to account for the observed variation of the optical rotation versus interlayer distance. Moreover, the behaviour of the monophosphonate intercalates (**DM25***) reinforces this conclusion (*cf.* Figure 2.14). In this case, the inorganic layers are not covalently bonded whatsoever. Yet, the possible coiling of the organic chains must be also disturbed by compression and disorder when the lamellae are scarcely separated either because there is no intercalation or short amines are intercalated. Nevertheless, with amines of sufficient length, the coiling can be freely expressed, and since the monophosphonate chains are only attached by only one end to the inorganic phase, a longer amine cannot produce chain stretching. Coiling cannot thus be destroyed by the longer amines as it certainly was in the case of the pillared **D25*** and **H25*** examples. This is compatible with the experimental observation of a large, nearly constant value of optical rotation from butyl- to tetradecylamine (Figure 2.14).

In summary, the tale behind the unexpected chirality behaviour of the accounted materials clearly points to the important fact that chiral properties can be created and amplified at the supramolecular level in the solid state and finely tuned by mild reactions in the solid–liquid interface. To the best of our knowledge, there are no examples in the literature achieving this effect in such a creative and clear-cut manner [82].

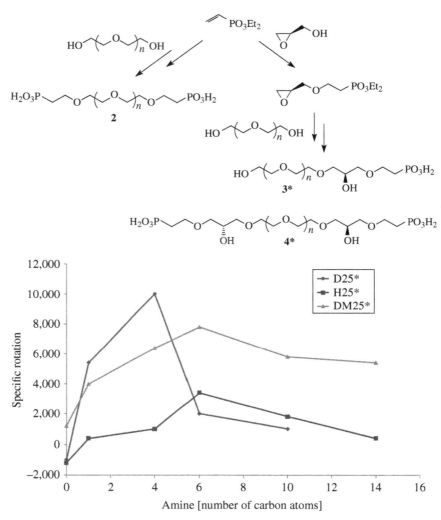

FIGURE 2.14 Variation of optical rotation upon amine intercalation in γ-ZrP exchanged at 25% with compounds **3*** (**DM25***, $n = 1$) and **4*** (**D25***, $n = 1$; **H25***, $n = 5$) of the scheme (see text).

The next logical step in this research is attempting enantioselective recognition. One of the most important challenges of modern organic, inorganic and analytical chemistry is the synthesis of microporous solids that selectively and reversibly enclose certain analytes. In recent years, layered metal phosphates and phosphonates have received considerable attention because the structures of these zeolite analogues can be tailor made at will. In terms of shape selectivity, enantioselective recognition from a racemic mixture is significantly more demanding than the detection of achiral molecules. Both enantiomers possess identical chemical functions and have to be

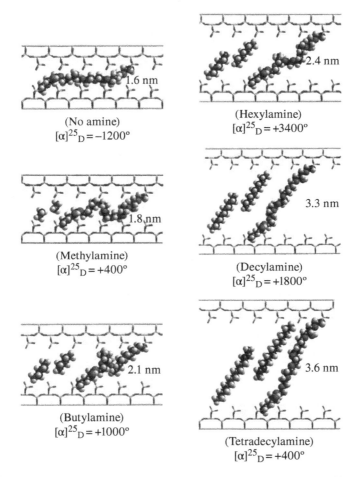

FIGURE 2.15 Calculated most stable conformations of a single column **4*** (scheme of Figure 2.14) attached to γ-ZrP (material **H25***) for the different interlayer distance (nm) induced by the corresponding intercalated amine shown.

resolved by multipoint binding to a chiral host following the lock-and-key principle. We reasoned that the pores within our chiral scaffolds could perform this function. Table 2.3 lists the results of the materials used to perform enantioselective molecular recognition experiments. The descriptors **D25*** and **H25*** have already been defined (Figure 2.14), and the acronyms **TARn*** refer to **γ-ZrP** exchanged to *n* % with the diphosphonates **5*** derived from enantiomerically pure tartaric acid as shown in the scheme of Table 2.3.

We have selected 1-phenetylamine (PEA) as the guest molecule, due to its simplicity and inexpensive availability as racemate or enantiomerically pure forms. Besides, the e.e. can be quickly and accurately measured by GC with the appropriate chiral stationary phase. We thus designed two different groups of experiments: one

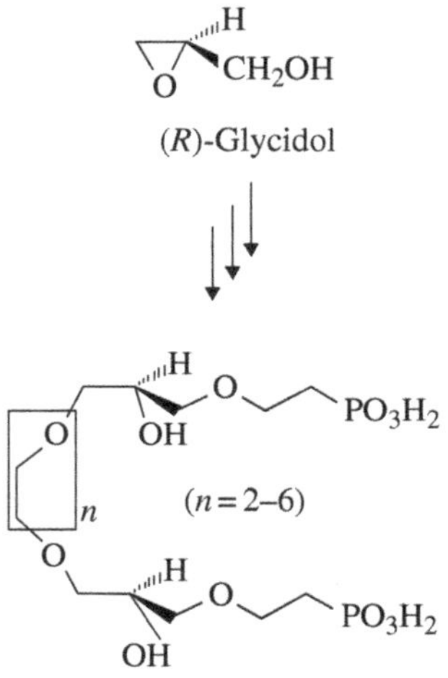

(R)-Glycidol

$(n = 2–6)$

SCHEME 2.1

TABLE 2.3   Enantiomeric excess measured in the mother liquors of the suspension of the indicated materials with (±)-PEA (see text)

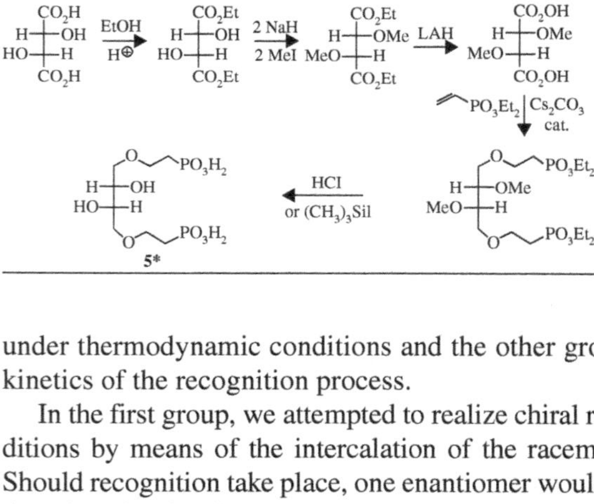

		e.e. (isomer)		PEA content per Zr
Product	24 h	48 h	
D25*	2.3 (R)	2.2 (R)	0.80
H25*	3.1 (S)	2.5 (S)	0.90
TAR25*	5.9 (S)	8.0 (S)	0.92
TAR35*	7.5 (S)	11.7 (S)	0.80
TAR70*	11.5 (S)	14.1 (S)	0.25

under thermodynamic conditions and the other group to qualitatively determine the kinetics of the recognition process.

In the first group, we attempted to realize chiral recognition under equilibrium conditions by means of the intercalation of the racemic amine into the chiral material. Should recognition take place, one enantiomer would predominantly intercalate versus the other, thus leading to a measurable e.e. in the mother liquors (MLs). Unfortunately, enantiomeric recognition was very small. The long diphosphonates derived from di- and hexaethyleneglycol (see Scheme 2.1) showed the lowest degree of enantioselective recognition. The **TAR** materials, derived from tartaric acid (Scheme 2.2), exhibited a sizeable recognition for the R-(+)-PEA isomer since its mirror-image isomer was found to be enriched in the MLs. In general, the higher the diphosphonate content of the material, the larger the observed recognition. Table 2.3 also displays the amine content of the resulting PEA-intercalated material of each experiment. It may be seen that all solids ended up with approximately one amine per remaining acidic phosphate. XRD analysis of these materials showed in general broad peaks (low crystallinity), but in average, an interlayer distance of 2.1 nm was measured, very similar to the basal spacing of pristine **γ-ZrP** intercalated with a double layer of PEA (2.15 nm). In the case of the **TAR** derivatives, this interlayer distance is close to the maximum possible calculated by molecular modelling (2.3 nm) for the most extended conformation of **TAR** columns. Therefore, it should be assumed that for PEA-intercalated **TAR70***, the sample showing the highest enantiomer recognition, there is also a double layer of amines in the scarce space left among the almost fully extended tartrate-derived columns. The highest preference for the *R* isomer of PEA was attained in this case where the amines should be in the closest possible contact with the tartrate columns. The higher proximity of the stereocenters to the phosphonate end of the chain in the **TAR** derivatives as compared to those derived from glycidol (Scheme 2.2) and the higher abundance of chains in **TAR70*** should make the void spaces more shape demanding to the host PEA isomer. The e.e. obtained for **TAR70*** is very promising in order to use this material as a stationary phase for chromatographic separations where the lock-and-key interactions should be magnified.

The kinetic type of experiment consisted of preparing two different binding phases, each of them enclosing one of the pure PEA enantiomers. These solids were suspended and stirred at room temperature in an aqueous solution, which contained two eq of the opposite enantiomer. By means of the e.e. evolution in the MLs over a period of time, it was possible to determine the rate at which the two enantiomers exchanged places from the interior of the solid material into the MLs (and vice versa). This enabled us to study whether one of the enantiomers was accommodated within the chiral cavities faster than the other, that is, whether or not the binding phases were capable of exerting kinetic chiral recognition. This is a competitive experiment, in which one enantiomer replaces the other previously accommodated inside. Thus, it was necessary to run independent assays with both enantiomers inside and outside in order to ascertain that the results were complementary, in case some kinetic recognition exists. We decided to use **D25***, which showed very little, if at all, thermodynamic recognition (Table 2.3).

Figure 2.16 shows the variation of e.e. over time in the MLs for **D25*–PEA(*S*)** and **D25*–PEA(*R*)** during 3 h. The dashed lines indicate the equilibrium situation (2 : 1 or ±33.3% e.e.), equivalent to a lack of chiral recognition, in which the PEA(*S*)-to-PEA(*R*) ratio would be the same inside and outside the material. The tale from the unexpected was that these levels were never reached. First of all, it should be noted that the in-/outflux of amines was relatively fast because after the first 5 min the e.e. of the MLs was very far from the initial expected ±100%. This is seen by the e.e. reduction to 55% of PEA(*R*) in the MLs after the first 5 min of incubation when starting with **D25*–PEA(*S*)**, whereas the e.e. decrease was reduced to only 20% of PEA(*S*) in the ML during the same period when starting with **D25*–PEA(*R*)**. This suggests that PEA(*S*) is faster in leaving and slower in entering the chiral matrix than its PEA(*R*) counterpart.

At the end of the experiment (3 h), e.e.'s were 45 and −60% from **D25*–PEA(*S*)** and **D25*–PEA(*R*)**, respectively, quite far from the ±33.3% indicative of no recognition. In other words, after 3 h, 1 eq of pure **D25*–PEA(*S*)** suspended in a solution of 2 eq of pure PEA(*R*) ended up with a 0.55 : 0.45 *R*/*S* ratio within the solid, whereas

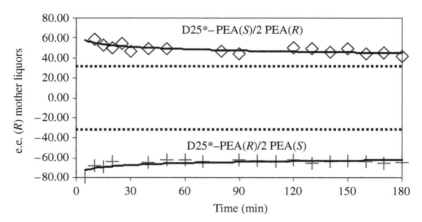

FIGURE 2.16 Variation of e.e. for *R* enantiomer of PEA over time for the dispersions of **D25*–PEA(*S*)** and **D25*–PEA(*R*)** in water acetone solution of the corresponding opposite enantiomer (see text).

1 eq of **D25*–PEA(*R*)** suspended in a solution of 2 eq of PEA(*S*) gave a 0.60 : 0.40 *R*/*S* ratio within the solid. Therefore, in the longer run of 3 h, the PEA(*S*) isomer showed a tendency to leave the inside of the solid host. On the other hand, the PEA(*R*) was much more reluctant to depart from the organic–inorganic host. These facts clearly show that PEA(*R*) is more stable than its enantiomer within **D25*** and that the chiral solid matrix displayed a kinetic and thermodynamic recognition towards PEA(*R*), which was not evident from the experiments with the racemate (*vide supra*).

There is a relatively simple explanation to these apparently contradictory results. When the material **D25*** was treated with racemic PEA, a negligible enantiomeric recognition was observed (Table 2.3). The initially deflated volume matrix, with no pre-existing conformational preference of the **3*** columns (*n* = 1; Scheme 2.1), expands to accommodate a double layer of racemic PEA. It is not possible to know whether the two enantiomers enter the same interlayer gallery or not. However, when **D25*** is treated with either PEA(*R*) or PEA(*S*), the chiral columns have the chance to express in full a particular conformation to best house the specific handedness of the amine. As a consequence, two supramolecularly diastereomeric composites are formed, which should differ in energy. The fact that PEA(*R*) was more reluctant to exit the solid matrix would suggest that **D25*–PEA(*R*)** is the most stable diastereomeric entity and that there seems to be a better complementary *matching* between the conformation/configuration of the chiral columns and PEA(*R*).

In summary, the experiments detailed earlier showed that the conformation of intrinsically chiral organic moieties can be heavily altered at the supramolecular level within a solid matrix, determining striking differences in the measured values of specific optical rotation. Very long ago, Brewster suggested that conformation and chiroptical properties should be closely intertwined [83]. While this was relatively easy to demonstrate in solution [84], the solid state has resisted up until now a clear-cut proof of it. Nonetheless, our research endeavours with chiral organic–inorganic materials ultimately were fruitful, and Brewster's assumption has thus been expressed in the solid state, in the new field of tailored organic–inorganic composites.

2.5.1.2 Gas and Vapour Storage

Recognition can be established on many different expressions. The aforementioned chiral recognition is perhaps the most spectacular in terms of structural shape requirements. However, other ways of recognition can be of fundamental importance for modern applications. This is another tale from the unexpected because the organic–inorganic scaffoldings based on γ-ZrP may be used for gas storage.

Hydrogen is a very appealing energy vector because the release of its energy does not involve the noxious carbon dioxide, provided its production is performed from renewable energy sources. Therefore, its environmentally friendly production and safe storage/transportation are key problems to be solved if hydrogen is to be efficiently used as the clean energy carrier of the future. Although there are already many reasonably useful technical approaches, neither of the two riddles is nowadays at a level of resolution, which would make the use of hydrogen routinely possible, though society is in a rush to achieve that.

Hydrogen storage may be attained by *physisorption* in porous matrices [85] among other procedures. Why not try appropriate matrices based on ZrP to serve the latter purpose? The thorough examination of the literature allows one to find numerous organic, inorganic and hybrid systems in which the storage of hydrogen has been tested, from the delusive carbon nanotubes to the very cleverly designed systems with almost unbelievable specific surface areas close to $5000\,m^2\,g^{-1}$ [86]. The analysis of the different available structures points to the conclusion that a large accessible volume is desirable but by means of micropores or ultramicropores and therefore presenting a large internal contact surface to the elusive hydrogen molecules. Some interesting theoretical studies and experimental ones on zeolites also point to the benefit of having polarizing centres, namely, small cations as, for instance, Li^+. Also, the presence of transition metals with open coordination sites appears to be important. We believe that our set of tools based on the chemistry in the solid–liquid interface displayed by **γ-ZrP** might allow us to design appropriate materials for hydrogen storage. Figure 2.17 summarizes our rationale and the best achieved results [87].

The attachment of polyphenyl or polyphenylethynyl diphosphonates to either α- or **γ-ZrP** led to materials with slit-like ultramicropores of different length. Further exchange reactions led to polar or non-polar groups at both ends of the pores. This set of materials allowed us to check for a large number of different arrangements. The best results were obtained with the material named **αT60Li**, α-ZrP with 60% of terphenyldiphosphonates [88], the remainder phosphates bearing Li^+ as counterions. At 800 Torr and 77 K, 1.7% w/w of hydrogen could be stored, and the Department of Energy's goal for 2010 (45 g of hydrogen per litre of material) was thus attained below 2 atm at 77 K (Figure 2 17).

Another important aspect of gas or vapour storage and/or detection is the seeking out of a 'chemical nose' [89] for the *in situ* sampling of certain volatile components

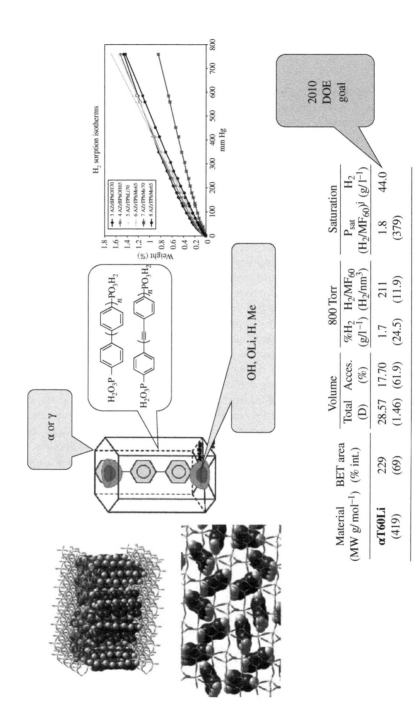

FIGURE 2.17 Phosphonate-exchanged ZrP materials for hydrogen storage (see text).

of, for instance, explosives. The design of simple, non-intrusive devices to detect dangerous substances within different objects like air cargo, luggage, and so on is an important activity of many security companies. Such a device could be based in the piezo-resistive detector indicated in Figure 2.18, where the micro cantilever resonates at a frequency that depends on its mass.

Should a porous material be deposited on it, its total mass would change when a given vapour is absorbed in the pores. The extent of the resonance frequency thus changes, and together with a simple calibration protocol, it enables the quantification of the absorbed component. To this effect, many porous matrices have been used as, for instance, porous and mesoporous silica, zeolites and ordered mesoporous materials. Modified **γ-ZrP** could be considered within the latter, yet it has never been tested for this purpose. Our material named **αT60**, that is, **α-ZrP** pillared with 60% of terphenyldiphosphonates, had the ability of reversibly detecting ca. 400 ppm of toluene vapour (used as a model for the detection of the aromatic compounds contained in certain explosives) in N_2 with a response time lower than 30 s. This is a very promising result upon which we are devoting further research in order to improve sensitivity and selectivity. What would the capabilities of ZrP phases pillared with the molecules of Figure 2.18 or with any other one think of? More tales from the unexpected no doubt will follow.

2.5.2 Dissymmetry and Luminescence Signalling

Important research is being directed to the design of molecular systems able to display the strong luminescence of lanthanide metals [90]. To accomplish it with efficiency, two stringent conditions have to be met: (1) a suitable organic chromophore should absorb light and efficiently transfer energy to the metal (*antenna effect*); (2) the coordination sphere of the metal should be free of water molecules because the OH oscillators easily quench metal emission. The pillared materials described in the previous section with polyethyleneoxa columns may constitute an excellent spider web to enshroud the oxygenphylic lanthanide metals and isolate them within the solid matrix. Unfortunately, the white powder of **γ-ZrP** does not absorb visible light and does it feebly in the UV region what is not adequate to excite the lanthanide metals. However, acid–base reactions with the surface phosphates may be taken as the driving force to intercalate suitable chromophores in the organic–inorganic lattice. Figure 2.19 schematizes part of the achieved results with this rationale.

In electronics' terms, the composite created from **γ-ZrP** topotactically exchanged with polyethyleneoxa diphosphonates and intercalated with 2,2′-bipyridyl behaves as a triple AND gate – for the metal luminescence to take place, it is necessary to gather together the three components: lanthanide metal, polyethyleneoxa columns and 2,2′-bipyridyl as sensitizer. These materials may find a number of different applications either in solid or in the solid–liquid interface. More sophisticated chromophores are being tried (Figure 2.20) and preliminary results showed the pursued strong luminescence of the metals [91]. The bis-triazolylpyridine diphosphonates have been provided with chirality (Figure 2.21) in order to check whether the resulting materials from the topotactic exchange are able to show circularly polarized

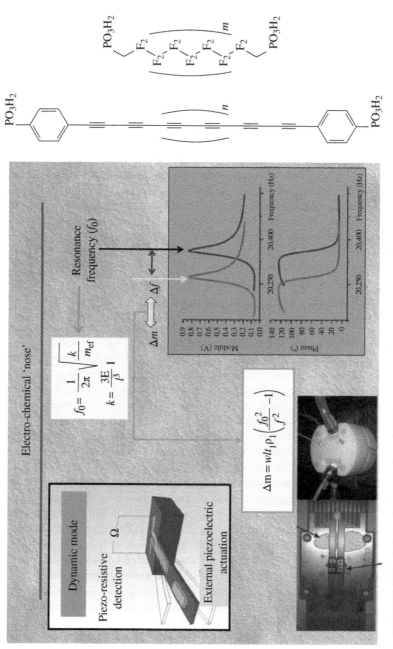

FIGURE 2.18 Schematics for the detection of vapours by porous materials based on chemical and piezoelectric means (see text).

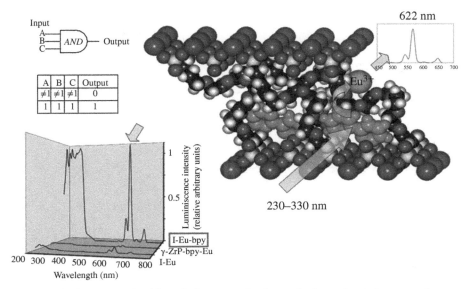

FIGURE 2.19 The lanthanide emission was only observed when polyethyleneoxa columns, the 2,2′-bipiridyl and the metal were confined within **γ-ZrP**.

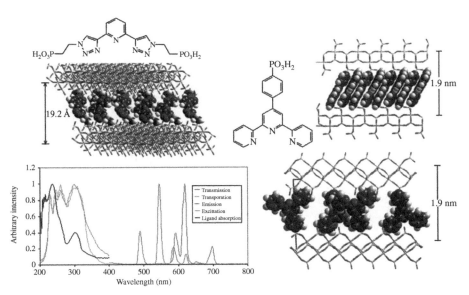

FIGURE 2.20 Exchanged **γ-ZrP** with chromophores derived from bis-triazolylpyridine and phenylterpyridine, showing the characteristic emission of Eu^{3+} and Tb^{3+}.

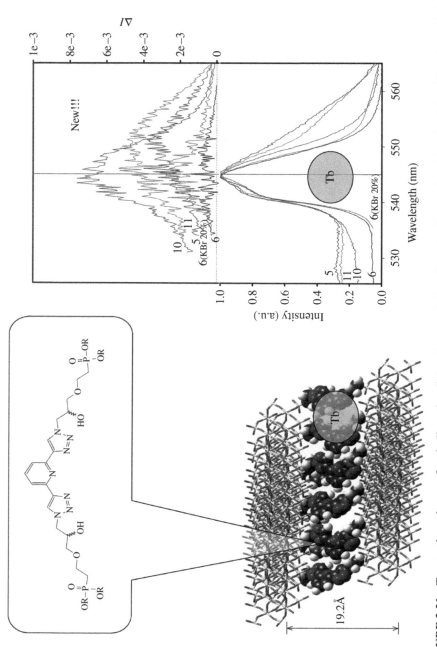

FIGURE 2.21 Topotactic exchange of optically active diphosphonates derived from the chromophore bis-triazolylpyridine, intercalation of Tb³⁺ ions and observation of CPL (see text).

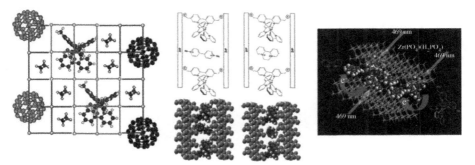

FIGURE 2.22 Idealized arrangement of electron donor (Ru^{II}(bpy)$_3$) and electron acceptors (fullerene derivatives or viologens) on the simplified surface of **γ-ZrP**.

luminescence (CPL) of lanthanides [92]. This property, seldom measured in the solid state, might be very important in the technology of optical handling of information. Work is under way.

2.5.3 Building DSSCs

There is no fundamental principle running against the possibility of attaining the proper arrangement of chemical components on a surface to accomplish long-lived photoinduced electron transfer [93], the first step in either the achievement of artificial photosynthesis or the construction of efficient solar cells. Yet again, inorganic **γ-ZrP** may constitute an excellent carving board to realize that goal. We have attached to it phosphonate derivatives of the Ru^{II}(bpy)$_3$ complex, well known by its excellent light-absorption and electron-donor properties, and placed it side by side to different electron-donor acceptors ranging from relatively simple viologens to more elaborated fullerene derivatives. Figure 2.22 contains some molecular models of these complex structures, which were positively characterized by the usual techniques [94].

The first important observation was that the phosphorescence emission of the Ru^{II}(bpy)$_3$ complex was heavily quenched by the presence of the electron-acceptor species, suggesting that the pursued electron transfer took place. Very recently, flash photolysis revealed some of these materials to have outstanding properties in that the separation of charges is long-lived and the initially thought inert inorganic layer resulted semiconducting, thus driving the separated electrons and/or holes along the material [95].

Additional results showed that these powders can be arranged as solar cells with promising efficiency [96]. Figure 2.23 shows the results obtained by the intercalation of the indicated dyad into **γ-ZrP**. The most important finding shown by Mott–Schottky measurements is that the inorganic salt is an n-type semiconductor. We have thus been able to make a photovoltaic cell based on zirconium phosphate without relatively simple chemistry. It should be highlighted that its efficiency is still low but, after a few attempts, only eight times inferior to the current titanium-based solar cells

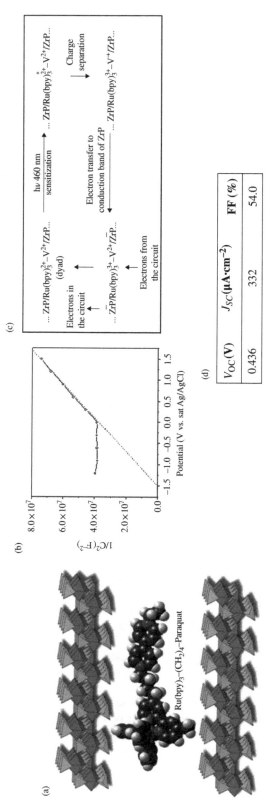

FIGURE 2.23 (a) Intercalation of the indicated dyad into **γ-ZrP**, (b) Mott–Schottky plot of the resultant material showing the *n*-type semiconductivity, (c) proposed mechanism for the generation of current and (d) efficiency values achieved by the material.

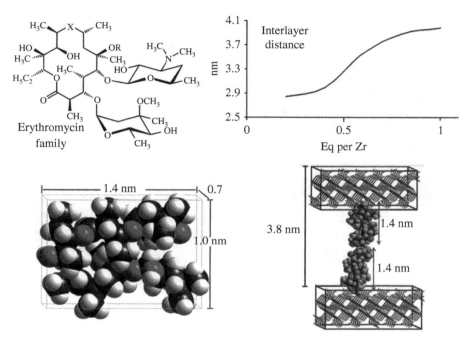

FIGURE 2.24 Intercalation of antibiotics of the erythromycin family (X = CO, R = H, erythromycin A; R = Me, clarithromycin; X = MeN–CH$_2$, R = H, azithromycin). The interlayer distance indicates the formation of mono- and bilayers of antibiotic between the inorganic lamellae at low and high stoichiometric ratio of the reactants, respectively. The X-ray crystal structure of clarithromycin and an ideal model of the bilayer are represented.

that have been optimized following the hard work of numerous well-reputed research groups and the result of several thousands of papers.

2.5.4 Molecular Confinement

Confinement of drugs into molecular matrices might be a good solution to their slow release at particular locations on the body [97]. We have preliminary studies where it can be shown that the erythromycin family antibiotics can intercalate into **γ-ZrP**, the driving force being the acid–base interaction between the amine group(s) of the organic compound and the acidic surface phosphates. The measurement of the interlayer distance versus the stoichiometry of the reactants (Figure 2.24) shows that mono- and bilayers of erythromycin derivatives may be enclosed between the inorganic lamellae. More detailed studies will follow.

On its part, dyes are an integral part of many technologies. For example, malachite green (MG) is used in Gram stain, which is a technique used to classify bacteria. Also, it is used for anti-fungal purposes. However, many dyes are toxic and carcinogenic and have to be handled with special care if serious environmental contamination is to be avoided. Purification of water and air resources that contain even traces of dyes is an important technological challenge and has attracted wide attention.

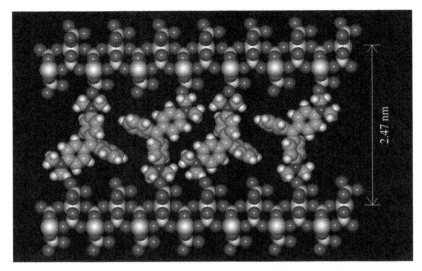

FIGURE 2.25 Possible arrangement of MG intercalated in **γ-ZrP** complying with the observed interlayer distance measured by XRD.

Dye-containing coloured water is of no use, but appropriately bleached solutions may still be used for washing, cooling, irrigation and cleaning purposes. Materials able to efficiently entrap dyes are thus quite desirable. Moreover, mineral-accommodated dyes or inclusion pigments are finding growing attention with regard to their potential applications as components of, for instance, optical devices. The strict orderly orientation of chromophores necessary in such devices can be achieved by the intercalation of the polar dye molecules into layered minerals. Wouldn't **γ-ZrP** have a role to play in this scenario? MG has been effectively intercalated into **γ-ZrP** as shown in Figure 2.25, and the resulting materials have been positively characterized [98].

These preliminary results are a good sign regarding the high prospects of using **γ-ZrP** in the uptake of different drugs and dyes like MG or its next-of-kin methyl violet and crystal violet. They also show the importance of **γ-ZrP** and its surfactant composites in the field of intercalation and inclusion chemistry of bioactive organic compounds within layered inorganic and organic–inorganic materials.

2.6   CONCLUSION AND PROSPECTS

Despite the limited length of this chapter, it is our belief that the gathered experimental facts eloquently show the limitless fields where the building of organic–inorganic 3D structures based on ZrP can play a crucial role either in basic or applied research. The number and quality of the rather complex composites that can be figured out from the combination of the inorganic matrix and any phosphonate is only limited by the imagination and chemical needs of the researcher. Moreover, the confinement of the organic molecules would confer them new properties at the supramolecular level: Tales from the unexpected.

FINAL COMMENTS AND ACKNOWLEDGEMENTS

This work would not have been possible along the last 15 years without the smart contribution at different levels of many people. Although always difficult, throughout these years, financial support has been obtained from governmental sources like the Ministries of Science and Education and Foreign Affairs, the Autonomous Community of Madrid and the European Union. However, it unfortunately ceased in 2010 for reasons difficult to understand save for the impact of the global economic crisis. On the contrary, private indirect funding from ERCROS Farmacia S.A. (Aranjuez, Spain) has been and is being generous. I wish to expressly thank Dr. Carmen Cruzado, its R&D manager, for her generosity in letting us develop this chemistry as an important bonus to our consulting work with the company. Quite recently, these activities are performed within the framework of the UAM–ERCROS Chair for Pharmaceutical Chemistry that I have the high honour to direct.

REFERENCES

1. R. Feynman, *Eng. Sci.*, **1960**, *23*, 22.

2. J.A. Wisner, *Nat. Chem.*, **2013**, *5*, 646–647; A. Priimagi, G. Cavallo, P. Metrangolo, G. Resnati, *Acc. Chem. Res.*, **2013**, *46*, 2686.

3. P. Dandekar, Z.B. Kuvadia, M.F. Doherty, *Ann. Rev. Mater. Res.*, **2013**, *43*, 359–386; C. Wang, D. Liu, W. Lin, *J. Am. Chem. Soc.*, **2013**,*135*, 13222–13234; G.R. Desiraju, *J. Am. Chem. Soc.*, **2013**, *135*, 9952–9967.

4. R. Ciriminna, A. Fidalgo, V. Pandarus, F. Beland, L.M. Ilharco, M. Pagliaro, *Chem. Rev.*, **2013**, *113*, 6592–6620; D.O.V. Uche, *Adv. App. Sci. Res.*, **2013**, *4*, 506–510.

5. N.S. Oltra, J. Swift, A. Mahmud, K. Rajagopal, S.M. Loverde, D.E. Discher, *J. Mater. Chem. B Mater. Biol. Med.*, **2013**, *1*, 5177–5185; C.A. Hunter, H.L. Anderson, *Angew. Chem. Int. Ed.*, **2009**, *48*, 7488; E. Kharlampieva, V. Kozlovskaya, S.A. Sukhishvili, *Adv. Mater.*, **2009**, *21*, 3053; Y. Liu, H. Yan, *Science*, **2009**, *325*, 685.

6. C.R. Evans, *J. Mater. Chem. C Mater. Opt. Elect. Dev.*, **2013**, *1*, 4190–4200; S. Cataldo, B. Pignataro, *Materials*, **2013**, *6*, 1159–1190.

7. M. Moliner, C. Martinez, A. Corma, *Chem. Mater.*, **2014**, *26*, 246; A. Dhakshinamoorthy, M. Alvaro, A. Corma, H. Garcia, *Dalton Trans.*, **2011**, *40*, 6344–6360; C. Martinez, A. Corma, *Coord. Chem. Rev.*, **2011**, *255*, 1558–1580.

8. P. Van Der Voort, E.F. Vansant, P. Cool (P. Somasundaran, ed.) in *Encyclopedia of Surface and Colloid Science (2nd Edition)* **2012**, 6162–6172.

9. U. Costantino, F. Costantino, F. Elisei, L. Latterini, M. Nocchetti, *Phys. Chem. Chem. Phys.*, **2013**, *15*, 13254–13269; S. Nishimura, A. Takagaki, K. Ebitani, *Green Chem.*, **2013**, *15*, 2026–2042.

10. M.A. Bizeto, A.L. Shiguihara, V.R.L. Constantino, *J. Mat. Chem.*, **2009**, *19*, 2512.

11. G. Alberti, S. Murcia-Mascaros, R. Vivani, *Mat. Sci. Forum.*, **1994**, *87*, 152; R. Vivani, G. Alberti, F. Costantino, M. Nocchetti, *Microporous Mesoporous Mater.*, **2007**, *107*, 58.

12. R. Vivani, G. Alberti, F. Costantino, M. Nocchetti, *Microporous Mesoporous Mater.*, **2007**, *107*, 58.

13. A. Clearfield, J.A. Stynes, *J. Inorg. Nucl. Chem.*, **1964**, *26*, 117.

14. A. Clearfield, G.D. Smith, *Inorg. Chem.*, **1969**, *8*, 431; J.M. Troup, A. Clearfield, *Inorg. Chem.*, **1977**, *16*, 3311–3314.

15. L. Sun, J.Y. O'Reilly, D. Kong, J.Y. Su, W.J. Boo, H.-J. Sue, A. Clearfield, *J. Coll. Interf. Sci.*, **2009**, *333*, 503.

16. A. Clearfield, *Prog. Inorg. Chem.*, **1998**, *47*, 371.

17. G.K.H. Shimizu, R. Vaidhyanathan, J.M. Taylor, *Chem. Soc. Rev.*, **2009**, *38*, 1430; T. Devic (C. Serre, V. Valtchev, S. Mintova, M. Tsapatsis, eds., Elsevier, Amsterdam) in *Ordered Porous Solids* **2009**, *77*; A. Clearfield, *Dalton Trans.*, **2008**, *44*, 6089.

18. M.P. Poojary, B. Zhang, A. Clearfield, *J. Chem. Soc. Dalton Trans.*, **1994**, *16*, 2453.

19. G. Alberti, M. Bartocci, M. Santarelli, R. Vivani, *Inorg. Chem.*, **1997**, *36*, 3574.

20. A. Clearfield, R.H. Blessing, J.A. Stynes, *J. Inorg. Nucl. Chem.*, **1968**, *30*, 2249.

21. N. Clayden, *J. Chem. Soc. Dalton Trans.*, **1987**, *9*, 1877.

22. D.M. Poojary, B. Shpeizer, A. Clearfield, *J. Chem. Soc. Dalton Trans. Inorg. Chem.*, **1995**, *1*, 111.

23. L. Sun, W.J. Boo, R.L. Browning, H.-J. Sue, A. Clearfield, *Chem. Mater.*, **2005**, *17*, 5606; L. Sun, J.Y. O'Reilly, D. Kong, J.Y. Su, W.J. Boo, H.-J. Sue, A. Clearfield, *J. Coll. Interf. Sci.*, **2009**, *333*, 503.

24. C.V. Kumar, A. Bhambhani, N. Hnatiuk (S.M. Auerbach, K.A. Carrado, P.K. Dutta, eds.) in *Handbook of Layered Materials* **2004**, 313.

25. T. Moller, N. Bestaoui, M. Wierzbicki, T. Adams, A. Clearfield, *Appl. Radiat. Isotopes*, **2011**, *69*, 947.

26. A. Diaz, B.M. Mosby, V.I. Bakhmutov, A.A. Marti, J. Batteas, A. Clearfield, *Chem. Mater.*, **2013**, *25*, 723–728.

27. V. Saxena, A. Diaz, A. Clearfield, J.D. Batteas, M.D. Hussain, *Nanoscale*, **2013**, *5*, 2328–2336; A. Diaz, V. Saxena, J. Gonzalez, A. David, B. Casanas, C. Carpenter, J.D. Batteas, J.L. Colon, A. Clearfield, H.M. Delwar, *Chem. Commun.*, **2012**, *48*, 1754–1756.

28. A. Diaz, A. David, R. Riviam, M.L. Gonzalez, A. Baez, S.E. Wark, P. Zhang, A. Clearfield, J.L. Colon, *Biomacromolecules*, **2010**, *11*, 2465–2470.

29. M.B. Santiago, G.A. Daniel, A. David, B. Casanas, G. Hernandez, A.R. Guadalupe, J.L. Colon, *Electroanalysis*, **2010**, *22*, 1097.

30. M.B. Santiago, M.M. Velez, S. Borrero, A. Diaz, C.A. Casillas, C. Hofmann, A.R. Guadalupe, J.L. Colon, L. Jorge, *Electroanalysis*, **2006**, *18*, 559.

31. S. Vecchio, R. Rocco, C. Ferragina, *J. Thermal Anal. Calorim.*, **2009**, *97*, 805.

32. Y. Du, Q. Pan, J. Li, J. Yu, R. Xu, *Inorg. Chem.*, **2007**, *46*, 5847.

33. H. Nakayama, *Phosph. Res. Bull.*, **2009**, *23*, 1.

34. A. Hayashi, H. Nakayama, M. Tsuhako, *Solid State Sci.*, **2009**, *11*, 1007.

35. S. Shi, Y. Peng, J. Zhou, *J. Nanosci. Nanotech.*, **2009**, *9*, 2746.

36. C. Ferragina, R. Di Rocco, P. Giannoccaro, P. Patrono, L. Petrilli, Lucantonio, *J. Incl. Phenom. Macro. Chem.*, **2009**, *63*, 1.

37. S. Shi, J. Zhou, R. Zong, J. Ye, Jianping, *J. Lumin.*, **2007**, *122–123*, 218.

38. M.H. Xiang, X.B. Shi, N. Li, K.A. Li, *Chin. Chem. Lett.*, **2007**, *18*, 89.

39. H.Y. Wang, W.D. Ji, D.X. Han, *Chin. Chem. Lett.*, **2008**, *19*, 1330.

40. B. Wozniak, W. Apostoluk, J. Wodka, *Solv. Extr. Ion Exch.*, **2008**, *26*, 699; B. Wozniak, W. Apostoluk, *Solv. Extr. Ion Exch.*, **2010**, *28*, 665–681.

41. A. Hayashi, Y. Yoshikawa, N. Ryu, H. Nakayama, M. Tsuhako, T. Eguchi, *Phosph. Res. Bull.*, **2008**, *22*, 48.

42. A. Bhambhani, C.V. Kumar, *Microporous Mesoporous Mater.*, **2008**, *110*, 517.

43. Y. Liu, C. Lu, W. Hou, J.-J. Zhu, *Anal. Biochem.*, **2008**, *375*, 27.

44. A. Bhambhani, C.V. Kumar, *Microporous Mesoporous Mater.*, **2008**, *109*, 223.

45. Z.P. Xu, Y. Jin, J.C. Diniz da Costa, G.Q. Lu, *Solid State Ionics*, **2008**, *178*, 1654.

46. E.J. Rivera, C. Figueroa, J.L. Colon, L. Grove, W.B. Connick, *Inorg. Chem.*, **2007**, *46*, 8569.

47. H.Y. Wang, D.X. Han, *Chin. Chem. Lett.*, **2007**, *18*, 764.

48. H. Hu, J.C. Martin, M. Zhang, C.S. Southworth, M. Xiao, Y. Meng, L. Sun, *RSC Adv.*, **2012**, *2*, 3810–3815.

49. L. Benes, K. Melanova, J. Svoboda, V. Zima, M. Kincl, *J. Phys. Chem. Solids*, **2007**, *68*, 803.

50. M.B. Santiago, C. Declet-Flores, A. Diaz, M.M. Velez, M.Z. Bosques, Y. Sanakis, J.L. Colon, *Langmuir*, **2007**, *23*, 7810.

51. S. Shi, R. Zong, Y. Liu, Y. Wang, J. Zhou, *Key Eng. Mater.*, **2007**, 336–338, 2589.

52. K. Melanova, L. Benes, V. Zima, J. Svoboda, M. Trchova, J. Dybal, *J. Incl. Phenom. Macro. Chem.*, **2007**, *58*, 95.

53. A.A. Marti, N. Rivera, K. Soto, L. Maldonado, J.L. Colon, *Dalton Trans.*, **2007**, *17*, 1713.

54. A.A. Marti, G. Paralitici, L. Maldonado, J.L. Colon, *Inorg. Chim. Acta*, **2007**, *360*, 1535.

55. S. Vecchio, R. Di Rocco, C. Ferragina, *Therm. Acta*, **2007**, *453*, 105.

56. C. Ferragina, R. Di Rocco, S. Foglia, L. Petrilli, *Coll. Surf. A Phys. Eng. Aspects*, **2007**, *293*, 114–122.

57. A. Hayashi, H. Katsuta, A. Tsuruta, H. Nakayama, M. Tsuhako, *Phosp. Res. Bull.*, **2004**, *17*, 107.

58. A. Bhambhani, C.V. Kumar, *Adv. Mater.*, **2006**, *18*, 939.

59. A. Hayashi, Y. Nakabayashi, J. Yasutomi, H. Nakayama, M. Tsuhako, *Bull. Chem. Soc. Jpn.*, **2006**, *79*, 262.

60. J. Liu, W.-J. Boo, A. Clearfield, H.-J. Sue, *Mater. Manuf. Proc.*, **2006**, *21*, 143.

61. N. Bestaoui, N.A. Spurr, A. Clearfield, *J. Mater. Chem.*, **2006**, *16*, 759.

62. B. Ha, K. Char, H.S. Jeon, *J. Phys. Chem. B*, **2005**, *109*, 24434.

63. R.A. Bermudez, R. Arce, J.L. Colon, *J. Photochem. Photobiol. A Chem.*, **2005**, *175*, 201.

64. M. Casciola, G. Alberti, A. Donnadio, M. Pica, F. Marmottini, A. Bottino, P. Piaggio, *J. Mater. Chem.*, **2005**, *15*, 4262.

65. S. Vecchio, R. Di Rocco, C. Ferragina, S. Materazzi, *Therm. Acta*, **2005**, *435*, 181.

66. H. Wang, D. Han, L. Na, K. Li, *J. Incl. Phenom. Macro. Chem.*, **2005**, *52*, 247.

67. S. Park, T.D. Chung, S.K. Kang, R.-A. Jeong, H. Boo, H.C. Kim, *Anal. Sci.*, **2004**, *20*, 1635.

68. A. Chaudhari, J. Thota, C.V. Kumar, *Microporous Mesoporous Mater.*, **2004**, *75*, 281.

69. C. Ferragina, R. Di Rocco, A. Fanizzi, P. Giannoccaro, L. Petrilli, *J. Therm. Anal. Calorim.*, **2004**, *76*, 871.

70. J. Wang, Y. Hu, B. Li, Z. Gui, Z. Chen, *Ultrason. Sonochem.*, **2004**, *11*, 301.

71. G. Alberti, U. Costantino, C. Dionigi, S. Murcia-Mascaros, R. Vivani, *Supramol. Chem.*, **1995**, *6*, 29; A. Clearfield, U. Costantino, *Compr. Supramol. Chem.*, **1996**, *7*, 107; P. Olivera-Pastor, P. Maireles-Torres, E. Rodriguez-Castellon, A. Jimenez-Lopez, T. Cassagneau, D.J. Jones, J. Roziere, *J. Chem. Mater.*, **1996**, *8*, 1758.

72. G. Alberti, E. Giontella, S. Murcia-Mascaros, *Inorg. Chem.*, **1997**, *36*, 2844.

73. G. Alberti, E. Giontella, S. Murcia-Mascarós, R. Vivani, *Inorg. Chem.*, **1998**, *37*, 4672.

74. S. Yamanaka, K. Yamasaka, M. Hattori, *J. Incl. Phenom.*, **1984**, *2*, 297.

75. E. Brunet, M. Huelva, R. Vazquez, O. Juanes, J.C. Rodriguez-Ubis, *Chem. Eur. J.*, **1996**, *2*, 1578.

76. E. Brunet, unpublished results.

77. E. Brunet, M. Huelva, J.C. Rodriguez-Ubis, *Tetrahedron Lett.*, **1994**, *35*, 8697.

78. G. Alberti, E. Brunet, C. Dionigi, O. Juanes, M.J. Mata, J.C. Rodriguez-Ubis, R. Vivani, *Angew. Chem. Int. Ed.*, **1999**, *38*, 3351.

79. E. Brunet, M.J. Mata, O. Juanes, J.C. Rodriguez-Ubis, *Angew. Chem. Int. Ed.* **2004**, *43*, 619; E. Brunet, M.J. Mata, H.M.H. Alhendawi, C. Cerro, M. Alonso, O. Juanes, J.C. Rodriguez-Ubis, *Chem. Mat.*, **2005**, *17*, 1424.

80. E. Brunet, *Chirality*, **2002**, *14*, 135.

81. E. Brunet, M.J. Mata, O. Juanes, H.M.H. Alhendawi, C. Cerro, J.C. Rodriguez-Ubis, *Tetrahedron Asym.*, **2006**, *17*, 347.

82. E. Brunet, O. Juanes, J.C. Rodriguez-Ubis, *J. Mex. Chem. Soc.*, **2009**, *53*, 154.

83. J.H. Brewster, *J. Am. Chem. Soc.*, **1959**, *81*, 5475–5483.

84. E.L. Eliel, E. Brunet, *J. Org. Chem.*, **1991**, *56*, 1668–1670.

85. E. Brunet, C. Cerro, O. Juanes, J.C. Rodriguez-Ubis, A. Clearfield, *J. Mat. Sci.*, **2008**, *43*, 1155.

86. J.G. Vitillo, L. Regli, S. Chavan, G. Ricchiardi, G. Spoto, P.D.C. Dietzel, S. Bordiga, A. Zecchina, *J. Am. Chem. Soc.*, **2008**, *130*, 8386, and references cited therein.

87. E. Brunet, H.M.H. Alhendawi, C. Cerro, M.J. Mata, O. Juanes, J.C. Rodriguez-Ubis, *Angew. Chem. Int. Ed.*, **2006**, *45*, 6918.

88. E. Brunet, H.M.H. Alhendawi, C. Cerro, M.J. Mata, O. Juanes, J.C. Rodriguez-Ubis, *Chem. Eng. J.*, **2010**, *158*, 333.

89. M. Urbiztondo, P. Pina, J. Santamaria (V. Valtchev, S. Mintova, M. Tsapatsis eds.) in *Ordered Porous Solids* **2009**, 387.

90. E. Brunet, O. Juanes, J.C. Rodriguez-Ubis, *Cur. Chem. Biol.*, **2007**, *1*, 11.

91. E. Brunet, M.J. Mata, O. Juanes, J.C. Rodriguez-Ubis, *Chem. Mat.*, **2004**, *16*, 1517; E. Brunet, H.M.H. Alhendawi, O. Juanes, L. Jimenez, J.C. Rodriguez-Ubis, *J. Mat. Chem.*, **2009**, *19*, 2494.

92. G. Muller, *Dalton Trans.*, **2009**, *44*, 9692.

93. P.G. Hoertz, T.E. Mallouk, *Inorg. Chem.*, **2005**, *44*, 6828.

94. E. Brunet, M. Alonso, M.J. Mata, S. Fernandez, O. Juanes, O. Chavanes, J.C. Rodriguez-Ubis, *Chem. Mat.*, **2003**, *15*, 1232; E. Brunet, M. Alonso, C. Cerro, O. Juanes, J.C. Rodriguez-Ubis, A.E. Kaifer, *Adv. Funct. Mat.*, **2007**, *17*, 1603.

95. E. Brunet, M. Alonso, M.C. Quintana, P. Atienzar, O. Juanes, J.C. Rodriguez-Ubis, H. Garcia, *J. Phys. Chem. C*, **2008**, *112*, 4029.

96. L. Teruel, M. Alonso, M.C. Quintana, A. Salvador, O. Juanes, J.C. Rodriguez-Ubis, E. Brunet, H. Garcia, *Phys. Chem. Chem. Phys.*, **2009**, *11*, 2922.

97. J. Salonen, A.M. Kaukonen, J. Hirvonen, V.-P. Lehto, *J. Pharm. Sci.*, **2008**, *97*, 632.

98. M.H. Hussein, J. Alhendawi, *Mater. Chem.*, **2011**, *21*, 7748.

3

PHOSPHONATES IN MATRICES

Konstantinos E. Papathanasiou and Konstantinos D. Demadis
Crystal Engineering, Growth and Design Laboratory, Department of Chemistry, University of Crete, Crete, Greece

3.1 INTRODUCTION: PHOSPHONIC ACIDS AS VERSATILE MOLECULES

Phosphonic acids constitute a special class of molecules within phosphorus-containing compounds [1]. The phosphonic acid group is a pentavalent, tetrahedral P atom connected via a double bond to O (P=O), while it forms two P–O single bonds with two OH groups (Figure 3.1). The P atom also has a single bond with carbon (P–C), the latter originating from an aliphatic or aromatic fragment.

The H atoms exhibit variable acidity. A thorough review has been published on the acid behaviour of several phosphonic acids [2]. In this review, the authors make available experimental data on stability constants of proton and metal complexes for 10 phosphonic acids. These are methylphosphonic acid, 1-hydroxyethane-1,1-diylbisphosphonic acid, dichloromethylenebisphosphonic acid, aminomethanephosphonic acid, *N*-(phosphono-methyl)glycine, imino-*N*,*N*-bis(methylenephosphonic acid), *N*-methylamino-*N*,*N*-bis(methylenephosphonic acid), nitrilotris(methylenephosphonic acid), 1,2-diaminoethane-*N*,*N*,*N'*,*N'*-tetrakis-(methylenephosphonic acid) and diethylenetriamine-*N*,*N*,*N'*,*N''*,*N''*-pentakis-(methylenephosphonic acid). The data were taken from papers published in the time frame 1950–1997. The acid–base behaviour of phosphonic acids will be discussed further in the following part of this chapter.

Phosphonic acids are widely used in a variety of applications. Their ability to prevent precipitation of alkaline-earth metal sparingly soluble salts at substoichiometric concentrations (threshold inhibition effect) finds wide application in chemical water treatment for scale inhibition [3]. Others are used extensively in laundry detergent formulations [4].

Tailored Organic–Inorganic Materials, First Edition. Edited by Ernesto Brunet, Jorge L. Colón and Abraham Clearfield.
© 2015 John Wiley & Sons, Inc. Published 2015 by John Wiley & Sons, Inc.

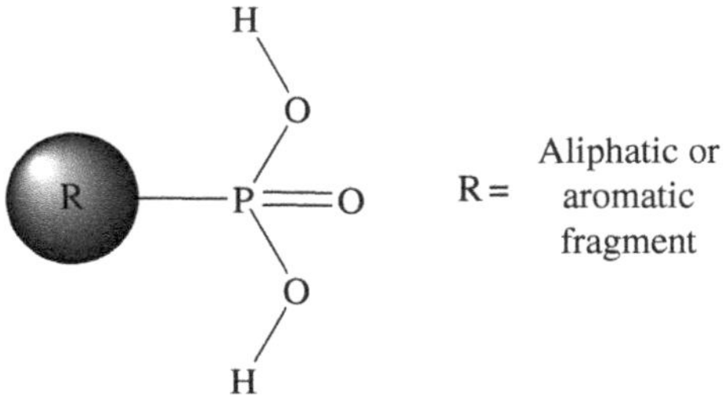

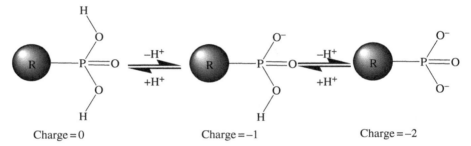

FIGURE 3.1 The chemical identity of the phosphonic acid group.

FIGURE 3.2 The two deprotonation processes in the phosphonic acid group.

Some are also used as corrosion inhibitors [5], in industrial cleaning [6] and in peroxy bleach stabilization [7]. Uses of organophosphonates span applications in flame-resistant polymers [8], photographic processing [9], ore flotation (aminophosphonic surfactants) [10], actinide separation processes [11] and analytical chemistry [12]. Recently, organophosphonates have been identified as promising reagents for the creation of the so-called 'structurally tailored' materials [13] and microporous materials [14], in catalysis [15] and in the electrochemical treatment of polluted soils [16].

The high biological activity of carboxyalkylphosphonates, aminoalkylphosphonates and alkylenediphosphonates makes them useful agents as components of microfertilizers and pesticides in agriculture [17], as well as drugs and diagnostic reagents in biology and medicine [18]. Annual industrial output of organophosphonates is in the thousands of tons [19].

3.2 ACID–BASE CHEMISTRY OF PHOSPHONIC ACIDS

As mentioned before, the phosphonic acid group exhibits variable acidity. There are two stepwise deprotonation processes, as shown in Figure 3.2.

The pK_a values depend on the backbone of the phosphonic acid molecule and the presence of other functional groups (e.g. amino, sulphonate, carboxylate, etc.).

We will refer to some notable examples of phosphonic acids and their acid–base behaviour. Hence, the schematic structures of some (poly)phosphonic acids are shown in Figure 3.3.

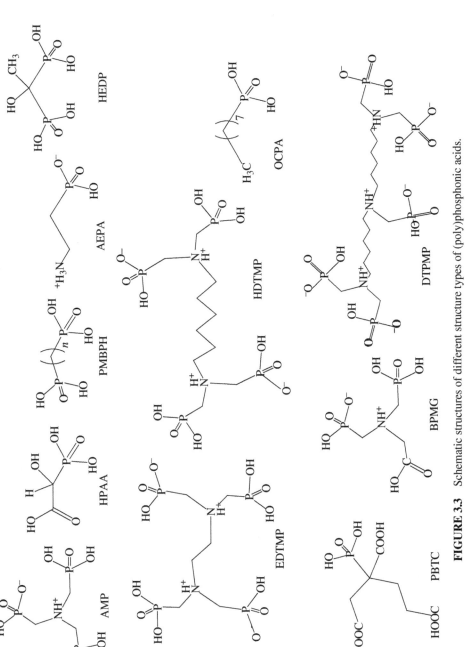

FIGURE 3.3 Schematic structures of different structure types of (poly)phosphonic acids.

The protonation constants of iminobis(methylenephosphonic acid) (IDPH, H_4idph, H_4L), N-methyliminobis(methylenephosphonic acid) (MIDPH, H_4midph, H_4L) and nitrilotris(methylenephosphonic acid) (NTPH, H_6ntph, H6L) were determined by ^{31}P NMR spectroscopy at 25 °C in 0.1 M KNO_3 at $11 < pH < 14$. For equilibrium $L + H \leftrightarrow HL$, log $K = 11.5$ (0.1), 12.2 (0.1) and 12.9 (0.1), respectively [20].

Protonation constants for three common polyphosphonic acids (AMP, HEDP and DTPMP) have been reported (see Table 3.1) [21].

An important observation is the very high acidity of the first acidic proton. Speciation graphs, like the one shown in Figure 3.4, are useful in gaining an overview of acid–base behaviour of phosphonic acids, particularly when they interact with metal ions of variable charge to form hybrid materials or salts, as we will see in the following paragraph.

TABLE 3.1   Protonation constants of AMP, HEDP and DTPMP (for chemical structures, see Figure 3.3)

	AMP[a]	HEDP[a]	DTPMP[b]
Log K_1	12.5 ± 0.2	11.0 ± 0.2	12.58
Log K_2	7.22 ± 0.03	6.9 ± 0.1	11.18
Log K_3	5.90 ± 0.02	2.7 ± 0.1	8.30
Log K_4	4.59 ± 0.03	1.6 ± 0.2	7.23
Log K_5	1.6 ± 0.3		6.23
Log K_6	0.5 ± 0.3		5.19
Log K_7			4.15
Log K_8			3.11
Log K_9			2.08
Log K_{10}			1.04

[a]Data taken from Ref. [22]. Conditions: $I = 0.1$ mol·l^{-1} (KNO_3), $T = 25 ± 0.5$ °C.
[b]Data taken from Ref. [23].

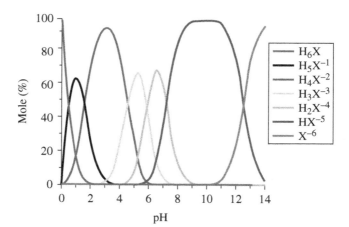

FIGURE 3.4 The distribution of AMP phosphonate ionic species as a function of pH. Reprinted with permission from Ref. [21], Copyright (2006) American Chemical Society.

TABLE 3.2   The stepwise protonation of compounds 1, 2, 3, 4 and 5, as reported by Vepsäläinen et al. [24]

	1 $n=2$	**5** $n=7$	**9** $n=11$
	2 $n=3$	**6** $n=8$	**10** $n=15$
	3 $n=4$	**7** $n=9$	
	4 $n=5$	**8** $n=10$	

Protonation Reaction	\multicolumn{6}{c}{log K}
	1	2	3	4	5	pKa
$L^{4-}+H^+ \leftrightarrow HL^{3-}$	12.86	12.13	12.05	11.94	11.65	pKa$_5$
$HL^{3-}+H^+ \leftrightarrow H_2L^{2-}$	10.04	10.69	10.78	10.86	10.67	pKa$_4$
$H_2L^{2-}+H^+ \leftrightarrow H_3L^-$	5.90	6.26	6.44	6.62	6.75	pKa$_3$
$H_3L^-+H^+ \leftrightarrow H_4L$	1.70	2.12	2.30	2.38	2.52	pKa$_2$
$H_4L+H^+ \leftrightarrow H_5L^-$	1.06	0.60	0.62	0.94	1.08	pKa$_1$

The group of Vepsäläinen has studied the protonation constants of a series of bis-phosphonic acids, of the 'dronate' family, used as therapeutics for osteoporosis [24]. The following general trend can be found in the results of this study: the lengthening of the CH_2 chain between BP and amino groups decreases the value of the first protonation constant (pKa$_5$, amino group) and increases the values of the other protonation constants (phosphonate groups) in most cases (Table 3.2).

3.3   INTERACTIONS BETWEEN METAL IONS AND PHOSPHONATE LIGANDS

The simplest organophosphonate, methylphosphonic acid, reveals higher ML complex stability than acetic acid (by about one order of magnitude) for both alkaline-earth and 3d metal ions. This fact demonstrates the almost equal importance for coordination compound stability of an increase in ligand basicity and an increase in the number of ionic and covalent bonds. Within the 3d elements, definite but small deviations from the Irving–Williams sequence can be seen: Cu > Mn > Co ~ Ni. Although the systematic error is higher than some of the differences in log K_{ML}, the relative error is expected to be small enough to make this a valid conclusion.

Bisphosphonates have significantly higher stability constants than those of monodentate alkylphosphonic acids owing to both higher basicity and bidentate coordination. The log K_{ML} values for HEDPA (1-Hydroxyethane-1,1-diyl)bis(phosphonic acid) and CMDPA (Dichloromethylenediphosphonic acid) are also much higher than those for the dicarboxy analogue malonic acid. For the pair HEDPA/CMDPA, there is a reasonable difference in stability that can be attributed to the electron-attracting effect of the two Cl atoms.

Aminomethylenephosphonic acids (AMPH (Aminomethanephosphonic acid), IDPH (Imino-N,N-bis(methylenephosphonic acid)), MIDPH (N-Methylamino-N, N-bis(methylenephosphonic acid)), NTPH (Nitrilotris(methylenephosphonic acid))) also demonstrate generally higher affinity to cations than their carboxy analogues (see Table 3.3) [25–30].

TABLE 3.3 Stability constants (log K_{ML}) of aminomethylenephosphonates and their carboxy analogues ($I = 0.1\,\text{M}$, $T = 25\,^\circ\text{C}$)

Cation	AMPH	Gly	MIDPH	MIDA	NTPH [25]	NTA	EDTPH [25–30]		EDTA
H^+	10.07	9.60	12.1	9.65	14.2	12.7	9.7	13.8 13.0	10.26
Mg^{2+}	1.99	—	5.1	3.44	9.0	7.5	5.43	9.1 8.4	8.69
	1.67	—	4.6	3.75	9.4	7.9	6.45	10.1 9.4	10.7
Ca^{2+}	1.34	—	4.0	2.85	8.0	6.5	5.0	8.3 7.6	8.6
	4.5	4.66	9.4	7.62	15.5	14.0	10.4	17.8 17.1	16.11
Sr^{2+}	5.3	5.8	9.5	8.73	13.2	11.7	11.5	17.1 16.4	18.6
Co^{2+}	8.10	8.2	14.2	11.09	18.7	17.2	13.0	23.9 23.2	18.8
	—	5.0	10.2	7.66	16.1	14.6	10.7	19.5 18.8	16.26

The complexation constants of AMP and HEDP with different metal cations were determined [31]. These two phosphonic acids are used in detergents as builders, because of their complexing properties, especially the complexation with calcium and heavy metal cations. The cations studied were those usually encountered in natural waters (Ca(II), Zn(II)) and anthropogenic heavy metals (Cu(II), Ni(II), Cd(II), Pb(II)).

The equilibrium constant values and the titration curves point out the following order:

AMP: Cu(II) > Zn(II) > Pb(II) > > Cd(II) > Ni(II) > > Ca(II)

HEDP: Cu(II) > Zn(II) > > Cd(II) > Ni(II) > Ca(II)

The affinity order for all the cations is the same for AMP and HEDP. The AMP complexing properties for each cation are higher than those of HEDP, particularly with copper and zinc (difference of six log units).

Stone et al. have studied the formation of metal ion-chelating agent complexes in aqueous solution in an excellent review [32]. They also studied and compared the adsorption of various phosphonates onto (hydr)oxide mineral surfaces (Figure 3.5). Interconnections between the coordination chemistry and chemical reactivity of phosphonates were also made.

Synthetic manipulation of organic platforms by introducing phosphonate groups often yields highly selective ligands. For example, Gałezowska et al. described the synthesis of two new ligands (L^2 and L^3) composed of ethylenediamine (EN), pyridyl (Py) moieties and phosphonic groups, designed to bind metal ions with the donor set based on four nitrogen atoms and two phosphonic units (Figure 3.6) [33]. These ligands were used to evaluate their coordination abilities towards Cu(II), Ni(II) and Zn(II) and compare those to ligand L^1.

Ligands L^1 and L^2 possessing the ethylenediamine core situated closely to the phosphonic acid units bind the Cu(II), Ni(II) and Zn(II) ions using the same donor set, two N atoms from ethylenediamine and two O atoms from phosphonate. The presence of the pyridyl moieties in L^3 leads to very effective ligand (pM 15.27) for

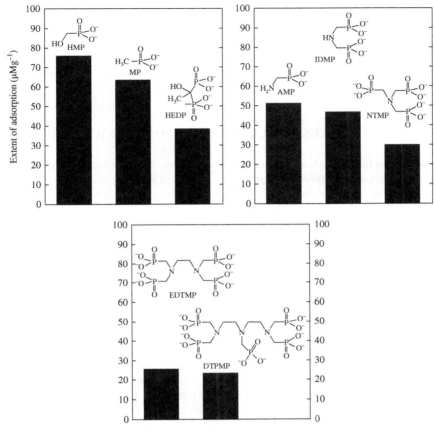

FIGURE 3.5 The adsorption of five phosphonate molecules onto goethite surfaces. Reprinted with permission from Ref. [32], Copyright (2002) American Chemical Society.

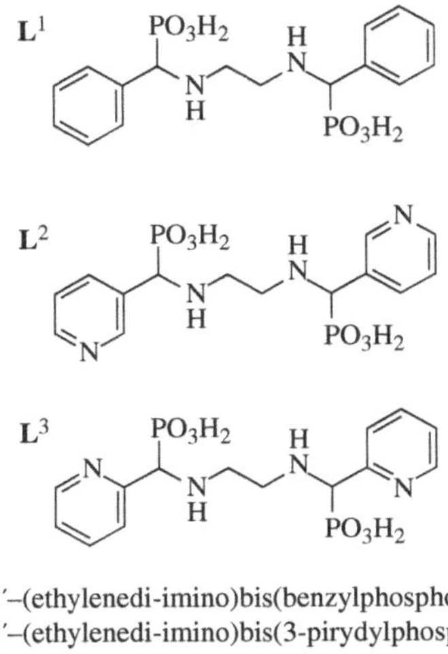

FIGURE 3.6 Chemical structures of studied ligands L^1, L^2, and L^3. Reprinted with permission from Ref. [33], Copyright (2009) Elsevier.

L^1 2,2′–(ethylenedi-imino)bis(benzylphosphonic-acid)
L^2 2,2′–(ethylenedi-imino)bis(3-pirydylphosphonic-acid)
L^3 2,2′–(ethylenedi-imino)bis(2-pirydylphosphonic-acid)

Cu(II) ions, with tetradendate coordination mode and axial involvement of two phos-phonate groups in Zn(II) and Ni(II) octahedral complexes. L^3 is able to involve a four-nitrogen donor system and additionally two phosphates and is unusually an effective ligand for both planar and octahedral complexes.

3.4   PHOSPHONATES IN 'ALL-ORGANIC' POLYMERIC SALTS

When phosphonic acids deprotonate, they can interact in solution or in the solid state with organic cations. The salts that form could be envisioned as 'all organic', since they do not contain metal ions. In this section, we will review some repre-sentative examples of materials that conform to the general type 'organic cation phosphonate'.

An example of an 'intramolecular' salt is the heterotopic phosphonic acid, 3-amino-5-(dihydroxyphosphoryl)benzoic acid, which was synthesized and structur-ally characterized [34] (Figure 3.7).

Long-chain monophosphonates with ammonium and ethylenediammonium cat-ions have been structurally characterized. Specifically, the crystal structure of ammonium 1-decylphosphonate and ethylenediammonium 1-decylphosphonate 1.5 hydrate have been reported [35]. The layered structure of the crystal of ammonium 1-decylphosphonate is dominated by the complex system of hydrogen bonds between phosphonate group and ammonium ion (Figure 3.8). The 1-decy-lphosphonate ions are situated in the head-to-head and tail-to-tail relation. The very wide amphiphilic layers (26.345(5) Å) parallel to the ab crystallographic plane are built of the hydrophilic central part and the hydrophobic external part. The interlayer contacts are of the very weak van der Waals type, and the aliphatic chains do not interdigitate.

The organization of the crystal of ethylenediammonium 1-decylphosphonate is very similar to that of the crystal of ammonium 1-decylphosphonate (Figure 3.9). The crystal structure of the latter is more compact than that of the former because the aliphatic chains interdigitate. The phosphonate groups interact with ethylenedi-ammonium cations through four strong hydrogen bonds.

Co-crystallization of melamine (ma) with m-sulphophenylphosphonic acid (sppH$_3$) from water in different molar ratios (2 : 1 and 4 : 1) yields [(maH)$_2$(sppH)] · 3H$_2$O and [(maH)$_3$(spp)(ma)] · 12H$_2$O, respectively [36]. Structure analysis reveals that two very intricate hydrogen-bonded networks are formed in them, with two or three protons of the m-sulphophenylphosphonic acid being transferred to melamine (Figure 3.10). The resultant (sppH)$^{2-}$ or (sppH)$^{3-}$ anion can form as many as 12 or 14 hydrogen bonds with melamine and water molecules, showing a very high hydrogen-bonding capability.

Hexamethylenediamine-N,N,N',N'-tetrakis(methylenephosphonic acid) (HDTMP) has been isolated as a crystalline solid with the ethylenediammonium (en) dication, as (en)(HDTMP) · 2H$_2$O. The crystal structure of the solid has been determined [37].

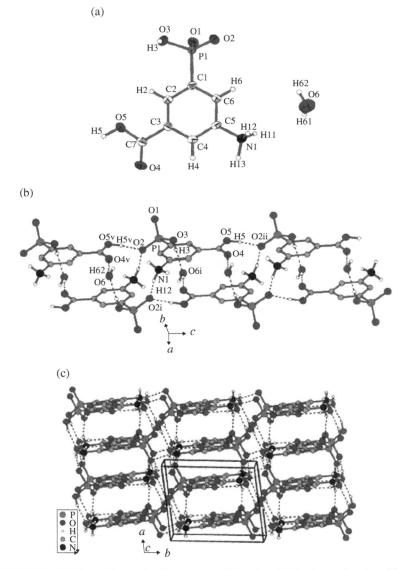

FIGURE 3.7 (a) View of the molecular structure of 3-amino-5-(dihydroxyphosphoryl)benzoic acid. (b) View of the O–H···O hydrogen-bonded chains of zwitterionic 3-amino-5-(dihydroxyphosphoryl)benzoic acid molecules. (c) Crystal packing of 3-amino-5-(dihydroxyphosphoryl)benzoic acid showing the layered structure. Dashed lines represent the O–H···O and N–H···O hydrogen bonds. Reprinted with permission from Ref. [34], Copyright (2013) Elsevier.

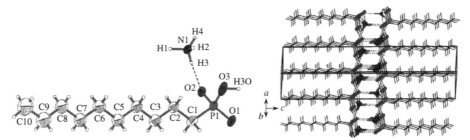

FIGURE 3.8 View of the molecular unit of ammonium 1-decylphosphonate (left) and the crystal packing (right). Reprinted with permission from Ref. [35], Copyright (2012) Elsevier.

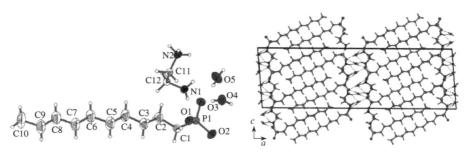

FIGURE 3.9 View of the molecular unit of ethylenediammonium 1-decylphosphonate 1.5 hydrate (left) and the crystal packing (right). Reprinted with permission from Ref. [35], Copyright (2012) Elsevier.

The molar ratio between the dication and dianion has been found $1 : 1$. The disposition of the ionic pair is shown in Figure 3.11. The presence of waters of crystallization, HDTMP^{2-} dianions and en^{2+} dications leads to a complicated network of hydrogen bonds, finally yielding a 2D layered structure (shown in Figure 3.12). They are described in detail in the following paragraph. The $-NH_3^+$ portion of the en molecule is hydrogen bonded with one of the two water molecules of crystallization (at a distance of 2.974 Å) and with three O–P moieties (all from different phosphonate groups) (at distances of 2.708, 2.804 and 2.984 Å).

The two waters of crystallization are located in the vicinity of the en^{2+} dication. One water molecule forms hydrogen bonds with one of the N–H$^+$ moieties of the phosphonate ligand (2.963 Å), with the ^+H_3N group of the en^{2+} cation (2.974 Å) and with the O atoms from a deprotonated phosphonate ligand, $^-$O–P (2.999 Å). The second water molecule participates in three hydrogen bonds, one with a –P–OH group (2.553 Å) and two hydrogen bonds with two deprotonated P–O$^-$ moieties (2.729 and 2.743 Å) from neighbouring phosphonate ligands. There are several other hydrogen bonds in the structure that are described in detail in the original paper.

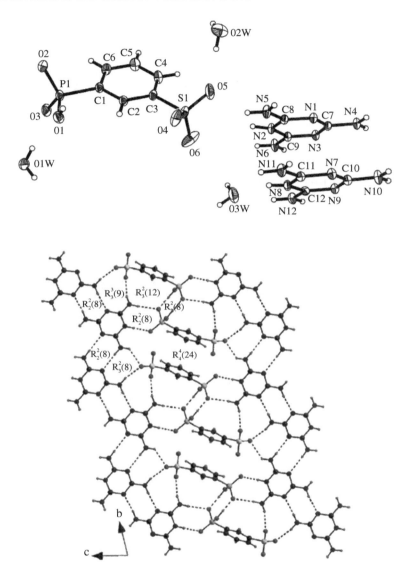

FIGURE 3.10 Asymmetric unit of [(maH)$_2$(sppH)]·3H$_2$O (left) and one-dimensional ladder-like chain formed by hydrogen bond interactions between the (sppH)$^{2-}$ and (maH)$^+$ ions in [(maH)$_2$(sppH)]·3H$_2$O (right). Reprinted with permission from Ref. [36], Copyright (2013) Elsevier.

Ethylenediamine-N,N'-tetrakis(methylenephosphonic acid) (EDTMP) is structurally related to HDTMP, except that the N atoms are connected by two methylene groups. A crystalline solid was isolated, which contains two ammonium cations per one EDTMP dianion [37]. In the structure of (NH$_4$)$_2$(EDTMP), there are discrete EDTMP^{2-} dianions and NH$_4^+$ cations (see Figure 3.13).

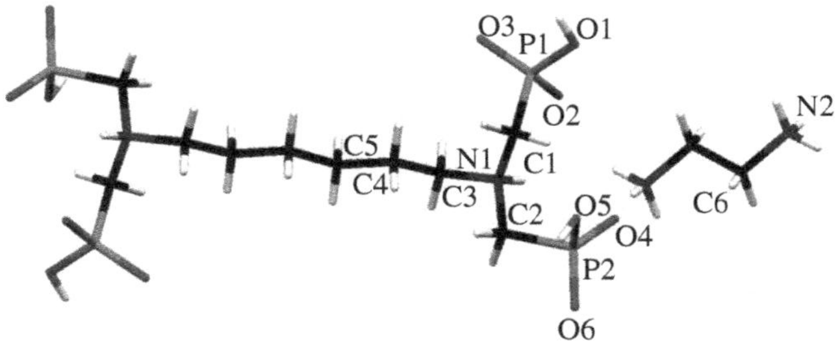

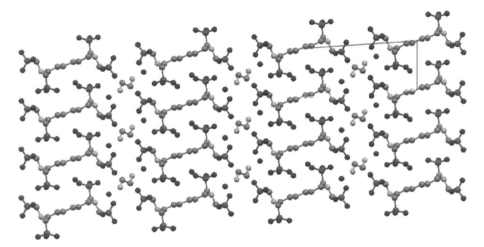

FIGURE 3.11 Structure of HDTMP²⁻ with the en²⁺ dications (the water molecules are omitted for clarity). Reprinted with permission from Ref. [37], Copyright (2009) Elsevier.

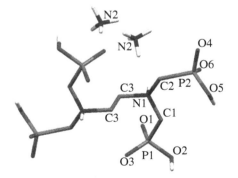

FIGURE 3.12 Layers of HDTMP²⁻ and en²⁺ dications (lower) down the *b*-axis. Hydrogen atoms are omitted for clarity.

FIGURE 3.13 Structure of EDTMP²⁻ with the two NH₄⁺ cations. Reprinted with permission from Ref. [37], Copyright (2009) Elsevier.

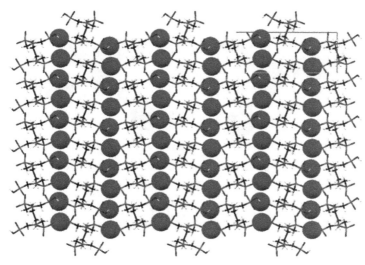

FIGURE 3.14   Layers of EDTMP^{2-} and NH$_4^+$ cations, shown as spheres, shown down the *a*-axis. Reprinted with permission from Ref. [37], Copyright (2009) Elsevier.

The presence of EDTMP^{2-} dianions and NH$_4^+$ cations leads to a complicated network of hydrogen bonds, finally yielding a 2D layered structure (see Figure 3.14).

Each of the ammonium cations participates in five hydrogen bonds, all with non-protonated P–O groups from different phosphonate ligands, with O⋯N distances ranging from 2.799 to 2.949 Å. The P(1)O$_3$H$^-$ phosphonate group participates in five hydrogen bonds. Protonated P–O(2)H participates in a hydrogen bond with the ammonium N(2) group of the NH$_4^+$ cation (2.929 Å) and the non-protonated group O(4)–P(2) from a neighbouring phosphonate (2.559 Å). The group P–O(1) forms only one H bond with the protonated N(1)H$^+$ group belonging to another EDTMP molecule (2.648 Å). Finally, P–O(3) forms two H bonds with two symmetry-related N(2) groups from two different en molecules (2.799 and 2.895 Å). The P(2)O$_3$H$^-$ phosphonate group also participates in five hydrogen bonds. Protonated P–O(5)H participates in a hydrogen bond with the non-protonated P(2)–O(6) group from a neighbouring phosphonate. At the same time, non-protonated P(2)–O(6) hydrogen bonds with the protonated P(2)–O(5)–H moiety with the same phosphonate group. This creates a hydrogen-bonded dimer, the structure of which is shown in Figure 3.15. A similar phosphonate hydrogen-bonded dimer has been observed in the structure of 2-phosphonobutane-1,2,4-tricarboxylic acid monohydrate (PBTC·H$_2$O) [38].

Tetraphosphonates are biomimetic hosts for bisamidinium cations in drugs such as pentamidine and DAPI (4′,6-diamidino-2-phenylindole). Similar to their insertion into DNA's minor groove, these drugs are often sandwiched by two tetraphosphonate hosts in a 2 : 1 ratio, as shown in Figure 3.16 [39].

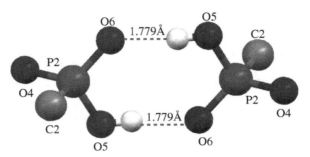

FIGURE 3.15   Structure of the hydrogen-bonded dimer in the structure of $(NH_4)_2(EDTMP)$.

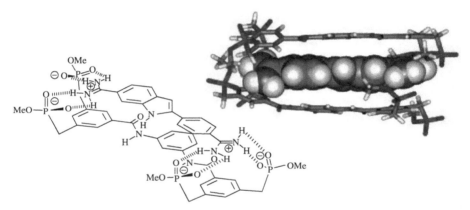

FIGURE 3.16   1 : 1 complex between a tetraphosphonate and DAPI according to Monte Carlo simulations in water. Reprinted with permission from Ref. [39], Copyright (2003) American Chemical Society.

3.5   PHOSPHONATES IN COORDINATION POLYMERS

The field of coordination polymers has exploded in the last decades. Since a substantial number of original research papers, reviews, book chapters and books have been published on the subject, we will attempt a concise look into this topic, with a focus on metal phosphonate-based coordination polymers. A concise source on the topic is a book by Clearfield and Demadis [40].

There is a battery of structurally characterized metal phosphonate materials of essentially all metal ions of the periodic table. We briefly note those of alkali metal ions [41–47]; alkaline-earth ions [48–60]; transition elements of the first period [61–70], the second period [71–80] and the third period [81–90]; lanthanides [91–100]; and actinides [101–105].

Among the plethora of anionic ligands used for the construction of inorganic–organic hybrids, polycarboxylates are predominant. Polyphosphonates have also attracted significant interest, because they exhibit a number of similarities but also differences to the carboxylates: (i) phosphonate building blocks possess three

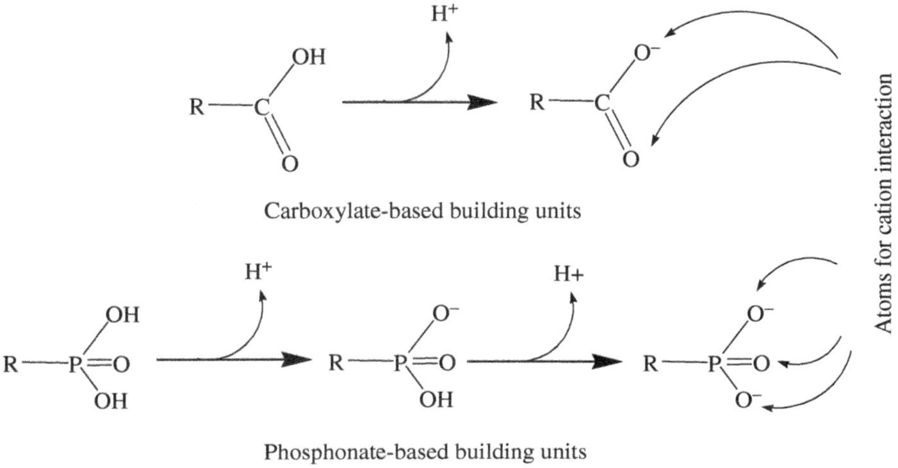

FIGURE 3.17 Structural and functional differences between carboxylic and phosphonic acids.

O atoms linked to the phosphorus atom in the coordinating moiety, compared to two O atoms in the case of carboxylates. This increases the possibilities for access to novel structures. (ii) The phosphonic acid moiety can be doubly deprotonated in two well-defined successive steps, depending on solution pH [2]. Carboxylic acid ligands can only be deprotonated once (see Figure 3.17).

Again, this allows access to a variety of potential novel phosphonate-containing structures, by simply varying the pH. (iii) The phosphonate group can be (potentially) doubly esterified, in contrast to the carboxylate group that can only be monoesterified [41, 64, 106, 107]. The introduction of at least one phosphonate ester in the building block is expected to enhance solubility (in the case of very insoluble materials) or, by virtue of its hydrolysis [63], to yield structural diversity in the end material. (iv) Synthesis of metal phosphonate materials can be carried out via a number of different routes that do not necessarily give products with the same structure. There is hence a greater potential of structural diversity in the products derived. Several of these methods lend themselves to a combinatorial approach allowing high-throughput screening of candidate materials to be achieved [108]. In this context, a recent review was published on 'non-carboxylate' metal–organic frameworks (MOFs) [109].

Because the focus of this chapter is not metal phosphonate frameworks, we will refer the reader to the relevant book mentioned earlier and its chapters for further details and literature [40].

3.6 PHOSPHONATE-GRAFTED POLYMERS

A phosphonate moiety can be grafted on a polymeric chain or matrix, by use of established organic synthetic methodology. We will briefly examine these grafting methods and also review the types of polymers that result.

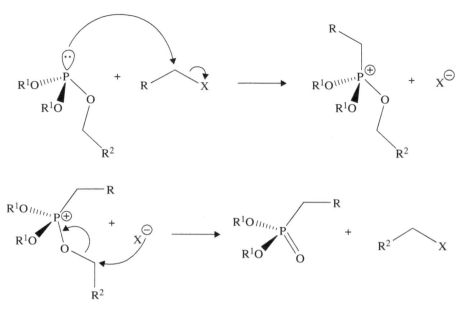

FIGURE 3.18 Mechanism of the Michaelis–Arbuzov reaction. Reprinted with permission from Ref. [40], Copyright (2012) Royal Society of Chemistry.

The Michaelis–Arbuzov reaction is a method for C–P bond formation, leading to a dialkoxyphosphonate. The reaction was discovered by Michaelis and investigated and developed by Arbuzov. It proceeds mainly between primary alkyl halides and trialkyl phosphite and is usually thermally initiated (see Figure 3.18) [110].

The Mannich-type condensation (occasionally called the Moedritzer–Irani reaction) [111] is a convenient approach for the synthesis of *N,N*-disubstituted aminomethylphosphonic acids or *N*-substituted iminobis(methylphosphonic acids). The reaction is conducted in highly acidic solutions. Zon et al. have proposed a mechanism for this reaction (Figure 3.19) [112].

Zon et al. explained that the first step of the reaction is a nucleophilic attack of *N,N*-dialkylamine nitrogen. Further rearrangement gives *N*-hydroxyethylamine, which in strong acidic condition undergoes elimination of water molecule yielding an imine salt. Phosphorous acid in acidic conditions behaves as a nucleophile and therefore attacks the electrophilic imine salt. The charged adduct is stabilized by loss of a proton to give *N,N*-disubstituted aminomethylphosphonic acid. *N*-substituted iminobis(methylphosphonic acid) is formed when one starts from primary alkyl-amine. In the following, we present an example of the formation of an aromatic tetraphosphonic acid ligand (Figure 3.20), which has been used before for the construction of metal phosphonate frameworks.

Below, we will present several examples of phosphonate incorporation into polymeric matrices. Biomaterials such as inulin, chitin, chitosan and their

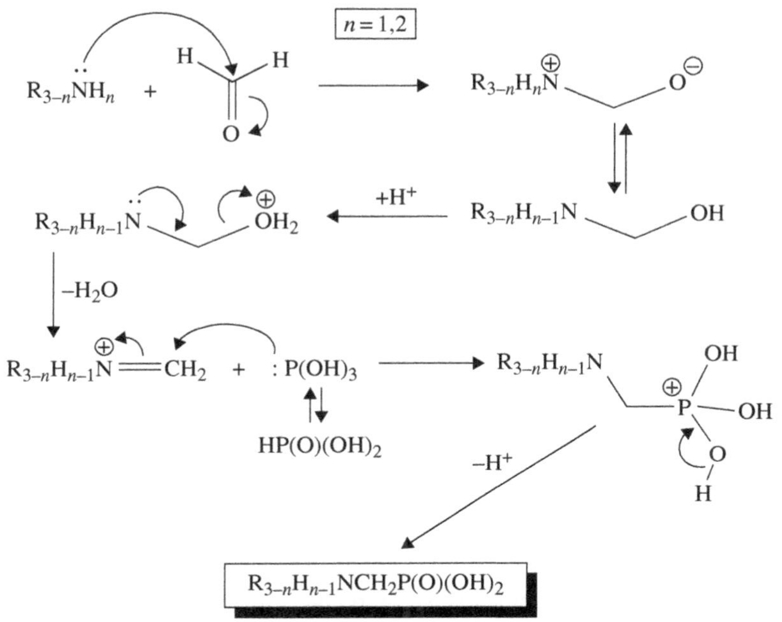

FIGURE 3.19 Mechanism of the Mannich-type (Moedritzer–Irani) reaction. Reprinted with permission from Ref. [40], Copyright (2012) Royal Society of Chemistry.

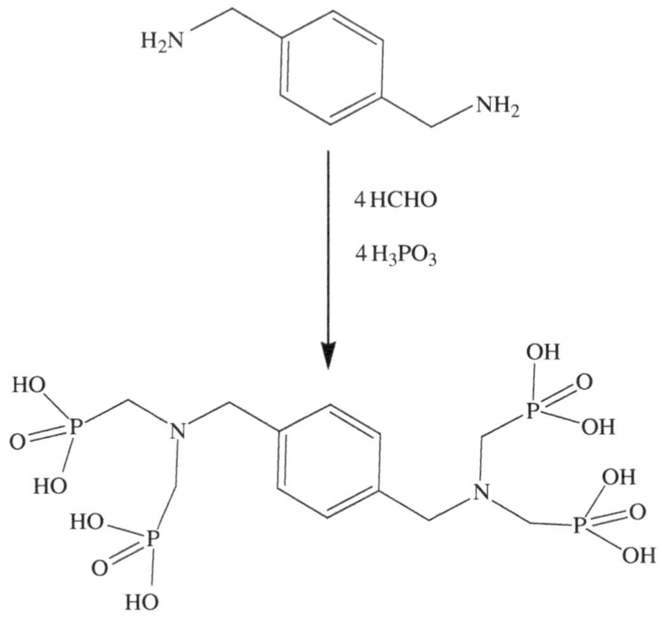

FIGURE 3.20 Synthesis of the xylene-diamine-tetrakis(methylenephosphonic acid) ligand.

derivatives have a significant and rapid development in recent years. They have become the focus of intense research because of an unusual combination of biological activities together with mechanical and physical properties. However, the applications of chitin and chitosan are limited due to insolubility issues in most solvents. The chemical modifications of chitin and chitosan are of keen interest because these modifications would not change the fundamental skeleton of chitin and chitosan but would keep the original physico-chemical and biochemical properties. They would also improve certain properties. The chemical modification of chitin and chitosan by phosphorylation is expected to be biocompatible and is able to promote tissue regeneration. Thus, we will start with these polymers.

A novel conjugate of a polysaccharide and a Gd(III) chelate with potential as contrast agent for magnetic resonance imaging (MRI) was synthesized [113]. The structure of the chelate was derived from H_5DTPA by replacing the central pendant arm by a phosphinic acid functional group, which was covalently bound to the polysaccharide inulin. On the average, each monosaccharide unit of the inulin was attached to approximately one chelate moiety. The ligand binds the Gd^{3+} ion in an octadentate fashion via three nitrogen atoms, four carboxylate oxygen atoms and one P–O oxygen atom, and its first coordination sphere is completed by a water molecule. This compound shows promising properties for application as a contrast agent for MRI thanks to a favourable residence lifetime of this water molecule (170 ns at 298 K), a relatively long rotational correlation time (866 ps at 298 K) and the presence of two water molecules in the second coordination sphere of the Gd^{3+} ion. The synthesis of this interesting polymer is shown in Figure 3.21.

A water-soluble chitosan derivative carrying phosphonic groups was synthesized using a one-step reaction [114, 115]. Detailed NMR studies permitted the identification of the structure by the substituent distribution of the product, which is partly N-monophosphonomethylated (0.24) and N,N-diphosphonomethylated (0.14) and N-acetylated (0.16) without modification of the initial degree of acetylation (Figure 3.22).

The introduction of an alkyl chain onto a water-soluble, modified chitosan (N-methylene phosphonic chitosan) allows the presence of hydrophobic and hydrophilic branches for the control of solubility properties [116]. A simple methodology for the preparation of a new chitosan derivative surfactant, N-lauryl-N-methylene phosphonic chitosan, was developed. The degree of lauryl substitution was estimated to be 0.33.

A simple methodology for the preparation of a new chitosan derivative called N-propyl-N-methylene phosphonic chitosan (PNMPC) was proposed. As before, the introduction of a propyl chain onto a modified chitosan (N-methylene phosphonic chitosan) offers the presence of hydrophobic and hydrophilic branches for the control of solubility properties of the new derivative. The degree of propyl substitution was 0.64. An SEM image of the solid polymer is shown in Figure 3.23 [117].

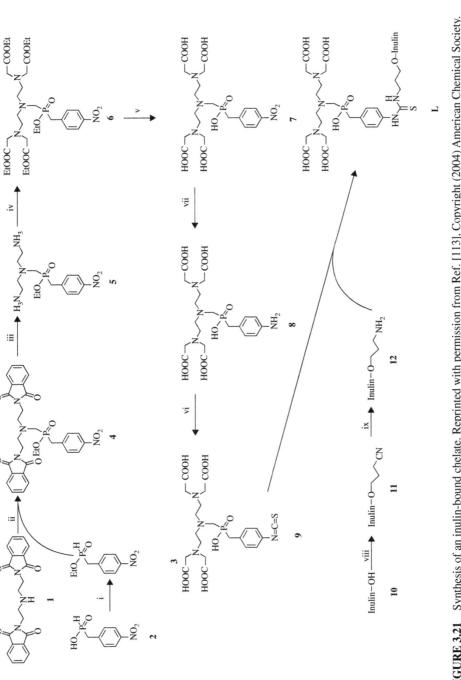

FIGURE 3.21 Synthesis of an inulin-bound chelate. Reprinted with permission from Ref. [113], Copyright (2004) American Chemical Society.

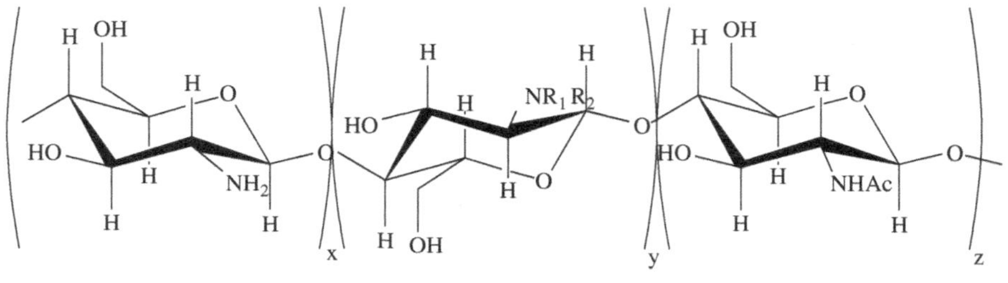

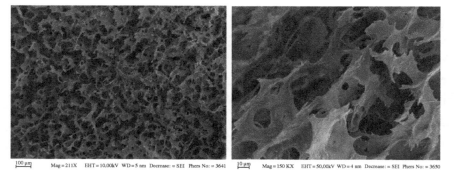

$$R_1 = H, R_2 = CH_2PO_3H_2$$
$$R_1 = R_2 = CH_2-PO_3H_2$$

FIGURE 3.22 Synthesis of phosphono-substituted (*N*-methylene phosphonic) chitosan. Reprinted with permission from Ref. [114], Copyright (2001) Elsevier.

FIGURE 3.23 SEM images of the *N*-propyl-*N*-methylene phosphonic chitosan (PNMPC). Reprinted with permission from Ref. [117], Copyright (2010) Elsevier.

Phosphorylation of chitosan but at the hydroxy and amino groups under the conditions of the Kabachnik–Fields reaction was reported (Figure 3.24) [118]. Conditions were found under which the reaction yields chitosan derivatives containing *N*-phosphonomethylated and chitosan phosphite fragments.

Recently, several derivatives of 2-(arylamino phosphonate)-chitosan (2-AAPCS) were prepared by different Schiff bases of chitosan reacted with dialkyl phosphite in benzene solution (Figure 3.25) [119]. The structures of the derivatives (2-AAPCS) were characterized by FTIR spectroscopy and elemental analysis. In addition, the anti-fungal activities of the derivatives against four kinds of fungi were evaluated. The results indicated that all the prepared 2-AAPCS had a significant inhibiting effect on the investigated fungi when the derivatives concentration ranged from 50 to $500\,g\,l^{-1}$. Furthermore, the anti-fungal activities of the derivatives increased with increasing molecular weight and concentration.

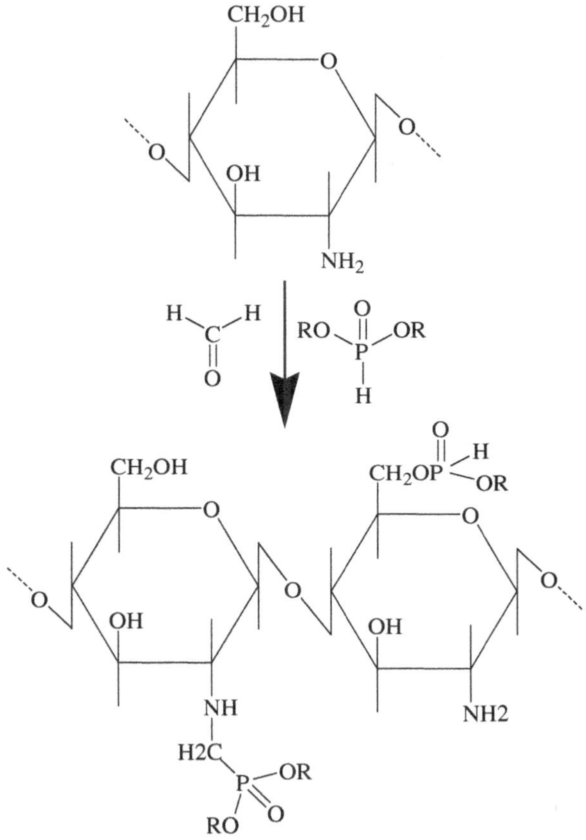

FIGURE 3.24 Kabachnik–Fields synthesis of chitosan derivatives containing *N*-phospho-nomethylated and chitosan phosphite fragments.

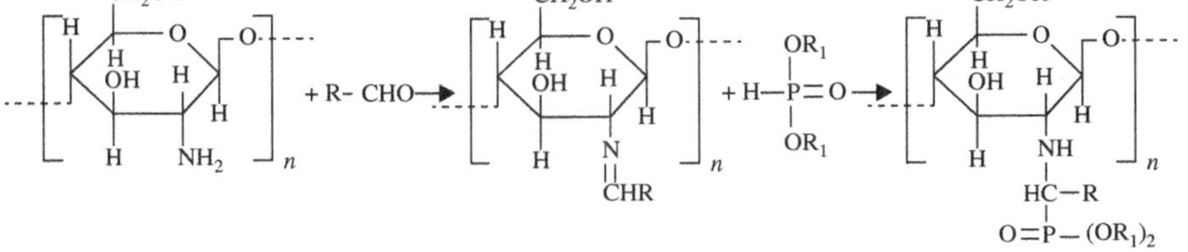

FIGURE 3.25 Synthetic pathway of 2-(-arylamino phosphonate)-chitosan. Reprinted with permission from Ref. [119], Copyright (2009) Elsevier.

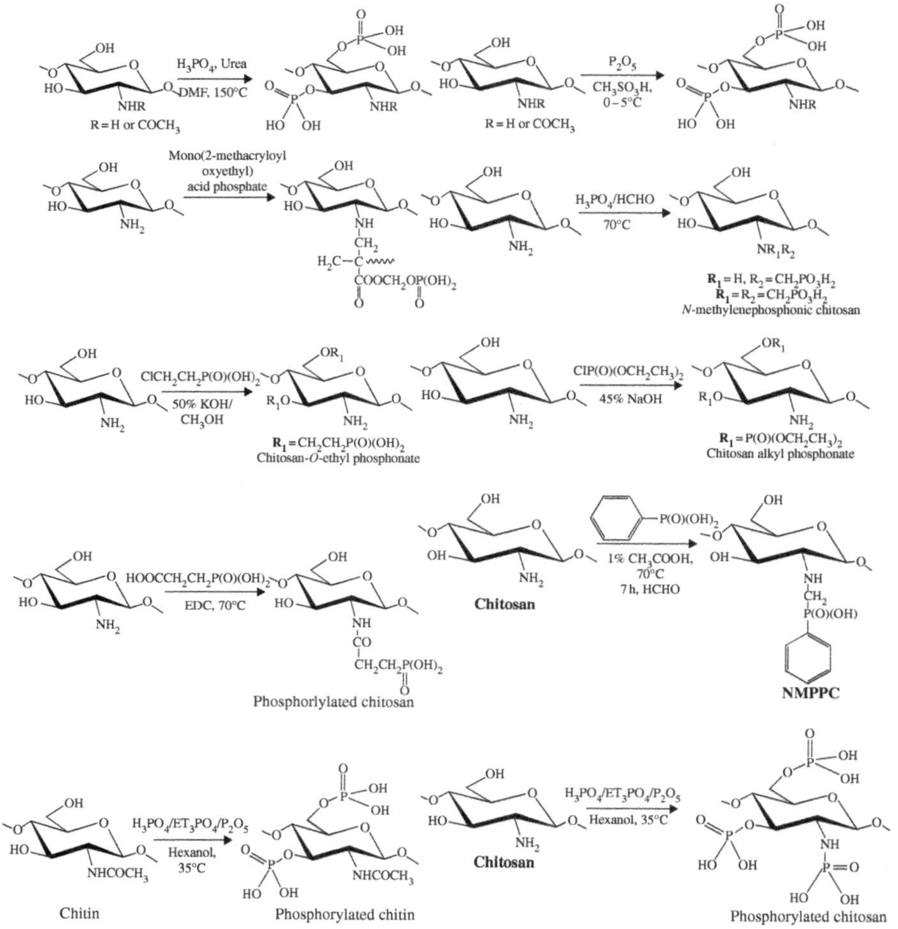

FIGURE 3.26 Preparation of phosphorylated chitin and chitosan with different methods. Reprinted with permission from Ref. [120], Copyright (2008) Elsevier.

Jayakumara et al. have presented the recent developments in the preparation of phosphorylated chitin and chitosan with different methods [120]. These are presented in Figure 3.26.

Chitosan was also phosphorylated by P_2O_5 in methanesulphonic acid, and the product, water-soluble phosphorylated chitosan, was characterized by P elemental analysis and IR and ^{31}P NMR spectroscopy (Figure 3.27) [121]. The phosphorylated chitosans were used to improve the mechanical properties of calcium phosphate cement (CPC) systems of two types: (1) monocalcium phosphate monohydrate (MCPM) and calcium oxide (CaO) and (2) dicalcium phosphate dihydrate (DCPD) and calcium hydroxide [Ca(OH)₂]. The results were successful (based on the compressive strength (CS) and Young's modulus of both CPC formulations). The results

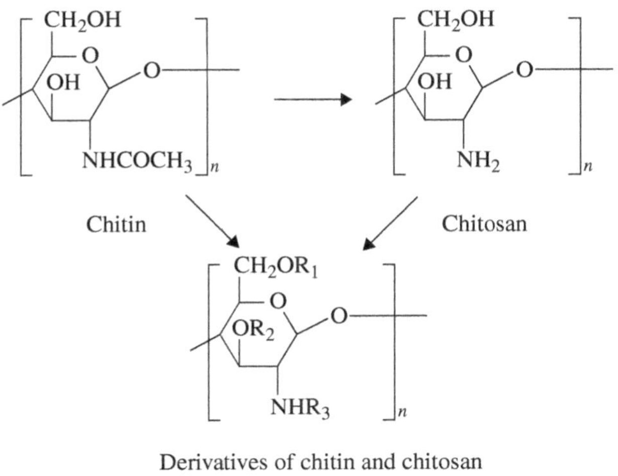

Derivatives of chitin and chitosan

FIGURE 3.27 Schematic structures of chitin, chitosan and derivatives (R_1, R_2, R_3 = $-COCH_3$, $-CH_3$, $-CH_2COOH$, $-SO_3H$, $-P(O)(OH)_2$). Reprinted with permission from Ref. [121], Copyright (2001) Elsevier.

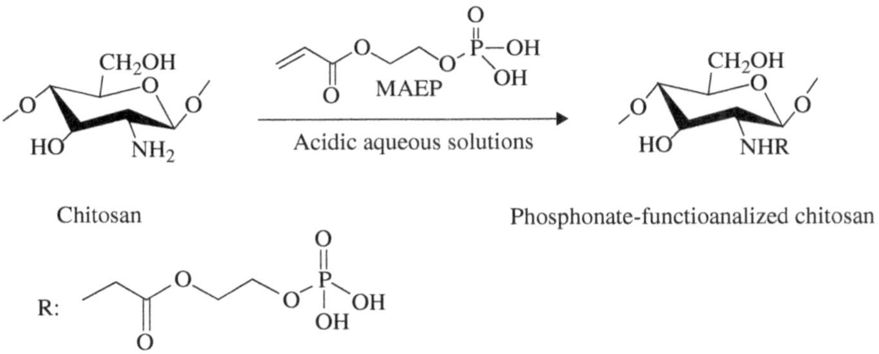

FIGURE 3.28 Synthesis of Phosphonate-functionalized chitosan. Reprinted with permission from Ref. [122], Copyright (2006) Elsevier.

indicated that P-chitosan-reinforced CPC have some good characteristics for clinical applications.

A series of phosphonate-functionalized pH-responsive chitosans were directly synthesized via Michael addition of chitosan with mono-(2-acryloyloxyethyl) phosphonate (Figure 3.28) [122]. The results indicated that the inter- or intra-chain electrostatic interactions of the phosphonate-functionalized chitosans could be controlled via adjusting the solution pH, leading to the reversible conformational and phase transitions of these chitosans.

New phosphorus-containing chitosan derivatives were prepared in good yield under mild conditions from 6-O-triphenylmethyl-chitosan and native chitosan. The

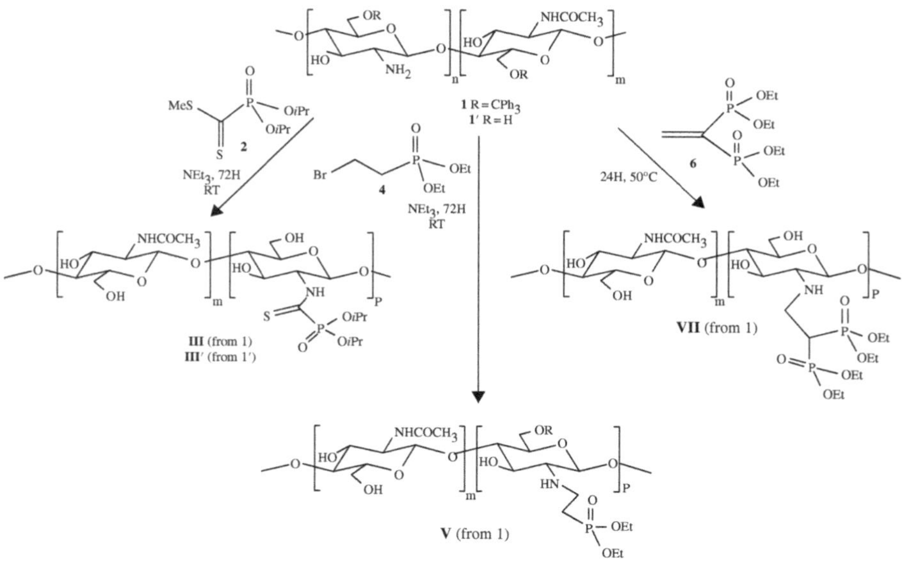

FIGURE 3.29 Synthesis of phosphorus-containing chitosans. Reprinted with permission from Ref. [123], Copyright (2009) Taylor & Francis.

three reactions used are thioacylation by a phosphonodithioester, alkylation by a halogeno-phosphonate and Michael addition using a tetraethyl vinylidenebisphosphonate (Figure 3.29) [123]. The modified chitosan derivatives were fully characterized, and their solubilities and thermal properties were evaluated.

Poly[(1-vinyl-1,2,4-triazole)-co-(vinylphosphonic acid)] (poly(VTAz/VPA)) hydrogels were prepared by ^{60}Co γ-irradiation of binary mixtures of 1-vinyl 1,2,4-triazole and vinylphosphonic acid in the presence of NaHCO$_3$. The polymers form hydrogels under certain conditions (Figure 3.30) [124].

The preparation and characterization of some chelating resins, phosphonate grafted on polystyrene-divinylbenzene supports, were reported [125]. The resins were prepared by an Arbuzov-type reaction between chloromethyl polystyrene-divinylbenzene copolymers and triethylphosphite, yielding the phosphonate ester copolymer (resin A) (Figure 3.31). This can be hydrolyzed by HCl to yield the phosphonate/phosphonic acid copolymer (resin B) (Figure 3.31). The phosphonate resins A and B were characterized by the determination of the phosphorus content, infrared spectroscopy and thermal analysis. The total sorption capacity of the phosphonate ester-functionalized resin (A) and phosphonate/phosphonic acid-functionalized resin (B) for divalent metal ions such as Ca^{2+}, Cu^{2+} and Ni^{2+} was studied in aqueous solutions. Resin A retains approximately 3.25 mg Ca^{2+}/g copolymer, 2.75 mg Cu^{2+}/g copolymer but no Ni^{2+} at pH = 1. On the other hand, resin B retains 8.46 mg Ca^{2+}/g copolymer, 7.17 mg Cu^{2+}/g copolymer but no Ni^{2+} at pH = 1. Efficient Ni^{2+} retention was observed at pH = 7 only for the phosphonate/phosphonic acid-functionalized resin (B) at the level of 19 mg Ni^{2+}/g polymer B. Polymer A was incapable of retaining Ni^{2+} at pH = 7.

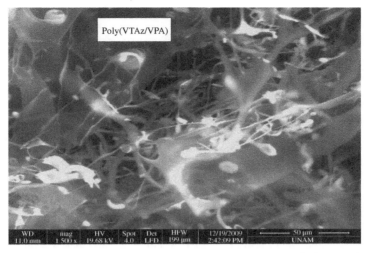

FIGURE 3.30   SEM image of poly(VTAz/VPA) hydrogels. Reprinted with permission from Ref. [124], Copyright (2013) Taylor & Francis.

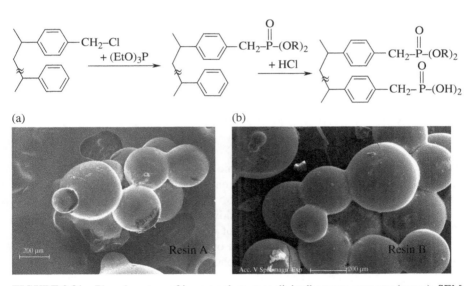

FIGURE 3.31   Phosphonate grafting on polystyrene-divinylbenzene supports (upper). SEM images of the produced resins (lower). The bars for both images are 200 μm. Reprinted with permission from Ref. [125], Copyright (2008) American Chemical Society.

The Mannich-type reaction was used to graft methylenephosphonic acid groups to polyethyleneimine (PEI) to produce an ion-exchange polymer, polyethyleneimine methylenephosphonic acid (PEIMPA) [126]. The removal of various heavy metal ions such as Cu^{2+}, Co^{2+}, Zn^{2+}, Ni^{2+} and Pb^{2+} from aqueous solutions by induced

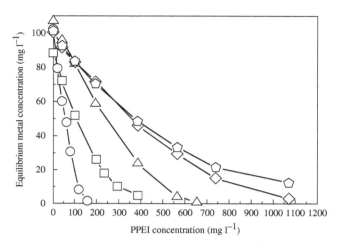

FIGURE 3.32   Removal of heavy metals by the PPEI–Ca^{2+} flocculant system. Circles, Pb; pentagons, cobalt; squares, Cu; triangles, zinc; rhombs, nickel. Reprinted with permission from Ref. [127], Copyright (2002) Taylor & Francis.

flocculation of phosphonomethylated polyethyleneimine (PPEI) heavy metal complex with Ca^{2+} ions was studied (Figure 3.32) [127]. Considerable floc formation accompanying metal sequestration was demonstrated, even at low initial concentration of the target metals. The PPEI–Ca^{2+} flocculant system was also effective for heavy metal scavenging purposes.

3.7   POLYMERS AS HOSTS FOR PHOSPHONATES AND METAL PHOSPHONATES

Phosphonates have been studied as enzyme inhibitors. Since this action requires incorporation of the phosphonate into an enzyme matrix, we will mention a few representative examples here.

   Phosphonates were found to competitively inhibit phosphotriesterase from *Pseudomonas* by chelating both zinc atoms of a binuclear metal centre in the enzyme's active site [128]. The inhibitory properties of a series of substituted phosphonates were also measured for the bacterial phosphotriesterase. The incorporation of fluorine, hydroxyl, thiol, carbonyl and carboxyl groups adjacent to the phosphoryl group was designed to assist in the direct coordination with one or both of the metal ions contained within the structure of the binuclear metal centre. Of the compounds tested, the diethyl thiomethylphosphonate was by far the most potent inhibitor identified.

   Foscarnet (phosphonoformate trisodium salt), an antiviral used for the treatment of HIV and herpesvirus infections, also acts as an activator or inhibitor of the metalloenzyme carbonic anhydrase (CA, EC 4.2.1.1). The interaction of the drug with 11 CA isozymes has been investigated kinetically, and the X-ray structure of its adduct with isoform I (hCA I–foscarnet complex) has been resolved (Figure 3.33) [129]. The first

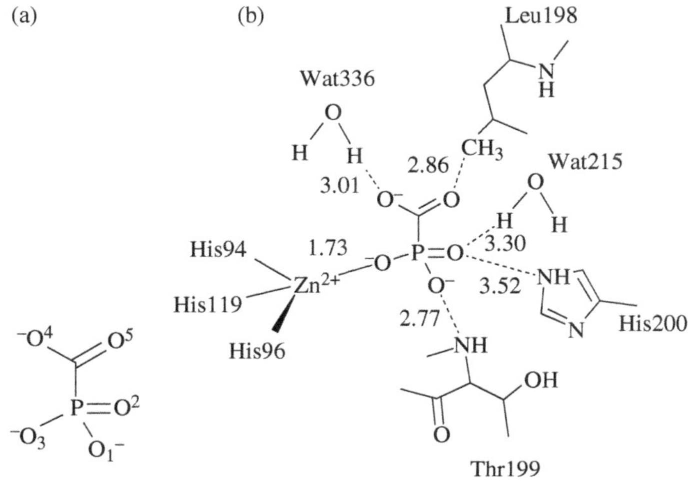

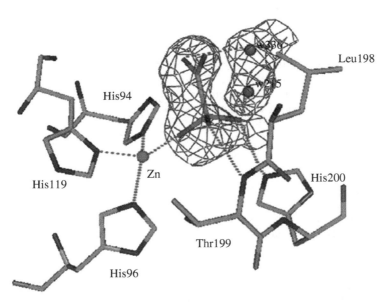

FIGURE 3.33 Scheme of the interactions between inhibitor (foscarnet, a) and the active site amino acid residues (b). Reprinted with permission from Ref. [129], Copyright (2007) Elsevier.

FIGURE 3.34 Electron density map showing the zinc ion coordinated by three histidine ligands and a phosphonate oxygen. Reprinted with permission from Ref. [129], Copyright (2007) Elsevier.

CA inhibitor possessing a phosphonate zinc-binding group was thus evidenced, together with the factors governing recognition of such small molecules by a metalloenzyme active site (Figure 3.34). Foscarnet is also a clear-cut example of a modulator of an enzyme activity, which can act either as an activator or inhibitor of a CA isozyme.

(a)   (b)

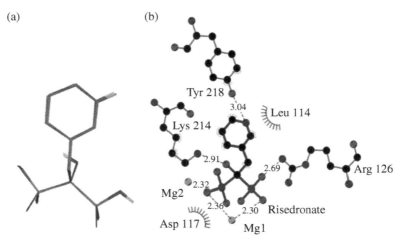

FIGURE 3.35   Docked structures and LigPlot interactions between bisphosphonate inhibitors and an avian FPPS. A, risedronate, 10 lowest energy conformations. LigPlot diagram showing the main interactions between risedronate and FPPS. Reprinted with permission from Ref. [130], Copyright (2004) American Chemical Society.

Oldfield et al. investigated the docking of a variety of inhibitors and substrates to the isoprene biosynthesis pathway enzymes farnesyl diphosphate synthase (FPPS), isopentenyl diphosphate/dimethylallyl diphosphate isomerase (IPPI) and deoxyxylulose-5-phosphate reductoisomerase (DXR) using the Lamarckian genetic algorithm program, AutoDock [130]. The structures of three isoprenoid diphosphates docked to the FPPS enzyme reveal strong electrostatic interactions with Mg^{2+}, lysine and arginine active site residues (Figure 3.35).

Similar results are obtained with the docking of four IPPI inhibitors to the IPPI enzyme. Bisphosphonate inhibitors are found to bind to the allylic binding sites in both eukaryotic and prokaryotic FPPSs, in good accord with recent crystallographic results (Figure 3.36). Overall, these results show for the first time that the geometries of a broad variety of phosphorus-containing inhibitors and substrates of isoprene biosynthesis pathway enzymes can be well predicted by using computational methods, which can be expected to facilitate the design of novel inhibitors of these enzymes.

Another detailed inhibition study of five carbonic anhydrase (CA, EC 4.2.1.1) isozymes with inorganic phosphates, carbamoyl phosphate, the antiviral phosphonate foscarnet as well as formate was reported [131]. The membrane-associated isozyme hCA IV was the most sensitive to the inhibition of bisphosphonates/phosphonates. Foscarnet was the best inhibitor of this isozyme highly abundant in the kidneys, which may explain some of the renal side effects of the drug.

Thus far, phosphonate incorporation of 'free' phosphonates (no metals) into organic matrices was presented. In the remaining part of this section, we will present the incorporation of metal phosphonates into selected organic and inorganic matrices.

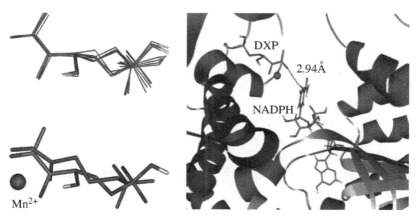

FIGURE 3.36   Docked structures of deoxyxylulose-5-phosphate (left) bound to NADPH-DXR (right). Reprinted with permission from Ref. [130], Copyright (2004) American Chemical Society.

Eddaoudi reported the successful growth of highly crystalline homogeneous MOF thin films of HKUST-1 and ZIF-8 on mesoporous silica foam, by employing a layer-by-layer (LBL) method [132]. The newly constructed hybrid materials, MOFs on mesoporous silica foam, were characterized and evaluated using different techniques including PXRD, SEM and TEM (Figure 3.37). This study confirms the unique potential of the LBL method for the controlled growth of desired MOF thin films on various substrates, which permits rational construction of hierarchical hybrid porous systems for given applications. The ability to control and direct the growth of MOF thin films on confined surfaces, using the stepwise LBL method, creates new opportunities for new prospective applications, such as hybrid systems construction of pure MOF-based membranes, as well as coating a variety of polymers for gas separation applications.

Hydrolysis and condensation reactions of diethylphosphato-ethyltriethoxysilane (SiP) and a mixture of SiP and tetraethoxysilane (TEOS) have been studied in ethanol and N-methylacetamide (NMA) as solvent. The reactions were investigated by high-resolution ^{29}Si NMR. The hydrolyzed and condensed species, from SiP and TEOS, were identified and quantified as a function of reaction time. The influence of the amide medium as well as the catalytic effect of the phosphonate function of SiP on TEOS hydrolysis was established [133].

A zinc silico-phenylphosphonate was synthesized [134]. The substitution of phosphorus by silicon induced a deficit of the positive charges that were balanced by cations located in the interlayer space. The XRD powder diffraction pattern corresponds to a lamellar structure and exhibits a series of sharp peaks assigned to the series $(00l)$ reflections with a d_{001} of 2 nm. SEM micrographs of this material show a morphology similar to the one observed for clay minerals and especially smectites 'gypsum-like morphology' (Figure 3.38). TEM images exhibit a rods' arrangement for the layered materials (Figure 3.38). Q4 sites were identified by ^{29}Si solid state NMR indicating a full polymerization of silica.

FIGURE 3.37 SEM images of the mesoporous silica foam, SEM (upper left) and TEM (upper right). SEM images of HKUST-1 grown on silica foam (middle left and right). SEM images of ZIF-8 grown on silica foam (lower left and right). Reprinted with permission from Ref. [132], Copyright (2012) Royal Society of Chemistry.

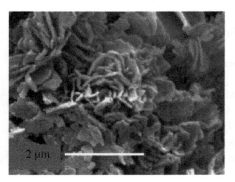 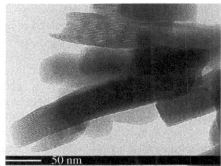

FIGURE 3.38 SEM (left) and TEM images (right) of the zinc silico-phenylphosphonate. Reprinted with permission from Ref. [134], Copyright (2007) Elsevier.

Thermally stable proton-conducting composite sheets 50–100 μm thick have been prepared from phosphosilicate gel (P/Si = 1 molar ratio) powders and polyimide precursor [135, 136]. Polyimide was selected as an organic polymer matrix because of its excellent thermal stability and good sheet-forming property. Proton conductivity, mechanical properties and chemical durability of the resultant composite sheets in the low and medium temperature range have been examined. In addition, a single test fuel cell has been fabricated using the composite sheet as an electrolyte.

3.8 APPLICATIONS

The importance of phosphonic acids and derivatives in the field of supramolecular chemistry, crystal growth and materials chemistry has been well recognized [137]. Besides basic chemistry, phosphonates play a significant role in several other technologically/industrially important areas, such as water treatment [138], oilfield drilling [139], minerals processing [140], corrosion control [141], metal complexation and sequestration [142], dental materials [143], bone targeting [144], cancer treatment [145] and so on.

3.8.1 Proton Conductivity

Solid-state ion conductors are an important class of materials because of their use as electrolytes in batteries and fuel cells, gas sensors and so on. The mechanism of ion conduction in solids depends on structural considerations. Specific important factors are concentration, mobility and charge of conductive ions. Control of the spatial distribution and dynamic behaviour of target ions in solids are also significant. Ion conductivity is also dependent on temperature because ions need to overcome the activation energy between the hopping sites in various structures. Commonly, organic polymers exhibit ion conductivity below 200 °C and inorganic materials (such as metal oxides and metal halides) above 400 °C.

In this section, we will deal with proton conductivity exhibited by phosphonate-containing materials (either organic or hybrid). A number of reviews have appeared on the subject [146].

A novel proton-conducting polymer blend was prepared by mixing poly(vinylphosphonic acid) (PVPA) with poly(1-vinylimidazole) (PVI) at various stoichiometric ratios via changing molar ratio of monomer repeating unit to achieve the highest protonation. The network was used for the immobilization of invertase, and then the enzyme activity was studied. The results reveal that the most stable and highly proton-conducting polymer network may play a pioneer role in the biosensor applications as given by FTIR, elemental analysis, impedance spectroscopy and storage stability experiments [147].

Sulphophenylphosphonic acid was used for the synthesis of two zirconium salts, $Zr(HO_3SC_6H_4PO_3)_2 \cdot 2H_2O$ and $Zr(HPO_4)_{0.7}(HO_3SC_6H_4PO_3)_{1.3} \cdot 2H_2O$ [148]. Powder patterns indicate that these layered compounds are structurally derived from alpha modification of zirconium phosphate monohydrate, in which the zirconium atoms are octahedrally coordinated by six oxygen atoms of the phosphate groups. In the case of phosphonates, the fourth oxygen atom of the phosphate group is replaced by an organic residue that points into the interlayer space. In $Zr(HPO_4)_{0.7}(HO_3SC_6H_4PO_3)_{1.3} \cdot 2H_2O$, the incorporation of the phosphate group causes structural disorder in the whole system, increases the amount of 'labile' protons and changes their behaviour. This is in agreement with the increased conductivity of this compound. The conductivity of both compounds increases from 0.028 to $0.063 \, S \cdot cm^{-1}$ in the relative humidity (RH) range of 50–90%.

Recently, two similar series of lanthanide carboxyphosphonates were synthesized and structurally characterized [149]. The ligand used was hydroxyphosphonoacetic acid (HPAA). The presence of 1D channels, filled with water molecules, in the crystal structures of both series (see Figure 3.39), suggests the possibility of proton conductivity behaviour.

Furthermore, there are certain structural features that make these good candidates as proton conductors at room temperature. These include the –POH groups pointing towards the interior of the channels, the network of hydrogen bonds within the channels and the proximity between the lattice water molecules. Therefore, conductivity studies have been carried out for one representative member of each series.

When GdHPA-II is exposed to the highest % RH value of 98%, a spike is observed that has an associated capacitance of ~1 μF. Since the spike is inclined to the Z′ axis by ~70°, it indicates a partial-blocking electrode response that allows limited diffusion; therefore, the conducting species must be ionic, that is, H^+ ions. The total pellet resistance, R_T, was obtained from the intercept of the spike and/or the arc (low-frequency end) on the Z′ axis. At 98% RH and $T = 21 \, °C$, σ_T was $3.2 \cdot 10^{-4} \, S \cdot cm^{-1}$.

Shimizu et al. have reported a Zn material with the ligand 1,3,5-benzenetriphosphonic acid [150]. At 90% RH and 85 °C, the proton conductivity reaches $2.1 \times 10^{-2} \, S \cdot cm^{-1}$. This is the highest proton conductivity reported for a phosphonate-based MOF.

(a)

(b)

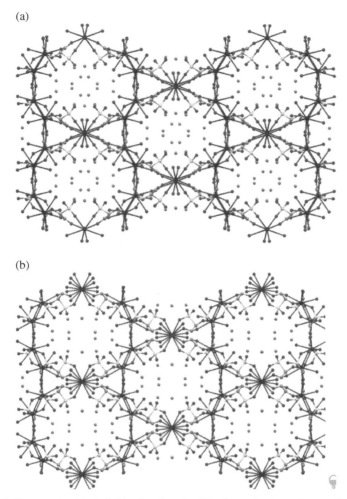

FIGURE 3.39 Lanthanide hybrids showing the 1D channels along the c-axis filled with lattice waters for: (a) La-HPAA-II and (b) La-HPAA-I. The lattice water molecules occupying the channels are as single dots. Reprinted with permission from Ref. [139], Copyright (2012) American Chemical Society.

A new flexible ultramicroporous solid, La(H$_5$DTMP)·7H$_2$O, has been crystallized at room temperature using the tetraphosphonic acid H$_8$DTMP, hexamethyl-enediamine-N,N,N',N'-tetrakis(methylenephosphonic acid) [151]. Its crystal structure, solved by synchrotron powder X-ray diffraction, is characterized by a 3D pillared open framework containing 1D channels filled with water (Figure 3.40).

Upon dehydration, a new related crystalline phase, La(H$_5$DTMP), is formed. Partial rehydration of La(H$_5$DTMP) led to La(H$_5$DTMP)·2H$_2$O. These new phases contain highly corrugated layers showing different degrees of conformational flexibility of

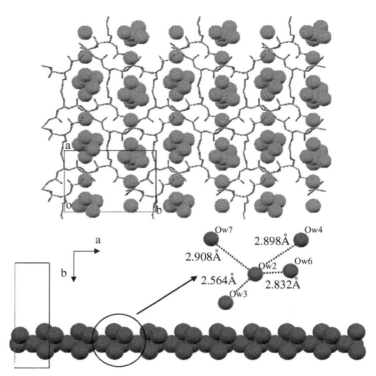

FIGURE 3.40 Upper: 3D framework for La(H$_5$DTMP)·7H$_2$O showing the one-dimensional (1D) channels running along the c-axis and defined by 30-membered rings. The water molecules within the channels are depicted as exaggerated spheres. Lower: 1D chain of hydrogen-bonded lattice water molecules within the channel along the c-axis. H-bonding distances in the 5-water cluster are shown in the inset.

the long organic chain. Impedance data indicates that proton conductivity takes place in La(H$_5$DTMP)·7H$_2$O with a value of 8·10^{-3} S·cm^{-1} at 286 K and 99% of RH.

A new multifunctional light hybrid, Mg(H$_6$ODTMP)·2H$_2$O(DMF)$_{0.5}$, was synthesized using the tetraphosphonic acid H$_8$ODTMP, octamethylenediamine-N,N,N',N'-tetrakis(methylenephosphonic acid), by high-throughput methodology. Its crystal structure, solved by synchrotron powder X-ray diffraction, is characterized by a 3D pillared open framework containing cross-linked 1D channels filled with water and DMF (Figure 3.41) [152]. Upon H$_2$O and DMF removal and subsequent rehydration, Mg(H$_6$ODTMP)·6H$_2$O is formed. These processes take place through crystalline–quasi–amorphous–crystalline transformations, during which the integrity of the framework is maintained. Impedance data indicates that Mg(H$_6$ODTMP)·6H$_2$O has high proton conductivity, $\sigma = 1.6 \times 10^{-3}$ S·cm^{-1} at $T = 292$ K at approximately 100% RH, with an activation energy of 0.31 eV.

A new 3D metal–organic framework, [La(H$_5$L)(H$_2$O)$_4$], L = 1,2,4,5-tetrakisphosphonomethylbenzene, was reported, which conducts protons above 10^{-3} S·cm^{-1} at

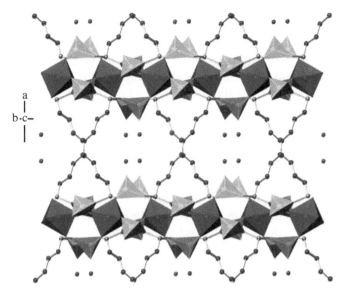

FIGURE 3.41 Crystal structure of Mg(H$_6$ODTMP)·2H$_2$O(DMF)$_{0.5}$. The framework constructed as MgO$_6$ and PO$_4$ polyhedra and ball-and-stick for the organic ligand, with lattice water molecules shown as red spheres, creating 1D channels vertical to the page. DMF molecules are not shown. Reprinted with permission from Ref. [152], Copyright (2012) American Chemical Society.

60 °C and 98% RH [153]. The MOF contains free phosphonic acid groups, shows high humidity stability and resists swelling in the presence of hydration. Channels filled with crystallographically located water and acidic groups are also observed.

Shimizu has published a comprehensive review on proton conductivity in metal phosphonate frameworks, which provides useful further reading [154].

3.8.2 Metal Ion Absorption

A great deal of research has been performed with a variety of materials that absorb metal ions from aqueous solutions, with obvious environmental implications [155]. In this section, we will present selected phosphonate-based systems that are capable for metal ion absorption.

A flexible open-framework sodium-lanthanum(III) tetrakis-phosphonate, NaLa[(HO$_3$P)$_2$CH–C$_6$H$_4$–CH(PO$_3$H)$_2$]·4H$_2$O, was published by Bein et al. [156]. They studied the exceptional ion-exchange selectivity between the Na$^+$ ions of the material and alkaline-earth, alkaline and selected transition metal ions. Exchange between the hosted Na$^+$ ions and other monovalent ions with ionic radii ranging from 0.76 A (Li$^+$) to 1.52 A (Rb$^+$) was accomplished. The divalent ions with approximately the same size did not appear to be exchanged.

Also, [Pb$_7$(HEDTP)(H$_2$O)]·7H$_2$O and [Zn(H$_4$EDTP)]·2H$_2$O [H$_8$EDTP = N,N,N',N'-ethylenediamine-tetrakis(methylenephosphonic acid)] showed very high adsorption

(>96 %) of Fe^{3+} ions in aqueous medium. Excellent ion adsorption performance was also found for other divalent ions such as Ca^{2+}, Cr^{2+}, Mn^{2+}, Cu^{2+}, Zn^{2+} and Cd^{2+} [90].

PEIMPA was investigated in liquid–solid extraction of a mixture of Cd(II), Co(II), Cu(II), Fe(III), Ni(II), Pb(II) and Zn(II) cations from a mineral residue of zinc ore dissolved in nitric acid. The selectivity of this polymer was studied as a function of pH. PEIMPA can adsorb much higher amounts of Fe ion than Cd, Co, Cu, Ni, Pb and Zn ions. The recovery of Fe(III) is almost quantitative [157].

A polyelectrolyte complex of PEIMPA (polyanion) and PEI (polycation) was found useful for the removal of various heavy metal ions such as Cu^{2+}, Co^{2+}, Zn^{2+}, Ni^{2+} and Pb^{2+} from aqueous solutions by the co-precipitation method. Heavy metal binding with PEIMPA was initially allowed to occur, and then upon equilibration, PEI was added to initiate precipitation of the polyelectrolyte complex together with the heavy metal ion. The PEIMPA–PEI system was found effective for heavy metal scavenging purposes even in the presence of high concentrations of non-transition metal ions like Na^+. Heavy metal concentration may be reduced beyond emission standards for industrial wastewaters. The PPEI–PEI polyelectrolyte complex was found to be more effective than traditional precipitation methods for the treatment of a representative electroless Ni plating waste solution [158].

PEIMPA was used as an effective sorbent for solid-phase extraction of Pb(II) ions from an aqueous solution [159]. Conditions for effective sorption are optimized with respect to different experimental parameters in a batch process. The results showed that the amount of extraction decreases with solution pH in the range between 3.5 and 5.8. The sorption capacity is $609\,mg \cdot g^{-1}$. Also, PEIMPA was tested in the recovery of Pb(II) from a synthesized binary solution of Pb(II)–Zn(II) and from real Zn(II)-electrolyzed wastewaters. The presence of Cd(II), Co(II), Cu(II), Fe(III), Ni(II) and Zn(II) in large concentrations has a significantly negative effect on extraction properties (Figure 3.42).

A very promising application especially for treatment of some technological solution and waste comes from some new Tin(IV) nitrilo(methylene)triphosphonates. These materials have been synthesized by a gel method in granular form (spherical beads) [160]. The ion-exchange behaviour between the ligands and alkali, alkaline-earth and some transition metal ions was studied. It is worth mentioning that even at very acidic conditions, the ion exchangers operate efficiently because of the highly acidic adsorption sites.

Some phosphonic acids have the ability to graft onto polymeric matrices. Popa et al. synthesized a group of materials that have been used successfully for the removal of metal ions from aqueous solutions [121]. An Arbuzov-type reaction between chloromethyl polystyrene-divinylbenzene copolymers and triethylphosphite took place and led to phosphonate ester copolymer (resin A). In the presence of HCl resin A hydrolyzes to yield the phosphonic acid copolymer (resin B). Then, the total sorption capacity in aqueous solutions of the two resins was studied for divalent metal ions such as Ca^{2+}, Cu^{2+} and Ni^{2+}. Resin A retains approximately 3.25 mg Ca^{2+}/g copolymer, 2.75 mg Cu^{2+}/g copolymer but no Ni^{2+} at pH = 1. On the other side, resin B retains 8.46 mg Ca^{2+}/g copolymer, 7.17 mg Cu^{2+}/g copolymer but no Ni^{2+} at pH = 1. Efficient Ni^{2+} retention observed at pH = 7 only for the resin B and at the level of

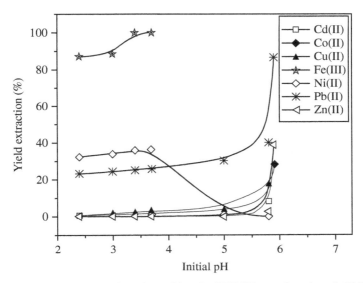

FIGURE 3.42   Recovery of indicated metal ions by PEIMPA, as a function of pH. Reprinted with permission from Ref. [159], Copyright (2009) Taylor & Francis.

19 mg Ni^{2+}/g polymer. The first polymer was incapable of retaining Ni^{2+} at pH = 7. A number of metal-binding modes were proposed for these phosphonate polymers (Figure 3.43).

A three-dimensionally ordered macroporous titanium phosphonate material was synthesized by an inverse opal method using 1-hydroxy ethylidene-1,1-diphosphonic acid (HEDP). This material was tested for the adsorption of Cu^{2+}, Cd^{2+} and Pb^{2+}, showing 10–20% removal efficiency, depending on the metal ion [161].

The same group reported an improvement by synthesizing organic–inorganic hybrid materials of porous titania–phosphonate using tetra- or penta-phosphonates, ethylenediamine-N,N'-tetrakis(methylenephosphonic acid) (EDTMP) and diethylenetriamine-pentakis(methylenephosphonic acid) (DTPMP) [162]. These were anchored to the titania network homogeneously. The synthesized titania–phosphonate hybrids possess irregular mesoporosity formed by the assembly of nanoparticles in a crystalline anatase phase. The titania–DTPMP sample achieved the highest adsorption capacity for each metal ion (Cu^{2+}, Cd^{2+} and Pb^{2+}), probably due to the phosphonate-binding sites than in titania–EDTMP: for Cd(II) up to 88.75 and 89.17%, respectively.

3.8.3   Controlled Release of Phosphonate Pharmaceuticals

Prodrugs of phosphonoformic acid (PFA), an antiviral agent used clinically as the trisodium salt (foscarnet), are of interest due to the low bioavailability of the parent drug, which severely limits its utility. Neutral PFA triesters are known to be susceptible to P–C bond cleavage under hydrolytic de-esterification conditions, and it was

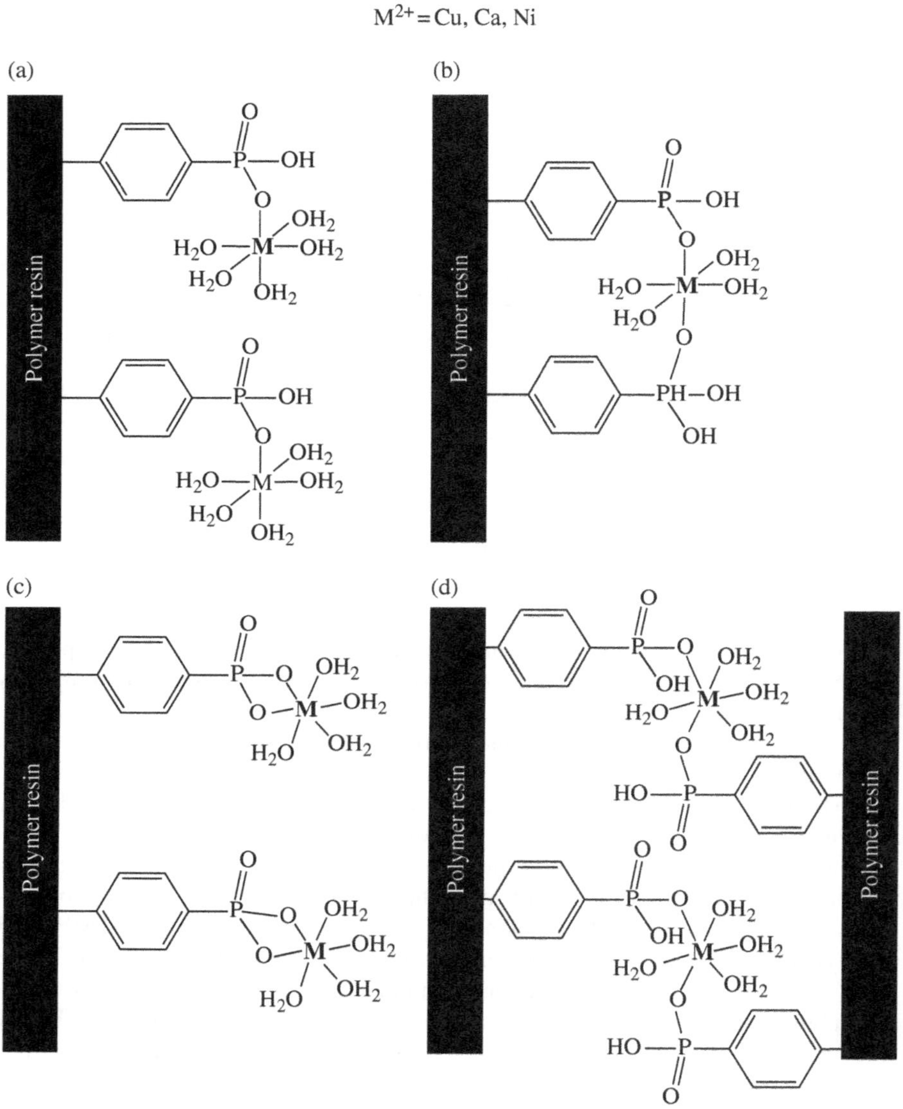

$M^{2+} = $ Cu, Ca, Ni

FIGURE 3.43 Various metal-binding modes of the phosphonate resins. (a) Monodentate terminal binding, (b) monodentate chelating binding, (c) bidentate chelating binding, and (d) monodentate bridging binding. Reprinted with permission from Ref. [125], Copyright (2008) American Chemical Society.

previously found that P,C-dimethyl PFA P–N conjugates with amino acid ethyl esters did not release PFA at pH 7 and could not be fully deprotected under either acid or basic conditions, which led, respectively, to premature cleavage of the P–N linkage (with incomplete deprotection of the PFA ester moiety) or to P–C cleavage [163].

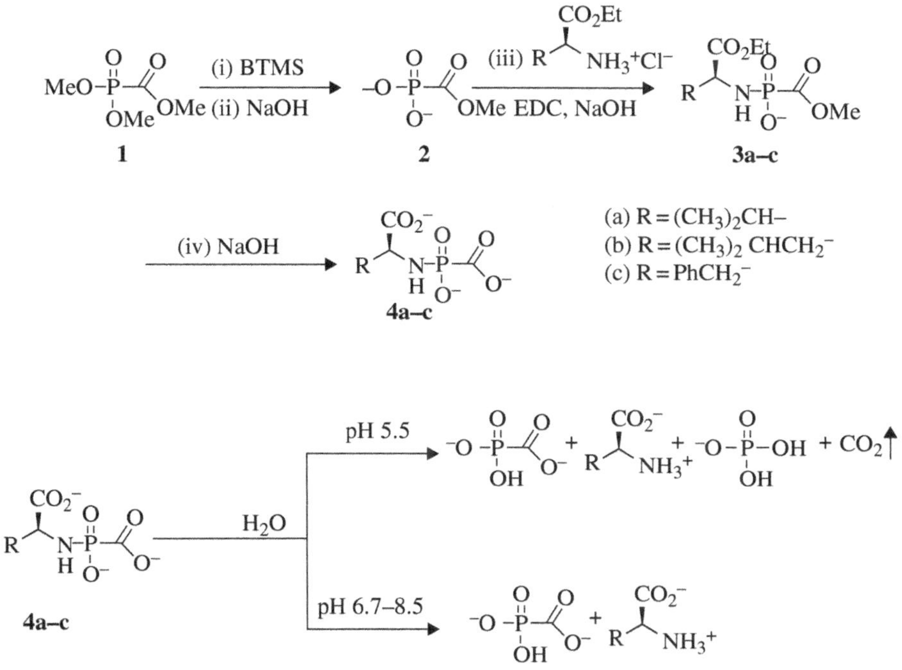

FIGURE 3.44 Hydrolytic processes of the conjugates at various pH values. Reprinted with permission from Ref. [163], Copyright (2004) Elsevier.

Fully deprotected PFA-amino acid P–N conjugates (compounds 4a–c in Figure 3.44) can be prepared via coupling of C-methyl PFA dianion 2 (in Figure 3.44) with C-ethyl-protected amino acids using aqueous EDC, which gives a stable monoanionic intermediate 3 (in Figure 3.44) that resists P–C cleavage during subsequent alkaline deprotection of the two carboxylate ester groups. At 37 °C, the resulting new PFA-amino acid (Val, Leu, Phe) conjugates (4a–c) undergo P–N cleavage near neutral pH, cleanly releasing PFA. A kinetic investigation of 4a hydrolysis at pH values 6.7, 7.2 and 8.5 showed that PFA release was first order in [4a] with respective $t_{1/2}$ values of 1.4, 3.8 and 10.6 h.

Foscarnet inhibits two RNA polymerases but with different patterns of inhibition. Influenza virus RNA polymerase is inhibited in a non-competitive manner with respect to nucleoside triphosphates, apart from GTP, which gives a mixed pattern of inhibition. It was also found that initiation of influenza virus mRNA synthesis by the polymerase, when primed by exogenous mRNA, could occur in the presence of foscarnet but that the elongation was inhibited [164]. This block of mRNA formation by foscarnet occurred during or after the synthesis of the 12-nucleotide-long conserved sequence, ending at a GMP, found at the 5′-end of the viral message (Figure 3.45).

Pamidronate, one of the therapeutic bisphosphonates for osteopenic diseases, directly inhibits bone healing in the traumatic defect model of the rabbit calvaria. The inhibition effect of pamidronate on bone healing was supported by radiographic and

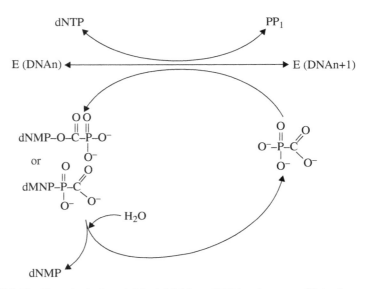

FIGURE 3.45   Hypothetical model for inhibition of DNA polymerase (E) by foscarnet. Two different reaction products can be envisioned for the degradative reaction (none was isolated). Reprinted with permission from Ref. [164], Copyright (1989) Elsevier.

histological analyses. Radiographic analysis showed that pamidronate combined with poly-L-lactide-*co*-glycolide (PLGA) had less bone formation than that of the defect-only or PLGA-only group. Histological analysis further confirmed that pamidronate inhibited bone healing [165].

The intercalation of 1-hydroxyethylidene-1,1-diphosphonic acid (HEDP), which is also a drug for osteoporosis, in layered double hydroxide (LDH) was examined with the goal of developing a novel drug delivery system of HEDP (Figure 3.46). To prevent side reactions, the intercalation reaction was carried out at pH 4–6. The uptake of HEDP was determined as $3.5 \, mmol \cdot g^{-1}$ of LDH, and the interlayer distance increased from 7.8 to $13.0 \, \text{Å}$. The HEDP-release profiles into K_2CO_3 aqueous solution and into various buffer solutions were also examined [166].

Calcification is the principal cause of the clinical failure of bioprosthetic heart valves (BHV), fabricated from glutaraldehyde-treated porcine valves of bovine pericardium. A study examined the dose–response of local controlled-release disodium HEDP therapy for BHV calcification and its mechanism of action [167]. The controlled release of HEDP from ethylene-vinyl acetate matrices was regulated by co-incorporation of the insert filler inulin, and subdermal calcification of BHV cusps was studied with the co-implantation of these matrices in rats (Figure 3.47). Subdermal BHV tissue calcification was inhibited *in vivo* for 7, 60 and 84 days, without any adverse effects. At 84 days, matrices (0.2, 2 and 20% w/w HEDP) co-implanted with BHV resulted in explant calcification levels of 210.4, 39.1 and 11.7 pg/mg in comparison to control values of 213.2 pg/mg (a level equivalent to that of clinically failed BHV). The diffusion coefficient of HEDP through BHV was

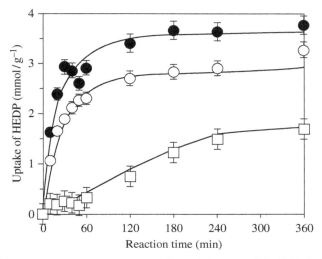

FIGURE 3.46 Uptake of HEDP by three different LDH materials (full circles LDH(Cl), empty circles LDH(CO$_3$), squares LDH calcined at 500 °C). Reprinted with permission from Ref. [166], Copyright (2003) Wiley.

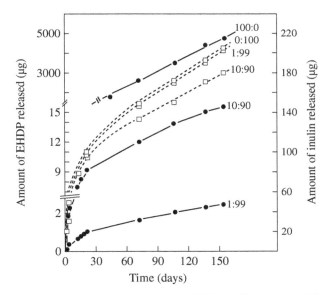

FIGURE 3.47 HEDP and inulin release profiles. HEDP–inulin ratios 0 : 100, 1 : 99, 10 : 90 and 100 : 0. Reprinted with permission from Ref. [167], Copyright (1986) Elsevier.

$0.8 \times 10^{-10}\,\mathrm{cm^2\,s^{-1}}$ reflecting low tissue permeability and high affinity. It was concluded that both *in vitro* and *in vivo* releases of HEDP from the 20% w/w HEDP matrices were suitable to inhibit BHV calcification and that this effect is most likely due to the interaction of HEDP with the BHV tissue surface.

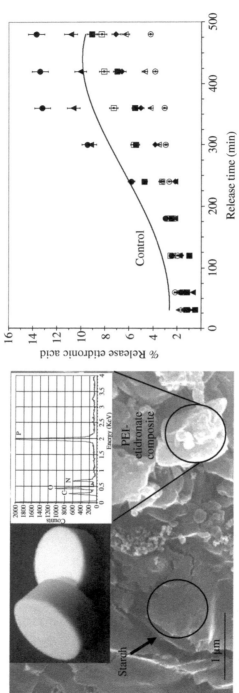

FIGURE 3.48 Left: typical tablet containing polymer–etidronate active ingredient (upper left). Surface particles of the PEI–etidronate composite are highlighted. EDS spectrum of the tablet showing the presence of P from etidronic acid and C, O, and N from the polymer. Right: controlled-release curves for the PEI–etidronate and CATIN–etidronate composites. For clarity purposes, the average release of free etidronic acid from a starch-only tablet (control) is presented as a continuous line. Filled symbols represent the release of etidronic acid from CATIN matrices and hollow symbols show the release from PEI matrices. Specifically: □ PEI (from pH 3), ○ PEI (from pH 4), △ PEI (from pH 5), ■ CATIN-3 (from pH 9), ◆ CATIN-2 (from pH 7), ● CATIN-3 (from pH 7), ▲ CATIN-1 (from pH 9). Reprinted with permission from Ref. [168], Copyright (2011) American Chemical Society.

The controlled release of etidronic acid (HEDP), immobilized onto cationic polymeric matrices, such as PEI or cationic inulin (CATIN), was studied [168]. Several CATIN–etidronate and PEI–etidronate composites were synthesized at various pH regions and characterized. Tablets with starch as the excipient containing the active ingredient (polymer–etidronate composite) were prepared (Figure 3.48 left), and the controlled release of etidronate was studied at aqueous solutions of pH 3 (to mimic the pH of the stomach) for 8 h. All studied composites showed a delayed etidronate release in the first 4 h (Figure 3.48 right), compared to the 'control' (a tablet containing only starch and etidronic acid, without the polymer).

3.8.4 Corrosion Protection by Metal Phosphonate Coatings

Many phosphonic acids are used as corrosion inhibitors in the interdisciplinary field of corrosion science. The phosphonate-based corrosion inhibitors are effective in decreasing metallic corrosion [169], especially near neutral conditions by the formation of poorly soluble metal phosphonate compounds with the existing metal ions of these aqueous solutions. The necessity to develop inhibitors that are free from carcinogenic, chromates [170], nitrates, nitrites and so on fuelled the interest in the area of phosphonic acids. Commonly, the procedure is as follows. The phosphonate is introduced into the system in the acid form or as an alkali metal soluble salt. It then rapidly forms stable complexes with metal cations, such as Ca^{2+}, Mg^{2+}, Sr^{2+} or Ba^{2+} that pre-exist in the process stream. Certain times, cations such as Zn^{2+} are purposely added to the system. A few representative examples follow.

For the protection of carbon steel, a combination of Zn^{2+} and HDTMP in a 1 : 1 molar ratio offers excellent corrosion protection [62]. The corrosion rate for the control (absence of any additives) is 7.28 mm/year, whereas the corrosion rate for the Zn-HDTMP-protected sample is 2.11 mm/year. That is a approximately 170% reduction in corrosion rate. The FTIR, XRF and EDS studies show that the 'anti-corrosion' film is a material composed of Zn^{2+} (externally added) and P (from HDTMP) in an approximate 1 : 4 ratio, as expected.

Other metal phosphonate systems have also been reported. They include Sr/Ba–HPAA [53, 54], Ca–HPAA [59], Sr/Ba–HDTMP [55], Ca–EDTMP [57], Ca–PBTC [51]. Demadis et al. have published a series of reviews on the subject, which offer additional information and further literature [171–175].

3.8.5 Gas Storage

The last decades have witnessed the explosion in the area of MOFs. Many microporous 3D framework structures, thermally stable up to 700 K, have been discovered. Among the most readily recognizable are certain carboxylate-based frameworks such as MOF-5 [176], MIL-100 [177] and HKUST-1 [178]. These materials show remarkable adsorption properties, very high surface areas, flexibility in their framework and unsaturated metal adsorption sites. There are also porous MOFs based on amines, such as zeolitic imidazoles [179], amino acids [180] and amino carboxylates [181].

Among the first organic–inorganic adsorbents investigated were also metal phosphonate materials, which offer an alternative set of chemical and structural possibilities, but until recently, none of them was found to possess pores larger than 6Å (in comparison to carboxylate-based materials that have pores up to approximately 20Å). The competitive property of phosphonates is that the $O_3P–C$ bond is stable at elevated temperatures. Examples include the syntheses of divalent metal piperazine-bis(methylenephosphonate)s of Co(II), Mn(II), Fe(II) and Ni(II) that showed remarkable pore volumes [182]. Other reactions with this ligand and Fe(II), Co(II) and Ni(II) acetates at pH values below 6.5 led to porous solids. The structure can be described as inorganic columns of helical chains of edge-sharing NiO_5N octahedra, which cause an hexagonal array of channels with a free diameter taking into account van der Waals radii of hydrogen atoms of approximately 10Å.

The nickel version of the aforementioned materials is the first fully crystalline phosphonate MOF with pores approaching 1nm and pore volume values observed for large pore zeolites [183]. This structure contains channels filled with physisorbed and chemisorbed water molecules. The framework has a honeycomb arrangement that comes from helical chains of edge-sharing NiO_5N octahedra and the ligand. Through the phosphonate oxygen atoms and the N atoms of piperazine, the ligand coordinated with the nickel atoms, a feature that imparts thermal stability up to 650K. Because of the reversible dehydration and rehydration and the changes in the local structure that temperature results, experiments showed that H_2 and CO are excellent probes of adsorption site at low temperatures, while CO_2, CH_3OH, CD_3CN and CH_4 are useful probes at room temperatures in terms of the fully dehydrated sample. In order to evaluate the adsorption properties of the materials for potential storage and separation, all the measurements were carried out at 303K.

Another use of phosphonic acids in the field of gas adsorption is featured in the materials UAM-150, UAM-151 and UAM-152 [184]. The reaction of the rigid 4,4′-biphenyldiphosphonic acid (BPDP), phosphorus acid and aluminium salts leads to amorphous products. The H_2 intake displayed by material UAM-152 is close to the highest reported for organic–inorganic materials at 77K and atmospheric pressure. From the study, the conclusion is reached that H_2 adsorption increases as the amount of phosphorous acid incorporated increases.

3.8.6 Intercalation

The field of intercalation chemistry is thriving, as more inorganic layered materials are discovered as potential 'hosts' to 'guest' molecules. The most important characteristics for the developing host materials are high thermal stability, resistance to chemical oxidation, selectivity to ions and molecules and the ability to expand their interlamellar space in the presence of guest molecules [184, 185].

The role of metal phosphonates in the field of intercalation is being thoroughly studied [186, 187]. The majority of lamellar metal monophosphonates have the general formula $M(O_3PR)_x \cdot nH_2O$, where M is the metal ion and R is the aliphatic or aromatic group. Molecules such as n-alkylmonoamines [188–192], n-alkyldiamines [193], aromatic amines (pyridines) [194] and dendritic polyamines [195] have been intercalated into lamellar metal phosphonates. The interaction of zirconium phenylphosphonate, $Zr(O_3PC_6H_5)_2$, with n-alkylmonoamines $R–(CH_2)_n–NH_2$ ($n = 0–6$), was studied by Ruiz and Airoldi [196]. Studies on the ability of pyridine (py) and a-, b- and c-picolines to intercalate into crystalline-hydrated barium phenylphosphonate, $Ba[(HO)O_2PC_6H_5]_2 \cdot 2H_2O$, were reported by Lazarin and Airoldi [194].

When $Zn(O_3PC_6H_5) \cdot H_2O$ is in contact with liquid amine, a replacement of the water molecule by amine molecule takes place. The product $Zn(O_3PC_6H_5) \cdot (RNH_2)$ is thermally stable with the amine occupying the same coordination site as the water molecule in the monohydrate hybrid material [197].

In the case of dehydration of $Cu(O_3PC_6H_5) \cdot H_2O$ and $Zn(O_3PCH_3) \cdot H_2O$, the adsorption of amine can take place. The Cu^{2+} centre in the product is in an unusual 5-coordinated environment, with a water molecule occupying the equatorial position of a distorted square pyramid [198].

Studies of Lima and Airoldi focused on the intercalation of crystalline calcium phenylphosphonate [199] and calcium methylphosphonate [200] with n-alkylmono-amines. The hydrated compound $Ca(HO_3PC_6H_5)_2 \cdot 2H_2O$ has two coordinated water molecules to the inorganic backbone, which in the presence of basic polar molecules are gradually replaced [69]. An increase in interlamellar distance takes place to accommodate the guest molecules in the interlayer space [201].

Poojary and Clearfield used powder X-ray data to study the intercalated products formed by propyl-, butyl- and pentylamine and zinc phenylphosphonates (hydrous and anhydrous structures) [202]. In all three intercalates, the zinc centre is found in a slightly distorted tetrahedron. The three coordination sites are occupied by three different oxygen atoms from three different phosphonate groups. The fourth site is occupied by the nitrogen atom of the amine molecule.

3.9 CONCLUSIONS

Phosphonates and a plethora of their derivatives play a significant role in tailoring the structures and properties of materials. In this chapter, we attempted to accumulate some basic information on phosphonates that may exist, 'trapped' or purposely incorporated into organic, biological or inorganic matrices. This research area is vast, but the information and associated literature may offer to the reader a starting point and a basic map for this exciting field.

ACKNOWLEDGMENTS

K.E. Papathanasiou thanks the Onassis Foundation for a doctoral scholarship. K.D. Demadis thanks the EU for funding the Research Program SILICAMPS-153, under the ERA.NET-RUS Pilot Joint Call for Collaborative S&T projects.

REFERENCES

1. M. Peruzzini, L. Gonsalvi (eds), *Phosphorus Compounds: Advanced Tools in Catalysis and Material Sciences*, Springer, Dordrecht, **2011**.
2. K. Popov, H. Rönkkömäki, L.H.J. Lajunen, *Pure Appl. Chem.* **2001**, *73*, 1641–1677.
3. K.D. Demadis, *Water Treatment Processes*, Nova Science Publishers, New York, **2012**.
4. Human & Environmental Risk Assessment on ingredients of European household cleaning products: Phosphonates, http://www.heraproject.com (accessed 4 September 2014).
5. V.S. Sastri, *Corrosion Inhibitors: Principles and Applications*, Chichester, John Wiley & Sons, Ltd, **1998**.
6. W.W. Frenier, S.J. Barber, *Chem. Eng. Progress* **1998**, *July*, 37–44.
7. E. Valsami-Jones, *Phosphorus in Environmental Technology: Principles and Applications*, IWA Publishing, London, **2004**.
8. (a) S. Failla, G. Consiglio, P. Finocchiaro, *Phosphorus Sulfur Silicon* **2011**, *186*, 983–988. (b) G.A. Consiglio, S. Failla, P. Finocchiaro, V. Siracusa, *Phosphorus Sulfur Silicon Relat. Elem.* **1998**, *134/135*, 413–418. (c) P. Finocchiaro, A.D. La Rosa, A. Recca, *Curr. Trends Polymer Sci.* **1999**, *4*, 241–246. (d) M. Frigione, A. Maffezzoli, P. Finocchiaro, S. Failla, *Adv. Polymer Tech.* **2003**, *22*, 329–342. (e) A.D. La Rosa, S. Failla, P. Finocchiaro, A. Recca, V. Siracusa, J.T. Carter, P.T. McGrail, *J. Polymer Eng.* **1999**, *19*, 151–160.
9. M. Hagiwara, S. Koboshi, H. Kobayashi, Pat. Eur. No. 293,729; *Chem. Abstr.* **1988**, *111*, 14329f.
10. M. Ash, I. Ash, *Handbook of Green Chemicals*, Synapse Information Resources Inc., New York, **2004**.
11. (a) J.D. Burns, T.C. Shehee, A. Clearfield, D.T. Hobbs, *Anal. Chem.* **2012**, *84*, 6930–6932. (b) H.H. Someda, A.A. El-Zahhar, M.K. Shehata, H.A. El-Naggar, *J. Radioanal. Nucl. Chem.* **1998**, *228*, 37–41.
12. T.N. Van der Walt, P.J. Fourie, *Appl. Radiat. Isot.* **1987**, *38*, 158.
13. E. Brunet, L. Jiménez, M.V. Rodriguez, V. Luu, G. Muller, O. Juanes, J.C. Rodríguez-Ubis, *Microporous Mesoporous Mater.* **2013**, *169*, 222.
14. M.T. Wharmby, P.A. Wright, in *Metal Phosphonate Chemistry: From Synthesis to Applications*. Clearfield, A., Demadis, K.D., Editors, Royal Society of Chemistry, London, **2012**, p. 317.
15. P. Silva, F. Vieira, A.C. Gomes, D. Ananias, J.A. Fernandes, S.M. Bruno, R. Soares, A.A. Valente, J. Rocha, F.A. Almeida Paz, *J. Am. Chem. Soc.* **2011**, *133*, 15120.
16. K. Popov, V. Yachmenev, A. Kolosov, N. Shabanova, *Colloids Surf. A* **1999**, *160*, 135.
17. R.L. Hilderbrand, *The Role of Phosphonates in Living Systems*, CRC Press, Boca Raton, FL, **1983**.
18. M. Notelovitz, *Osteoporosis: Prevention, Diagnosis and Management*, Professional Communications Inc., New York, **2008**.
19. B. Novack. *Wat. Res.* **1998**, *32*, 1271.
20. A. Popov, H. Ronkkomaki, K. Popov, L.H.J. Lajunen, A. Vendilo, *Inorg. Chim. Acta* **2003**, *353*, 1.
21. E. Ruiz-Agudo, C. Rodriguez-Navarro, E. Sebastian-Pardo, *Cryst. Growth Des.* **2006**, *6*, 1575.
22. V. Deluchat, J.C. Bollinger, B. Serpaud, C. Caullet, *Talanta* **1997**, *44*, 897–907.
23. M.B. Tomson, A.T. Kan, J.E. Oddo, *Langmuir* **1994**, *10*, 1442.

24. A.-L. Alanne, H. Hyvönen, M. Lahtinen, M. Ylisirniö, P. Petri Turhanen, E. Kolehmainen, S. Peräniemi, J. Vepsäläinen, *Molecules* **2012**, *17*, 10928.

25. K. Popov, E. Niskanen, H. Ronkkomaki, L.H.J. Lajunen. *New J. Chem.* **1999**, *23*, 1209.

26. K. Sawada, T. Miyagawa, T. Sakaguchi, K. Doi, *J. Chem. Soc. Dalton Trans.* **1993**, 3777.

27. K. Sawada, T. Araki, T. Suzuki, K. Doi, *Inorg. Chem.* **1989**, *28*, 2687.

28. K. Sawada, T. Araki, T. Suzuki, *Inorg. Chem.* **1987**, *26*, 1199.

29. R. Motekaitis, I. Murase, A.E. Martell, *Inorg. Chem.* **1976**, *15*, 2303.

30. R. Motekaitis, I. Murase, A.E. Martell, *Inorg. Nucl. Chem. Lett.* **1971**, *7*, 1103.

31. V. Deluchat, B. Sepraud, E. Alves, C. Caullet, J.-C. Bollinger, *Phosphorus Sulfur Silicon*, **1996**, 109–110, 209.

32. A.T. Stone, M.A. Knight, B. Nowack, in *Chemicals in the Environment*, ACS Symposium Series 806, Lipnick, R.L., Mason, R.P., Phillips, M.L., Pittman, C.U. Jr., Editors, **2002**, pp. 59–94.

33. J. Gałezowska, P. Kafarski, H. Kozłowski, P. Młynarz, V.M. Nurchi, T. Pivetta, *Inorg. Chim. Acta* **2009**, *362*, 707.

34. P. Garczarek, J. Janczak, J. Zon, *J. Mol. Struct.* **2013**, *1036*, 505.

35. D. Boczula, A. Cały, D. Dobrzynska, J. Janczak, J. Zon, *J. Mol. Struct.* **2012**, *1007*, 220.

36. Z.-Y. Du, C.-C. Zhao, Z.-G. Zhou, K.-J. Wang, *J. Mol. Struct.* **2013**, *1035*, 183.

37. K.D. Demadis, E. Barouda, H. Zhao, R.G. Raptis, *Polyhedron* **2009**, *28*, 3361.

38. K.D. Demadis, R.G. Raptis, P. Baran, *Bioinorg. Chem. Appl.* **2005**, *3*, 119.

39. T. Grawe, G. Schäfer, T. Schrader, *Org. Lett.* **2003**, *5*, 1641.

40. A. Clearfield, K.D. Demadis (eds), *Metal Phosphonate Chemistry: From Synthesis to Applications*, Royal Society of Chemistry, London, **2012**.

41. C.-Y. Cheng, K.-J. Lin, *Acta Cryst.* **2006**, *C62*, m363.

42. D. Vega, D. Fernaandez, J.A. Ellena, *Acta Cryst.* **2002**, *C58*, m77.

43. N.B. Padalwar, C. Pandu, K. Vidyasagar, *J. Solid State Chem.* **2013**, *203*, 321.

44. A.A. Ayi, A.D. Burrows, M.F. Mahon, V.M. Pop, *J. Chem. Crystallogr.* **2011**, *41*, 1165.

45. T.L. Kinnibrugh, N. Garcia, A. Clearfield, *J. Solid State Chem.* **2012**, *187*, 149.

46. T. Lis, *Acta Cryst.* **1997**, *C53*, 28.

47. K.D. Demadis, P. Baran, *J. Solid State Chem.* **2004**, *177*, 4768.

48. K.D. Demadis, J.D. Sallis, R.G. Raptis, P. Baran, *J. Am. Chem. Soc.* **2001**, *123*, 10129.

49. K.D. Demadis, S.D. Katarachia, *Phosphorus Sulfur Silicon* **2004**, *179*, 627.

50. K.D. Demadis, S.D. Katarachia, H. Zhao, R.G. Raptis, P. Baran, *Cryst. Growth Des.* **2006**, *6*, 836.

51. K.D. Demadis, P. Lykoudis, R.G. Raptis, G. Mezei, *Cryst. Growth Des.* **2006**, *6*, 1064.

52. E. Barouda, K.D. Demadis, S. Freeman, F. Jones, M.I. Ogden, *Cryst. Growth Des.* **2007**, *7*, 321.

53. K.D. Demadis, M. Papadaki, R.G. Raptis, H. Zhao, *J. Solid State Chem.* **2008**, *181*, 679.

54. K.D. Demadis, M. Papadaki, R.G. Raptis, H. Zhao, *Chem. Mater.* **2008**, *20*, 4835.

55. K.D. Demadis, E. Barouda, R.G. Raptis, H. Zhao, *Inorg. Chem.* **2009**, *48*, 819.

56. K.D. Demadis, Z. Anagnostou, H. Zhao, *ACS-Appl. Mater. Interf.* **2009**, *1*, 35.

57. K.D. Demadis, E. Barouda, N. Stavgianoudaki, H. Zhao, *Cryst. Growth Des.* **2009**, *9*, 1250.

58. E. Akyol, M. Öner, E. Barouda, K.D. Demadis, *Cryst. Growth Des.* **2009**, *9*, 5145.

59. K.D. Demadis, M. Papadaki, I. Cisarova, *ACS-Appl. Mater. Interf.* **2010**, *2*, 1814.

60. R.M.P. Colodrero, A. Cabeza, P. Olivera-Pastor, J. Rius, D. Choquesillo-Lazarte, J.M. García-Ruiz, M. Papadaki, K.D. Demadis, M.A.G. Aranda, *Cryst. Growth Des.* **2011**, *11*, 1713.

61. K.D. Demadis, S.D. Katarachia, M. Koutmos, *Inorg. Chem. Comm.* **2005**, *8*, 254.

62. K.D. Demadis, C. Mantzaridis, R.G. Raptis, G. Mezei, *Inorg. Chem.* **2005**, *44*, 4469.

63. S. Lodhia, A. Turner, M. Papadaki, K.D. Demadis, G.B. Hix, *Cryst. Growth Des.* **2009**, *9*, 1811.

64. K.D. Demadis, M. Papadaki, M.A.G. Aranda, A. Cabeza, P. Olivera-Pastor, Y. Sanakis, *Cryst. Growth Des.* **2010**, *10*, 357.

65. R.M.P. Colodrero, A. Cabeza, P. Olivera-Pastor, D. Choquesillo-Lazarte, J.M. Garcia-Ruiz, A. Turner, G. Ilia, B. Maranescu, K.E. Papathanasiou, K.D. Demadis, et al., *Inorg. Chem.* **2011**, *50*, 11202.

66. J. Weber, G. Grossmann, K.D. Demadis, N. Daskalakis, E. Brendler, M. Mangstl, J. Schmedt auf der Guenne, *Inorg. Chem.* **2012**, *51*, 11466.

67. Y.-S. Ma, Y. Song, Y.-Z. Li, L.-M. Zheng, *Inorg. Chem.* **2007**, *46*, 5459.

68. S. Kunnas-Hiltunen, E. Laurila, M. Haukka, J. Vepsäläinen, M. Ahlgrén, Z. Anorg, *Allg. Chem.* **2010**, *636*, 710.

69. G.B. Hix, K.D.M. Harris, *J. Mater. Chem.* **1998**, *8*, 579.

70. K. Gholivand, A.R. Farrokhi, Z. Anorg, *Allg. Chem.* **2011**, *637*, 263.

71. N. Calin, S.C. Sevov, *Inorg. Chem.* **2003**, *42*, 7304.

72. E. Dumas, C. Sassoye, K.D. Smith, S.C. Sevov, *Inorg. Chem.* **2002**, *41*, 4029.

73. C. du Peloux, A. Dolbecq, P. Mialane, J. Marrot, F. Sécheresse, *J. Chem. Soc. Dalton Trans.* **2004**, 1259.

74. H. Li, L. Zhang, G. Li, Y. Yu, Q. Huo, Y. Liu, *Microporous Mesoporous Mater.* **2010**, *131*, 186.

75. J.-G. Mao, Z. Wang, A. Clearfield, *J. Chem. Soc. Dalton Trans.* **2002**, 4457.

76. R.-L. Sang, L. Xu, *Chem. Commun.* **2008**, 6143.

77. S.-F. Tang, X.-B. Pan, X.-X. Lu, X.-B. Zhao, *J. Solid State Chem.* **2013**, *197*, 139.

78. Z.-Y. Du, C.-C. Zhao, L.-J. Dong, X.-Y. Deng, Y.-H. Sun, *J. Coord. Chem.* **2012**, *65*, 813.

79. C.R. Samanamu, E.N. Zamora, L.A. Lesikar, J.-L. Montchamp, A.F. Richards, *CrystEngComm* **2008**, *10*, 1372.

80. D.S. Sagatys, C. Dahlgren, G. Smith, R.C. Bott, J.M. White, *J. Chem. Soc. Dalton Trans.* **2000**, 3404.

81. J.A. Fry, C.R. Samanamu, J.-L. Montchamp, A.F. Richards, *Eur. J. Inorg. Chem.* **2008**, 463.

82. W.P. Power, M.D. Lumsden, R.E. Wasylishen, *Inorg. Chem.* **1991**, *30*, 2997.

83. P. Vojtíšek, J. Rohovec, I. Lukeš, *Collect. Czech. Chem. Commun.* **1997**, *62*, 1710.

84. J.-G. Mao, Z. Wang, A. Clearfield, *J. Chem. Soc. Dalton Trans.* **2002**, 4541.

85. S.-M. Ying, J.-G. Mao, *Eur. J. Inorg. Chem.* **2004**, 1270.

86. D.M. Poojary, B. Zhang, A. Cabeza, M.A.G. Aranda, S. Bruque, A. Clearfield, *J. Mater. Chem.* **1996**, *6*, 639.

87. A. Cabeza, M.A.G. Aranda, S. Bruque, *J. Mater. Chem.* **1999**, *9*, 571.

88. F.-Y. Yi, T.-H. Zhou, J.-G. Mao, *J. Mol. Struct.* **2011**, *987*, 51.

89. C. Lei, J.-G. Mao, Y.-Q. Sun, *J. Solid State Chem.* **2004**, *177*, 2449.

90. J. Wu, H. Hou, H. Han, Y. Fan, *Inorg. Chem.* **2007**, *46*, 7960.

91. F. Costantino, P.L. Gentili, N. Audebrand, *Inorg. Chem. Comm.* **2009**, *12*, 406.

92. G. Cao, V.M. Lynch, J.S. Swinnea, T.E. Mallouk, *Inorg. Chem.* **1990**, *29*, 2112.

93. J. Legendziewicz, P. Gawryszewska, E. Gałdecka, Z. Gałdecki, *J. Alloys Compd.* **1998**, 275–277, 356.

94. S.M. Ying, J.Q. Liu, S.L. Cai, F. Zhong, G.P. Zhou, *Acta Cryst.* **2007**, E63, m415.

95. S.M. Ying, X.R. Zeng, X.N. Fang, X.-F. Li, D.-S. Liu, *Inorg. Chim. Acta* **2006**, *359*, 1589.

96. P. Silva, J.A. Fernandes, F.A. Almeida Paz, *Acta Cryst.* **2012**, E68, m294.

97. L. Tei, A.J. Blake, C. Wilson, M. Schröder, *Dalton Trans.* **2004**, 1945.

98. R.C. Wang, Y. Zhang, H. Hu, R.R. Frausto, A. Clearfield, *Chem. Mater.* **1992**, *4*, 864.

99. F. Serpaggi, G. Ferey, *Inorg. Chem.* **1999**, *38*, 4741.

100. T.-H. Zhou, F.-Y. Yi, P.-X. Li, J.-G. Mao, *Inorg. Chem.* **2010**, *49*, 905.

101. A.N. Alsobrook, E.V. Alekseev, W. Depmeier, T.E. Albrecht-Schmitt, *J. Solid State Chem.* **2011**, *184*, 1195.

102. D.M. Poojary, D. Grohol, A. Clearfield, *Angew. Chem. Int. Ed.* **1995**, *34*, 1508.

103. A.N. Alsobrook, B.G. Hauser, J.T. Hupp, E.V. Alekseev, W. Depmeier, T.E. Albrecht-Schmitt, *Chem. Commun.* **2010**, *46*, 9167.

104. K.E. Knope, C.L. Cahill, *Eur. J. Inorg. Chem.* **2010**, 1177.

105. Z. Liao, J. Ling, L.R. Reinke, J.E.S. Szymanowski, G.E. Sigmona, P.C. Burns, *Dalton Trans.* **2013**, *42*, 6793.

106. R. Fu, S. Xia, S. Xiang, S. Hu, X. Wu, *J. Solid State Chem.* **2004**, *177*, 4626.

107. R.M.P. Colodrero, P. Olivera-Pastor, A. Cabeza, M. Papadaki, K.D. Demadis, M.A.G. Aranda, *Inorg. Chem.* **2010**, *49*, 761.

108. P.M. Forster, N. Stock, A.K. Cheetham, *Angew. Chem. Int. Ed.* **2005**, *44*, 7608.

109. S. Natarajan, P. Mahata, *Curr. Opin. Solid State Mater. Sci.* **2009**, *13*, 46.

110. A.K. Bhattacharya, G. Thyagarajan, *Chem. Rev.* **1981**, *81*, 415.

111. K. Moedritzer, R.R. Irani, *J. Org. Chem.* **1966**, *31*, 1603.

112. J. Zon, P. Garczarek, M. Bialek, in *Metal Phosphonate Chemistry: From Synthesis to Applications.* Clearfield, A., Demadis, K.D., Editors, Royal Society of Chemistry, London, **2012**, pp. 170–191.

113. P. Lebduskova, J. Kotek, P. Hermann, L. Vander Elst, R.N. Muller, I. Lukes, J.A. Peters, *Bioconjug. Chem.* **2004**, *15*, 881.

114. A. Heras, N.M. Rodriguez, V.M. Ramos, E. Agullo, *Carbohydr. Polym.* **2001**, *44*, 1.

115. V.M. Ramos, N.M. Rodrıguez, M.F. Dıaz, M.S. Rodrıguez, A. Heras, E. Agullo, *Carbohydr. Polym.* **2003**, *52*, 39.

116. V.M. Ramos, N.M. Rodrıguez, M.S. Rodrıguez, A. Heras, E. Agullo, *Carbohydr. Polym.* **2003**, *51*, 425.

117. A. Zuñiga, A. Debbaudt, L. Albertengo, M.S. Rodríguez, *Carbohydr. Polym.* **2010**, *79*, 475.

118. G.L. Matevosyan, S. Yukha, P.M. Zavlin, *Rus. J. Gen. Chem.* **2003**, *73*, 1725.

119. Z. Zhong, P. Li, R. Xing, X. Chen, S. Liu, *Int. J. Biol. Macromol.* **2009**, *45*, 255.

120. R. Jayakumara, N. Selvamurugan, S.V. Nair, S. Tokura, H. Tamura, *Int. J. Biol. Macromol.* **2008**, *43*, 221.

121. X. Wang, J. Ma, Y. Wang, B. He, *Biomaterials* **2001**, *22*, 2247.

122. H. Kang, Y. Cai, J. Deng, H. Zhang, Y. Tang, P. Liu, *Eur. Polym. J.* **2006**, *42*, 2678.

123. F. Lebouc, I. Dez, M. Gulea, P.-J. Madec, P.-A. Jaffres, *Phosphorus Sulfur Silicon* **2009**, *184*, 872.

124. A. Cengiz, N.P. Bayramgil, *Soft Mater.* **2013**, *11*, 476.

125. A. Popa, C.-M. Davidescu, N. Petru, I. Gheorghe, A. Katsaros, K.D. Demadis, *Ind. Eng. Chem. Res.* **2008**, *47*, 2010.

126. R.R. Navarro, S. Wada, K. Tatsumi, *Sep. Sci. Technol.* **2003**, *38*, 2327.

127. R.R. Navarro, K. Tatsumi, *Sep. Sci. Technol.* **2002**, *37*, 203.

128. S.B. Hong, F.M.J. Raushel, *Enzymol. Inhib.* **1997**, *12*, 191.

129. C. Temperini, A. Innocenti, A. Guerri, A. Scozzafava, S. Rusconi, C.T. Supuran, *Bioorg. Med. Chem. Lett.* **2007**, *17*, 2210.

130. F. Cheng, E. Oldfield, *J. Med. Chem.* **2004**, *47*, 5149.

131. S. Rusconi, A. Innocenti, D. Vullo, A. Mastrolorenzo, A. Scozzafava, C.T. Supuran, *Bioorg. Med. Chem. Lett.* **2004**, *14*, 5763.

132. O. Shekhah, L. Fu, R. Sougrat, Y. Belmabkhout, A.J. Cairns, E.P. Giannelis, M. Eddaoudi, *Chem. Commun.* **2012**, *48*, 11434.

133. P. Van Nieuwenhuyse, V. Bounor-Legare, F. Boisson, P. Cassagnau, A. Michel, *J. Non-Cryst. Solids* **2008**, *354*, 1654.

134. M. Jaber, O. Larlus, J. Miehe-Brendle, *Solid State Sci.* **2007**, *9*, 144.

135. A. Matsuda, N. Nakamoto, K. Tadanaga, T. Minami, M. Tatsumisago, *Solid State Ionics* **2003**, 162–163, 247.

136. A. Matsuda, T. Kanzaki, K. Tadanaga, M. Tatsumisago, T. Minami, *Solid State Ionics* **2002**, 154–155, 687.

137. (a) A. Clearfield, *Chem. Mater.* **1998**, *10*, 2801. (b) K. Maeda, *Microporous Mesoporous Mater.* **2004**, *73*, 47. (c) K.D. Demadis, in *Solid State Chemistry Research Trends*, Buckley, R.W., Editor, Nova Science Publishers, New York, **2007**, p. 109. (d) A. Clearfield, in *Progress in Inorganic Chemistry*, Karlin, K.D., Editor, John Wiley & Sons, Inc., New York. **1998**, p. 371. (e) A. Vioux, L., Le Bideau, P., Hubert Mutin, D., Leclercq, *Top. Curr. Chem.* **2004**, *232*, 145. (f) A.K. Cheetham, G. Ferey, T. Loiseau, *Angew. Chem. Int. Ed.* **1999**, *38*, 3268. (g) B.A. Breeze, M. Shanmugam, F. Tuna, R.E.P. Winpenny, *Chem. Commun.* **2007**, 5185. (h) U. Costantino, M. Nocchetti, R. Vivani, *J. Am. Chem. Soc.* **2002**, *124*, 8428.

138. K.D. Demadis, *Phosphorus Sulfur Silicon* **2006**, *181*, 167.

139. (a) S.J. Dyer, C.E. Anderson, G.M. Graham, *J. Pet. Sci. Eng.* **2004**, *43*, 259. (b) J.E. Oddo, M.B. Tomson, M.B. *Appl. Geochem.* **1990**, *5*, 527. (c) J.J. Xiao, A.T. Kan, M.B. Tomson, *Langmuir* **2001**, *17*, 4668. (d) S.J. Friedfeld, S. He, M.B. Tomson, *Langmuir* **1998**, *14*, 3698. (e) V. Tantayakom, H.S. Fogler, P. Charoensirithavorn, S. Chavadej, *Cryst. Growth Des.* **2005**, *5*, 329. (f) F.H. Browning, H.S. Fogler, *AIChE J.* **1996**, *42*, 2883. (g) R. Pairat, C. Sumeath, C., F.H. Browning, H.S. Fogler, *Langmuir* **1997**, *13*, 1791. (h) V. Tantayakom, H.S. Fogler, F.F. de Moraes, M. Bualuang, S. Chavadej, P. Malakul, *Langmuir* **2004**, *20*, 2220.

140. (a) A.-L. Penard, F. Rossignol, H.S. Nagaraja, C. Pagnoux, T. Chartier, *Eur. J. Ceram. Soc.* **2005**, *25*, 1109. (b) M.J. Pearse, *Miner. Eng.* **2005**, *18*, 139.

141. (a) I. Sekine, T. Shimode, M. Yuasa, *Ind. Eng. Chem. Res.* **1992**, *31*, 434. (b) B. Mosayebi, M. Kazemeini A. Badakhshan, *Br. Corr. J.* **2002**, *37*, 217. (c) Yu, I. Kouznetsov, *Prot.*

Met. **2001**, *37*, 434. (d) Yu, V. Balaban-Irmenin, A.M. Rubashov, N.G. Fokina, *Prot. Met.* **2006**, *42*, 133. (e) J.L. Fang, Y. Li, X.R. Ye, Z.W. Wang, Q. Liu, *Corrosion* **1993**, *49*, 266. (f) A. Paszternák, S. Stichleutner, I. Felhősi, Z. Keresztes, F. Nagy, E. Kuzmann, A. Vértes, Z. Homonnay, G. Pető, E. Kálmán, *Electrochim. Acta* **2007**, *53*, 337.

142. (a) *Biogeochemistry of Chelating Agents,* Nowack, B., Van Briessen, J.M., Editors, ACS Symposium Series, ACS, Washington, DC, **2003**, Vol. 910. (b) T.P. Knepper, *Trends Anal. Chem.* **2003**, *22*, 708.

143. (a) K. Miyazaki, T. Horibe, J.M. Antonucci, S. Takagi, L.C. Chow, *Dent. Mater.* **1993**, *9*, 46. (b) M. Atai, M. Nekoomanesh, S.A. Hashemi, S. Amani, S. *Dent. Mater.* **2004**, *20*, 663. (c) J.W. Nicholson, G. Singh, *Biomaterials* **1996**, *17*, 2023. (d) H. Tschernitschek, L. Borchers, W. Geurtsen, W. *J. Prosth. Dent.* **2006**, *96*, 12.

144. (a) M. Bottrill, L. Kwok, N.J. Long, N.J. *Chem. Soc. Rev.* **2006**, *35*, 557. (b) I.G. Finlay, M.D. Mason, M. Shelley, *Lancet Oncol.* **2005**, *6*, 392. (c) V. Kubicek, J. Rudovsky, J. Kotek, P. Hermann, L. Vander Elst, R.N. Muller, Z.I. Kolar, H.T. Wolterbeek, J.A. Peters, I. Lukeš, *J. Am. Chem. Soc.* **2005**, *127*, 16477. (d) H. Kung, R. Ackerhalt, M. Blau, *J. Nucl. Med.* **1978**, *19*, 1027.

145. (a) S.S. Padalecki, T.A. Guise, *Breast Cancer Res.* **2001**, *4*, 35. (b) V. Stresing, F. Daubiné, I. Benzaid, H. Mönkkönen, P. Clézardin, *Cancer Lett.* **2007**, *257*, 16. (c) R. Layman, K. Olson, C. Van Poznak, *Hematol. Oncol. Clinics North Am.* **2007**, *21*, 341.

146. S. Horike, D. Umeyama, S. Kitagawa, *Acc. Chem. Res.* **2013**, *46*, 2376 doi:10.1021/ar300291s. (b) K.-D. Kreuer, *Chem. Mater.* **1996**, *8*, 610. (c) M. Yoon, K. Suh, S. Natarajan, K. Kim, *Angew. Chem. Int. Ed.* **2013**, *52*, 2688. (d) J. Lee, O.K. Farha, J. Roberts, K.A. Scheidt, S.T. Nguyen, J.T. Hupp, *Chem. Soc. Rev.* **2009**, *38*, 1450.

147. S. Isikli, S. Tuncagil, A. Bozkurt, L. Toppare, *J. Macromol. Sci. Part A* **2010**, *47*, 639.

148. V. Zima, J. Svoboda, K. Melánová, L. Beneš, M. Casciola, M. Sganappa, J. Brus, M. Trchová, *Solid State Ionics* **2010**, *181*, 705.

149. R.M.P. Colodrero, A. Cabeza, P. Olivera-Pastor, E.R. Losilla, K.E. Papathanasiou, N. Stavgianoudaki, J. Sanz, I. Sobrados, D. Choquesillo-Lazarte, J.M. García-Ruiz, et al., *Chem. Mater.* **2012**, *24*, 3780.

150. S. Kim, K.W. Dawson, B.S. Gelfand, J.M. Taylor, G.K.H. Shimizu, *J. Am. Chem. Soc.* **2013**, *135*, 963.

151. R.M.P. Colodrero, P. Olivera-Pastor, E.R. Losilla, M.A.G. Aranda, M. Papadaki, A. McKinlay, R.E. Morris, K.D. Demadis, A. Cabeza, *Dalton Trans.* **2012**, *41*, 4045.

152. R.M.P. Colodrero, P. Olivera-Pastor, E.R. Losilla, D.H. Alonso, M.A.G. Aranda, L. Leon-Reina, J. Rius, K.D. Demadis, B. Moreau, D. Villemin, et al., *Inorg. Chem.* **2012**, *51*, 7689.

153. J.M. Taylor, K.W. Dawson, G.K.H. Shimizu, *J. Am. Chem. Soc.* **2013**, *135*, 1193.

154. (a). G.K.H. Shimizu, in *Metal Phosphonate Chemistry: From Synthesis to Applications,* Clearfield, A., Demadis, K.D., Editors, Royal Society of Chemistry, London, **2012**, pp. 493–524. (b) P. Ramaswamy, N.E. Wong, G.K.H. Shimizu, *Chem. Soc. Rev.* **2014**, 43, 5913.

155. M. Arkas, D. Tsiourvas, C.M. Paleos, *Macromol. Mater. Eng.* **2010**, *295*, 883–898.

156. M. Plabst, L.B. McCusker, T. Bein, *J. Am. Chem. Soc.* **2009**, *131*, 18112.

157. O. Abderrahim, M.A. Didi, B. Moreau, D. Villemin, *Solv. Extract. Ion Exch.* **2006**, *24*, 943.

158. R.R. Navarro, S. Wada, K. Tatsumi, *J. Hazard. Mater.* **2005**, *B123*, 203.

159. O. Abderrahim, M.A. Didi, D. Villemin, *Anal. Lett.* **2009**, *42*, 1233.

160. A.I. Bortun, L.N. Bortun, A. Clearfield, S.A. Khainakov, J.R. García, *Solv. Extract. Ion Exch.* **1998**, *16*, 651.

161. T.-Y. Ma, X.-J. Zhang, G.-S. Shao, J.-L. Cao, Z.-Y. Yuan, *J. Phys. Chem. C* **2008**, *112*, 3090.

162. X.-J. Zhang, T.-Y. Ma, Z.-Y. Yuan, *J. Mater. Chem.* **2008**, *18*, 2003.

163. M.S. Marma, B.A. Kashemirov, C.E. McKenna, *Bioorg. Med. Chem. Lett.* **2004**, *14*, 1787.

164. B. Oberg, *Pharm. Ther.* **1989**, *40*, 213.

165. J.-Y. Choi, H.-J. Kim, Y.-C. Lee, B.-O. Cho, H.-S. Seong, M. Cho, S.-G. Kim, *Oral Surg. Oral Med. Oral Pathol. Oral Radiol. Endod.* **2007**, *103*, 321.

166. H. Nakayama, K. Takeshita, M. Tsuhako, *J. Pharm. Sci.* **2003**, *92*, 2419.

167. G. Golomb, R. Langer, F.J. Schoen, M.S. Smith, Y.M. Choi, R.J. Levy, *J. Control. Release* **1986**, *4*, 181.

168. K.D. Demadis, I. Theodorou, M. Paspalaki, *Ind. Eng. Chem. Res.* **2011**, *50*, 5873.

169. G. Gunasekaran, R. Natarajan, V.S. Muralidharan, N. Palaniswamy, B.V. Appa Rao, *Anti-Corr. Meth. Mater.* **1997**, *44*(4), 248.

170. S.M. Cohen, *Corrosion* **1995**, *51*, 71.

171. K.D. Demadis, N. Stavgianoudaki, in *Metal Phosphonate Chemistry: From Synthesis to Applications*. Clearfield, A., Demadis, K.D., Editors, The Royal Society of Chemistry, London, **2012**, pp. 438–492.

172. K.D. Demadis, M. Papadaki, D. Varouchas, in *Green Corrosion Chemistry and Engineering: Opportunities and Challenges*, Sharma, S.K., Editor, Wiley-VCH Verlag GmbH & Co., Germany, **2012**, pp. 243–296.

173. M. Papadaki, K.D. Demadis, *Comments Inorg. Chem.* **2009**, *30*, 89.

174. K.D. Demadis, G. Angeli, in *Mineral Scales in Biological and Industrial Systems*, Z. Amjad, Editor, Taylor & Francis, New York, **2013**, pp. 343–360.

175. K.D. Demadis, in *Solid State Chemistry Research Trends*, Buckley, R.W., Editor, Nova Science Publishers, Inc., New York, **2007**, pp. 109–172.

176. H. Li, M. Eddaoudi, M. O'Keeffe, O.M. Yaghi, *Nature* **1999**, *402*, 276.

177. G. Ferey, C. Mellot-Draznieks, C., Serre, F. Millange, *Acc. Chem. Res.* **2005**, *38*, 217.

178. S.S.Y. Chui, J.P.H. Charmant, A.G. Orpen, I.D. Williams, *Science* **1999**, *283*, 1148.

179. K.S. Park, Z. Ni, A.P. Côté, J.Y. Choi, R.D. Huang, F.J. Uribe-Romo, H.K. Chae, M. O'Keeffe, O.M. Yaghi, *Proc. Natl. Acad. Sci. U.S.A.* **2006**, *103*, 10186.

180. R. Vaidhyanathan, D. Bradshaw, J.N. Rebilly, J.P. Barrio, J.A. Gould, N.G. Berry, M.J. Rosseinsky, *Angew. Chem. Int. Ed.* **2006**, *45*, 6495.

181. R. Matsuda, R. Kituara, S. Kitagawa, Y. Kubota, R.V. Belosludov, T.C. Kobayashi, H. Sakamoto, T. Chiba, M. Takata, Y. Kawazoe, et al., *Nature* **2005**, *436*, 238.

182. J.A. Groves, S.R. Miller, S.J. Warrender, C. Mellot-Draznieks, P. Lightfoot, P.A. Wright, *Chem. Commun.* **2006**, 3305.

183. S.R. Miller, G.M. Pearce, P.A. Wright, F. Bonino, S. Chavan, S. Bordiga, I. Margiolaki, N. Guillou, G. Ferey, S. Bourrelly, et al., *J. Am. Chem. Soc.* **2008**, *130*, 15967.

184. E. Brunet, C. Cerro, O. Juanes, J.C. Rodriguez-Ubis, A. Clearfield, *J. Mater. Sci.* **2008**, *43*, 1155.

185. K. Peeters, P. Grobet, E.F. Vansant, *J. Mater. Chem.* **1996**, *6*, 239.

186. R. Hoppe, G. Alberti, U. Contantino, C. Dionigi, G.S. Ekloff, R. Vivani, *Langmuir* **1997**, *13*, 7252.

187. S.L. Suib, *Chem. Rev.* **1993**, *93*, 803.

188. G. Cao, H.-G. Hong, T. Mallouk, *Acc. Chem. Res.* **1992**, *25*, 420.

189. A.M. Lazarin, C. Airoldi, *J. Incl. Phenom. Macrocycl. Chem.* **2005**, *51*, 33.

190. C.B.A. Lima, C. Airoldi, *Solid State Sci.* **2002**, *4*, 1321.

191. B. Zhang, D.M. Poojary, A. Clearfield, G. Peng, *Chem. Mater.* **1996**, *8*, 1333.

192. P. Gendraud, M.E. de Roy, J.P. Besse, *Inorg. Chem.* **1996**, *35*, 6108.

193. T. Kijima, S. Watanabe, M. Machida, *Inorg. Chem.* **1994**, *33*, 2586.

194. A.M. Lazarin, C. Airoldi, *J. Chem. Thermodyn.* **2005**, *37*, 243.

195. T. Kijima, K. Ohe, F. Sasaki, M. Yada, M. Machida, *Bull. Chem. Soc. Jpn.* **1998**, *71*, 141.

196. V.S.O. Ruiz, C. Airoldi, *Thermochim. Acta* **2004**, *420*, 73.

197. Y. Zhang, K.J. Scott, A. Clearfield, *J. Mater. Chem.* **1995**, *5*, 315.

198. Y.P. Zhang, A. Clearfield, *Inorg. Chem.* **1992**, *31*, 2821.

199. C.B.A. Lima, C. Airoldi, *Thermochim. Acta* **2003**, *400*, 51.

200. C.B.A. Lima, C. Airoldi, *Int. J. Inorg. Mater.* **2001**, *3*, 907.

201. F. Fredoueil, D. Massiot, P. Janvier, F. Gingl, M.B. Doeuff, M. Evain, A. Clearfield, B. Bujoli, *Inorg. Chem.* **1999**, *38*, 1831.

202. D.M. Poojary, A. Clearfield, *J. Am. Chem. Soc.* **1995**, *117*, 11278.

4

HYBRID MATERIALS BASED ON MULTIFUNCTIONAL PHOSPHONIC ACIDS

Aurelio Cabeza, Pascual Olivera-Pastor
and Rosario M. P. Colodrero

*Departamento de Química Inorgánica, Cristalografía y Mineralogía,
Universidad de Málaga, Málaga, Spain*

4.1 INTRODUCTION

The versatility of phosphorus organochemistry has led to a huge expansion of the metal phosphonate chemistry during the last decades [1–14]. Although many alkyl and aryl phosphonic acids tend monotonically to form pillared layered networks with a diversity of metal ions, attaching functional groups to the organic moiety has been proven to be an effective strategy to synthesize a great variety of non-layered networks. A number of these compounds possess structures with accessible internal surface and/or superficial characteristics usable for multiple applications [15].

Polyfunctional phosphonic acids can be relatively easy to prepare by well-established methods, the majority relying on the Michaelis–Arbuzov reaction or the Mannich reaction [16]. The phosphonic acid moiety itself has three O atoms capable of coordinating to metal ions and different states of protonation that influences its coordinative properties. Incorporating additional functional groups increases the number of possible modes of coordination and structural arrangements. So, metal phosphonates containing groups such as carboxylate, amine, hydroxyl, ether, pyridine and so on have been successfully prepared. More recently, tri- and tetraphosphonate and even more highly functionalized ligands have been used to create a rich diversity of architectures and topologies, from 0D molecular to three-dimensional

Tailored Organic–Inorganic Materials, First Edition. Edited by Ernesto Brunet, Jorge L. Colón and Abraham Clearfield.

(3D) open frameworks. Furthermore, employing auxiliary ligands not only extends beyond the structural and functional complexity but also may facilitate, in some cases, adjusting of desired features in the targeted compounds [17].

Manipulating the connectivity of the phosphonate moiety, $H_2O_3P–R–X$, is a key factor to manage the final arrangement of the phosphonate hybrid materials. By changing the nature and/or the length of the R spacer and rational choice of the X functional end, the cohesion of the network can be modified in different ways by varying the number of its potential binding sites. The next sections will show the role of different functional groups in configuration of metal phosphonate frameworks.

This chapter will present a survey of recent developments of polyfunctionalized phosphonic acids relevant to the chemistry of metal phosphonates. A classification according to the number of phosphonate groups in the ligand is established, and some of the most relevant properties, related to the nature of the functionality inserted, are briefly discussed. We pay attention to polyfunctionalized phosphonates that have been used for different applications, including proton conductivity, gas adsorption and catalytic properties; some of them are shown simultaneously by a same solid. Due to the large number of materials reported and for reasons of simplicity, we will focus on crystalline metal phosphonates in absence of auxiliary ligands, the role of which have been recently reviewed [17].

4.2 STRUCTURAL TRENDS AND PROPERTIES OF FUNCTIONALIZED METAL PHOSPHONATES

4.2.1 Monophosphonates

One advantage of phosphonate over phosphate ligands in the design of metal derivatives is that the former can be chemically decorated by various types of functional groups. In this section, we discuss the structural effects of carboxyl, hydroxyl and amino groups as the most prominent functionalities attached to the phosphonate moiety.

4.2.1.1 Metal Alkyl- and Aryl-Carboxyphosphonates The use of bifunctional carboxyphosphonate units $[O_3P–R–COO]^{3-}$ has led to many new materials. These contain tuneable organic units (R = alkyl or aryl) bound to the phosphonate and carboxylate groups. Among them, carboxymethylphosphonic and 2-carboxyethylphosphonic acids have been widely employed as ligands for a number of metal ions. The structural variety of these metal carboxyphosphonates ranges from one-dimensional (1D) networks to 3D open frameworks, the most common architectures being, however, layered and pillared layered structures.

The structural diversity of manganese carboxyphosphonates illustrates very well how simple changes in the synthesis conditions and the organic units may affect the metal–ligand connectivity and, hence, the topology of the hybrid solid. So, $Mn_3(O_3PCH_2COO)_2$ and $Mn_3(O_3PCH_2CH_2COO)_2$ [18] show an extensive change in the crystal structure by simple extension of the organic moiety from CH_2 to CH_2CH_2.

The former exhibits a 3D network, while the latter forms a pillared structure. The structure of $Mn_3(O_3PCH_2COO)_2$ is made up of ribbons of $Mn(1)O_6$ and $Mn(2)O_5$ polyhedra, which are connected via the $Mn(3)O_5$ units to form layers in the *ab* plane. These layers are connected by the carboxylate groups. $Mn_3(O_3PCH_2CH_2COO)_2$ forms a pillared structure, where edge-sharing pairs of MnO_5 polyhedra are corner-linked to the MnO_6 octahedra forming layers in the *bc* plane. The carboxyethylphosphonate ions act as bridging as well as chelating ligands, but the chelation mode is different from that of carboxymethylphosphonate ions. As found in other compounds, the organic groups connect adjacent layers by the phosphonate and carboxylate ends, giving rise to the formation of a pillared layered structure (Figure 4.1). On the other hand, $Mn^{II}(O_3PCH_2CH_2COOH) \cdot H_2O$ and $Mn^{III}(OH)$ $(O_3PCH_2CH_2COOH) \cdot H_2O$ have inorganic layered structures with the carboxylic groups pointing towards the interlayer space and the organic moieties arranged in bilayer or monolayer configurations, respectively [19]. Altogether, the manganese carboxyphosphonates show a clear trend to adopt higher dimensionality upon increasing the extent of deprotonation of the ligand and, hence, the availability of the donor atoms to coordinate the metal ions.

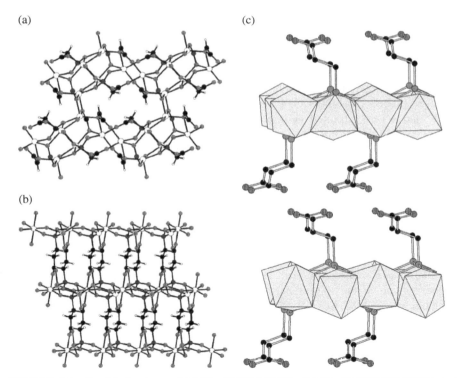

FIGURE 4.1 Crystal structures of (a) $Mn_3(O_3PCH_2COO)_2$, (b) $Mn_3(O_3PCH_2CH_2COO)_2$ and (c) $Mn^{II}(O_3PCH_2CH_2COOH) \cdot H_2O$. C, black spheres; Mn, white spheres; MnO_6, light-grey polyhedra; O, grey spheres; P, medium-sized dark-grey spheres. Adapted from Refs. [18] and [19].

The compounds $Al(O_3PCH_2COO) \cdot 3H_2O$ [20] and $Zr(O_3PCH_2COOH)_2$ [21] are representative examples of layered M(III) and M(IV) carboxyphosphonates. In the structure of crystalline aluminium carboxymethylphosphonate, aluminium atoms in the same layer are coordinated by both the carboxylate and phosphonate groups. Al^{3+} is located in an octahedral environment formed by three carboxymethylphosphonate ions, one chelating bidentate plus two monodentate modes, and two water molecules. The chelating bidentate linkage is made through one phosphonate oxygen and another carboxylate oxygen, while in monodentate fashion one phosphonate oxygen links the Al^{3+} ion. Although the ligand is found fully deprotonated, the trivalent aluminium atom satisfies its coordination in a layered arrangement with the presence of coordinated water. The layers are held together by hydrogen bonding between interlayer lattice water, coordinated water and the unbound oxygen of the carboxylate group.

Many metal carboxyethylphosphonates show a pillared layered arrangement. In zinc compounds, tetrahedral or both tetrahedral and octahedral coordination environments can exist [22]. The structure of $Zn(O_3PCH_2CH_2CO_2H) \cdot 1.5H_2O$ is built up from ZnO_4 and RPO_3 tetrahedra forming layers. Each organic group of one layer connects to the adjacent layer through one of the oxygen atoms of the carboxylic group. Alternating Zn and P tetrahedra in the layer form eight-membered rings, in which the water molecules are located. In a more dense structure, with composition $Zn_3(O_3PCH_2CH_2CO_2)_2$, ZnO_4 and ZnO_6 polyhedra coexist within layers. Metal–ligand connectivity is higher than in the hydrated framework, and there is no space for water molecules. The layers are strongly linked via the carboxylate end that uses its two oxygen atoms to connect to both sites of Zn^{2+}. Hypothetically, this second structure looks as if it was formed from the first one by exchange of a zinc complex species, $[Zn_2(O_3PCH_2CH_2CO_2)]^+$, for a H^+. On the other hand, Cd^{2+} also exhibits five- and six-coordinated geometries in $Cd_2(OH)(O_3PCH_2CH_2CO_2)$ [23] but a very distorted sixfold coordination in $Cd_3(O_3PCH_2CH_2CO_2)_2 \cdot 2H_2O$ [24]. The crystal structure of $Pb_5(O_3PCH_2CH_2CO_2)_2(O_3PCH_2CH_2COOH)_2$ [25] contains three independent Pb^{2+} ions, with seven- and eightfold coordination, and two types of carboxyethylphosphonate ions, one bonding adjacent layers and the other one bounded to a single inorganic layer through both carboxylate and phosphonate groups. For trivalent metal ions, the crystal structure of $NH_4[Fe_2(OH)\{O_3P(CH_2CH_2CO_2\}_2]$ was solved. The inorganic layers are made up of corner-sharing distorted $Fe(III)O_6$ octahedra, and the bifunctional ligand $[O_3P(CH_2)_2CO_2]^{3-}$ is pillaring adjacent inorganic layers [26]. The crystal structure of $Al_3(OH)_3(O_3PCH_2CH_2CO_2)_2 \cdot 3H_2O$ [19] is pillared as is that of $Mn_3(O_3PCH_2CH_2COO)_2$ [18], but in this case, the inorganic layers contain chains of aluminium octahedra running parallel to each other. In the structure of $Pr(O_3PCH_2COO)$ [27], praseodymium atoms are sevenfold coordinated by five oxygens from four phosphonate groups and two oxygens from two carboxylate groups. One of the phosphonate groups exhibits a chelating bidentate coordination mode. Isostructural compounds were prepared for yttrium and the entire series of the lanthanide elements.

Tridimensional open frameworks are characteristics of compounds such as $Co_3(O_3-PCH_2CH_2COO)_2 \cdot 6H_2O$ [28], $Cu_3[O_3PCH_2CO_2]_2 \cdot 2H_2O$ [29], $Pb_3(O_3PCH_2COO)_2$ [30] and $[Fe^{II}_3(OH)_2(H_2O)_4(O_3PCH_2CH_2COOH)_2]$ [31]. The copper phosphonate contains distorted CuO_6 octahedra and distorted $Cu(H_2O)O_4$ trigonal bipyramids in which the planar oxygens of the CuO_6 octahedra are shared with the trigonal bipyramidal CuO_5 units to form $Cu_3(H_2O)_2O_{10}$ trimers (Figure 4.2a). Successive trimers are linked by two phosphonate units forming chains, and these chains are then cross-linked by further phosphonate and carboxylate connections to form a dense 3D framework.

The structure of $Pb_3(O_3PCH_2CH_2COO)_2$ (Figure 4.2b) consists of three crystallo-graphically independent lead atoms having three different coordination geometries: tetrahedral PbO_3, trigonal bipyramidal PbO_4 and octahedral PbO_5 (the lone pairs occupying the fourth, fifth and sixth coordination sites, respectively). There are two 2-carboxyethylphosphonate ions in the asymmetric unit. While carboxylate groups act both as monodentate and bridging units, the connectivity of the phosphonate groups was found to be unusually high, some oxygens linking simultaneously three Pb^{2+} ions. These complex connectivities give rise to an interesting 3D open frame-work with several types of channels. The structure of $[Fe^{II}_3(OH)_2(H_2O)_4(O_3PCH_2CH_2COOH)_2]$ [31] is built up from linear trimers of edge-sharing iron(II) octahedra linked by the organic moieties. The protonation of the carboxylic group and the presence of O–OH sharing edges inside the trimers were the only differences found with the compound $Co_3(O_3PCH_2CH_2COO)_2 \cdot 6H_2O$.

Uranyl (UO_2^{2+}) carboxyphosphonates are typically layered with anionic net-works separated by either alkali metal or organoammonium cations. The addition of transition metals allows for the lower-dimensional features to be connected into higher-dimensional frameworks [32]. An exception is the compound $Cs_3[(UO_2)_4(PO_3CH_2CO_2)_2(PO_3–CH_2CO_2H_{0.5})_2]_3 \cdot nH_2O$ that without transition metals also adopts a 3D network structure with large channels housing the Cs^+ ions [33].

Although reports on monodimensional structures are still scarce, some attempts at obtaining metal carboxyphosphonates with low dimensionality were made recently.

(a)   (b)

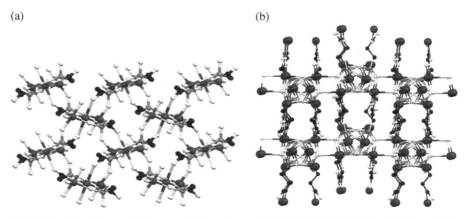

FIGURE 4.2 View of the crystal structures of (a) $Cu_3[O_3PCH_2CO_2]_2 \cdot 2H_2O$ and (b) $Pb_3(O_3PCH_2CH_2COO)_2$. C, black spheres; H, not shown; M, dark-grey spheres; O, grey spheres; P, medium-sized dark-grey spheres.

Ayi et al. [34] have prepared hydrothermally $(NH_4)_2Al(H_{1/2}O_3PCH_2CO_2)_2$. The structure of this compound consists of a unique aluminium(III) ion at the inversion centre with octahedral geometry. The aluminium atom is sixfold coordinated to four different carboxymethylphosphonates, where two moieties chelate to the aluminium centre through one oxygen atom from each phosphonate and carboxylate group. The neighbouring phosphonate groups chelating to equivalent aluminium atoms occupy the two remaining coordination sites forming $\{Al_2O_4P_2\}$ dimeric units with an eight-membered ring. These units compose linear chains running along the a-axis.

Substitution of more rigid aryl for alkyl groups further extends the coordinative variability of the carboxyphosphonate ligands giving rise to new topologies of metal carboxyphosphonates. So, the solvothermal reaction of p-phosphonobenzoic acid $(HOOC–C_6H_4–PO_3H_2,$ pbc) with Zn^{2+} leads to the compound $Zn_2(pbc)_2 \cdot H_2dma \cdot 3H_2O$ (dma = dimethylamine), which presents a 3D zeolite-like open framework [35]. The structure contains two crystallographically independent Zn^{2+} as well as pbc^{3-} ions. Zn(1) is tetrahedrally linked to four pbc ligands through the oxygen atoms of two PO_3 and two carboxylate groups. On the other hand, Zn(2) is coordinated to four phosphonate oxygen atoms. Each pbc^{3-} ion coordinates to four Zn atoms through three phosphonate oxygen atoms and one carboxylate oxygen atom, thus leaving the other carboxylate oxygen atom uncoordinated. The connectivity between the ZnO_4 and O_3PC tetrahedra gives rise to four-membered rings, which are connected edgewise to form zigzag chains. Further interconnection of the chains results in a 3D open framework with large intersected channels. The co-solvent, DMF, is partly decomposed to dma, which acts as a structure-directing agent; these molecules are protonated and located into the channels together with the water molecules. Relative to other metal carboxyphosphonates, such as $Na[Zn(O_3PC_2H_4COO)] \cdot H_2O$ [36], which are constructed by thick inorganic layers connected to flexible organic ligands, the structure of $Zn_2(pcb)_2 \cdot H_2dma \cdot 3H_2O$ shows significantly larger intersected channels, upheld by the rigid benzene-ring ligands.

The system $MnCl_2/H_2O_3PCH_2–C_6H_4–COOH/NaOH$ was investigated in detail using a systematic methodology (high throughput) [37]. The effect of the gradual deprotonation of the phosphonocarboxylic acid on the structure was demonstrated, with increasing pH of the reaction mixture favouring the synthesis of solids with higher dimensionality. So, the compound $Mn[HO_3PCH_2–C_6H_4–COOH]_2 \cdot 2H_2O$, obtained at low pH, exhibits a 1D chain structure of alternating corner-sharing MnO_6 and O_3PC polyhedra, whereas the compound $Mn[O_3PCH_2–C_6H_4–COOH] \cdot H_2O$ having the phosphonic acid group fully deprotonated shows a two-dimensional (2D) structure. The same structural trends were observed for the compounds $Cu[HO_3PCH_2–C_6H_4–COOH]_2 \cdot 2H_2O$ and $Cd[O_3PCH_2–C_6H_4–COOH] \cdot H_2O$.

By using the high-throughput approach [38], several crystalline phases of zinc carboxy-aryl-phosphonate were discovered. The synthesized compounds $Zn(HO_3PCH_2C_6H_4COOH)_2$, $Zn(O_3PCH_2C_6H_4COOH)$ and $Zn_3(O_3PCH_2C_6H_4COO)_2 \cdot 3H_2O$ exhibited monodimensional, bidimensional and tridimensional structures, respectively, showing thus a gradual evolution to more condensed structures with increasing deprotonation of the ligand.

4.2.1.2 *Hydroxyl-Carboxyphosphonates*   The use of the inexpensive racemic ligand, 2-hydroxyphosphonoacetic acid $H_2O_3PCH(OH)CO_2H$ (H_3HPA), featuring three close connected sub-functional groups has afforded a vast family of multifunctional metal phosphonates, with structures ranging from a discrete cluster to intricate 3D frameworks [39–45]. A common characteristic of this ligand is the formation of multiple chelating rings, displaying numerous coordination modes with alkaline, alkaline earth and transition metal ions. Reactions of metal(II) salts and H_3HPA show a similar trend as that reported for other metal carboxyphosphonates, with low-dimensional solids being synthesized generally at low pH and room temperature, whereas hydrothermal reactions result in 2D [41] or 3D frameworks [42, 43]. The latter may even incorporate various alkaline cations for charge compensation.

In the basic building unit of 1D solids $[M(II)(H_1HPA)(H_2O)_2]$ [42], the M(II) cations are located in a distorted octahedral environment formed by two chelating bidentate H_1HPA^{2-} ligands and two water molecules. Each ligand links two M(II) cations through two pairs of oxygens, one belonging to the phosphonate and carboxylate groups and another one to the carboxylate and the hydroxyl groups. This coordination mode generates five- and six-membered chelate rings (Figure 4.3a). A maximum of two lattice water molecules can be additionally hosted between the chains. The loss of the metal-bound water molecules by heating yields 3D solids by cross-linking between the coordination polymers (Figure 4.3b). The connectivity of the phosphonate group with respect to the metal cation substantially changes in the tridimensional structure, generating edge-sharing M_2O_{10} dimers and a third 12-membered ring.

Hydrothermal reactions at 140–180 °C lead usually to the formation of layered compounds $[M(II)(H_1HPA)(H_2O)_2]$ (M = Mn, Fe, Co, Zn) [41] showing different metal–ligand connectivity with respect to the 1D solids. In these layered compounds, the M(II) cations are located in a distorted octahedral environment formed by a chelating bidentate H_1HPA linked through carboxylate and hydroxyl oxygens, two oxygens of two phosphonate groups and two water molecules. Each H_1HPA ligand is, therefore, linked to three M(II) cations. Wriggled chains are formed by interconnected $[MnO_6]$ octahedra and $[CPO_3]$ tetrahedral. These chains then connect to each other by carboxylate and hydroxyl groups into a wave-like hybrid layer.

Anionic frameworks with fully deprotonated trianionic HPA^{3-} have been reported with divalent metal ions [41, 46–48]. In these structures, an extra organic or inorganic cation is also incorporated for charge balance. The resulting 3D bimetallic frameworks show chain connectivity different from the other ones already described. The structural changes result in a more open framework to host the incorporation of charge-compensating ions (Figure 4.3c). The Zn^{2+} derivatives feature a variable coordination around the M(II) centre, depending on the size of the companion cation, but the tetrahedral environment leads to a more open 3D framework, able to host bigger alkali cations. This is achieved by leaving the hydroxyl group unbound [42].

Large divalent M(II) ions, such as Sr(II) and Ba(II) [39], showing a high coordination number (nine-coordinated environment), adopt a 3D architecture with 2-hydroxyphosphonoacetate, despite of having the same elemental composition as the layered compounds with six-coordinated M(II) ions.

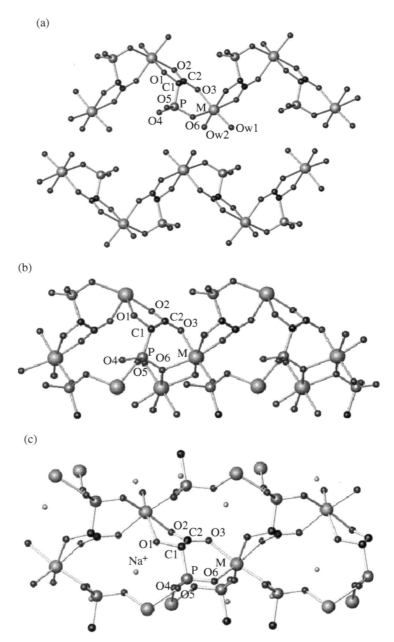

FIGURE 4.3 Comparative view of the chains: (a) M(II)(H$_1$HPA)(H$_2$O)$_2$, (b) M(II)(H$_1$HPA) and (c) NaM(II)(HPA). C, black spheres; M, big light-grey spheres; Na$^+$, small grey spheres; O, grey spheres; P, medium-sized dark-grey spheres. Adapted from Ref. [42].

In the 3D structure of $Ba(H_1HPA)(H_2O)_2$, which is isostructural with the Sr(II) derivative, all oxygen atoms of the three functional groups of the ligand (phosphonate, carboxylate and hydroxyl) are bounded to metal ions in such a way that each H_1HPA^{2-} links to five symmetry-equivalent, nine-coordinated Ba^{2+} centres. One oxygen atom of the carboxylate group is bridging two metal centres, and two water molecules complete the coordination sphere. The five- and six-membered chelate rings featuring other previously discussed structures are also preserved in this 3D framework.

The metal–ligand connectivity of Ca^{2+} and Cd^{2+} with 2-hydroxyphosphonoacetate, relative to the disposition of the five- and six-membered chelate rings, is quite different from that described earlier for other M^{2+} ions. This change in connectivity leads to completely different topologies. So, in the structure of the molecular (0D) $Ca_3(HPA)_2(H_2O)_{14}$ trimer [40], two HPA^{3-} ligands bridge three Ca^{2+} ions forming six-membered chelate rings with the central Ca^{2+} and five-membered chelate rings with the peripheral Ca^{2+}centres. The central Ca^{2+} centre is found in an octahedral environment, whereas the peripheral Ca^{2+} centres are in an eight-coordinated environment $(CaO_8–CaO_6–CdO_8)$ (Figure 4.4).

Following a systematic study of the $Ca^{2+}–H_3HPA$ system, two new compounds, namely, $Ca(H_1HPA)\cdot3H_2O$ (2D) and $Ca_5(HPA)_2(H_1HPA)_2\cdot6H_2O$ (3D), were obtained at room temperature and hydrothermally at $180\,^\circ C$, respectively [49]. The main structural feature of these compounds is that the frameworks are constructed from the trimeric Ca–HPA–Ca–HPA–Ca secondary building units (SBUs) that were isolated as a 'stand-alone' species at neutral pH. Both the 2D and the 3D compounds are thought to be formed by interaction of the cationic trimeric bricks with anionic monomeric species. However, the incorporation of alkaline cations, such as Li^+ or Na^+, disrupts this common structural feature, leading to new highly dense bimetallic frameworks. Trimeric building blocks $CdO_7–CdO_6–CdO_7$ are also used to generate the 2D framework of the compound $Na_2[Cd_2(HPA)_2(H_2O)_3]\cdot2H_2O$ [50]. In this structure, hydrated Na^+ ions, instead of H^+, are used to compensate the negative charge of the layer.

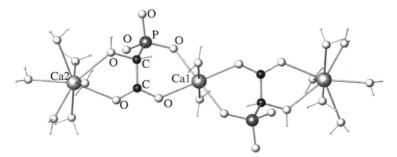

FIGURE 4.4 Trimer structure of $Ca_3(O_3PCH–(OH)COO)_2\cdot14H_2O$, showing the five- and six-membered rings. C, black spheres; Ca, big light-grey spheres; H, small white spheres; O, white spheres; P, medium-sized dark-grey spheres. Adapted from Ref. [40].

With the lanthanide (Ln^{3+}) ions, a full new family of 3D Ln metal–organic frameworks (MOFs) (Ln = La, Ce, Pr, Sm, Eu, Gd, Tb and Dy) based on the ligand 2-hydroxyphosphonoacetate could be obtained by slow crystallization at ambient conditions, from very acidic aqueous solutions [51]. This simple procedure is in contrast with preparation of many 3D solids, which usually requires hydrothermal/solvothermal conditions [52]. Formulated as $Ln_3(H_{0.75}HPA)_4 \cdot xH_2O$ ($x = 15$–16), these compounds crystallize in the orthorhombic system, and two polymorphs could be identified. They also exhibit the trimeric SBU structural feature, LnO_9–LnO_8–LnO_9. This trimeric unit shows a pronounced bending in the Ln–Ln–Ln angle attributed to the higher coordination of the central Ln^{3+} centre. On the other hand and on the basis of the presence of a chiral carbon centre in the HPA ligand, the central lanthanide ion in the trimeric SBU shows a preference for the S isomer, thus exerting chirality to this unit. The 3D framework may be envisaged as a layered structure formed by {Ln2(P1)} units and pillared along the b-axis by Ln1{P2}2 units (numbers 1 and 2 indicating two different ligand environments (P) and central and peripheral Ln^{3+}, respectively). This kind of connectivity gives rise to the open frameworks shown in Figure 4.5. The structures enclose water-filled 1D channels and exhibit remarkable crystalline-to-amorphous-to-crystalline transformations [51].

Even though 2-hydroxyphosphonoacetate is a racemic ligand, 3D homochiral manganese phosphonate with right-handed helical chains, $[enH_2]_{0.5}Mn_2[(H_1HPA)(HPA)]$ (en = ethylenediamine) could be synthesized by hydrothermal reaction in the presence of ethylenediamine as template [45]. These results open the way for preparing other novel 3D open-framework metal phosphonates with homochiral motifs by spontaneous resolution on crystallization from the racemic phosphonic acids (Figure 4.6).

Studies on the system M(II)/2-hydroxyphosphonoacetate have been recently extended with the synthesis of new Zn(II), Co(II) and Ba(II) derivatives having secondary ligands or bimetallic networks [53]. From this study, five new coordination modes were discovered for 2-hydroxyphosphonoacetate. It follows that the high adaptability of this ligand in different frameworks offers great chances for getting new hybrid solids with tuneable properties.

(a)   (b)

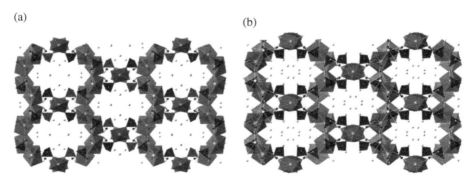

FIGURE 4.5 c-axis views (b-axis horizontal) of the crystal structures corresponding to both polymorphic forms of $Ln_3(H_{0.75}HPA)_4 \cdot xH_2O$: (a) LaHPA-I and (b) LaHPA-II. LnO_8/LnO_9, dark-grey polyhedra; O, white spheres; PO_3C, grey tetrahedral.

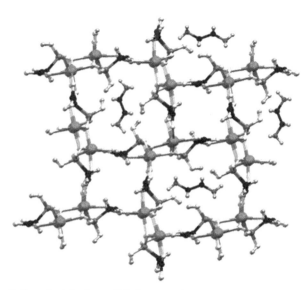

FIGURE 4.6 Ball-and-stick view of 3D framework structure showing the free achiral channels in $[enH_2]_{0.5}Mn_2[(H_1HPA)(HPA)]$. C, black spheres; H, small white spheres; Mn, big-sized
grey spheres; N, small-sized grey spheres; O, grey spheres; P, medium-sized dark-grey spheres.

4.2.1.3 Nitrogen-functionalized phosphonates The inclusion of amine groups in the
phosphonate molecule not only offers new binding sites but also introduces new functionalities into the hybrid solids. Phosphonic acids containing one or two amine groups
usually exist as zwitterions, but direct coordination to the metal may occur, depending on
the metal and the synthesis experimental conditions. To illustrate the different structural
roles of the amine group, we consider here various representative examples.

The compound $Zn(O_3PC_2H_4NH_2)$ is a 3D phosphonate with a channel-type
arrangement and amino ends present in the cavities [54]. The structure contains tetra-
coordinated zinc atoms bounded to three oxygens of three different phosphonate
groups and the nitrogen of an additional phosphonate unit. The zinc atoms are nearly
coplanar and arranged in 16-membered rings, constructed by corner sharing of 4
ZnO_3N tetrahedra and 4 PO_3C tetrahedra. The sheet linkage of these rings results in
eight-membered rings (Zn–O–P–O–Zn–O–P–O), with an arrangement very similar
to that of $Zn(O_3PC_2H_4CO_2H) \cdot 1.5H_2O$. The stacking of the 16-membered rings forms
infinite elliptical tunnels with windows of approximate dimensions $3.6 \times 5.3\,\text{Å}$. The
inorganic layers are connected together via O–P–C–C–N links to form a 3D network.
No guest molecules were trapped in the structure.

A number of 'macrocyclic leaflets' with a 1D chain structure as well as other
structural moieties resulted from reactions of metal(II) salts with phosphonic acids
attached with an aza-crown ether [55–57].

N-(Phosphonomethyl)iminodiacetic acid (H_4PMIDA) is a well-known organic
linker due to its use in preparation of the herbicide glyphosate. It has been found to
be able to adopt various kinds of coordination modes under different reaction

conditions. The structures of the corresponding hybrid solids depend on the extent of deprotonation of the ligand as well as on the nature of the metal ions. Under acidic conditions, only the phosphonic group is coordinated to the metal ions, and carboxylic groups and nitrogen are only involved in hydrogen bonding. Consequently, the synthesized compounds tend to be low-dimensional solids. Good examples are the Zr–PMIDA compounds [58]. If phosphoric acid was present in the reaction mixture, a layered compound with composition $Zr_2(PO_4)(H_2PMIDA)(HPMIDA)$ was obtained. Without addition of phosphoric acid, a linear chain compound was isolated. Each phosphonate group bridges with two and three Zr(IV) ions in the chain and layer Zr(IV) compounds, respectively. The chelation coordination mode is adopted when the solution is less acidic and the metal ions used have an affinity for the nitrogen atom, such as Co(II) and Zn(II) ions. Tetradentate chelation was firstly observed in the compounds $[Co_2(PMIDA)(H_2O)_5] \cdot H_2O$ and $[Zn_2(PMIDA)(CH_3CO_2H)] \cdot 2H_2O$ [59]. In both cases, the structure contains two crystallographic independent metal centres. One of the two metal atoms is chelated by a PMIDA ligand in a tetradentate fashion by three oxygen atoms and a nitrogen atom. The other metal atom having a different coordination environment acts as a bridging centre, extending, thus, the structure into infinite layers. If these second metal centres were not present, only discrete mononuclear complexes or 1D polymers would instead be solely formed [60]. At first sight, construction of these structures may be envisaged as a result of the interconnection of two different building blocks: anionic mononuclear chelate species and cationic metal complexes, the latter acting as the bridging centres.

The tetradentate chelation, characterized by the formation of three 5-membered chelate rings, appears to be a common coordination mode of H_4PMIDA with a number of metal ions (Scheme 4.1). So, the high-symmetry 3D structure of $Pb_2\{PMIDA\} \cdot 1.5H_2O$ [61] contains two crystallographic Pb^{2+} ions with severely distorted octahedral geometries. Pb(1) is six-coordinated by tetradentate chelation by a PMIDA ligand and two phosphonate oxygen atoms from neighbouring Pb(PMIDA) units. Pb(2) is six-coordinated by four carboxylate and two phosphonate oxygen atoms. These two different types of Pb^{2+} ions are interconnected through bridging carboxylate and phosphonate groups, resulting in a network with micropores, whose cavity is filled by lattice water molecules.

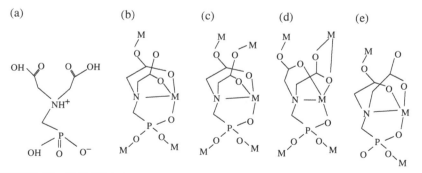

(a)   (b)   (c)   (d)   (e)

SCHEME 4.1 H_4PMIDA (a) and its coordination modes in 2D [Co(II) (b) and Zn (c)] and 3D [Pb(II) (d) and Ln(III) (e)] networks. Adapted from Refs. [59–61].

Similarly, hybrid structures of rare earth elements are 3D frameworks as a result of combining large coordination numbers of the cations and high ligand connectivity. In the frameworks of the seven isostructural compounds $Ln(HPMIDA)(H_2O)_2 \cdot H_2O$ (Ln = Gd, Tb, Dy, Y, Er, Yb and Lu) [62], the lanthanide ion is eight-coordinated by a chelating $HPMIDA^{3-}$ anion in a tetradentate fashion, one carboxylate oxygen and one phosphonate oxygen from two other $HPMIDA^{3-}$ anions and two aqua ligands. The Ln(HPMIDA) chelating units are interconnected via bridging carboxylate and phosphonate groups into a 3D network with helical tunnels containing lattice water. Each Ln atom is linked by three HPMIDA ligands, and each HPMIDA ligand bridges to three Ln atoms. Tetragonal right-handed and octagonal left-handed helices are present in this net; however, the whole net is racemic because the helices are alternatively arranged.

Hybrid solids with chiral networks have attracted attention due to their potential applications in enantioselective separation and catalysis. It was shown that using enantiomerically pure phosphonate ligands as chiral sources is an effective route of producing chiral metal phosphonates. Several chiral metal phosphonates based on crown ether derivatives have been reported [63], and more recently, the use of a proline-derived enantiopure phosphonic acid, 1-phosphonomethylproline [(S)-HO$_2$CC$_4$H$_7$NCH$_2$PO$_3$H$_2$], afforded various homochiral layered and 3D compounds [64, 65] (Scheme 4.2).

The framework of the homochiral 3D zinc phosphonate $Zn_2[(S)$-O$_3$PCH$_2$ NHC$_4$H$_7$CO$_2$]$_2$ [63] consists of alternately arranged left- and right-handed helices that are connected through 1-phosphonomethylproline ligands. The L- and R-handed helical chains consist of distinct pairs of ZnO$_4$ and O$_3$PC tetrahedral. Two types of microchannels with dimensions of 4×6 and 5×12 Å are generated, the amino and partially uncoordinated carboxylate groups being directed into the channels (Figure 4.7).

4.2.1.4 Metal Phosphonatosulphonates

Metal phosphonates derived from reactions of metal salts with phosphonatosulphonic acids have come into the focus of interest recently [66, 67]. The compounds based on rigid organic building units, such as m- and p-sulphophenylphosphonic acid, show a large structural variety ranging from isolated cluster to 3D frameworks. More recently, hybrid compounds with flexible phosphonosulphonic acids have also been reported [68, 69]. By using the flexible linker H_2O_3P–C$_4$H$_8$–SO$_3$H, a series of isostructural rare earth compounds

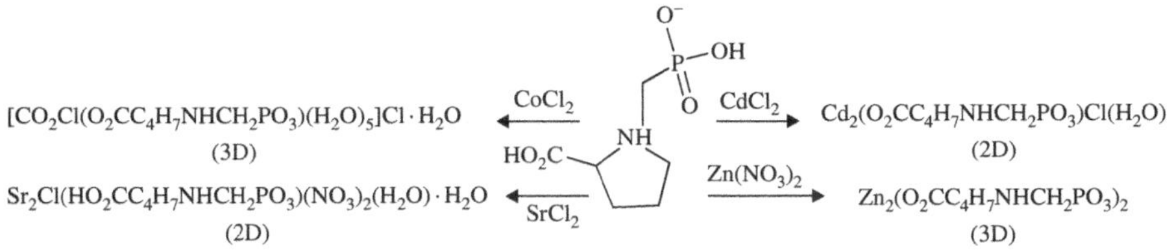

[CO$_2$Cl(O$_2$CC$_4$H$_7$NHCH$_2$PO$_3$)(H$_2$O)$_5$]Cl · H$_2$O $\xleftarrow{\text{CoCl}_2}$ Cd$_2$(O$_2$CC$_4$H$_7$NHCH$_2$PO$_3$)Cl(H$_2$O)$\xrightarrow{\text{CdCl}_2}$
(3D)   (2D)

Sr$_2$Cl(HO$_2$CC$_4$H$_7$NHCH$_2$PO$_3$)(NO$_3$)$_2$(H$_2$O) · H$_2$O $\xleftarrow{\text{SrCl}_2}$ Zn$_2$(O$_2$CC$_4$H$_7$NHCH$_2$PO$_3$)$_2$$\xrightarrow{\text{Zn(NO}_3)_2}$
(2D)   (3D)

SCHEME 4.2 Chiral compounds derived from 1-phosphonomethylproline. Adapted from Ref. [65].

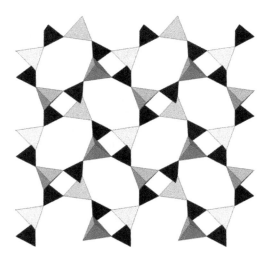

FIGURE 4.7 Polyhedral view of an inorganic layer in $Zn_2[(S)–O_3PCH_2NHC_4H_7CO_2]_2$ along the a-axis. S-phosphonomethylproline ligands were omitted for clarity. CPO_3; black tetrahedral; ZnO_4, grey tetrahedral. Adapted from Ref. [63].

$Ln(O_3P–C_4H_8–SO_3)(H_2O)$ ($Ln = La–Gd$) were obtained from high-throughput experiments [70]. The structures are built up from chains of edge-sharing LnO_8 polyhedra that are connected by the phosphonate and sulphonate groups into layers. These layers are linked by the $–(CH_2)_4$ group to form a 3D framework. Each rare earth ion is connected to five $O_3P–C_4H_8–SO_3^{3-}$ ions through five P–O–Ln and two S–O–Ln bonds. The full coordination sphere is completed by an H_2O molecule. Furthermore, two phosphonate oxygen atoms act by bridging two metal centres, which gives rise to the formation of the chains. As discussed in Section 4.3.3, the presence of the sulphonic group usually confers high proton-conducting capabilities to the solids.

4.2.2 Diphosphonates

Metal diphosphonates constitute the most widespread and versatile class of polyfunctional metal phosphonates, not only for the diversity of coordination modes but also for the variability of the organic moieties of the organodiphosphonate ligands $[O_3P–R–PO_3]^{4-}$. From a historical perspective, the leitmotif of metal diphosphonates has been the preparation of porous materials having large surface areas for diverse applications. Despite many of the synthesized materials are not porous and not considered as MOFs, it has been demonstrate that phosphonate-based MOFs can be prepared by appropriate selection of the organic linkers and synthesis methods [15]. The work on diphosphonate MOFs has been recently summarized in comprehensive [5, 9, 13] and systematic reviews [12], and therefore, we mainly consider here progress made dating from 2005. Particularly, we focus on crystalline functionalized metal diphosphonates.

4.2.2.1 Aryldiphosphonates: 1,4-Phenylenebisphosphonates and Related Materials

Independently on the oxidation state of the metal ion, the majority of crystalline metal 1,4-phenylene(bisdiphosphonates), 1,4-H_4PBP, show a typical pillared layered structure, with the phenyl groups propping apart the inorganic layers [5, 71–78]. Generally, the closeness of the phenylene pillars makes the interlayer region crowded and hydrophobic, although separation of the organic pillars to a certain extent can be achieved by using small spacer ligands. This idea, firstly developed on semi-crystalline zirconium bisphosphonates by the Dines's, Clearfield's and Alberti's groups [2–5, 79], has been recently applied to the crystalline compound [Al$_2$(O$_3$PC$_6$H$_4$PO$_3$)(H$_2$O)$_2$F$_2$·2H$_2$O] [80]. So, the presence of F$^-$ ions in the coordination sphere of Al^{3+} gives rise to small cavities filled with lattice water within the interlayer region by lowering the connectivity of the [O$_3$PC] tetrahedra to the metal ion. However, this feature was not enough to develop permanent porosity in the solid. Adelani and Albrecht-Schmitt have also demonstrated that the rigid 1,4-benzenebisphosphonates can be used in the presence of fluoride ions [81] or monovalent/divalent metal cations [81, 82] or in conjunction with a variety of organic templates as structure-directing agents [83, 84] to produce a high diversity of topologies in uranyl diphosphonates, with the lattice water and/or the template molecules having a structuring role. Pillared layered [82, 83, 85], nanotubules [81], layered [84] and even pillared layered 3D frameworks [83, 84] could be formed.

The influence of the relative position of the {O$_3$P–} groups, that is, *ortho-*, *meta-* and *para-*arrangements, has been studied for vanadium diphosphonates [77]. While the crystal structure of [VO(p-HO$_3$PC$_6$H$_4$PO$_3$H)] consists of {VO(HO$_3$PR)$_2$} layers linked through the {p-C$_6$H$_4$} groups, [V$_2$O$_2$(H$_2$O)$_2$(m-O$_3$PC$_6$H$_4$PO$_3$)]·1.5H$_2$O shows a complex 3D structure built up from chains of corner-sharing VO$_6$ octahedra, which are linked through (m-O$_3$PC$_6$H$_4$PO$_3$)$^{4-}$ groups into V–P–O layers, which are in turn connected through mononuclear square-pyramidal VO$_5$ units. However, when the phosphonic groups were in *ortho-*position, a bidimensional compound, [V$_2$O$_2$(H$_2$O)$_2$(o-O$_3$PC$_6$H$_4$PO$_3$)]·2H$_2$O, resulted. In this compound, the layers are formed by chains of corner-sharing VO$_6$ octahedra. Adjacent chains are linked through 1,2-phenyldiphosphonate groups to VO$_5$ square-pyramidal units, which in turn serve to link neighbouring {VO(O$_3$PC$_6$H$_4$PO$_3$)}$_n^{2n-}$ chains.

On the other hand, the functionalization of the ligand 1,4-H_4PBP in the 2- and 5-positions of the ring with hydroxyl groups has been found to impart porosity to the network. Thus, a new 3D microporous material with permanent porosity, Zn(H$_2$DHBP)·(DMF)$_2$ (DMF=dimethylformamide) was prepared using as precursor the ligand 2,5-dihydroxy-1,4-benzenediphosphonate (H$_4$DHBP) [86]. The solid is composed of 1D chains of distorted ZnO$_4$ tetrahedra bridged by the phosphonate ligand (Figure 4.8). Only two O atoms from each phosphonate group are bounded to the Zn centres, so each R–PO$_3^{2-}$ is functioning like a 'charge-assisted carboxylate' in the formation of the robust infinite SBU. Furthermore, the hydroxyl group is not coordinated to the zinc atom, and therefore, it only plays a steric role preventing formation of a regular layered motif.

Recently, bidentate phosphonate monoesters, analogues in dentation to dicarboxylate linkers in MOFs, have been employed to obtain 3D MOF materials [87, 88].

(a)   (b)

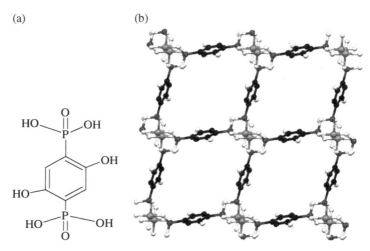

FIGURE 4.8 (a) A molecule of the ligand H_4DHBP. (b) View looking down the c-axis showing the network of pores for [Zn(DHBP)] (DMF)$_2$. C, black spheres; H, template molecules not shown; O, grey spheres; P, medium-sized dark-grey spheres; Zn, grey spheres. Adapted from Ref. [86].

These compounds seem to be extremely sensitive both to the flexing of the structure and to the orientation of the alkyl tether. Thus, while Cu[BDP–R] (H_2BDP–R = 1,4-benzenediphosphonate-bis(monoalkyl ester), R = Me, Et) with R = Et is non-porous, the methyl derivative, with an isomorphous framework, shows permanent porosity. As for Zn(H_2DHBP)·(DMF)$_2$, their crystal structures are formed by 1D chains of very distorted CuO$_4$ tetrahedra linked by fully deprotonated BDP–Et dianions forming a rhombohedral grid. The alkyl groups of the phosphonate esters protrude into the pores restricting the aperture of the channels [87]. However, the zinc derivative, Zn(BDP–Et), shows a bidimensional framework. In this solid, the corrugate sheets are also built up from 1D Zn(R–PO$_3$Et) chains bridged by the ligand and where the ethylene groups of adjacent sheets interact with each other in an unusual zigzag manner that increases the intermolecular contacts (Figure 4.9) [88].

One of the multiple strategies pursued to generate pores in crystalline metal phosphonates has been to modify the length of the carbon chain to increment and modulate the size of the pores. In this sense, p-xylenediphosphonic acid, also named 1,4-phenylenebis(methylene)diphosphonic acid (p-H_4pmd), has been widely employed with transition (V^{4+} [89], Mn^{2+} [90], Ni^{2+} [90], Cd^{2+} [90], Co^{2+} [91], Zn^{2+} [92, 93] and Cu^{2+} [94]), p-block (Ga^{3+} [95], Sn^{2+} [96] and Pb^{2+} [97]) and lanthanide (La^{3+} [98], Ce^{3+} [99] and Pr^{3+} [99]) metals.

All M$_2$[p-C$_6$H$_4$(CH$_2$PO$_3$)$_2$]·2H$_2$O (M = Mn [90], Ni [90], Cu [94], Zn and Cd [90]) show pillared networks where the inorganic layers have similar structural characteristics to those encountered in the layered structure of the corresponding phenylphosphonates. In all of them, the inorganic layers are pillared by the –CH$_2$C$_6$H$_4$CH$_2$– units, although stacking of the phenyl rings between adjacent layers differs substantially.

(a)   (b)

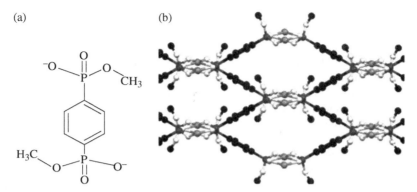

FIGURE 4.9 (a) A molecule of the ligand $H_2BDP–Me$. (b) Crystal structure of Zn{BDP–Me} showing the hydrophobic channels. C, black spheres; H, not shown; O, grey spheres; P, medium-sized dark-grey spheres; Zn, grey spheres.

Different topologies have been reported for $V^{IV}O(H_2O)(p-C_6H_4(CH_2PO_3H)_2)$ [89], $Cu[m-C_6H_4(CH_2PO_3H)_2] \cdot 3H_2O$ [100] and $Zn[p-C_6H_4(CH_2PO_3H)_2$ [92]], all of them presenting monoprotonated phosphonate groups. $V^{IV}O(H_2O)(p-C_6H_4(CH_2PO_3H)_2)$ [89] shows a bidimensional structure formed by hybrid organic–inorganic sheets containing corner-sharing octahedra. $Cu[m-C_6H_4(CH_2PO_3H)_2] \cdot 3H_2O$ [100] also exhibits a 2D framework, but in this case, it contains infinite hybrid organic–inorganic layers built up from 1D chains of edge-sharing copper octahedra connected by the O_3PC groups of the $m-C_6H_4(CH_2PO_3H)_2^{2-}$ ligands. The interlayer space is occupied by the water molecules involved in a network of hydrogen bonds between adjacent layers. However, the Zn derivative, synthesized from tetraethyl p-xylylenediphosphonate under solvothermal conditions, shows a 3D open framework. This solid has the peculiarity of containing 1D linear inorganic chains formed by corner-sharing $[ZnO_4]$ and $[PO_3C]$ tetrahedral, which are linked through the $p-C_6H_4(CH_2PO_3H)_2^{2-}$ units creating a 3D structure [92]. As in other divalent metals, Pb(II) [97] and Sn(II) [96] derivatives of p-xylenediphosphonic acid also exhibit pillared-layered or bidimensional structures, respectively, but the effect of lone pair in the latter ones is noted by a distinctive metal–ligand connection.

Several p-xylenediphosphonates of trivalent cations have been reported. For instance, $Ga_2[p-C_6H_4(CH_2PO_3)_2](H_2O)_2F_2$ [95] shows a dense packing of the organic portions with layers, containing linear chains of corner-sharing GaO_4F_2 octahedra, linked by the diphosphonate groups. The incorporation of phosphite, as spacer group, in its structure gave solids of composition $Ga_2[p-(C_6H_4)(CH_2PO_3)_2]_{1-x}HPO_3)_{2x}\}(H_2O)_2F_2$, ($0 < x < 0.146$) with larger apertures randomly arranged throughout the framework. For lanthanide derivatives of the same ligand, a structural diversity, from 1D to 3D, has been found by using microwave or one-pot methodologies [98]. With both methods, conversions of 1D to 3D networks occurred. However, conducting the reaction in ionic liquids composed of choline chloride and malonic acid for Ce(III) and Pr(III) derivatives yielded non-porous Ce(III) and Pr(III) derivatives [99].

By using ligands such as 4,4'-biphenylenebis(phosphonic) acid, $[H_2O_3P(C_6H_4)_2PO_3H_2]$, and 4-(4'-phosphonophenoxy)phenylphosphonic acid, $[H_2O_3P–C_6H_4]_2–O$, combined with divalent, trivalent or tetravalent metals [5, 75, 77, 93, 101, 102], non-porous pillared structures or bidimensional networks resulted. The main differential feature between them was that the bidimensional ones contained partially protonated phosphonate groups. A completely different network is observed for the uranyl compound $M^I_2[(UO_2)_2F(HO_3P(C_6H_4)_2PO_3H)(HO_3P(C_6H_4)_2PO_3)]\cdot 2H_2O$ ($M^I = Cs^+$, Rb^+) [103]. Differently from other uranyl phosphonates showing circular nanotubules [104–107], this solid presents a 3D open framework formed by the packing of elliptical nanotubules. Its structure is composed of corner-sharing uranyl dimers of pentagonal bipyramids UO_6F, which are linked through a fluoride anion. The PO_3 moieties help to create the dimers and also bridge the dimers to each other, creating 1D chains extended along the length of the nanotubules. The resulting chains are connected by the biphenyl rings generating tubular channels filled with water molecules (Figure 4.10). The negative charge of these nanotubules is compensated by the Cs^+/Rb^+ cations located between them, imparting, thus, a high ion-exchange capacity.

Trivalent and tetravalent metal phosphonates of $[H_2O_3P(C_6H_4)_2PO_3H_2]$ and $[H_2O_3P–C_6H_4]_2–O$ are invariably semi-crystalline pillared layered solids [13, 108–112]. By using different small spacer molecules, some of these compounds were made microporous solids [5, 113]. Another synthetic strategy to get porous materials consisted in using different solvent mixtures containing DMSO or short-chain aliphatic alcohols [110–112]. In general, these semi-crystalline solids show narrow pore size distributions, ranging between 10 and 20 Å. Owing to the lack of

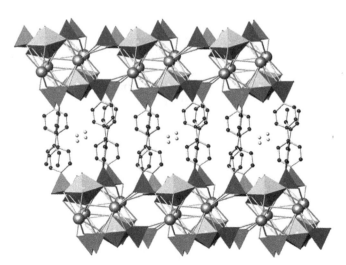

FIGURE 4.10   Layered crystal structure of $M^I_2[(UO_2)_2F(HO_3P(C_6H_4)_2PO_3H)(HO_3P(C_6H_4)_2PO_3)]\cdot 2H_2O$ ($M^I = Cs^+$, Rb^+). C, black spheres; H, small white spheres; M^I, medium-sized grey spheres; O, grey spheres; PO_3C, grey polyhedra; UO_7, grey polyhedra. Adapted from Ref. [103].

long-range order in their structures, these compounds have been coined globally as unconventional MOFs (UMOFs), as contrasted with the highly ordered structures that define conventional MOFs [109].

4.2.2.2 1-Hydroxyethylidinediphosphonates

Among functionalized diphosphonic acids, 1-hydroxyethane-1,1-diphosphonic acid (etidronic acid) $H_2O_3P–CRR'–PO_3H_2$ (R=OH, R'=CH$_3$) (H$_4$HEDP) is one of the most thoroughly studied. This acid and some of its salts have been widely used as a strong chelating agent for water treatment, detergents, cosmetics and pharmaceutical treatments. In comparison with methylenediphosphonic acid [114, 115], functionality of the R tether in H$_4$HEDP extends significantly its coordination ability. The hydroxyl group is capable to bind to metal atoms or participate in hydrogen bonds with a number of s, p, d and f metal ions. A survey in Cambridge Structural Database (V5.34) [116] reveals more than a hundred of metal-H$_4$HEDP derivatives and a large structural diversity, from 0D to 3D frameworks, although the majority of these solids show low dimensionality. There are several reviews where the coordination modes, topologies and complexation abilities of this diphosphonic acid have already been examined [12, 117, 118]. In this section, we particularly pay attention to metal derivatives of H$_4$HEDP with tridimensional topologies.

A wide range of structures can be formed depending on the synthesis conditions used for the preparation [119, 120]. However, only a few transition metal- and lanthanide-HEDP phosphonates exhibit 3D networks when the metal centres and etidronic acid are used solely as primary building units [121–123]. One of the first compounds reported with a 3D open framework was $Na_2Cu_{15}(HEDP)_6(OH)_2(H_2O)$, which constitutes a rare example of mixed-valence copper phosphonates [124].

The framework structures formed from different combinations of Ln^{3+} ions and etidronic acid exhibit often zeolite-type behaviour (presence of channels with charge-compensating extra-framework cations that may or not be hydrated), for instance, the two series of chiral $[NH_4][Ln(HEDP)(H_2O)] \cdot 3H_2O$ (Ln=La–Nd) and achiral $[NH_4]$ $[Ln(HEDP)(H_2O)]$ (Ln = Sm, Eu, Gd, Tb, Dy, Er, Y) [125] or the series of compounds $Na_4[Ln_2(HEDP)_2(H_2O)_2] \cdot nH_2O$ [126].

The crystal structure of the chiral isomorphous compounds $[NH_4][Ln(HEDP)(H_2O)] \cdot 3H_2O$ (Ln=La–Nd) [125] consists of infinite chains of CeO$_8$ polyhedra and CPO$_3$ tetrahedra, where each HEDP^{4-} ligand connects four Ln(III) atoms via all its phosphonate oxygen atoms and the hydroxyl group. Such chains are linked to each other, forming a 3D framework structure via CPO$_3$ tetrahedra. This connectivity creates two types of small channels along the a- and b-axis, respectively, where the lattice water molecules and charge-compensating NH$_4$$^+$ ions are located with extensive hydrogen-bond interactions (Figures 4.11a).

Lanthanide atoms with lower ionic radii (Ln=Sm, Eu, Gd, Tb, Dy, Er, Y) crystallize with similar composition but without lattice waters, $[NH_4][Ln(HEDP)(H_2O)]$. These solids exhibit achiral 3D network topologies quite similar to those obtained for the bigger lanthanide cations but with the channels occupied only by NH$_4$$^+$ ions (Figure 4.11b). The absence of lattice waters inside the channels implies that the NH$_4$$^+$ ion acts as a template [125].

(a)

(b)

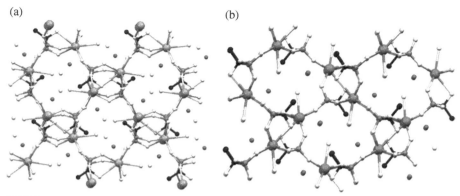

FIGURE 4.11 Ball-and-stick views of the framework for (a) chiral $[NH_4][La(HEDP)(H_2O)] \cdot 3H_2O$ and (b) achiral $[NH_4][Er(HEDP)(H_2O)]$. C, black spheres; H, not shown; Ln, big-sized grey spheres; N, small-sized grey spheres; O, grey spheres; P, medium-sized dark-grey spheres.

The framework-type structures $Na_2[Y(HEDP)(H_2O)_{0.67}]$ and $Na_4[Ln_2(HEDP)_2(H_2O)_2] \cdot H_2O$ (Ln = La, Ce, Nd, Eu, Gd, Tb and Er) and layered networks $[Eu(H_2-HEDP)(H_2O)_2] \cdot H_2O$, $Na_{0.9}[Nd_{0.9}Ge_{0.10}(H_1-HEDP)(H_2O)_2]$, $[Ln(H_2-HEDP)(H_2O)] \cdot H_2O$ (Ln = Y, Tb) and $[Yb(H_2-HEDP)]H_2O$ are, to a certain extent, interconvertible by hydrothermal treatments with HCl or NaCl [126].

Basically, the crystal structure of $Na_4[Ln_2(HEDP)_2(H_2O)_2] \cdot H_2O$ is composed of LnO_8 polyhedra and $HEDP^{5-}$ anions, where each $HEDP^{5-}$ links four metal centres in bidentate, tridentate and doubly monodentate fashions. The 3D framework is formed by 1D inorganic–organic chains, built up of $[Eu(HEDP)(H_2O)]$ fragments linked together in a zigzag-type fashion. Adjacent chains are interconnected via shared oxygen atoms of the phosphonate groups creating an anionic porous framework, with two types of perpendicular channels filled with Na^+ and both free and lanthanide-coordinated water molecules, which are removed reversibly by calcination at 300 °C. In layered $[Y(H_2hedp)(H_2O)] \cdot 3H_2O$, non-coordinated water molecules present in the interlayer space form a $(H_2O)_{13}$ cluster through hydrogen-bond interactions, which is the basis of an unprecedented 2D water network [126].

$(H_3O)_2\{[(UO_2)(H_2O)]_3(HEDP)_2\} \cdot 2H_2O$ shows interesting selective ion-exchange properties [127]. Its crystal structure is composed of three UO_7 pentagonal bipyramids and four PO_3C tetrahedra as SBUs. These SBUs are connected to each other through corner sharing, which creates an anionic 3D open framework with two types of channels occupied by water molecules. Hydronium ions act as charge-compensating cations. These ions can be easily exchanged by monovalent cations, which occupy the sites preferentially according to their sizes.

4.2.2.3 R-Amino-N,N-bis(methylphosphonates) and R-N,N'-bis(methylphosphonates)

Ligands containing nitrogen atoms have been employed as versatile building blocks for the preparation of wide variety of functionalized metal phosphonates. The corresponding phosphonic acids exist in solution as zwitterions, which are a common feature of phosphonic acids containing the amine function.

As already commented for amino-monophosphonate ligands, the coordination modes adopted by the amino-bisphosphonate ligands are strongly influenced by the extent of protonation of the molecule, but, additionally in this case, the organic substituent plays also a subtle role in the structures and, even more, may affect markedly the properties of the resulting metal phosphonates.

R-Amino-N,N-bis(methylphosphonates) The large number of crystal structures described for R-amino-N,N-bis(methylphosphonates) responds to the different coordination modes that this ligand can adopt depending on the synthesis conditions and the features of the metal centres. To illustrate this variety of coordination modes, Scheme 4.3 depicts those shown by the iminobis(methylenephosphonic acid) with divalent transition metal.

Slight differences in synthesis conditions yield compounds with different stoichiometries. On one hand, in the compounds $M(II)[NH_2(CH_2PO_3H)_2]_2(H_2O)_2$ (M=Co [128], Mn [129], Ni [130]), the cations are in a six-coordinated environment, and the ligand adopts a bidentate bonding mode to produce 1D chains. On the other hand, in the compounds $M(II)[NH_2(CH_2PO_3H)(CH_2PO_3)](H_2O)_3$ (M=Co, Mn) [131], the ligand is bridging four cobalt(II) ions in a layered architecture or is linking a Mn(II) ion bidentately as well as bridging monodentately three other Mn(II) ions in a 3D framework.

The compound $Cu[NH(CH_2PO_3H)_2]$ shows, however, a linear chain structure composed of distorted square pyramid of Cu(II) bridged by two phosphonate oxygen

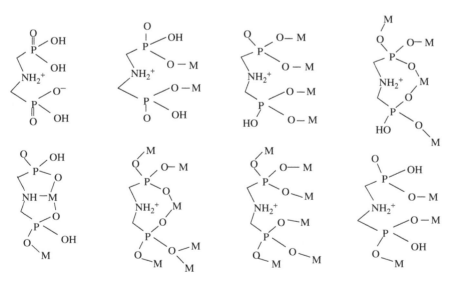

SCHEME 4.3 Ligand $NH(CH_2PO_3H_2)_2$ and its coordination modes in divalent transition metal derivatives.

atoms forming Cu_2O_2 dimeric units, which are further interconnected via phosphonate groups. In this case, the phosphonate anion chelates one copper(II) ion tridentately (two oxygen and the nitrogen atom) and also bridges two other Cu(II) ions [131].

Removing of more protons from the phosphonate ligand results in new coordination modes and crystal structures, as observed for the compounds $M_3[NH_2(CH_2PO_3)_2]_2$ (M = Cu, Co) [129]. The Cu(II) derivative was also obtained indirectly by refluxing $CuCl_2$ with nitrilotris(methylene)triphosphonic acid, $[N(CH_2PO_3H_2)_3]$ [132]. Both compounds exhibit a 3D framework, but with distinct topologies due to variations in the coordination mode of the ligand. In both frameworks, the ligand links six metal ions in two different ways. In the cobalt complex, one Co^{2+} ion is chelated by two oxygens of different phosphonate groups, and one of the oxygens of a phosphonate group bridges two Co^{2+} ions. In the copper derivative, the six oxygens of the two phosphonate groups link six Cu^{2+} ions. The crystal structure of $Cu_3[NH_2(CH_2PO_3)_2]_2$ is built up from two distinct copper environments, one a tetragonally elongated tetrahedron and a second distorted square plane, which are bridged by one terminal oxygen atom. The dimers Cu_2O_9 are interconnected by phosphonate groups, resulting in the formation of 3D networks [129, 132].

Additional coordination modes of this trifunctional phosphonate anion have been reported for Co(II) and Zn(II) derivatives [133, 134]. In $M_3(NH_2(CH_2PO_3)_2]_2$ (M = Zn, Co), a non-porous 3D network is formed by the cross-linking of chains of tetrahedrally coordinated metal atoms.

Co(II)-iminobis(methylphosphonate) compounds clearly illustrate the influence of the synthesis conditions [135, 136] in determining the composition and structure of the resulting solids. Two Co(II) derivatives, one of them having two polymorphs, were synthesized [133]. These compounds were of low dimensionality (1D or 2D), and in all networks, there is an octahedrally coordinated cobalt centre that is slightly distorted.

Many other R-amino-N,N-bis(methylenephosphonates) of divalent and trivalent metals, where R can be n-alkyl, carboxyalkyl, benzyl, and so on, have been described with layered structures [137–144]. In these compounds, different arrangements in the interlayer region as well as in the inorganic layers are found, as a consequence of the presence of additional functions such as carboxylic [145] or carboxypyridine groups [146]. For instance, N-methyl-iminobis(methylenephosphonates) with divalent cations basically gives three types of structures: layered, as $M[NHCH_3(HO_3PCH_2)_2]_2 \cdot 2H_2O$ (M = Mn [147], Cd [148]); with a double-chain structure, as $Zn[NHCH_3(HO_3PCH_2)$ $(O_3PCH_2)](H_2O)$ [148] and $Pb[NHCH_3(HO_3PCH_2)_2]$ [61]; or with a 3D network, as $Zn_3[NHCH_3(O_3PCH_2)_2]_2$ [147] and $Zn_3[NHCH_3(O_3PCH_2)_2]_2$. The same types of frameworks have also been reported for other derivatives with longer chains.

Recently, Costantino et al. [149] have analysed the family of Zr alkylamino-N,N-bismethylenephosphonates to establish the role that the length of the chain of the n-alkyl fraction may play as a structural director. In this study, the systems $ZrF_2[R–NH(O_3PCH_2)_2]$ and $R = –CH_3$, $–CH_2–CH_3$, $–(CH_2)_n–CH_3$ and $n = 2–9$ were analysed [149, 150]. All compounds were synthesized following equivalent synthetic procedures, so the structural and compositional differences between them were not a consequence of the synthesis conditions. From a structural point of view, the results obtained from this study suggest that the self-assembly of alkyl chains may act as the orienting factor during the formation of the solid. Thus, the Zr alkylamino-bismethyle-nephosphonates with 3–10 carbon atoms in the chain show layered frameworks, with the chains situated in the interlayer space in an interdigitated fashion (Figure 4.12). This layered arrangement is likely due to the effect of the non-polar interactions among the organic groups. However, the methylamino- and ethylamino-zirconium derivatives show different 3D structures. Thus, zirconium methylamino-*N,N*-bismethylenephos-phonate exhibits a pillared layered structure where the inorganic layers are composed of two types of chains of ZrO_6 or ZrO_5F polyhedra and PO_3C tetrahedral, which are running concurrently along the same crystallographic direction but with different orientations. Neighbouring chains are interconnected through the small organic groups, defining small 1D tunnels occupied by the methyl groups. On the other hand, the ethylamino derivative, $Zr[(HO_3PCH_2)(O_3PCH_2)NHC_2H_5]_2$, shows a 3D network with ellipsoidal cavities where the organic moieties are located. So, both compounds with short organic chains tend to maintain these non-polar moieties confined in spaces with adequate dimensions (Figure 4.12) [149].

A special case is the series of luminescent lanthanide squarato-aminophospho-nates that constitutes a rare example of *in situ* organic reactions in the chemistry of metal phosphonates [140, 151]. This family of compounds results from the *in situ* condensation reaction of the aminodiphosphonic acid $(H_2O_3PCH_2)_2NCH_2C_6H_4COOH$ with squaric acid, initially used as a second linker. Under hydrothermal conditions

(a)

(b)

(c)

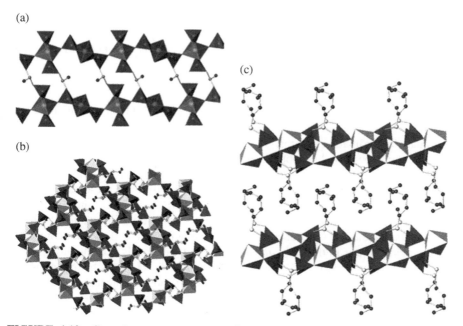

FIGURE 4.12 Crystal structures corresponding to Zr alkylamino-bismethylenephospho-nates with alkyl: (a) methyl, (b) ethyl and (c) pentyl.

and in the presence of lanthanide nitrate salts, these two ligands react, producing a new multifunctional monophosphonic acid, $HOOC-C_6H_4-CH_2-N(C_4O_3H)$ $(CH_2PO_3H_2)$, which gives a series of isostructural Ln(III) (Ln = La, Pr, Eu, Gd, Tb, Er) [152]. The resulting lanthanide derivatives show a pillared layered architecture, where the cross-linking between adjacent layers takes place via the coordination of one oxygen atom of the carboxylate group of one layer to the lanthanide cation of another layer.

R-N,N'-bis(methylenephosphonates): Piperazinyldiphosphonates   In this group, the most relevant results have been obtained with ligands containing the piperazine ring. In contrast to the mostly bidimensional metal 1,4-phenylene(bisdiphosphonate) solids, crystalline open and microporous framework metal phosphonates have been obtained using N,N'-piperazinebis(methylenephosphonic acid), $H_2O_3PCH_2N(C_4H_8)$ $NCH_2PO_3H_2$, and related phosphonic acids. This markedly different behaviour is mainly due to the versatile metal-binding and hydrogen-bonding capabilities of the last ligands. The piperazine-based bisphosphonic linkers can make use of one, two or three phosphonate O atoms to bind to the metal cations, and additionally, piperazinyl N atoms may also be available to coordination. On the contrary, if the N atom is protonated, it is involved in H-bond arrangements. Moreover, they are flexible and can adopt different configurations, which extends beyond its coordinative potential.

The work on metal bisphosphonates formed with the ligand N,N'-piperazinebis(m ethylenephosphonic acid) has been recently reviewed by Wharmby and Wright [153] and partially included in other reviews [11, 13]. Therefore, here, we only discuss a few representative examples to illustrate the structural trends in this family of compounds, some of which are considered archetype of phosphonate MOFs.

In compounds of alkaline earth metals with MO_8 and MO_9 polyhedra M = Sr [154] and Ba [155], the bigger cation tends to form a more complex network. In fact, the Ba(II) derivative shares some structural features with bulky lanthanide derivatives as $Nd_2(O_3PCH_2NH(C_4H_8)NHCH_2PO_3)_3 \cdot 9H_2O$ [156].

Other transition metals such as V [157], Cu [158] and Zn [159] exhibit also a layered topology, but the 2D architectures differ from that of Ba as the metal ions present smaller coordination spheres.

Zirconium piperazinyl phosphonates represent an example of the effect of pH on building up networks. A layered compound, $ZrF_2(O_3PCH_2NH(C_4H_8)NHCH_2PO_3)$, containing ZrO_4F_2 octahedra was obtained at pH = 2 [160] in the presence of HF. On the other hand, increasing the synthesis pH up to 7 gave rise to a 3D framework solid, $Zr_2H_4(O_3PCH_2NH(C_4H_8)NHCH_2PO_3]_3 \cdot 9H_2O$, made of infinite inorganic chains of ZrO_6 octahedra and PO_3C tetrahedra, running along the *c*-axis direction, connected by piperazine groups in the *ab* plane, and generating channels running along the *c*-axis. In this framework, each Zr atom is bounded to six O atoms belonging to six different phosphonate groups, in such a way that each phosphonate tetrahedra bridges two Zr atoms of a same chain. The third oxygen atom of the phosphonate group is pointed towards the channels formed by the interconnection of adjacent chains. However, no permanent porosity was measured upon removing water inside cavities due to contraction of the channels after dehydration (Figure 4.13).

(a) (b)

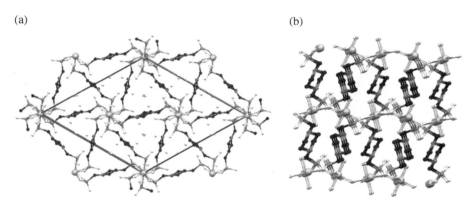

FIGURE 4.13 3D frameworks for (a) $Zr_2H_4(O_3PCH_2NH(C_4H_8)NHCH_2PO_3]_3 \cdot 9H_2O$ and (b) $ZrF_2(O_3PCH_2NH(C_4H_8)NHCH_2PO_3)$ view along the c-axis. C, black spheres; H, not shown; N, small-sized grey spheres; O/F small white spheres; P, medium-sized dark-grey spheres; Zr, big-sized grey spheres.

Outstanding structural changes occur by fine-tuning of the synthesis conditions in piperazinyl derivatives of divalent metals. The framework of the compounds $M(O_3PCH_2NH(C_4H_8)NHCH_2PO_3) \cdot xH_2O$ (Mn [161], Fe [162], Co [161], Zn [163]), synthesized at pH values below 6.5, is composed of chains containing MO_4 tetrahedra and bridging PO_3C groups. This connectivity results in formation of narrow channels that contain water molecules, but the dehydrated form shows no permanent porosity for larger molecules. On the other hand, raising, in most cases, the synthesis pH yields a second family of M(II) derivatives (M=Mg, Mn, Fe, Co, Ni) with formula $M_2(H_2O)_2(O_3PCH_2N(C_4H_8)NCH_2PO_3) \cdot xH_2O$ (STA-12) [162, 164, 165]. The framework is built from helical chains of edge-sharing MO_5N octahedra, one of the oxygens corresponding to coordinated water. Each chain is connected to three other chains by the piperazinyl moieties to form large hexagonal channels (Figure 4.14). Inside the cavities, surface groups are forming H-bond networks with the lattice water. Upon dehydration, the structure remains porous, and the coordinated water can be removed, thus leaving coordinatively unsaturated Lewis acid sites. All these characteristics convert the structure-type STA-12 in the first-reported large-pore, fully ordered phosphonate MOF [165].

Going beyond, Wright and co-workers have applied the concept of *isoreticular synthesis*, used for MOF materials, to enhance opening of the framework. In this way, reaction of the ligand *N,N'*-4,4'-bipiperidinebis(methylenephosphonic acid) with cobalt(II) or Ni(II) results in a new MOF bisphosphonate, named STA-16, with the same topology than STA-12 but with pores doubling the diameter of those of STA-12 (1.8 nm) (see Figure 4.14). As in STA-12, as-synthesized STA-16 [166] contains lattice and bound water molecules that can be removed and give rise to permanent microporosity and the appearance of coordinatively unsaturated sites.

Another isostructural family of 3D connected microporous phosphonate frameworks is MIL-91 [153, 167], $M(OH)(O_3PCH_2NHC_4H_8NHCH_2PO_3) \cdot nH_2O$ (MIL-91), whose

(a)   (b)

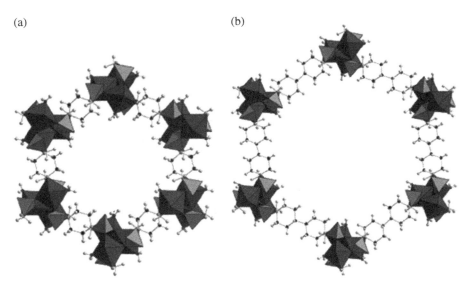

FIGURE 4.14   Crystal structures of STA-12 (a) and STA-16 (b). Adapted from Ref. [166].

(a)   (b)

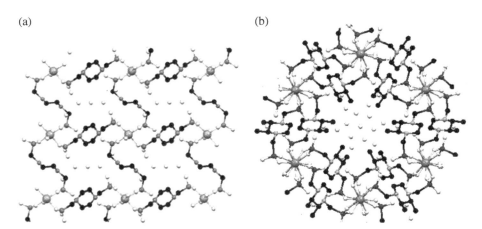

FIGURE 4.15   3D porous frameworks with regular open channels for (a) MIL-91 and (b) STA-13. C, black spheres; H, not shown; M, big-sized grey spheres; N, small-sized grey spheres; O small white spheres; P, medium-sized dark-grey spheres.

structure consists of corner-sharing MO_6 octahedra chains linked together in two directions by the piperazinyl ligand (Figure 4.15a). It must be noted that the non-coordinated P=O group of each phosphonate group stabilizes the framework by H bonding with the protonated N atom of the piperazinyl ring. The same structural arrangement is also adopted by Ti^{4+} where $[TiO]^{2+}$ cations substitute the $[MOH]^{2+}$ units in the structure. The resulting frameworks contain 1D channels with 4Å free diameter, upon removal of solvent, and a Langmuir surface area of approximately $500\,m^2\,g^{-0}$ [167].

The versatility of the N,N-piperazinebisphosphonate ligand, and its methylated derivatives, results evident from the structural diversity presented with lanthanide metal ions. All compounds studied exhibit 3D open frameworks built up from 1D chains of isolated metal polyhedrons bridged by the piperazinyl moieties [156, 168, 169], but different structural types are obtained depending on the cation size, pH, metal salt used and the presence or absence of methyl groups on the piperazine ring. The differences among these structural types rely basically on variable local environments of metal ions (six-coordinated for smaller and seven- or eight-coordinated for the larger ones) and coordination modes and configurations, chair or boat, of the ligand. Synthesis in the presence of added alkali metal hydroxides results in deprotonated ligands and charge-balancing cations in the cavities, whereas lower pH synthesis leaves protons on the ligand that participates in H-bond networks. Reversible dehydration but no porosity for gases, such as nitrogen, was observed.

Crystallization of yttrium and other trivalent metal cations of similar radius with racemic N,N'-2-methylpiperazinebis(methylenephosphonic acid) results in the permanently porous lanthanide phosphonate framework STA-13, $Y_2(O_3PCH_2NHC_4H_7(CH_3)$ $NHCH_2PO_3)_3 \cdot 7H_2O$ [169], formed by 1D M^{3+} phosphonate chains (Figure 4.15b). Interconnection of these chains forms a hexagonal array of unidirectional tunnels filled with water molecules. Upon dehydration, the free diameter of the tunnels is approximately 3Å. Building up of the 3D porous framework STA-13 requires the presence of the R- and S-enantiomers of the ligand, as the use of the pure R-enantiomer only led to a non-porous structure, by steric effect.

4.2.3 Polyphosphonates

One more step intending to obtain phosphonate-based open structures has been the use of flexible or rigid polyphosphonic acids containing the phosphonate moieties far away in divergent positions [11, 14, 170–176]. Among them, tri- and tetraphosphonates have been the most employed ligands. More phosphonic groups in the polyphosphonate ligand usually lead to amorphous materials [14].

4.2.3.1 Functionalized Metal Triphosphonates

Benzenetriphosphonates and Related Triphosphonates 1,3,5-Benzenetriphosphonic acid (H_6BTP) is the phosphonic acid analogue of trimesic acid, broadly used for the preparation of MOF materials [177–181]. H_6BTP reacts with copper salts for the formation of 3D supramolecular layered architectures such as $\{Cu_6[C_6H_3(PO_3)_3]_2$ $(H_2O)_8\} \cdot 5.5H_2O$ [171].

With Zn^{2+}, the ligand H_6BTP reacts in different ways to form a number of hybrid materials with layered or 3D architectures. Taylor et al. obtained an atypical layered zinc triphosphonate with composition $Zn_3[C_6H_3(PO_3)_3](H_2O)_2 \cdot 2H_2O$ (PCMOF-3) [182]. Its structure is constituted by columns of $[C_6H_3(PO_3)_3]^{6-}$ molecules cross-linked by Zn^{2+} ions to form neutral layers (see Figure 4.16). Within a layer, molecules of $[C_6H_3(PO_3)_3]^{6-}$ form π-stacked dimers perpendicular to the layer where the phosphonate oxygens of each dimer coordinate to six tetrahedral zinc centres. In the interlayer, lattice waters forms a strong H-bond network with phosphonate oxygen atoms and coordinated water

FIGURE 4.16 Structure of PCMOF-3 showing two layers stacked to show the channels occupied by water molecules. Light tetrahedra represent fully phosphonate-ligated Zn1 centres, and dark tetrahedra represent bis(aquo)-ligated Zn2 centres. Adapted from Ref. [182].

molecules. As discussed in the following text, this network of highly ordered water molecules is the basis of high proton conductivity with very low activation energy.

Two Zn-BTP derivatives with 3D frameworks are $Zn_{2.5}(H)_{0.4-0.5}(C_6H_3O_9P_3)(H_2O)_{1.9-2}$ $(NH_4)_{0.5-0.6}$ and $Zn_{2.5}(H)_{0.75}(C_6H_3O_9P_3)(H_2O)_2(CH_3NH_3)_{0.25}$ [183]. These materials show isostructural frameworks built from inorganic–organic layers linked by ZnO_4 tetrahedra. The negative charge of the framework is compensated by NH_4^+ or $CH_3NH_3^+$ cations located together with free water molecules in the internal cavities. In addition, NH_4^+ can be exchanged for lithium ions, and reversible dehydration–rehydration takes place through intermediate structural changes.

Other 3D open frameworks respond to the formula $ZnH_4BTP–M$ (M = K, Rb and Cs) [170, 184]. The framework contains 2D zigzag channels and interlayer cage-like spaces. In some cases, remarkable cation-exchange selectivity for Rb^+ and Cs^+ was observed.

Up to now, only a lanthanum-BTP layered material has been reported. Its crystal structure contains 2D undulating layers in which a pair of parallel aligned organic moieties bridges three different LnO_8 polyhedra. The layers, held together by hydrogen bonds through P–OH groups, can be exfoliated up to the level of nanosheets. Furthermore, its Eu- and Tb-doped derivatives, in both solid and dispersed states, show photoluminescence properties [185].

Further separation of the phosphonic groups is achieved by using the analogue of H_6BTP, 1,3,5-tris(4-phosphonophenyl)benzene $[C_6H_3(C_6H_4PO_3H_2)_3]$ [186]. However, to our knowledge, only three metal derivatives containing this ligand have been reported. $[V_3O_3(OH)\{C_6H_3(C_6H_4PO_3H)_3\}_2]\cdot 7H_2O$ [77] presents a 2D framework that results from the linking of trinuclear $[V_3O_3(OH)(HO_3PR)_6]$ clusters through the backbone of the triphosphonate ligand (Figure 4.17). The layers are ruffled in profile and stack such that the vanadate clusters of adjacent layers nestle above and below the large cavities of a given layer. The water molecules of crystallization reside in the galleries of approximate dimensions $5.0\times 14.0\text{Å}$ in the interlayer region.

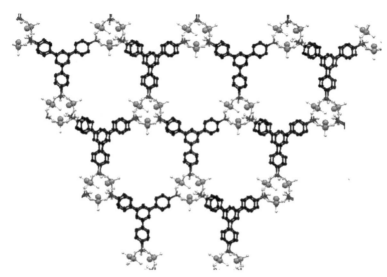

FIGURE 4.17 Bidimensional packing of $[V_3O_3(OH)\{C_6H_3(C_6H_4PO_3H)_3\}_2]\cdot 7H_2O$. C, black spheres; H, not shown; M, big-sized grey spheres; O small white spheres; P, medium-sized dark-grey spheres. Adapted from Ref. [77].

Designing porous frameworks based on other non-porous structures is a new strategy in metal phosphonate chemistry that has been very recently developed by Shimizu and co-workers. They have obtained a highly robust and porous Sn(IV) triphosphonate framework based on the crystal structure of $Sr_2(H_2L)$ [187], where L refers to 1,3,5-tris(4-phosphonophenyl)benzene. The structure of $Sr_2(H_2L)$ is twofold interpenetrated, and each net consists of 1D columns of phosphonate-bridged Sr^{2+} ions that link honeycomb assemblies of ligand molecules. The inter-penetrated nets are held together through van der Waals interactions and π-stacking between the central aromatic rings of the ligand molecules. The Sn(IV) derivative, CALF-28, is a permanently porous solid with BET surface area of $502\,m^2g^{-1}$ and a pore size distribution with maximum at $8.5\,Å$. It is also stable to high levels of moisture [188].

The inclusion of methylene groups into a phosphonate ligand may lead to significant structural variations on the final networks. This effect has been studied using as tripodal ligand benzene-1,3,5-tris(methylenephosphonic acid) (H_6BTMP) and 2,4,6-trimethylbenzene-1,3,5-tris(methylenephosphonic acid) (Me_3-H_6BTMP) with copper and cobalt salts [189]. Their crystal structures are highly correlated with the conformations of the flexible trisphosphonate ligand. The ligands H_6BTMP and Me_3-H_6BTMP are capable, to a certain extent, of adjusting themselves sterically, owing to the flexibility of 'phosphonic-arm' units. Each phosphonic group in H_6BTMP is linked by a C–C bond, which allows rotation about the single bond. This rotation leads to conformational changes, namely, the formation of a *cis,cis,cis*-form or a *cis,trans,trans*-form (Scheme 4.4).

SCHEME 4.4   *cis,cis,cis*- and *cis,trans,trans*-forms for the H_6BTMP ligand. Adapted from Ref. [189].

(a)   (b)

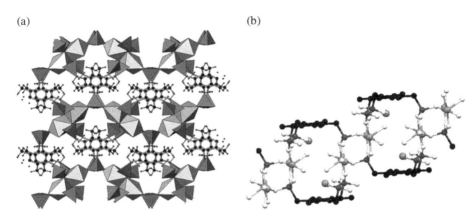

FIGURE 4.18   (a) Crystal structure for $[Co_6(Me_3-BTMP)_2(H_2O)_4]$ and (b) metalloprismatic cage of $[Cu_4(Me_3-H_2btmp)_2(H_2O)_4]$. C, black spheres; CoO_6, grey polyhedra; H, not shown for (b); M, big-sized grey spheres; O small white spheres; P, medium-sized dark-grey spheres; PO_3C, dark-grey tetrahedral. Adapted from Ref. [189].

This flexibility allows the formation of cage-array structures and 3D networks as have been described for $[Cu_3(BTMP)(H_2O)_{3.6}] \cdot H_2O$, $[Co_6(Me_3-BTMP)_2(H_2O)_4]$ and $[Cu_4(Me_3-H_2BTMP)_2(H_2O)_4]$. The crystal structures of the first two solids show a *cis,trans,trans*-conformation of $BTMP^{6-}$ and Me_3-BTMP^{6-}, respectively (Figure 4.18a). However, $[Cu_4(Me_3-H_2BTMP)_2(H_2O)_4]$ shows a 2D framework with an oak barrel shape of $M_4(Me_3-BTMP)_2$-type metalloprismatic cage structure with the Me_3-BTMP linkers in a *cis,cis,cis*-conformation (Figure 4.18b). As a result of these arrangements, the copper derivatives exhibit antiferromagnetic exchange between Cu^{2+} ions, whereas the complex $Co-Me_3-H_4BTMP$ shows a non-zero-spin ground state [189].

The reaction of H_6BMT with lanthanum salts using a microwave-assisted synthesis leads to $[La(H_4BMT)(H_5BMT)(H_2O)_2] \cdot 3H_2O$ and $[La_2(H_3BMT)_2(H_2O)_2] \cdot H_2O$ derivatives. While low temperatures and short reaction times yield the 1D compound $[La(H_4BMT)(H_5BMT)(H_2O)_2] \cdot 3H_2O$ [190], longer times and higher temperatures (>150°C) lead to non-porous 3D $[Ln_2(H_3BMT)_2(H_2O)_2] \cdot H_2O$ (Ln=La, Ce, Pr, Nd, $La_{0.95}Eu_{0.05}$ and $La_{0.95}Tb_{0.05}$) materials [191] with photoluminescent and catalytic properties. In this framework, the ligand acquires a *cis,trans,trans*-configuration, and reversible dehydration–rehydration is not accompanied by transient structural modifications.

Amino-tris-(methylenephosphonates) Amino-tris-(methylenephosphonate) or H_6AMP is one of the most common aminomethylenephosphonates and a very effective scale inhibitor [192]. It has been used with a large variety of metals: mono- [193], di- [194–199] and trivalent (Al [132], Ln [200, 201]). In general, divalent cations react with the AMP ligand displacing two protons to form compounds with the stoichiometry $M^{II}[HN(CH_2PO_3H)_3] \cdot 3H_2O$. All of them, even when combinations of several metals were used, yielded the same structure formed by 1D helical coordination polymers. This fact suggests that it is the hydrogen bonding that controls the formation of the framework [173, 197].

Reactions of H_6AMP with trivalent metal ions result in a non-porous 3D framework, $Al[N(CH_2PO_3H)_3] \cdot H_2O$ [132], or in 2D networks, $[Ln(H_3AMP)] \cdot 1.5H_2O$ (Ln = La, Pr, Nd, Sm and Eu) [201] or $La(H_3AMP)$ [200]. The crystal structure of the latter is formed by LaO_9 polyhedra, where the La^{3+} centre is coordinated to seven phosphonate groups arising from four AMP^{3-} anionic ligands. Interestingly, the thermal transformation of $La(H_3AMP)$ results in new 3D or layered phases with concomitant *in situ* polymerization of the ligand. Moreover, $La(H_3AMP)$ behaves as an effective heterogeneous catalyst, and its Eu- and Tb-doped derivatives show photoluminescent properties [200]. A different connectivity is observed in the structure of $[Pr(H_3AMP)]_3 \cdot 1.5H_2O$, where the layers are composed of 1D chains formed by edge-shared Ln^{3+} polyhedra interconnected via a phosphonate group.

4.2.3.2 Functionalized Metal Tetraphosphonates

The use of tetraphosphonic acids has expanded further the chemistry of metal phosphonates to include novel multifunctional hybrid metal derivatives. It was initially hypothesized that the use of such multidentate building blocks would induce formation of modular open frameworks that could be tailored by the right choice of interlinking organic groups [202]. Depending on the organic moiety, tetraphosphonate ligands can be broadly divided in two categories, rigid and flexible linkers. Rigidity in the organic portion may be a necessary condition in order to induce permanent porosity and robustness for better pore size control [203]. On the other hand, imparting structural flexibility may be a desirable goal if adaptability to distinct guest molecules is required [172, 204–208].

Rigid Tetraphosphonates Examples of rigid ligands are those containing aromatic rings in their backbones and those having the adamantane structural type that has an extended, rigid tetrahedral configuration. Combinations of xylylenediamine-based ligands, such as 1,4- and 1,3-bis(aminomethyl)benzene-N,N'-bis(methylenephosphonic acid), with divalent cations are known [206, 209–213]. In the series of compounds $M(II)$-(x-$C_{12}H_{18}O_{12}N_2P_4$)$\cdot 2H_2O$ ($M(II) = Mg$, Ca, Mn, Co, Ni, Zn, Cd), where x stands for *meta*- or *para*-position of the diaminetetraphosphonate groups, the dimensionality of the solids was found to be dependent on the relative position of the diaminetetraphosphonate moiety and the size of the metal ion. So, the compounds $M(II)$-(p-$C_{12}H_{18}O_{12}N_2P_4$)$\cdot 2H_2O$ ($M(II) = Mg$, Mn, Co, Ni and Zn) present 1D structure, and the tetraphosphonate ligand basically acts as a bridge between two metal octahedra. However, combining ligand p-$C_{12}H_{18}O_{12}N_2P_4$ with large cations, such as Ca^{2+} and Cd^{2+}, led to pillared layered networks. By contrast, the use of m-$C_{12}H_{20}O_{12}N_2P_4$ ligand results in a unique pillared layered network type, independently on the size of the cations. The inorganic layers of this network show the same metal–ligand connectivity as that found for Ca-(p-$C_{12}H_{18}O_{12}N_2P_4$)·$2H_2O$ [211, 212].

Constantino et al. prepared three homologous copper(II) diaminetetraphosphonates containing different tetraphosphonic building blocks, one with a flexible alkyl chain ($-CH_2-N-C_6H_{12}-N-CH_2-$, H_6HDTMP) and two other derived from the *para*- and *meta*-xylylenediamine [209]. These three building blocks contribute to the formation of hybrid inorganic–organic anionic layers. The former structures, $Cu_2(H_2O)_2(L)·Cu(H_2O)_6·2H_2O$ (L = H_4HDTMP or p-$C_{12}H_{18}O_{12}N_2P_4$), are isotypical, and their hybrid layers are made up of copper square pyramids and tetrahedral phosphonic groups. The negative charge of the layers is compensated by unusual hexaaqua copper(II) cations placed into the interlayer region and linked to the layers by means of a H-bond network between water molecules and CPO_3 groups in the sheets. By contrast, in $Cu_2(H_2O)_2(m$-$C_{12}H_{18}O_{12}N_2P_4)·Cu(H_2O)_3·3H_2O$, the interlayer copper ions are three-hydrated and are directly connected to the deprotonated PO groups. The N-chelating mode of the copper atoms belonging to the layer is common to the three compounds, which probably contributes to the structural isotypism of the sheets.

Other rigid tetraphosphonic ligands are 1,2,4,5-tetrakisphosphonomethylbenzene, 1,2,4,5-$(H_2O_3PCH_2)_4C_6H_2$, 1,4-phenylenebis(methylidyne)tetrakis(phosphonic acid) and dendritic-adamantane-tetraphosphonic acid. Cadmium and copper derivatives of 1,2,4,5-tetrakisphosphonomethylbenzene show similar 3D open frameworks formed by dimers [Cd_2O_{10}] or [Cu_2O_8] polyhedra connected to other units through tetraphosphonate ions [$HO_3PCH_2)_4C_6H_2$]$^{4-}$ [214, 215].

The lanthanum derivate [$La(H_5L)(H_2O)_4$] (PCMOF-5, L = 1,2,4,5-tetrakisphosphonomethylbenzene) adopts a modified pillared layered motif containing diprotic phosphonic acid groups within the framework [174]. Its crystal structure is made up of hydrophilic metal phosphonate layers pillared by hydrophobic organic layers. The 1D La-phosphonate chains are connected into a 3D framework through three of the four phosphonate groups on the ligand, leaving the fourth group, a diprotic phosphonic acid, uncoordinated and protruding into the interlayer region (see Figure 4.19).

As a result of the special features shown by the hydrated channels generated from this arrangement, PCMOF-5 exhibits high proton conductivity, as discussed in Section 4.3.3. 1,4-Phenylenebis(methylidyne)tetrakis(phosphonic acid), $(H_2PO_3)_2$ $CH-C_6H_4-CH(PO_3H_2)_2$ or $C_8H_{14}O_{12}P_4$, has been investigated as a building block of hybrid lanthanide derivatives (Ln = La, Nd, Gd, Dy). Fourteen new lanthanide tetraphosphonates were isolated and categorized into three structural and compositional types: (I) Ln($C_8H_{11}O_{12}P_4$)·$4H_2O$ (Ln = La, Nd, Gd and Dy), (II) $Ln_2(C_8H_8O_{12}P_4)·8H_2O$ (Ln = La, Nd, Gd and Dy) and (III) NaLn($C_8H_{10}O_{12}P_4$)·$4H_2O$ (Ln = La, Nd, Gd and Dy) [175]. Structure types I and II correspond to 2D networks, while compounds with structure type III present 3D open frameworks with a high ion-exchange selectivity for monovalent metal cations (Li$^+$, K$^+$, Rb$^+$ and Cs$^+$). The flexible anionic framework of these materials is constructed from Ln^{3+} ions bridged by the phosphonate groups of the [$(PO_3H)_2CH-C_6H_4-CH(PO_3H)_2$]$^{4-}$ ligands to form $-Ln-O-P-O-$ Ln$-$ chains, which are linked via the organic spacer to form a 3D framework. As can be seen in Figure 4.20, this connectivity creates rhombic channels where

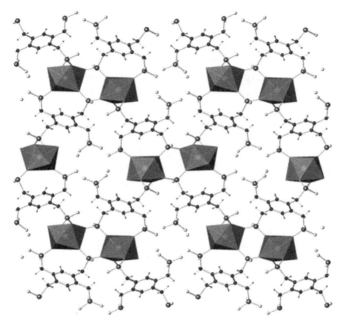

FIGURE 4.19 Crystal structure of [La(H$_5$L)(H$_2$O)$_4$] (PCMOF-5, L = 1,2,4,5-tetrakisphos-phonomethylbenzene) viewed along a-axis and showing the diprotic phosphonic acid groups and the uncoordinated water molecules inside the pores. Adapted from Ref. [174].

(a)   (b)

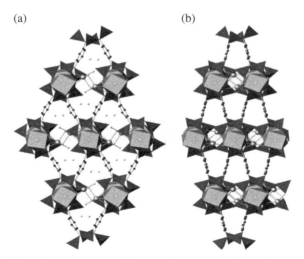

FIGURE 4.20 Views along the c-axis of the crystal structures of (a) hydrated and (b) anhydrous NaLa(C$_8$H$_{10}$O$_{12}$P$_4$)·4H$_2$O. Adapted from Ref. [216].

the $-Ln-O-P-O-Ln-$ chains are located at the corner of each rhombus and the free space is filled with Na^+ ions and water molecules [175]. A comparison of the monovalent metal ion-exchanged compounds, $M(I)La(C_8H_{10}O_{12}P_4)\cdot 4H_2O$, revealed that both the ionic radius and the enthalpy of hydration of the guest cation affect the equilibrium between the expanded (hydrated) and the contracted (dehydrated) forms and that the framework adapts specifically to the size of the guest cation [175, 216]. Additional sorption property studies of $M(I)La(C_8H_{10}O_{12}P_4)\cdot 4H_2O$ with N_2, Kr and H_2O showed a preference of the network for water molecules [217].

The extended tetrahedral phosphonate ligand, 1,3,5,7-tetrakis(4-phosphonatophenyl)adamantane, H_8TPPhA, was designed to direct the formation of open frameworks upon metal complexation [218]. $Cu_3-(H_3TPPhA\cdot(OH)\cdot(H_2O)_3)\cdot H_2O\cdot MeOH$ has a structure based on the diamondoid net formed by H_6TPPhA and trimetallic copper cluster as a second type of tetrahedral node, giving open channels. Upon desolvation, the structure contracts to a less ordered but structurally related phase with a CO_2 surface area of $198\,m^2\cdot g^{-1}$ and an average pore width of $5.0\,Å$. This solid also showed a low H_2 storage capacity ($0.24\,wt\%$) [176].

The crystalline derivate La_2-HTPPhA $\cdot 6H_2O$ is built upon binuclear $\{La_2(H_2O)_4(\mu\text{-}PO(OH)O_2)_2(-PO(OH)O_2)_3\}$ units, where the lanthanum ions are bridged by two $-PO(OH)O_2$ moieties and supported by two additional hydrogen bonds between pairs of axially coordinated water molecules [219]. Derivatives of Ti and V were mesoporous solids (surfaces areas of 557 and $118\,m^2\cdot g^{-1}$ from N_2 isotherms, respectively) with low crystallinity [202, 220].

Flexible Tetraphosphonates   Flexible tetraphosphonic acids with alkyl chains separating two aminobismethylenephosphonic groups $\{(H_2O_3PCH_2)_2N-R-N(CH_2PO_3H_2)_2$ $[R=(CH_2)_2, H_8EDTMP, (CH_2)_4, H_8TDTMP, (CH_2)_6, H_8HDTMP, etc.]\}$ are versatile building blocks that generate a variety of structures with different topologies, including closely packed 3D [38, 211, 215, 221] and pillared layered networks [210] and some hybrids with layered structures having the phosphonic groups within the layers [202]. When displaying microporosity, the dimensions of the tunnels may be tailored, to a certain extent, by the right choice of the interlinking organic group [222]. H_8XDTMP can potentially dissociate its eight protons (two H^+ from four $-PO_3H_2$) depending on the solution pH. The basic N groups remain protonated up to pH of approximately 10. Thus, these ligands are better described as 'zwitterionic' ligands with the negative charges located on monodeprotonated phosphonate groups and the positive charges on the N atoms [223].

Medical applications and photocatalytic activity [14, 224, 225] have been found for some hybrid materials containing the ligand H_8EDTMP [210, 224, 225], but they are amorphous or have not solved the crystal structure. By contrast, in 2007, Wu et al. [222] published two crystalline derivatives, $Pb_7[(H_1EDTMP)_2(H_2O)]\cdot 7H_2O$ and $[Zn_2(H_4EDTP)]\cdot 2H_2O$, with highly selective ferric ion sorption and exchange. In $Pb_7[(H_1EDTMP)_2(H_2O)]\cdot 7H_2O$, the interconnection of Pb^{II} ions with the phosphonate groups and nitrogen atoms of H_1EDTMP^{7-} results in a 3D microporous network with tunnels along the a- and b-axes. The tunnels along the a-axis are made of helical chains, which are created by twisted H_1EDTMP^{7-} ligands connected by lead atoms.

The cavity of the tunnels is occupied by the lattice water molecules. The tunnels along the b-axis are created by the packing of 24-membered rings, and each ring is formed by 6 lead atoms and 6 phosphonate groups. In $[Zn_2(H_4EDTP)] \cdot 2H_2O$, the H_4EDTP^{4-} anion bridges with eight Zn(II) ions, while the nitrogen atoms do not coordinate with any metal ion. Each Zn(II) ion is coordinated by four phosphonate oxygen atoms in a distorted tetrahedral geometry. This resulted in a zinc phosphonate layer, which shows a square-wave-like architecture along the a-axis, with cavities along the c-axis being occupied by the lattice water molecules interlinked through hydrogen bonds. The layers are connected by hydrogen bonds between uncoordinated phosphonate oxygen atoms.

The compounds $M(II)-H_6EDTMP \cdot 3H_2O$ (M = Ca, Sr) are coordination polymers with H_6EDTMP^{2-} acting as both chelating and bridging ligands. The main structural feature of these solids is that the $-PO_3H^-$ group coordinates to M^{2+} through only one of its three oxygens. At each side of the tetraphosphonate molecule, the $-PO_3H^-$ groups form 11-membered chelate rings with the M^{2+} that renders the whole tetraphosphonate a bridge between two M^{2+} centres. Thus, the 1D polymer can be envisioned as a ribbon in a 'wave-like' motion composed of $M-H_6EDTMP-M$ dimers [223].

The reaction of N,N,N',N-butylenediamine-tetrakis(methylenephosphonic acid), H_8TDTMP, with a number of divalent transition metals (Mg, Ca, Mn, Fe, Co, Ni, Zn and Cd) was studied by Stock et al. [38, 221] employing a high-throughput system. The solids, with general formula $M-H_6TDTMP$, are isostructural and present a 3D structure, which is composed of M^{2+} ions and zwitterions $[(HO_3PCH_2)_2(H)$ $N(CH_2)_4N(H)(CH_2PO_3H)_2]^{2-}$, where each phosphonate group and each N atom are protonated by one H atom. The M^{2+} ions are surrounded by six phosphonate groups forming layers with eight-membered rings composed of alternating MO_6 and PO_3C polyhedra. These layers are connected in a diagonal fashion by the organic groups, $-(CH_2)_2N(CH_2)_4N(CH_2)_2-$ blocking the eight-membered ring apertures. Vivani et al. [202] prepared a Zr hybrid, $Zr-H_6TDTMP \cdot 4H_2O$, with layered structure that has the phosphonic groups within the layers (see Figure 4.21a). Thus, each layer is formed by the connection, through the butylenediamino groups, of inorganic 1D polymeric units running along the c-axis. These units are constituted of a row of ZrO_6 octahedra connected to the phosphonate tetrahedra. Butyl chains covalently link these units from opposite parts, along the a-axis, generating intralayer channels running along the c-axis. These channels are partially occupied by water molecules. By using cyclohexyl-N,N,N',N-diaminotetraphosphonate as building blocks, $Zr(PO_3CH_2)_2N-C_6H_{10}-$ $N(CH_2PO_3)_2Na_2H_2 \cdot 5H_2O$ (ZrChDTMP) was obtained [226]. This solid exhibits a 3D open-framework structure made of inorganic polymeric units, with the same connectivity between ZrO_6 octahedra and PO_3C tetrahedra with that of $Zr-H_6TDTMP \cdot 4H_2O$, bridged by cyclohexyl groups in a 'brickwall-like' building texture, as shown in Figure 4.21b. As for $Zr-H_6TDTMP \cdot 4H_2O$, there are 1D cavities decorated with polar and acids P=O and P–OH groups and occupied by lattice water molecules and Na cations. Recently, its ion-exchange and conduction properties have been explored [202, 226]. This acid has also been reacted with lanthanide ions, such as gadolinium [227] and cerium [228], to obtain materials with luminescent properties.

(a)   (b)

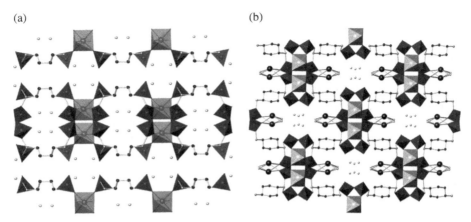

FIGURE 4.21 Polyhedra views of the crystal structures of (a) $Zr(PO_3CH_2)_2N-C_4H_8-N$ $(CH_2PO_3)_2 \cdot 4H_2O$ and (b) $Zr(PO_3CH_2)_2N-C_6H_{10}-N(CH_2PO_3)_2Na_2H_2 \cdot 5H_2O$ (ZrChDTMP) viewed along the c-axis. C: small black spheres; O, small white spheres; PO_3C, grey tetrahedral; ZrO_6, dark-grey octahedra. Adapted from Refs. [202] and [226].

With the linker HDTMP, a variety of structures from 1D to 3D architectures have been reported [229–234]. CaH_6HDTMP [172] presents a layered structure composed of a six-coordinated Ca^{2+} centre exclusively surrounded by phosphonate oxygen atoms from the H_6HDTMP^{2-} ligand embedded within the layers to give infinite chains along the a-axis. The lattice water molecules in both compounds are situated between the hybrid layers. An important feature is that the layers are only held together through hydrogen bonds. Moreover, this material is the first structurally characterized example of 2D materials that can reversibly store and remove host molecules while the network adjusts to these external stimuli by subtle changes within the framework (see Figure 4.22).

$LaH_5HDTMP \cdot 7H_2O$ [231] is characterized by a 3D pillared open framework, composed of corrugated layers in the plane bc and interconnected by the organic linker (Figure 4.23). Each layer is built by chains of LaO_6 polyhedra, which are built exclusively by six phosphonate oxygen atoms from six different H_5HDTMP^{3-} anions, interconnected by bridging phosphonate tetrahedra, resulting in 8- and 16-membered rings. The 1D channels formed are filled by seven lattice water molecules. The anhydrous material shows high affinity for water, but the rehydrate stoichiometries depend upon the experimental conditions.

$ZnH_6HDTMP \cdot H_2O$ [233] exhibits also chains of MO_6 polyhedra, but with different arrangement. An interesting feature is that one of the oxygen atoms of the coordination sphere of Zn^{2+} is a protonated phosphonate oxygen. Two Zn^{2+} centres and the amino-bis(methylenephosphonate) portions of H_6HDTMP form an 18-membered ring, and there is a concentric 8-membered ring formed by the same Zn^{2+} centres, and the protonated methylenephosphonate arm involved in the long $Zn \cdots HO$ interaction. Finally, a magnesium derivative with the ligand octamethylenediamine-N,N,N',N'-tetrakis(methylenephosphonic acid) has been

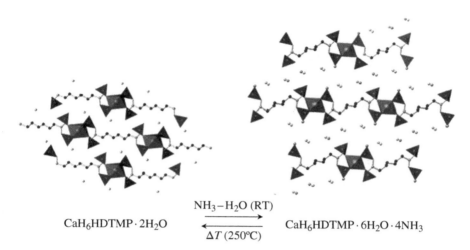

$$\text{CaH}_6\text{HDTMP}\cdot 2\text{H}_2\text{O} \quad \underset{\Delta T\,(250^\circ\text{C})}{\overset{\text{NH}_3-\text{H}_2\text{O (RT)}}{\rightleftharpoons}} \quad \text{CaH}_6\text{HDTMP}\cdot 6\text{H}_2\text{O}\cdot 4\text{NH}_3$$

FIGURE 4.22 Crystal structures of two members of the family CaH_6HDTMP showing the breathing effects during the adsorption/desorption of guest species. C, black spheres; CaO_6, dark-grey polyhedra; N, medium-sized grey spheres; O/NH_3, white spheres; PO_3C, grey polyhedra. Adapted from Ref. [172].

(a)   (b)

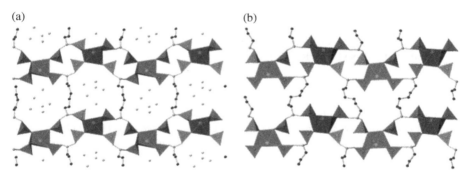

FIGURE 4.23 Crystal structures of (a) $\text{LaH}_5\text{HDTMP}\cdot 7\text{H}_2\text{O}$ and (b) LaH_5HDTMP showing the structural changes during the dehydration processes. C, black spheres; LaO_6, dark-grey polyhedra; N, medium-sized grey spheres; O, white spheres; PO_3C, grey polyhedra. Adapted from Ref. [231].

prepared, $\text{MgH}_6\text{ODTMP}\cdot 2\text{H}_2\text{O(DMF)}_{0.5}$ [235]. As shown in Figure 4.24, its crystal structure is characterized by a novel pillared open framework that contains cross-linked 1D channels filled with water and DMF. The inorganic layer is built in the bc plane by MgO_6 octahedra arranged in square polygonal fashion. The Mg^{2+} ions located in the corners of each square polygon are interconnected by two amino-bis(methylenephosphonate) moieties and two phosphonate groups corresponding to four different ligands, forming one central 16-membered ring and two 12-membered rings. Neighbouring layers are connected along the

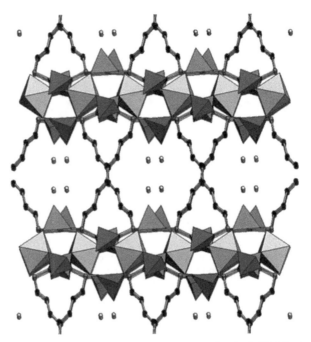

FIGURE 4.24 Pillared framework for Mg(H$_6$ODTMP)·2H$_2$O·0.5DMF view along the c-axis. C, black spheres; MgO$_6$, grey octahedra; N, small grey sphere; O, white spheres; PO$_3$C grey tetrahedral. Adapted from Ref. [235].

a-axis by the organic groups –(CH$_2$)$_2$NH–(CH$_2$)$_8$–NH–(CH$_2$)$_2$– in a cross-diagonal fashion. Removal of H$_2$O and DMF molecules from the channels and further rehydration lead to a new crystalline phase with higher water content, MgH$_6$ODTMP·6H$_2$O. These processes take place through crystalline–quasi-amorphous–crystalline transformation, with the integrity of the framework being maintained.

4.3 SOME RELEVANT APPLICATIONS OF MULTIFUNCTIONAL METAL PHOSPHONATES

The use of functionalized phosphonic acid has notoriously boosted the metal phosphonate chemistry, giving rise to novel and versatile structures for multiple purposes. A number of possible applications, such as gas adsorption, catalysis, ion exchange, proton conductivity, magnetic and photoluminescence properties, corrosion and so on, have recently been reported in the book *Metal Phosphonate Chemistry: From Synthesis to Applications* [15] and merit attention in other chapters of this book. So, this section is restricted to significant advances in the fields of gas adsorption, catalysis and proton conductivity in order to highlight the connections between structure and properties.

4.3.1 Gas Adsorption

The number of crystalline functionalized metal phosphonates with gas adsorption properties is much lesser than in the case of carboxylate and N-donor MOF materials. Until now, only a few phosphonate MOF materials showing high surface areas have been reported. Nevertheless, numerous UMOFs are known to possess high adsorption capabilities [13, 14, 109, 236, 237]. Crystalline phosphonate MOFs based on permanently porous N,N'-piperazinebis(methylenephosphonic acid) (STA series) and MIL-91materials have been the reference candidates to investigate the adsorption of different gases with environmental importance.

The most porous and stable STA-12 structure, the Ni form, has been investigated by measuring the CO_2 adsorption selectivity over CH_4 and CO. From 1 : 1 CO_2/CH_4 mixtures, selectivities to CO_2 of 6.5 (at 1 bar) and 24.5 (at 5 bar) were determined. These values are higher than those over MOFs without open metal sites, such as ZIF-8, but are lower than the selectivity values of CPO-27 materials [165].

N_2 and CO_2 isotherms for the STA-13 structure were of type I, indicating pore filling of $3.5 \, mmol \cdot g^{-1}$ for N_2 (corresponding to a pore volume of $0.12 \, cm^3 \cdot g^{-1}$) and $3.9 \, mmol \cdot g^{-1}$ for CO_2. Although the porosity of STA-13 is less than that observed for some other bisphosphonates such as MIL-91and STA-12(Ni), the STA-13 structure type is the first example of a lanthanide bisphosphonate with permanent porosity [169].

The 2D nickel phosphonate $\{[Ni(Hptz)_2] \cdot 7H_2O\}_n$, ($H_2ptz=4$-(1,2,4-triazol-4-yl) phenylphosphonic acid) [238], obtained by crystal transformation of the compound $[Ni_3(Hptz)_6(H_2O)_6] \cdot 9H_2O$ shows a BET surface area of $434 \, m^2 \cdot g^{-1}$ and a CO_2 uptake of $3.6 \, mmol \cdot g^{-1}$ (15.8 wt% at 298 K and 1 bar). The high adsorption enthalpy was considered to be due to strong interactions established between pending phosphonate groups ($-PO_3H$) with CO_2. A high selective separation of CO_2 over N_2 (114 : 1) was also observed.

Other remarkable examples of metal phosphonates showing significant CO_2 uptakes are those based on 1,4-benzenediphosphonate-bis(monoalkyl ester), for instance, Zn(1,4-benzenediphosphonate-bis(monoethyl ester)) [88], with layered structure and BET of $264 \, m^2 \cdot g^{-1}$ and a total uptake of $2.8 \, mmol \cdot g^{-1}$ at 273 K and 1 bar. Cu(1,4-benzenediphosphonate-bis(monomethyl ester) also exhibits CO_2 adsorption, $1.38 \, mmol \cdot g^{-1}$ at 273 K and 1 bar, although with a high isosteric heat of adsorption of $45 \, kJ \cdot mol^{-1}$ [87]. Neither of them shows significant adsorption of other adsorbates such as N_2, H_2 or CH_4.

As an example of flexible structures with selective adsorption of CO_2, we mention the magnesium tetraphosphonate $MgH_6ODTMP \cdot 2H_2O \cdot (DMF)_{0.5}$ [235]. This solid, upon dehydration, contains ultramicropores and shows a specific surface for CO_2 of $180 \, m^2 \cdot g^{-1}$ (at 273 K, 1 bar), with a maximum CO_2 uptake of $1.2 \, mmol \cdot g^{-1}$ at 900 kPa. Furthermore, it exhibits high selectivity for CO_2 in CO_2/CH_4 mixtures and low heat of adsorption of CO_2.

4.3.2 Catalysis and Photocatalysis

Phosphonate MOFs represent a high-metal-containing alternative to the frequently used zeolites in heterogeneous catalysis. To ascertain its feasibility as catalyst, a Co-based STA-12 material [STA-12(Co)] was investigated in the aerobic epoxidation of olefins [239]. It shows a high activity as compared with the Co-doped zeolite catalysts that are typically used in this reaction. The reaction was found to proceed mainly heterogeneously, and the catalyst was reusable with only a small loss of activity.

Recently, several articles have appeared reporting the use of lanthanide-functionalized organophosphonates as heterogeneous catalysts, many of them also exhibiting interesting photoluminescent features. Thus, $[La(H_4BMT)(H_5BMT)(H_2O)_2] \cdot 3H_2O$ shows a high activity as a heterogeneous catalyst in the methanolysis of styrene oxide with nearly complete conversions after 30 min of reaction without the need for catalyst regeneration [190]. Low yields were obtained, however, with other lanthanum phosphonates. In the cyclodehydration of xylose, a conversion closed to 85% was found for the catalyst $[Y(H_2CMP)(H_2O)]$ [142], a compound closely related to the family of layered rare earth carboxymethyliminodimethylphosphonates.

A 3D porous rutile-type zinc(II)-phosphonocarboxylate, $\{[Zn_3(pbdc)_2] \cdot 2H_3O\}_n$ (H_4pbdc = 5-phosphonobenzene-1,3-dicarboxylic acid), showed excellent size-selective properties for Friedel–Crafts benzylation. Its catalytic activity should be related to the presence of Brönsted acid centres in specific sites provided by the hybrid structure [240].

A semi-crystalline porous nickel tetraphosphonate based on the hexamethylenediamine-N,N,N',N'-tetrakis(methylphosphonic acid) ligand has been recently described as having interesting adsorption and catalytic properties [241]. It exhibits excellent catalytic activity and selectivity for the reduction of nitrobenzenes to the respective anilines in the presence of $NaBH_4$ as the reducing agent.

With regard to photocatalysis, there are only few works reporting the use of functionalized metal phosphonates. Titanium phosphonates, obtained by surfactant-assisted synthesis [242–244], were shown to be useful catalysts for the photodegradation of rhodamine B under solar light-analogous radiation. Crystalline lanthanide-containing organodiphosphonate-functionalized polyoxomolybdate cages have been found as effective catalysts in decolouration of rhodamine B and other dyes under UV irradiation, although mineralization data were not given [245].

To our best of knowledge, only a layered crystalline iron(II) hydroxyacetophosphonate, $Fe(HO_3PCHOHCOO) \cdot 2(H_2O)$ (FeHPAA), has been used as a Fenton photocatalyst for phenol oxidation under UVA radiation [246]. The photocatalyst showed a high stability under irradiation with UVA light, and leaching of Fe was sustained throughout the reaction, favouring thus a mixed heterogeneous–homogeneous catalytic process. About 90% of phenol mineralization was achieved at 80 min, under optimized conditions. This activity was quite higher than that displayed by the standard photocatalyst TiO_2–P25 (from Degussa) (see Figure 4.25). The bulk structure of FeHPAA remains unchanged after the phenol mineralization, but important chemical modifications occur on the surface of the catalyst upon pre-activation under UVA and after phenol photodegradation, as revealed by X-ray photoelectron spectroscopy. These modifications favour the availability of superficial Fe species and enhance phenol mineralization.

4.3.3   Proton Conductivity

Proton conductors are used in a variety of electrochemical devices [247]. Regarding proton exchange membrane fuel cells (PEMFC), proton conductor performance

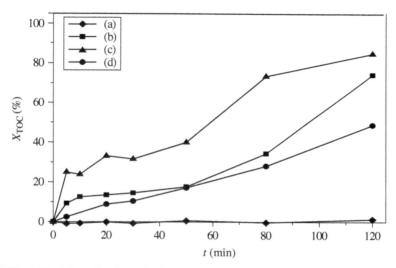

FIGURE 4.25 Mineralization of phenol under UVA irradiation for different systems: (a) without catalyst, (b) with as-prepared FeHPAA catalyst, (c) with as-prepared FeHPAA catalyst previously activated by UVA irradiation for 30 min and (d) with standard TiO_2 catalyst (Degussa-P25) under exactly the same conditions. Adapted from Ref. [246].

remains a critical issue [248]. Apart from perfluorinated polymers, such as Nafion, which work efficiently in the presence of water at temperatures lower than 80 °C, development of new proton conductors having high proton conductivity (over 0.1 Scm^{-1}) and operating in a wider temperature range (25–300 °C) is required to increase the efficiency of fuel cells.

Recently, MOF and more generally coordination polymers are receiving great attention as possible candidates to meet the necessary performance requirements and, by the way, taking advantage of their crystallinity, to give an insight into molecular-level transport mechanisms [249–253]. Proton-conducting MOFs are classified into two categories: water-mediated and anhydrous or non-water-mediated proton conductors. In the latter case, H^+ conductivity can be achieved by accumulating protonic organic molecules in the pores [251, 252]. For water-mediated proton conductors, the *vehicle-* and *Grotthuss-type mechanisms* have been described. The first type of mechanism, implying diffusion of hydrated protons or proton-containing groups (e.g. H_3O^+, NH_4^+), is recognized by high activation energies (0.5–0.9 eV). The Grotthuss-type mechanism, involving an ongoing exchange of covalent and hydrogen bonds between O and H atoms, is characterized by activation energies of 0.1–0.4 eV [250].

Metal phosphonates may offer acidic sites for proton exchange, some degree of structural adaptability, high crystallinity, pores filled with guest molecules (H_2O, heterocyclics, etc.) that act as proton carriers and, finally, some robust metal phosphonates that are susceptible of post-synthesis modifications, such as sulphonation, to increase the proton conductivity [253].

Table 4.1 displays the conductivity properties of selected amorphous and crystalline functionalized metal phosphonates. Conductivity values for metal phosphonates/polymer

TABLE 4.1 Proton Conductivities For Functionalized Metal Phosphonates

Proton-conducting phosphonate systems	Highest H+ σ (Scm^{-1})	Ea (eV)	T (°C)	Relative humidity (%)	Reference
BaH$_2$[C$_6$H$_3$PO$_3$(SO$_3$)$_2$]^a	1.2×10^{-1}		100	80	254
	3.0×10^{-3}		20	75	
Zr(HPO$_4$)$_{0.7}$ (HO$_3$SC$_6$H$_4$PO$_3$)$_{1.3}$ 2H$_2$O^a	7.0×10^{-2}	0.15	150	75	255
(ZrPL)	6.3×10^{-2}	—	100	90	
PCMOF-2.5	2.1×10^{-2}	0.21	85	90	256
Zr(O$_3$PC$_6$H$_4$SO$_3$H)$_{0.73}$ (O$_3$PCH$_2$OH)$_{1.27}$ 7.5H$_2$O^a	1.6×10^{-2}	0.22	RT	90	257
Zr(O$_3$PC$_6$H$_4$SO$_3$H)$_2$·3.6H$_2$O (EWS-3-89 (Zr))^a	1.0×10^{-2}	0.28	5	85	258
Zr meta-sulphophenylphosphonate^a	1.0×10^{-2}	0.21	20	90	259
	5.0×10^{-2}	—	100	95	
Ce(O$_3$PC$_6$H$_4$SO$_3$H)$_2$ 3.86H$_2$O^a	1.0×10^{-2}	0.21	100	100	260
LaH$_5$DTMP·7H$_2$O	8×10^{-3}	0.25	24	98	231
[UO$_2$(HO$_3$PC$_6$H$_5$)$_2$(H$_2$O)]$_2$ 8H$_2$O	3.0×10^{-3}	0.36	RT	80	261
PCMOF-5	4.0×10^{-3}	0.16	60	98	174
Ti[HO$_2$CCH$_2$N(CH$_2$PO$_3$)$_2$]$_{0.28}$[HO$_2$CCH$_2$N–(CH$_2$PO$_3$H$_2$)CH$_2$PO$_3$]$_{1.43}$ 0.6H$_2$O^a	3.7×10^{-4}	—	RT	100	262
MgH$_6$ODTMP·6H$_2$O	1.6×10^{-3}	0.31	19	98	235
Mn-(m-C$_{12}$H$_{18}$O$_{12}$N$_2$P$_4$)·2H$_2$O	1.5×10^{-3}	—	140	100	212
GdHPA-II	3.2×10^{-4}	0.23	25	98	52
ZrH$_4$(O$_3$PCH$_2$)$_2$N–C$_6$H$_{10}$–N (O$_3$CH$_2$P)$_2$·5.5H$_2$O (ZrChDTMP_lp@H)	1×10^{-4}	0.09	80	95	226
Mg-(p-C$_{12}$H$_{18}$O$_{12}$N$_2$P$_4$)·2H$_2$O	9.37×10^{-5}	0.55	24	98	213
Zn-(m-C$_{12}$H$_{18}$O$_{12}$N$_2$P$_4$)·H$_2$O	9.08×10^{-5}	0.25	24	98	213
Zn$_3$(1,3,5-benzenetriphosphonate) (H$_2$O)$_2$ 2H$_2$O (PCMOF-3)	5.0×10^{-5}	0.17	25	98	182
Zn-(p-C$_{12}$H$_{18}$O$_{12}$N$_2$P$_4$)·2H$_2$O	4.3×10^{-5}	—	140	100	212
Zr(O$_3$PC$_6$H$_4$SO$_3$H)$_{0.85}$ (O$_3$PC$_2$H$_5$)$_{1.15}$^{a}	1.2×10^{-5}	0.64	180	—	263
ZrH$_2$(O$_3$PCH$_2$)$_2$N–C$_6$H$_{10}$–N (O$_3$CH$_2$P)$_2$ 4.5H$_2$O^a (ZrChDTMP_np@H)	6.6×10^{-6}	—	80	95	226
LaHPA-I	5.6×10^{-6}	0.20	21	98	52
Zr(O$_3$P–(CH$_2$)$_2$–NH$_2$)$_2$ (ZrAP)^a	2.0×10^{-6}	0.86	200	—	264
Zr(O$_3$P–(CH$_2$)$_n$–COOH)$_2$, $n = 5$^{a}	1.4×10^{-7}	1.37	200	—	265
Zr(O$_3$P–(CH$_2$)$_n$–COOH)$_2$, $n = 3$^{a}	1.6×10^{-8}	1.08	200	—	265
Zr(O$_3$P–(CH$_2$)$_n$–COOH)$_2$, $n = 1$^{a}	2.0×10^{-9}	0.79	200	—	265
Zr(O$_3$PCH$_2$OH)$_2$^{a}	1.0×10^{-8}	—	RT	90	256
Ba$_2$[C$_6$H$_3$PO$_3$(SO$_3$)$_2$]^a	1.0×10^{-8}	—	100	70	254

^{a}Amorphous/semi-crystalline materials.

composites or pure phosphonic acids have been omitted and will not be discussed here. Table 4.1 shows that a large number of metal phosphonates showing significant proton conductivity are multifunctionalized structures. Materials exhibiting the highest proton conductivity are those containing sulphonic groups, incorporated by post-synthesis sulphonation, or forming part of the phosphonate moiety. The absence of crystallinity makes it difficult to establish a clear relationship between their crystal structures and conductivity mechanisms.

To the best of our knowledge, proton conductivity data on crystalline function-alized metal phosphonates are still scarce in the literature. Three of the compounds with high proton conductivity, $LaH_5DTMP \cdot 7H_2O$ [231], $MgH_6ODTMP \cdot 6H_2O$ [235] and ZrChDTMP [226] phases, contain the N,N,N',N'-diaminotetramethyle-nephosphonate moiety, which is characterized by imparting different degrees of flexibility in the structures. Structurally speaking (see Section 4.2.3.2.2), all of them exhibit pillared layered or 3D open frameworks with 1D channel decorated with polar $P{=}O$ and/or P–OH groups. These cavities are regularly occupied with lattice water molecules that establish H-bond interactions with each other and with the non-coordinated phosphonate groups, creating an extended H-bond network through the channels. This network favours the exchange of the protons, which explains the relative low activation energies measured, typical of a Grotthuss mechanism. In contrast to the behaviour shown by compounds ZrChDTMP, $LaH_5DTMP \cdot 7H_2O$ and $MgH_6ODTMP \cdot 6H_2O$ experience a diminution in their proton conductivities upon dehydration. These differences in behaviour may be explained in terms of framework flexibility. A high flexibility, as that in La(III) and Mg(II) derivatives, implies that water removal can be accompanied by concomitant structural changes. More rigidity, however, as that in ZrChDTMP derivatives, leaves the crystal structure almost intact upon dehydration without loss in proton conductivity or even enhancing it with increasing temperature up to a limit, at which H-bond networks are broken. Weaker hydrogen-bond interactions in ZrChDTMP_np@H with respect to ZrChDTMP_lp@H would explain lower values in proton conductivity for the former one.

PCMOF-5 [174] constitutes an example of the importance of having phosphonic groups for the proton-conducting mechanism. This lanthanum tetraphosphonate rep-resents the first coordination polymer with a 3D structure containing an uncoordi-nated diprotic phosphonic acid group that points towards the hydrated channels (see Section 4.2.3.2.1). The presence of these acidic groups and the hydrated nature of the channel (Figure 4.26) provide a chance of generating well-defined hydrogen-bond pathways, which, in practice, justifies the low activation energy measured (0.16 eV). High proton conductivity is also observed for the layer PCMOF-3 [182], which dis-plays a highly ordered H-bond network in the interlayer region, through interactions of water molecules with free P–OH groups, to form a highly efficient Grotthuss proton transfer pathway.

Values of proton conductivity similar to those of PCMOF-5 were reported for a gadolinium multifunctional phosphonates, GdHPA-II [51], containing 2-hydroxy-phosphonoacetic acid as ligand. However, a polymorphic phase of GdHPA-II,

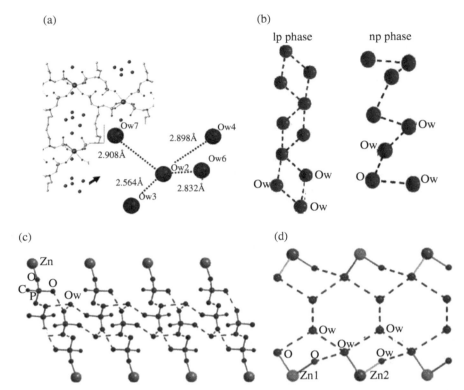

FIGURE 4.26 Hydrogen-bond networks inside the channels of (a) LaH$_3$DTMP, (b) ZrChDTMP (both phases), (c) PCMOF-5 and (d) PCMOF-3. Adapted from Refs. [187], [223], [236] and [241].

LaHPA-I [51], shows conductivity values approximately 2 orders of magnitude lower than GdHPA-II, despite both compounds display microporous frameworks, with 1D channels decorated with P–OH groups and filled with water molecules (see Section 4.2.1.2). It is thought that the different arrangement of the water molecules inside the channels, both metal-bonding and non-coordinated water, likely leads to markedly different pathways for the proton transferences, as can be suggested by their proton conductivities.

Functionalized metal phosphonates of low dimensionality, such as M–(m-C$_{12}$H$_{18}$O$_{12}$N$_2$P$_4$)·H$_2$O (M = Mn and Zn) and M–(p-C$_{12}$H$_{18}$O$_{12}$N$_2$P$_4$)·2H$_2$O (M = Mg and Zn) [219, 220], exhibit also significant proton conductivities Their crystal structures also contain H-bounded water molecules that are filling the space between the hydrophilic portions of the layers, for M–(m-C$_{12}$H$_{18}$O$_{12}$N$_2$P$_4$)·H$_2$O, or chains, for M–(p-C$_{12}$H$_{18}$O$_{12}$N$_2$P$_4$)·H$_2$O. The layered Mn–(m-C$_{12}$H$_{18}$O$_{12}$N$_2$P$_4$)·H$_2$O shows a high proton conductivity at 140 °C and 100% RH, but it decays abruptly by decreasing 10% the RH. The monodimensional solids M–(m-C$_{12}$H$_{18}$O$_{12}$N$_2$P$_4$)·H$_2$O show much lower values of conductivities. Furthermore, Zn–(p-C$_{12}$H$_{18}$O$_{12}$N$_2$P$_4$)·H$_2$O shows high activation barrier (0.55 eV), which suggests that a combined hopping/diffusion mechanism prevails over the Grotthuss mechanism for these compounds.

From this survey, we conclude that phosphonate-based proton conductors exhibiting good performances could be prepared without strong requirements about dimensionality of the solid or post-synthesis treatments. One important condition is possession of an extended H-bond network, the existence of which is, by nature, very common in these materials.

4.4 CONCLUDING REMARKS

Recent progress in the preparation and properties of multifunctional metal phosphonates, including some important applications, has been surveyed in this chapter. The significant efforts devoted to developing new strategies of synthesis and solid functionalization have resulted in a huge range of metal phosphonate materials, potentially useful for diverse applications. The intrinsic difficulty of designing phosphonate-based solids, as opposed to carboxylate-based MOFs, has been recognized for a long time. Yet, recent developments in multifunctional ligands design together with the implementation of new systematic methods of synthesis are steadily leading to a better control in basic structural features of metal phosphonates, such as dimensionality and network characteristics, such as structural rigidity/flexibility as well as the presence of specific functionalities. So, for instance, phosphonate MOFs exhibiting gas storage capabilities and catalytic properties have recently been prepared by design. Another area of research of growing interest is the potential use of these materials as proton conductors, based on the fact that metal phosphonate structures present commonly extended H-bond networks.

ACKNOWLEDGEMENTS

We thank the Junta de Andalucía and MINECO (Spain) for the financial support through the research projects FQM-1656, MAT2010-15175 and MAT2013-41836-R.

REFERENCES

1. Burwell, D., Thompson, M.E. (**1992**) ACS Symp. Ser. 499, 166–177.
2. Alberti, G. (**1996**) in Comprehensive Supramolecular Chemistry, Vol. 7, Eds. G. Alberti, T. Bein, Pergamon, Oxford, 151–187.
3. Alberti, G., Costantino, U., Dionigi, C., Murcis-Mascaro`s, S., Vivani, R. (**1995**) Supramol. Chem. 6, 29–40.
4. Clearfield, A., Costantino, U. (**1996**) in Comprehensive Supramolecular Chemistry, Vol. 7, Eds. G. Alberti, T. Bein, Pergamon, Oxford, 107–150.
5. Clearfield, A. (**1998**) Prog. Inorg. Chem. 47, 371–510.
6. Cao, G., Hong, H.-G., Mallouk, T.E. (**1992**) Acc. Chem. Res. 25, 420–427.

7. Thompson, M. (**1994**) *Chem. Mater. 6*, 1168–1175.

8. Clearfield, A. (**1996**) *Curr. Opin. Solid State Mater. Sci.* **1**, 268–278; (**2002**) *Curr. Opin. Solid State Mater. Sci. 6*, 495–506.

9. Maeda, K. (**2004**) *Microporous Mesoporous Mater. 73*, 47–55.

10. Demadis, K.D. (**2007**) in *Solid State Chemistry Research Trends*, Ed. R.W. Buckley, Nova Science Publication, Hauppauge, NY, 109–172.

11. Shimizu, G.K.H., Vaidhyanathan, R., Taylor, J.M. (**2009**) *Chem. Soc. Rev. 38*, 1430–1449.

12. Matczak-Jon, E., Videnova-Adrabinska, V. (**2005**) *Coord. Chem. Rev. 249*, 2458–2488.

13. Gagnon, K.J., Perry, H.P., Clearfield, A. (**2012**) *Chem. Rev. 112*, 1034–1054.

14. Ma, T.-Y., Yuan, Z.-Y. (**2011**) *ChemSusChem 4*, 1407–1419.

15. Clearfield, A., Demadis, K. (**2012**) in *Metal Phosphonate Chemistry: From Synthesis to Applications*. RSC, Cambridge, Chapter 1, p. 10.

16. Zon, J., Garczarek, P., Bialek, M. (**2012**) in *Metal Phosphonate Chemistry: From Synthesis to Applications*, Eds. A. Clearfield, K. Demadis. RSC, Cambridge, Chapter 6, pp. 170–191.

17. Mao, J.-G. (**2012**) in *Metal Phosphonate Chemistry: From Synthesis to Applications*, Eds. A. Clearfield, K. Demadis. RSC, Cambridge, Chapter 5, pp. 133–169.

18. Stock, N., Frey, S.A., Stucky, G.D., Cheetham, A.K. (**2000**) *J. Chem. Soc. Dalton Trans.*, 4292–4296.

19. Gómez-Alcantara, M.M., Aranda, M.A.G., Olivera-Pastor, P., Beran, P., García-Muñoz, J.L., Cabeza, A. (**2006**) *Dalton Trans.*, 577–585.

20. Hix, G.B., Wragg, D.S., Wright, P.A., Morris, R.E. (**1998**) *J. Chem. Soc. Dalton Trans.*, 3359–3362.

21. Dines, M.B., DiGiacomo, P.M. (**1981**) *Inorg. Chem. 20*, 92–97.

22. Drumel, S., Janvier, P., Barboux, P., Bujoli-Doeuff, M., Bujoli, B. (**1995**) *Inorg. Chem. 34*, 148–156.

23. Fredoueil, F., Evain, M., Massiot, D., Bujoli-Doeuff, M., Janvier, P., Clearfield, A., Bujoli, B. (**2002**) *J. Chem. Soc. Dalton Trans.*, 1508–1512.

24. Bestaoui, N., Ouyang, X., Fredoueil, F., Bujoli, B., Clearfield, A. (**2005**) *Acta Cryst. B 61*, 669–674.

25. Gómez-Alcantara, M.M., Cabeza, A., Aranda, M.A.G., Guagliardi, A., Mao, J.G., Clearfield, A. (**2004**) *Solid State Sci. 6*, 479–487.

26. Anillo, A., Altomare, A., Moliterni, A.G.G., Bauer, E.M., Bellitto, C., Colapietro, M., Portalone, G., Righini, G. (**2005**) *J. Solid State Chem. 178*, 306–313.

27. Serpaggi, F., Férey, G. (**1999**) *Inorg. Chem. 38*, 4741–4744.

28. Distler, A., Sevov, S.C. (**1998**) *Chem. Commun.* 959–960.

29. Tan, J.C., Merrill, C.A., Orton, J.B., Cheetham, A.K. (**2009**) *Acta Mater. 57*, 3481–3496.

30. Ayyappan, S., Delgado, G.D., Cheetham, A.K., Férey, G., Rao, C.N.R. (**1999**) *J. Chem. Soc. Dalton Trans.* 2905–2907.

31. Riou-Cavellec, M., Sanselme, M., Noguès, M., Grenèche, J.-M., Férey, G. (**2002**) *Solid State Sci. 4*, 619–625.

32. Knope, K.E., Cahill, C.L. (**2008**) *Inorg. Chem. 47*, 7660–7672.

33. Alsobrook, A.N., Albrecht-Schmitt, T.E. (**2009**) *Inorg. Chem. 48*, 11079–11084.

34. Ayi, A.A., Kinnibrugh, T.L., Clearfield, A. (**2011**) *Dalton Trans. 40*, 12648–12650.

35. Chen, Z., Zhou, Y., Weng, L., Yuan, C., Zhao, D. (**2007**) *Chem. Asian J. 2*, 1549–1554.

36. Zhang, X.M. (**2004**) *Eur. J. Inorg. Chem.* 544–548.

37. Stock, N., Bein, T. (**2005**) *J. Mater. Chem. 15*, 1384–1391.

38. Stock, N., Bein, T. (**2004**) *Angew. Chem. Int. Ed. 43*, 749–752.

39. Demadis, K.D., Papadaki, M., Raptis, R.G., Zhao, H. (**2008**) *Chem. Mater. 20*, 4835–4846.

40. Demadis, K.D., Papadaki, M., Císarova, I. (**2010**) *Appl. Mater. Interfaces 2*, 1814–1816.

41. Fu, R., Xiang, S., Zhang, H., Zhang, J., Wu, X. (**2005**) *Cryst. Growth Des. 5*, 1795–1799.

42. Colodrero, R.M.P., Olivera-Pastor, P., Cabeza, A., Papadaki, M., Demadis, K.D., Aranda, M.A.G. (**2010**) *Inorg. Chem. 49*, 761–768.

43. Demadis, K.D., Papadaki, M., Aranda, M.A.G., Cabeza, A., Olivera-Pastor, P., Sanakis, Y. (**2010**) *Cryst. Growth Des. 10*, 357–364.

44. Lodhia, S., Turner, A., Papadaki, M., Demadis, K.D., Hix, G.B. (**2009**) *Cryst. Growth Des. 9*, 1811–1822.

45. Dong, D.-P., Sun, Z.-G., Tong, F., Zhu, Y.-Y., Chen, K., Jiao, C.-Q., Wang, C.-L., Li, C., Wang, W.-N. (**2011**) *CrystEngComm 13*, 3317–3320.

46. Fu, R., Zhang, H., Wang, L., Hu, S., Li, Y., Huang, X., Wu, X. (**2005**) *Eur. J. Inorg. Chem.* 3211–3213.

47. Lai, Z., Fu, R., Hu, S., Wu, X. (**2007**) *Eur. J. Inorg. Chem.*, 5439–5446.

48. Cui, L., Sun, Z., Chen, H., Meng, L., Dong, D., Tian, C., Zhu, Z., You, W. (**2007**) *J. Coord. Chem. 60*, 1247–1254.

49. Colodrero, R.M.P., Cabeza, A., Olivera-Pastor, P., Papadaki, M., Rius, J., Choquesillo-Lazarte, D., García-Ruiz, J.M., Demadis, K.D., Aranda, M.A.G. (**2011**) *Cryst. Growth Des. 11*, 1713–1722.

50. Sun, Z., Chen, H., Liu, Z., Cui, L., Zhu, Y., Zhao, Y., Zhang, J., You, W., Zhu, Z. (**2007**) *Inorg. Chem. Commun. 10*, 283–286.

51. Colodrero, R.M.P., Papathanasiou, K.E., Stavgianoudaki, N., Olivera-Pastor, P., Losilla, E.R., Aranda, M.A.G., León-Reina, L., Sanz, J., Sobrados, I., Choquesillo-Lazarte, D., et al. (**2012**) *Chem. Mater. 24*, 3780–3792.

52. Stock, N., Biswas, S. (**2012**) *Chem. Rev. 112*, 933–969.

53. Fu, R., Hu, S., Wu, X. (**2011**) *J. Solid State Chem. 184*, 945–952.

54. Drumel, S., Janvier, P., Deniaud, D., Bujoli, B. (**1995**) *J. Chem. Soc. Chem. Commun.* 1051–1052.

55. Sharma, C.V.K., Clearfield, A. (**2000**) *J. Am. Chem. Soc. 122*, 1558–1558.

56. Clearfield, A., Sharma, C.V.K., Zhang, B. (**2001**) *Chem. Mater. 13*, 3099–3112.

57. Mao, J.-G., Wang, Z., Clearfield, A. (**2002**) *Inorg. Chem. 41*, 3713–3720.

58. Poojary, D.M., Zhang, B., Clearfield, A. (**1994**) *Angew. Chem. Int. Ed. Engl. 33*, 2324–2326.

59. Mao, J.-G., Clearfield, A. (**2002**) *Inorg. Chem. 41*, 2319–2324.

60. Almeida Paz, F.A., Rocha, J. (**2012**) In *Metal Phosphonate Chemistry: From Synthesis to Applications*, Eds. A. Clearfield, K. Demadis. RSC, Cambridge, 551–585.

61. Mao, J.G., Wang, Z., Clearfield, A. (**2002**) *Inorg. Chem. 41*, 6106–6111.

62. Tang, S.F., Song, J.-L., Li, X.-L., Mao, J.-G. (**2006**) *Crystal Growth Des. 6*, 2322–2326.

63. Ngo, H.L., Lin, W. (**2002**) *J. Am. Chem. Soc. 124*, 14298–14299.

64. Shi, X., Zhu, G., Qiu, S., Huang, K., Yu, J., Xu, R. (**2004**) *Angew. Chem. Int. Ed. 43*, 6482–6485.

65. Yang, B.-P., Mao, J.-G. (**2007**) *J. Mol. Struct. 830*, 78–84.

66. Sonnauer, A., Stock, N. (**2008**) *Eur. J. Inorg. Chem.*, 5038–5045.

67. Sonnauer, A., Stock, N. (**2005**) *Inorg. Chem. 44*, 5882–5889.

68. Sonnauer, A., Näther, C., Höppe, H.A., Senker, J., Stock, N. (**2007**) *Inorg. Chem. 46*, 9968–9974.

69. Sonnauer, A., Feyand, M., Stock, N. (**2008**) *Cryst. Growth Des. 9*, 586–592.

70. Feyand, M., Näther, C., Rothkirch, A., Stock, N. (**2010**) *Inorg. Chem. 49*, 11158–11163.

71. Ayi, A.A., Burrows, A.D., Mahon, M.F., Pop, V.M. (**2011**) *J. Chem. Crystallogr. 41*, 1165–1168.

72. Poojary, D.M., Zhang, B.L., Clearfield, A. (**1998**) *Anal. Quimica Int. Ed. 94*, 401–405.

73. Cao, D.-K., Gao, S., Zheng, L.-M. (**2004**) *J. Solid. State. Chem. 177*, 2311–2315.

74. Subbiah, A., Bhuvanesh, N., Clearfield, A. (**2005**) *J. Solid. State. Chem. 178* 1321–1325.

75. Poojary, D.M., Zhang, B.L., Bellinghausen, P., Clearfield, A. (**1996**) *Inorg. Chem. 35*, 4942–4949.

76. Amghouz, Z., García-Granda, S., García, J.R., Clearfield, A., Valiente, R. (**2011**) *Cryst. Growth Des. 11*, 5289–5297.

77. Ouellette, W., Wang, G., Liu, H., Yee, G.T., O'Connor, C.J., Zubieta, J. (**2009**) *Inorg. Chem. 48*, 953–963.

78. Adelani, P.O., Albrecht-Schmitt, T.E. (**2012**), *J. Solid State Chem. 192*, 377–384.

79. Dines, M.B., Digiacomo, P.M., Callahan, K.P., Griffith, P.C., Lane, R.H., Cooksey, R.E. (**1982**) Ed. J.S. Miller. *ACS Symposium Series 192*; American Chemical Society, Washington, DC, Chapter 12, p. 223.

80. Attfield, M.P., Mendieta-Tan, C., Telchaddera, R.N., Roberts, M.A. (**2012**) *RSC Adv. 2*, 10291–10297.

81. Adelani, P.O., Albrecht-Schmitt, T.E. (**2010**) *Angew. Chem. Int. Ed. 49*, 8909–8911.

82. Adelani, P.O., Albrecht-Schmitt, T.E. (**2012**) *Cryst. Growth Des. 12*, 5800–5805.

83. Adelani, P.O., Albrecht-Schmitt, T.E. (**2009**) *Inorg. Chem. 48*, 2732–2734.

84. Adelani, P.O., Albrecht-Schmitt, T.E. (**2011**), *J. Solid State Chem. 184*, 2368–2373.

85. Adelani, P.O., Albrecht-Schmitt, T.E. (**2011**), *J. Solid State Chem. 84*, 2368–2373.

86. Liang, J., Shimizu, G.K.H. (**2007**) *Inorg. Chem. 46*, 10449–10451.

87. Iremonger, S.S., Liang, J., Vaidhyanathan, R., Martens, I., Shimizu, G.K.H., Daff, T.D., Zein-Aghaji, M., S. Yeganegi, S., Woo, T.K. (**2011**) *J. Am. Chem. Soc. 133*, 20048–20051.

88. Iremonger, S.S., Liang, J., Vaidhyanathan, R., Shimizu, G.K.H. (**2011**) *Chem. Commun. 47*, 4430–4432.

89. Belier, F., Riou-Cavellec, M., Vichard, Riou, D.D. (**2000**) *C. R. Acad. Sci. Paris 3*, 655–660.

90. Stock, N., Bein, T. (**2002**) *J. Solid State Chem. 167*, 330–336.

91. Li, H., Zhu, G.S., Guo, X.D., Sun, F.X., Ren, H., Chen, Y., Qiu, S.L. (**2006**) *Eur. J. Inorg. Chem.* 4123–4128.

92. Xu, X., Wang, P., Hao, R., Gan, M., Sun, F., Zhu, G.-S. (**2009**) *Solid State Sci. 11*, 68–71.

93. Poojary, D.M., Zhang, B., Bellinghausen, P., Clearfield, A. (**1996**) *Inorg. Chem. 35*, 5254–5263.

94. Riou, D., Belier, F., Serre, C., Nogues, M., Vichard, D., Ferey, G. (**2000**) *Int. J. Inorg. Mater. 2*, 29–33.

95. Harvey, H.G., Herve, A.C., Hailes, H.C., Attfield, M.P. (**2004**) *Chem. Mater. 16*, 3756–3766.

96. Stock, N., Guillou, N., Bein, T., Fére, G. (**2003**) *Solid State Sci. 5*, 629–634.

97. Irran, E., Bein, T., Stock, N. (**2003**) *J. Solid State Chem. 173*, 293–298.

98. Vilela, S.M.F., Mendes, R.F., Silva, P., Fernandes, J.A., Tome, J.P.C., Almeida Paz, F.A. (**2013**) *Cryst. Growth Des. 13*, 543–560.

99. Shi, F.-N., Trindade, T., Rocha, J., Almeida Paz, F.A. (**2008**) *Cryst. Growth Des. 8*, 3917–3920.

100. Taddei, M., Costantino, F., Vivani, R., Sangregorio, C., Sorace, L., Castelli, L. (**2012**) *Cryst. Growth Des. 12*, 2327–2335.

101. Gomez-Alcantara, M.M., Cabeza, A., Martinez-Lara, M., Aranda, M.A.G., Suau, R., Bhuvanesh, N., Clearfield, A. (**2004**) *Inorg. Chem. 43*, 5283–5293.

102. Zhang, B., Poojary, D.M., Clearfield, A. (**1998**) *Inorg. Chem. 37*, 1844–1852.

103. Adelani, P.O., Albrecht-Schmitt, T.E. (**2011**) *Inorg. Chem. 50*, 12184–12191.

104. Poojary, D.M., Cabeza, A., Aranda, M.A.G., Bruque, S., Clearfield, A. (**1996**) *Inorg. Chem. 35*, 1468–1473.

105. Aranda, M.A.G., Cabeza, A., Bruque, S., Poojary, D.M., Clearfield, A. (**1998**) *Inorg. Chem. 37*, 1827–1832.

106. Mihalcea, I., Henry, N., Loiseau, T. (**2011**) *Cryst. Growth Des. 11*, 1940–1947.

107. Alekseev, E.V., Krivovichev, S.V., Depmeier, W. (**2008**) *Angew. Chem. Int. Ed. 47*, 549–551.

108. Wang, Z., Heising, J.M., Clearfield, A. (**2003**) *J. Am. Chem. Soc. 125*, 10375.

109. Clearfield, A. (**2008**) *Dalton Trans.*, 6089–6102.

110. Gomez-Alcantara, M.M., Cabeza, A., Olivera-Pastor, P., Fernandez-Moreno, F., Sobrados, I., Sanz, J., Morris, R.E., Clearfield, A., Aranda, M.A.G. (**2007**) *Dalton Trans.*, 2394–2404.

111. Cabeza, A., Gomez-Alcantara, M.M., Olivera-Pastor, P., Sobrados, I., Sanz, J., Xiao, B., Morris, R.E., Clearfield, A., Aranda, M.A.G. (**2008**) *Microporous Mesoporous Mater. 114*, 322–336

112. Gomez-Alcantara, M.M., Cabeza, A., Moreno-Real, L., Aranda, M.A.G., Clearfield, A. (**2006**) *Microporous Mesoporous Mater. 88*, 293–303.

113. Dines, M.B., Digiacomo, P.M., Callahan, K.P., Griffith, P.C., Lane, R.H., Cooksey, R.E. (**1982**) Ed. J.S. Miller. *ACS Symposium Series 192*; American Chemical Society, Washington, DC, Chapter 12.

114. Lohse, D.L., Sevov, S.C. (**1997**) *Angew. Chem. Int. Ed. Engl. 36*, 1619–1621.

115. Gao, Q., Guillou, N., Nogues, M., Cheetham, A.K., Ferey, G. (**1999**) *Chem. Mater. 11*, 2937–2947.

116. Allen, F.H. (**2002**) *Acta Cryst. B 58*, 380–388.

117. Sergienko, V.S. (**2001**) *Russian J. Coord. Chem. 27*, 681–710.

118. Sergienko, V.S. (**2000**) *Crystallogr. Rep. 45*, 69–76.

119. Song, H.-H., Yin, P., Zheng, L.-M., Korp, J.D., Jacobson, A.J., Gao, S., Xin, X.-Q. (**2002**) *J. Chem. Soc. Dalton Trans.*, 2752–2759.

120. Stavgianoudaki, N., Papathanasiou, K.E., Colodrero, R.M.P., Choquesillo-Lazarte, D., Garcia-Ruiz, J.M., Cabeza, A., Aranda, M.A.G., Demadis, K.D. (**2012**) *Cryst. Eng. Commun. 14*, 5385–5389.

121. Jingyang, B.N., Xiaoqing, Z., Donghui, Y., Junwei, Z., Pengtao, M., Ulrich, K., Jingping, W. (**2012**) *Chem. Eur. J. 18*, 6759–6762.

122. Hong-Yue, B.W., Weiting, Y., Zhong-Ming, S. (**2012**) *Cryst. Growth Des. 12*, 4669–4675.

123. Zongbin, B.W., Zhongmin, L., Peng, T., Lei, X., Haibin, S., Xinhe, B., Xiumei, L., Xianchun, L. (**2006**) *Cryst. Res. Technol. 41*, 1049–1054.

124. Zheng, L.-M., Duan, C.-Y., Ye, X.-R., Zhang, L.-Y. L.-Y., Wang, C., Xin, X.-Q. (**1998**) *J. Chem. Soc. Dalton Trans.*, 905–908.

125. Dong, D.-P., Liu, L., Sun, Z.-G., Jiao, C.-Q., Liu, Z.-M., Li, C., Zhu, Y.-Y., Chen, K., Wang, C.-L. (**2011**) *Cryst. Growth Des. 11*, 5346–5354.

126. Shi, F.N., Cunha-Silva, L., Ferreira, R.A.S., Mafra, L., Trindade, T., Carlos, L.D., Paz, F.A.A., Rocha, J. (**2008**) *J. Am. Chem. Soc. 130*, 150–167.

127. Yang, W., Wu, H.-Y., Wang, R.-X., Pan, Q.-J., Sun, Z.-M., Zhang, H. (**2012**) *Inorg. Chem. 51*, 11458–11465.

128. Jankovics, H., Dashalakis, M.C., Raptopoulou, P., Terzis, A., Tangoulis, V., Giapintzakis, J., Kiss, T., Slifoglou, A. (**2002**) *Inorg. Chem. 41*, 3366–3374.

129. Kong, D., Li, Y., Ouyang, X., Prosvirin, A.V., Zhao, H., Ross, J.H., Dunbar, K.R., Clearfield, A. (**2004**) *Chem. Mater. 16*, 3020–3031.

130. Menelaou, M., Dakanali, M., Raptopoulou, C.P., Drouza, C., Lalioti, N., Salifoglou, A. (**2009**) *Polyhedron 28*, 3331–3339.

131. Yanga, B.-P., Prosvirinb, A.V., Zhaob, H.-H., Mao, J.-G. (**2006**) *J. Solid State Chem. 179*, 175–185.

132. Cabeza, A., Bruque, S., Guagliardi, A., Aranda, M.A.G. (**2001**) *J. Solid State Chem. 160*, 278–286.

133. Gagnon, K.J., Prosvirin, A.V., Dunbar, K.R., Teat, S.J., Clearfield, A. (**2012**) *Dalton Trans. 41*, 3995–4006.

134. Turner, A., Jaffres, P.A., MacLean, E.J., Villemin, D., McKee, V., Hix, G.B. (**2003**) *Dalton Trans.*, 1314–1319.

135. Cabeza, A., Aranda, M.A.G. (**2012**) *Metal Phosphonate Chemistry: From Synthesis to Applications*, Eds. A. Clearfield, K. Demadis. RSC, Cambridge, 107–132.

136. Maniam, P., Stock, N. (**2012**) *Metal Phosphonate Chemistry: From Synthesis to Applications*, Eds. A. Clearfield, K. Demadis. RSC, Cambridge, 87–106.

137. Vivani, R., Alberti, G., Costantino, F., Nocchetti, M. (**2008**) *Microporous Mesoporous Mater. 107*, 58–70.

138. Meng, L., Li, J., Sun, Z., Zheng, X., Chen, H., Dong, D., Zhu, Y., Zhao, Y., Zhang, J. (**2008**) *J. Coord. Chem. 61*, 2478–2487.

139. Ying, S.-M., Mao, J.-G. (**2006**) *J. Mol. Struct. 783*, 13–20.

140. Tang, S.-F., Song, J.-L., Mao, J.-G. (**2006**) *Eur. J. Inorg. Chem.*, 2011–2019.

141. Bauer, S., Stock, N. (**2007**) *Angew. Chem. Int. Ed. 46*, 6857–6860.

142. Cunha-Silva, L., Lima, S., Ananias, D., Silva, P., Mafra, L., Carlos, L.D., Pillinger, M., Valente, A.A., Paz, F.A.A., Rocha, J. (2009) *J. Mater. Chem. 19*, 2618–2632.

143. Wang, C.-L., Li, J., Sun, Z.-G., Hua, R.-N., Zhu, Y.-Y., Chen, K., Jiao, C.-Q., Li, C., Zhen, M.-J., Chu, W., et al. (2012) *Z. Anorg. Allg. Chem. 638*, 111–115.

144. Yang, B.-P., Prosvirin, A.V., Guo, Y.-Q., Mao, J.-G. (2008) *Inorg. Chem. 47*, 1453–1459.

145. Bauer, S., Bein, T., Stock, N. (2006) *J. Solid State Chem. 179*, 145–155.

146. Costantino, F., Sassi, P., Geppi, M., Taddei, M. (2012) *Cryst. Growth Des. 12*, 5462–5470.

147. Mao, J.G., Wang, Z.K., Clearfield, A. (2002) *Inorg. Chem. 41*, 2334–2340.

148. Mao, J.G. Wang, Z.K., Clearfield, A. (2002) *J. Chem. Soc. Dalton. Trans.* 4457–4463.

149. Taddei, M., Vivania, R., Costantino, F. (2013) *Dalton Trans. 42*, 9671–9678.

150. Vivania, R., Costantino, F., Taddei, M. (2012) *Metal Phosphonate Chemistry: From Synthesis to Applications*, Eds. A. Clearfield, K. Demadis. RSC, Cambridge, 45–86.

151. Bauer, S., Bein, T., Stock, N. (2005) *Inorg. Chem. 44*, 5882–5889.

152. Song, J.-L., Yi, F.-Y., Mao, J.-G. (2009) *Crystal Growth Des. 9*, 3273–3277.

153. Wharmby, M.T., Wright, P. (2012), *Metal Phosphonate Chemistry: From Synthesis to Applications*, Eds. A. Clearfield, K. Demadis. RSC, Cambridge, 317–343.

154. Ma, K.-R., Wei, C.-L., Zhang, Y., Kan, Y.-H., Cong, M.-H., Yang, X.-J. (2013) *J. Spectrosc. 2013*, Article ID 378379, 9 pages.

155. Du, Z.-Y., Xie, Y.-R., Wen, H.-R. (2009) *Inorg. Chim. Acta 362*, 351–354.

156. Mowat, J.P.S., Groves, J.A., Wharmby, M.T., Miller, S.R., Li, Y., Lightfoot, P., Wright, P.A. (2009) *J. Solid State Chem. 182*, 2769–2778.

157. Soghomonian, V., Diaz, R., Haushalter, R.C., O'Connor, C.J., Zubieta, J. (1995) *Inorg. Chem. 34*, 4460–4466.

158. Wang, Y., Bao, S.-S., Xu, W., Chen, J., Gao, S., Zheng, L.-M. (2004) *J. Solid State Chem. 177*, 1297–1301.

159. Zhang, N., Huang, C.-Y., Sun, Z.-G., Zhang, J., Liu, L., Lu, X., Wang, W.-N., Tong, F. (2010) *Z. Anorg. Allg. Chem. 636*, 1405–1409.

160. Taddei, M., Costantino, F., Vivani, R. (2010) *Inorg. Chem. 49*, 9664–9670.

161. LaDuca, R., Rose, D., DeBord, J.R.D., Haushalter, R.C., O'Connor, C.J., Zubieta, J. (1996) *J. Solid State Chem. 123*, 408–412.

162. Groves, J.A., Miller, S.R., Warrender, S.J., Mellot-Draznieks, C., Lightfoota, P., Wright, P.A. (2006) *Chem. Commun.*, 3305–3307.

163. Grovew, J.A., Wright, P.A., Lightfoot, P. (2005) *Dalton Trans.*, 2007–2010.

164. Miller, S.R., Pearce, G.M., Wright, P.A., Bonino, F., Chavan, S., Bordiga, S., Margiolaki, I., Guillou, N., Ferey, G., Bourrelly, S., et al. (2008) *J. Am. Chem. Soc. 130*, 15967–15981.

165. Wharmby, M.T., Pearce, G.M., Mowat, J.P.S., Griffin, J.M., Ashbrook, S.E., Wright, P.A., Schilling, L.-H., Lieb, A., Stock, N., Chavan, S., et al. (2012) *Microporous Mesoporous Mater. 157*, 3–17.

166. Wharmby, M.T., Mowat, J.P.S., Thompson, S.P., Wright, P.A. (2011) *J. Am. Chem. Soc. 133*, 1266–1269.

167. Serre, C., Groves, J.A., Lightfoot, P., Slawin, A.M.Z., Wright, P.A., Stock, N., Bein, T., Haouas, M., Taulelle, F., Ferey, G. (2006) *Chem. Mater. 18*, 1451–1457.

168. Grovew, J.A., Stephens, N.F., Wright, P.A., Lightfoot, P. (**2006**) *Solid State Sci. 8*, 397–403.

169. Wharmby, M.T., Miller, S.R., Groves, J.A., Margiolaki, I., Ashbrook, S.E., Wright, P.A. (**2010**) *Dalton Trans. 39*, 6389–6391.

170. Maeda, K., Hatasawa, H., Nagayoshi, N., Matsushima, Y. (**2011**) *Chem. Lett. 40*, 215–217.

171. Kong, D., Zoñ, J., McBee, J., Clearfield, A. (**2006**) *Inorg. Chem. 45*, 977–986.

172. Colodrero, R.M.P., Cabeza, A., Olivera-Pastor, P., Infantes-Molina, A., Barouda, E., Demadis, K.D., Aranda, M.A.G. (**2009**) *Chem. Eur. J. 15*, 6612–6618.

173. Clearfield, A. (**2002**) *Curr. Opin. Solid State Mater. Sci. 6*, 495–506.

174. Taylor, J.M., Dawson, K.W., Shimizu, G.K.H. (**2013**) *J. Am. Chem. Soc. 135*, 1193–1196.

175. Plabst, M., Bein, T. (**2009**) *Inorg. Chem. 48*, 4331–4341.

176. Taylor, J.M., Mahmoudkhani, A.H., Shimizu, G.K.H. (**2007**) *Angew. Chem. Int. Ed. 46*, 795–798.

177. Habib, H.A., Sanchiz, J., Janiak, C. (**2008**) *Dalton Trans. 13*, 1734–1744.

178. Chen, W., Wang, J.-Y., Chen, C., Yue, Q., Yuan, H.-M., Chen, J.-S., Wang, S.-N. (**2003**) *Inorg. Chem. 42*, 944–946.

179. Seo, Y.-K., Hundal, G., Jang, I.T., Hwang, Y.K., Jun, C.-H., Chang, J.-S. (**2009**) *Microporous Mesoporous Mater. 119*, 331–337.

180. Plater, M.J., Foreman, M.R.S.J., Howie, R.A., Skakle, J.M.S., Coronado, E., Gomez-Garcia, C.J., Gelbrich, T., Hursthouse, M.B. (**2001**) *Inorg. Chim. Acta 319*, 159–175.

181. Davies, R.P., Less, R.J., Lickiss, P.D., White, A.J.P. (**2007**) *Dalton Trans. 24*, 2528–2535.

182. Taylor, J.M., Mah, R.K., Moudrakovski, I.L., Ratcliffe, C.I., Vaidhyanathan, R., Shimizu, G.K.H. (**2010**) *J. Am. Chem. Soc. 132*, 14055–14057.

183. Kinnibrugh, T.L., Ayi, A.A., Bakhmutov, V.I., Zoń, J., Clearfield, A. (**2013**) *Cryst. Growth Des. 13*, 2973–2981.

184. Maeda, K., Takamatsu, R., Mochizuki, M., Kawawa, K., Kondo, A. (**2013**) *Dalton Trans. 42*, 10424–10432.

185. Arika, T., Kondo A., Maeda, K. (**2013**) *Chem. Commun. 49*, 552–554.

186. Beckmann, J., Rüttinger, R., Schwich, T. (**2008**) *Cryst. Growth Des. 8*, 3271–3276.

187. Vaidhyanathan, R., Mahmoudkhani, A.H., Shimizu, G.K.H. (**2009**) *Can. J. Chem. 87*, 247–253.

188. Mah, R.K., Lui, M.W., Shimizu, G.K.H. (**2013**) *Inorg. Chem. 52*, 7311–7313.

189. Yang, C.-I., Song, Y.-T., Yeh, Y.-J., Liu, Y.-H., Tseng, T.-W., Lu, K.-L. (**2011**) *Cryst. Eng. Comm. 13*, 2678–2686.

190. Vilela, S.M.F., Firmino, A.D.G., Mendes, R.F., Fernandes, J.A., Ananias, D., Valente, A.A., Ott, H., Carlos, L.D., Rocha, J., Tomé, J.P.C., et al. (**2013**) *Chem. Commun. 49*, 6400–6402.

191. Vilela, S.M.F., Ananias, D., Gomes, A.C., Valente, A.A., Carlos, L.D., Cavaleiro, J.A.S., Rocha, J., Tomé, J.P.C., Almeida Paz, F.A. (**2012**) *J. Mater. Chem. 22*, 18354–18371.

192. Demadis, K.D., Ketsetzi, A. (**2007**) *Sep. Sci. Technol. 42*, 1639–1649.

193. Martínez-Tapia, H.S., Cabeza, A., Bruque, S., Pertierra, P., García-Granda, S., Aranda, M.A.G. (**2000**) *J. Solid State Chem. 151*, 122–129.

194. Demadis, K.D., Katarachia, S.D. (**2004**) *Phosphorus Sulfur Silicon 179*, 627–648.

195. Demadis, K.D., Katarachia, S.D., Raptis, R.G., Zhao, H., Baran, P. (**2006**) *Cryst. Growth Des. 6*, 836–838.

196. Cabeza, A., Aranda, M.A.G., Bruque, S. (**1999**) *J. Mater. Chem. 9*, 571–578.

197. Sharma, C.V.K., Clearfield, A., Cabeza, A., Aranda, M.A.G., Bruque, S. (**2001**) *J. Am. Chem. Soc. 123*, 2885–2886.

198. Cabeza, A., Ouyang, X., Sharma, C.V.K., Aranda, M.A.G., Bruque, S., Clearfield, A. (**2002**) *Inorg. Chem. 41*, 2325–2333.

199. Mao, J.G., Wang, Z., Clearfield, A. (**2002**) *New J. Chem. 26*, 1010–1014.

200. Silva, P., Vieira, F., Gomes, A.C., Ananias, D., Fernandes, J.A., Bruno, S.M., Soares, R., Valente, A.A., Rocha, J., Almeida Paz, F.A. (**2011**) *J. Am. Chem. Soc. 133*, 15120–15138.

201. Cunha-Silva, L., Mafra, L., Ananias, D., Carlos, L.D., Rocha, J., Almeida Paz, F.A. (**2007**) *Chem. Mater. 19*, 3527–3538.

202. Vivani, R., Costantino, F., Costantino, U., Nocchetti, M. (**2006**) *Inorg. Chem. 45*, 2388–2390.

203. Vasylyev, M.V., Wachtel, E.J., Popovitz-Biro, R., Neumann, R. (**2006**) *Chem. Eur. J. 12*, 3507–3514.

204. Horike, S., Shimomura1 S., Kitagawa, S. (**2009**) *Nat. Chem. 1*, 695–704.

205. Férey, G., Serre, C. (**2009**) *Chem. Soc. Rev. 38*, 1380–1399.

206. Taddei, M., Costantino, F., Ienco, A., Comotti, A., Dau, P.V., Cohen, S.M. (**2013**) *Chem. Commun. 49*, 1315–1317.

207. Choi, H.-S., Suh, M.P. (**2009**) *Angew. Chem. Int. Ed. 48*, 6865–6869.

208. Dalgarno, S.J. (**2010**) *Chem. Commun. 46*, 538–540.

209. Costantino, F., Bataille, T., Audebrand, N., Le Fur, E., Sangregorio, C. (**2007**) *Cryst. Growth Des. 7*, 1881–1888.

210. Bligh, S.W.A., Harding, C.T., Mcewen, A.B., Sadler, P.J. (**1994**) *Polyhedron 12*, 1937–1943.

211. Stock, N., Stoll, A., Bein, T. (**2004**) *Microporous Mesoporous Mater. 69*, 65–69.

212. Feyand, M., Seidler, C.F., Deiter, C. Rothkirch, A., Lieb, A., Wark, M., Stock, N. (**2013**) *Dalton Trans. 42*, 8761–8770.

213. Colodrero, R.M.P, Angeli, G., Bazaga-García, M., Pastor-Olivera, P., Villemin, D., Losilla, E.R., Martos, E.Q., Hix, G.B., Aranda, M.A.G., Demadis, K.D., et al. (**2013**) *Inorg. Chem. 52*, 8770–8783.

214. Stock, N., Guillou, N., Senker, J., Férey, G., Bein, T. (**2005**) *Z. Anorg. Allg. Chem. 631*, 575–581.

215. Kaempfe, P., Stock, N. (**2008**) *Z. Anorg. Allg. Chem. 634*, 714–717.

216. Plabst, M., McCusker, L.B., Bein, T. (**2009**) *J. Am. Chem. Soc. 131*, 18112–18118.

217. Plabst, M., Köhn, R., Bein, T. (**2010**) *Cryst. Eng. Comm. 12*, 1920–1926.

218. Jones, K.M.E., Mahmoudkhani, A.H., Chandler, B.D., Shimizu, G.K.H. (**2006**) *Cryst. Eng. Comm. 8*, 303–305.

219. Boldog, I., Domasevitch, K.V., Baburin, I.A., Ott, H., Gil-Hernández, B., Sanchiz, J., Janiak, C. (**2013**) *Cryst. Eng. Comm. 15*, 1235–1243.

220. Vasylyev, M., Neumann, R. (**2006**) *Chem. Mater. 18*, 2781–2783.

221. Stock, N., Rauscher, M., Bein, T. (2004) *J. Solid State Chem.* *117*, 642–647.

222. Wu, J., Hou, H., Han, H., Fan, Y. (2007) *Inorg. Chem.* *46*, 7960–7970.

223. Demadis, K.D., Baroudi, E., Stavgianoudaki, N., Zhao, H. (2009) *Cryst. Growth Des.* *9*, 1250–1253.

224. Su, M., Qiu, Y., Jia, W. (2005) *Adv. Ther.* *22*, 297–306.

225. Mathew, B., Chakraborty, S., Das, T., Sarma, H.D., Banerjee, S., Samuel, G., Venkatesh, M., Pillai, M.R.A. (2004) *Appl. Radiat. Isot.* *60*, 635–642.

226. Costantino, F., Donnadio, A., Casciola, M. (2012) *Inorg. Chem.* *51*, 6992–7000.

227. Ying, S.-M., Zeng, X.-R., Fang, X.-N., Li, X.-F., Liu, D.-S. (2006) *Inorg. Chim. Acta* *359*, 1589–1593.

228. Costantino, F., Ienco, A., Gentili, P.L., Presciutti, F. (2010) *Cryst. Growth Des.* *10*, 4831–4838.

229. Demadis, K.D., Barouda, E., Raptis, R.G., Zhao, H. (2009) *Inorg. Chem.* *48*, 819–821.

230. Zheng, G.-L., Ma, J.-F., Yang, J. (2004) *J. Chem. Res.* *6*, 387–388.

231. Colodrero, R.M.P., Olivera-Pastor, P., Losilla, E.R., Aranda, M.A.G., León-Reina, L., Papadaki, M., McKinlay, A.C., Morris, R.E., Demadis, K.D., Cabeza, A. (2012) *Dalton Trans.* *41*, 4045–4051.

232. Demadis, K.D., Mantzaridis, C., Raptis, R.G., Mezei, G. (2005) *Inorg. Chem.* *44*, 4469–4471.

233. Demadis, K.D., Mantzaridis, C., Lykoudis, P. (2006) *Ind. Eng. Chem. Res.* *45*, 7795–7800.

234. Demadis, K.D., Stavgianoudaki, M. (2012) in *Metal Phosphonate Chemistry: From Synthesis to Applications*, Eds. A. Clearfield, K. Demadis. RSC, Cambridge, 438–492.

235. Colodrero, R.M.P., Olivera-Pastor, P., Losilla, E.R., Hernández-Alonso, D., Aranda, M.A.G., León-Reina, L., Rius, J., Demadis, K.D., Moreau, B., Villemin, D., et al. (2012) *Inorg. Chem.* *51*, 7689–7698.

236. Brunet, E., Alhendawi, H.M.H., Cerro, C., de la Mata, M.J., Juanes, O., Rodriguez-Ubis, J.C. (2010) *Chem. Eng. J.* *158*, 333–344.

237. Brunet, E., Cerro, C., Juanes, O., Rodriguez-Ubis, J.C., Clearfield, A. (2008) *J. Mater. Sci.* *43*, 1155–1158.

238. Zhai, F., Zheng, Q., Chen, Z., Ling, Y., Liu, X., Weng, L., Zhou, Y. (2013) *Cryst. Eng. Commun.* *15*, 2040–2043.

239. Beier, M.J., Kleist, W., Wharmby, M.T., Kissner, R., Kimmerle, B., Wright, P.A., Grunwaldt, J.-D., Baiker, A. (2012) *Chem. Eur. J.* *18*, 887–898.

240. Liao, T.-B., Ling, Y., Chen, Z.-X., Zhou, Y.-M., Weng, L.-H. (2010) *Chem. Commun.* *46*, 1100–1102.

241. Dutta, A., Patra, A.K., Bhaumik, A. (2012) *Microporous Mesoporous Mater.* *155*, 208–214.

242. Ma, T.Y., Zhang, X.J., Yuan, Z.Y. (2009) *Microporous Mesoporous Mater.* *123*, 234–242.

243. Ma, T.Y., Lin, X.Z., Yuan, Z.Y. (2010) *J. Mater. Chem.* *20*, 7406–7415.

244. Ma, T.Y., Lin, X.Z., Yuan, Z.Y. (2010) *Chem. Eur. J.* *16*, 8487–8494.

245. Niu, J., Zhang, X., Yang, D., Zhao, J., Ma, P., Kortz, U., Wang, J. (2012) *Chem. Eur. J.* *18*, 6759–6762.

246. Bazaga-Garcia, M., Cabeza, A., Olivera-Pastor, P., Santacruz, I., Colodrero, R.M.P., Aranda, M.A.G. (**2012**) *J. Phys. Chem. C 116*, 14526–14533.

247. Ikawa, H. (**1992**) in *Proton Conductors*, Ed. P. Colomban. Cambridge University Press, Cambridge, 511–515.

248. Hickner, M.A., Ghassemi, H., Kim, Y.S., Einsla, B.R., McGrath, J.E. (**2004**) *Chem. Rev. 104*, 4587–4612.

249. Li, S.-L., Xu, Q. (**2013**) *Energ. Environ. Sci. 6*, 1656–1683.

250. Jiménez-García, L., Kaltbeitzel, A., Enkelmann, V., Gutmann, J.S., Klapper, M., Müllen, K. (**2011**) *Adv. Funct. Mater. 21*, 2216–2224.

251. Yoon, M., Suh, K., Natarajan, S., Kim, K. (**2013**) *Angew. Chem. Int. Ed. 52*, 2688–2700.

252. Horike, S., Umeyama, D., Kitagawa, S. (**2013**) *Acc. Chem. Res. 46*, 2376–2384.

253. Shimizu, G.K.H. (**2012**) *Metal Phosphonate Chemistry: From Synthesis to Applications*, Eds. A. Clearfield, K. Demadis. RSC, Cambridge, 493–524.

254. Adani, F., Casciola, M., Jones, D.J., Massinelli, L., Montoneri, E., Roziere, L., Vivani, R. (**1998**) *J. Mater. Chem. 8*, 961–964.

255. Zima, V., Svoboda, J., Melanova, K., Benes, L., Casciola, M., Sganappa, M., Brus, J., Trchova, M. (**2010**) *Solid State Ionics 181*, 705–713.

256. Kim, S., Dawson, K.W., Gelfand, B.S., Taylor, J.M., Shimizu, G.K.H. (**2013**) *J. Am. Chem. Soc. 135*, 963–966.

257. Alberti, G., Casciola, M., Palombari, R., Peraio, A. (**1992**) *Solid State Ionics 58*, 339–344.

258. Stein, E.W., Clearfield, A., Subramanian, M.A. (**1996**) *Solid State Ionics 83*, 113–124.

259. Alberti, G., Boccaili, L., Casciola, M., Massinelli, L., Montoneri, E. (**1996**) *Solid State Ionics 84*, 97–104.

260. Jang, M.Y., Park, Y.S., Yamazaki, Y. (**2003**) *Electrochemistry 8*, 691–694.

261. Grohol, D., Subramanian, M.A., Poojary, D.M., Clearfield, A. (**1996**) *Inorg. Chem. 35*, 5264–5271.

262. Jaimez, E., Hix, G.B., Slade, R.C.T. (**1997**) *J. Mater. Chem. 7*, 475–479.

263. Alberti, G., Casciola, M., Costantina, U., Peraio, A., Montoneri, E. (**1992**) *Solid State Ionics 50*, 315–322.

264. Casciola, M., Costantino, U., Peraio, A., Rega, T. (**1995**) *Solid State Ionics 77*, 229–233.

265. Alberti, G., Costantino, U., Casciola, M., Vivani, R., Peraio, A. (**1991**) *Solid State Ionics 46*, 61–68.

5

HYBRID MULTIFUNCTIONAL MATERIALS BASED ON PHOSPHONATES, PHOSPHINATES AND AUXILIARY LIGANDS

FERDINANDO COSTANTINO[1,2], ANDREA IENCO[2] AND MARCO TADDEI[1]

[1]*Dipartimento di Chimica, Biologia e Biotecnologie, University of Perugia, Perugia, Italy*
[2]*CNR – ICCOM, Sesto Fiorentino, Firenze, Italy*

5.1 INTRODUCTION

The research and development of new functional materials with tailored structure and reactivity has exponentially grown in the last decade owing to the high demand of sustainable processes such as clean energy production and storage that require the employment of low cost, versatile and extremely efficient compounds [1].

Hybrid inorganic–organic materials are a vast class of functional solids (surfaces, micro- and nano-crystalline materials, mesoporous substrates, etc.) in which the inorganic and the organic parts intimately contribute to the overall chemical reactivity [2].

Among them, coordination polymers (CPs) occupy a relevant place because they are crystalline solids with various dimensionalities, whose structure can be easily tuned with the rational choice of the ligands and metals used in their synthesis [3].

CPs with 3D structure and permanent porosity, easily accessible by the most part of common gases, are also called Metal-Organic Frameworks (MOFs). To date, the chemistry of MOFs has been mainly based on the use of polycarboxylates and transition metals, yielding highly porous compounds employed for many applications, such as gas sorption, catalysis, molecular recognition, conductivity, photocatalysis and so on [4].

Tailored Organic–Inorganic Materials, First Edition. Edited by Ernesto Brunet, Jorge L. Colón and Abraham Clearfield.
© 2015 John Wiley & Sons, Inc. Published 2015 by John Wiley & Sons, Inc.

Phosphonates are the other important class of linkers used for the construction of functional CPs. As a matter of fact, these ligands possess a very high coordination variability compared to carboxylates; this aspect can be considered as both an advantage and a drawback, because the structural control, intended as the easiness to obtain isoreticular compounds and to tailor the functionality, is hardly achievable. Despite these limitations, the number of reported CPs based on phosphonates in literature is extremely high: a recent book reports the state of the art of this fascinating chemistry [5].

A third class of less known ligands, although used for the synthesis of many CPs, is that of phosphinates [6]. Phosphinic ligands, compared to phosphonic ones, have a lower coordination capability due to the presence of two R groups attached to the central P atom. For that reason, the dimensionality of the phosphinate-based CPs is often lower than that of phosphonate-based CPs, and only few examples of 3D frameworks can be found in the literature. On the other hand, phosphinate CPs display a high thermal stability and, depending on the type of coordinating metals, they can also have interesting magnetic or optical properties.

The structures and properties of many metal phosphonates and phosphinates CPs have been reported in some recent reviews and book chapters. However, there is a considerable number of other CPs containing also a second auxiliary ligand like aza-heterocycles and/or carboxylates that have never been summarized in a dedicated review.

In this chapter, we will report a survey of the CPs based on transition metal phosphonates and phosphinates where an auxiliary ligand is also present. These auxiliary ligands will be divided into two classes: the N- and the O-donors. Among the N-donors, we will focus on the ligands that normally do not contribute to extend the dimensionality, like the chelating ones (2,2'-bipyridine, 1,10-phenantroline, terpyridine and related molecules) and those that are able to connect different metals, thus strongly influencing the overall connectivity (4,4'-bipyridine, imidazole and related molecules). Among the O-donors, we will mainly focus on carboxylic ligands that are able to connect the metal ions in different manners, thus strongly contributing to determine the structure and the reactivity.

In order to narrow the range of compounds to report, we decided to include in this chapter only the CPs where the phosphonate/phosphinate and the ancillary ligand are two distinct species, considering the species bearing different ligating groups (such as carboxyphosphonates) as the phosphorate ligand.

It is important to say that this field of research is still dominated by a crystal engineering approach, aimed at understanding the relationships between the molecular structure of the building blocks and the global crystal structure of the resulting CP, a pivotal requirement if the task is the preparation of tailor-made materials. In particular, there is a strong interest around the preparation of phosphonate- or phosphinate-based MOFs, especially because of the high water stability of compounds containing these ligands compared to the classical carboxylate-based MOFs, but examples of CPs displaying other interesting properties can be found in the literature.

Since this book is dedicated to functional materials, only the compounds having properties that could be directly ascribable to the presence of the secondary ligand (which could act as both a functional group and a spacer) will be discussed in detail. We will mainly focus on gas absorption, optical and magnetic properties primarily on the open-framework materials. The chapter will be divided in two parts: CPs constructed from phosphonates and auxiliary ligands and CPs constructed from phosphinates and auxiliary ligands.

5.1.1 Phosphonates and Phosphinates as Ligands for CPs: Differences in Their Coordination Capabilities

A phosphonic acid can be considered as a phosphoric acid [H_3PO_4 or $OP(OH)_3$] where one of the three hydroxyl groups is replaced with a generic R substituent, thus obtaining a general formula $OPR(OH)_2$ (or alternatively RPO_3H_2). Phosphinic acids possess two organic R substituents in place of two –OH groups, and therefore, the generic formula can be written as $OPR^1R^2(OH)$ (or alternatively $R^1R^2PO_2H$), in the case when the two substituent groups are different.

The molecular structure of phosphoric, phosphonic and phosphinic acids is shown in Scheme 5.1.

The synthetic methods and strategies normally used for the preparation of the most common phosphonates and phosphinates are discussed in detail in the recent book dedicated to metal phosphonate chemistry and in a review hence the interested readers can refer to them for any further insight [5a, 6].

It is well established that phosphonic ligands, compared to carboxylates, allow to obtain a higher number of structural motifs when employed for the synthesis of CPs, owing to their higher number of oxygen binding sites for metal ions and to the tetrahedral geometry of the PO_3C groups. The phosphonic tetrahedron possesses two acidic P–OH groups with two different acidic constants pK_a1 and pK_a2, which, depending on the electrophilic character of the attached R group, can vary from 0.5 to 3 for pK_a1 and from 5 to 9 for pK_a2. The third vertex of the tetrahedron is occupied by a P=O group that is able to coordinate to the metal ions as well as the deprotonated P–O$^-$ groups.

When a phosphonic group is completely deprotonated, it can bind up to six different metal ions. Also, the tetrahedral geometry favours the formation of extended inorganic units (like one-dimensional (1D) chains and two-dimensional (2D) layers) instead of isolated secondary building units (SBUs). Phosphonates can either link the metal ions in an endo-bidentate (chelating) mode or in an exo-bidentate fashion through the formation of bridges of the type M1–O–P–O–M2, the latter being the most common.

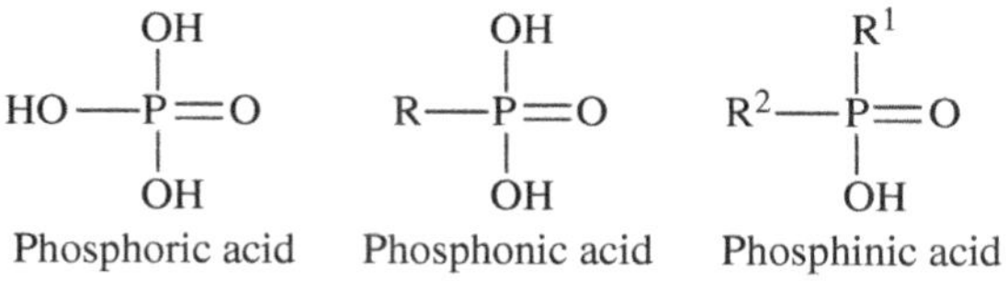

OH	OH	R^1
HO—P=O	R—P=O	R^2—P=O
OH	OH	OH
Phosphoric acid	Phosphonic acid	Phosphinic acid

SCHEME 5.1 Molecular structure of phosphoric, phosphonic and phosphinic acids.

Phosphinic ligands, due to the presence of a second R group linked to the central P atom, only have one acidic P–OH group, whose acidity varies in the 1.3–2.5 pK_a range.

The geometry of a phosphinate shares some similarities with that of a carboxylate, because both of them have only one negative charge and the two oxygen atoms (one P=O and one P–OH) have a comparable geometry. However, differently from carboxylates the P has a sp^3 hybridization with a O–P–O angle of about 110°, and, more importantly, the steric hindrance of the second R group normally has a strong influence on the type and directionality of the connection with the metal ions. The R groups frequently found in these ligands are quite voluminous (phenyls or alkyl chains), as phosphinates containing small groups (H and CH$_3$) are normally very reactive and therefore unstable. These steric limitations act as structure-orienting factors in the synthesis of CPs because they induce the formation of inorganic units with low dimensionality (especially 1D). The insertion of an auxiliary co-ligand normally induces drastic changes in the connectivity and in the dimensionality of the CPs, in a way that will be described in the next paragraphs.

5.1.2   The Role of the Auxiliary Ligands

5.1.2.1   N-Donors   The molecular structure of the N-donor heterocyclic coligands reported in this chapter is shown in Scheme 5.2.

They can roughly be divided in two categories: directional (4,4'-bipy, imidazole and related molecules) and chelating (2,2'-bipy, 1,10-phenanthroline, terpyridine and related molecules) ligands. These species have good affinity towards the most part of transition bi- and trivalent metals as well as lanthanide ions and have been extensively employed for the synthesis of many CPs based on phosphonates. They normally play an important role in the structural modulation due to their high directionality and/or chelating properties, whereas they have no influence on the charge balance, mainly acting as neutral ligands (although some examples of anionic species can be found, such as bpytrz and im in Scheme 5.2). An example of coordination modes of these co-ligands with metal ions is shown in Scheme 5.3. The chelating molecules act in different ways: the 2,2'-bipyridine (Scheme 5.3a) occupies two neighbouring positions of an octahedron, whereas the terpyridine (Scheme 5.3b) contains three N atoms and therefore occupies three positions in a *mer* arrangement. The 4,4'-bipyridine (Scheme 5.3c) is able to connect two different metals, and, in many phosphonate-based materials, it has found to create infinite 1D chains. The other positions are normally occupied by the oxygen atoms of the phosphonic groups or by water molecules. Moreover, the instauration of non-covalent interactions, like π···π stacking and CH···π bondings, could also contribute to the structural diversity. Depending on the type of metal and phosphonate used, they can also lead to the formation of constant structural motifs, like polymeric or discrete SBU, that can allow the design of isoreticular or tailor-made compounds. Some important examples will be described in detail in the next paragraphs. However, it is a common experience,

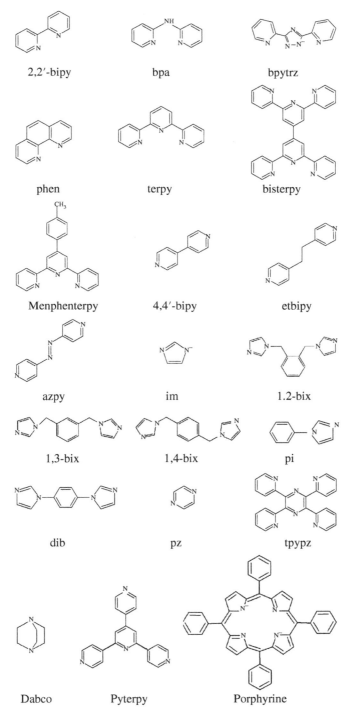

SCHEME 5.2 Molecular structure of the main N-donor heterocyclic co-ligands reported in this chapter.

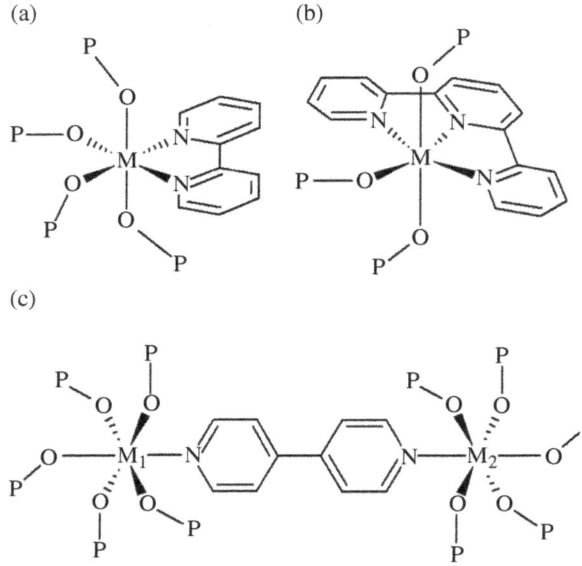

SCHEME 5.3 Coordination modes with metal ions of the N-donor heterocyclic co-ligands: equatorial (a and b) and bridging (c) modes are reported.

testified by a high number of examples, that small changes in the molecular structure of these ligands (a different substituent or different length of the alkyl chains separating the heterocycles) drastically influence the overall connectivity, thus leading to the attainment of compounds showing important differences in their structure and dimensionality and making the rational design of isoreticular compounds a challenge.

5.1.2.2 O-Donors
The molecular structure of the O-donor co-ligands that will be described in this chapter is reported in Scheme 5.4.

With respect to the N-donors, the number of O-donor ligands employed as auxiliary groups in the synthesis of phosphonate and phosphinate CPs is much lower. Carboxylic ligands show good affinity towards the most part of transition metals, and they are more competitive with phosphonates and phosphinates if compared to N-donors, when employed in the same synthesis. Their role in orienting different structural arrangements is similar to that of N-donor molecules. The main difference resides in the fact that they do not just contribute to fill the coordination sphere of the metal ions, but they also act as anionic ligands, thus also contributing to counterbalance the charge of the metal cation and pushing some P–O groups to be free from the engagement with the metal. Therefore, the use of these co-ligands frequently induces the formation of CPs with free polar groups that can also be available for the ion exchange reactions or post-synthetic modifications.

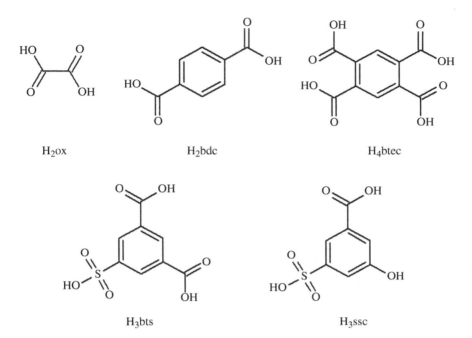

SCHEME 5.4   Molecular structure of the O-donor co-ligands reported in this chapter.

5.2   CPs BASED ON PHOSPHONATES AND N-DONOR AUXILIARY LIGANDS

A large number of mixed-linker phosphonate CPs with N-donor co-ligands and oxo-molybdate clusters have been synthesized in the recent past by the Zubieta group. A detailed and exhaustive description of these compounds has already been reported in a dedicated chapter on the recent metal phosphonate chemistry book [7]. Therefore, we will not treat about these materials, and the interested readers can find all the information on the aforementioned book.

Herein, we report on the state of the art of the CPs based on phosphonates and N-donor ancillary ligands, subdividing them in agreement with the nature of the ancillary ligand.

5.2.1   2,2′-Bipyridine and Related Molecules

This section reports on the CPs based on 2,2′-bipy and other ancillary ligands structurally related to it, such as phen, bpa and bpytrz. Table 5.1 reports all the CPs that we could find in the literature, along with their space group symmetry and dimensionality.

TABLE 5.1 CPs Based on 2,2′-bipy and Related Ancillary Ligands

Compound	Space group	Dimensionality	Reference
[Cu(2,2′-bipy)(H$_2$O)(VO)(O$_3$PCH$_2$PO$_3$)] (1)	Orthorhombic $Pca2_1$	2D	[8]
[Cu(2,2′-bipy)(VO)(O$_3$PCH$_2$CH$_2$PO$_3$)] (2)	Monoclinic $P2_1/n$	2D	[8]
[Cu(2,2′-bipy)(VO)(O$_3$PCH$_2$CH$_2$CH$_2$PO$_3$)]·H$_2$O (3)	Monoclinic $P2_1/c$	2D	[8]
Zn$_2$(O$_3$PC$_6$H$_5$)$_2$(2,2′-bipy) (4)	Unknown	2D	[9]
Zn$_2$(O$_3$PCH$_2$C$_6$H$_5$)$_2$(2,2′-bipy) (5)	Unknown	2D	[9]
[{Ni(2,2′-bipy)$_3$}{Ni(2,2′-bipy)$_2$(H$_2$O)}{(Mo$_5$O$_{15}$)(4,4′-dbp)}]·(4.75H$_2$O)] (6)	Monoclinic	1D	[10]
4,4′-dbp = 4,4′-dimethylenebiphenyldiphosphonic acid	$P2_1/c$		
[Cu(2,2′-bipy)]VO$_2$(O$_3$PC$_6$H$_5$)(HO$_3$PC$_6$H$_5$) (7)	Monoclinic $P2_1/c$	1D	[11]
[Cu$_3$(2,2′-bipy)$_3$](H$_2$O)V$_4$O$_9$(O$_3$PC$_6$H$_5$)$_4$ (8)	Monoclinic $P2_1/n$	1D	[11]
[Cu(2,2′-bipy)]$_2$V$_3$O$_6$(O$_3$PC$_6$H$_5$)$_3$(HO$_3$PC$_6$H$_5$) (9)	Triclinic P-1	1D	[11]
[Cu(2,2′-bipy)]VO(O$_3$PC$_6$H$_5$)$_2$ (10)	Monoclinic $P2_1/c$	1D	[11]
[Cu(2,2′_-bipy)]V$_2$O$_4$(O$_3$PCH$_3$)$_2$·H$_2$O (11)	Monoclinic Cc	1D	[11]
Cu(2,2′-bipy)(VO$_2$)$_2$(O$_3$PCH$_2$PO$_3$)H$_2$O (12)	Orthorhombic $Pbca$	2D	[12]
[Cu(2,2′-bipy)(VO$_2$)(O$_3$P(CH$_2$)$_2$PO$_3$H)]·1.5H$_2$O (13)	Monoclinic $C2/c$	2D	[12]
Cu(2,2′-bipy)(HO$_3$PCH$_2$CH$_2$PO$_3$H) (14)	Monoclinic $P2_1/n$	2D	[13]

Compound	Crystal system	Dimensionality	Reference
$[Cu(2,2'\text{-bipy})_2(o\text{-}O_3PC_6H_4PO_3)]\cdot 8H_2O$ (**15**)	Monoclinic $C2/c$	1D	[13]
$Zn_3(LH)_2(2,2'\text{-bipy})_3\cdot 18H_2O$ (**16**) H_4L = 1-amino-1-phenylmethane-1,1-diphosphonic acid	Monoclinic $C2/c$	1D	[14]
$[M(2,2'\text{-bipy})_2(p\text{-xdpaH}_4)](p\text{-xdpaH}_2)$ (**17**) M = Co^{2+}, Ni^{2+}	Monoclinic $C2/c$	1D	[15]
p-xdpaH$_4$ = α,α'-p-xylylenediphosphonic acid; $Cu(2,2'\text{-bipy})(HO_3PC_{10}H_6CO_2)$ (**18**) $H_2O_3PC_{10}H_6CO_2H$ = 2,6-carboxynaphthalene phosphonic acid	Triclinic $P\text{-}1$	1D	[16]
$[Cu(2,2'\text{-bipy})]_2V_2O_4F_2(p\text{-}O_3PC_6H_4PO_3)$ (**19**)	Monoclinic $P2_1/n$	2D	[17]
$[Cu(2,2'\text{-bipy})VO_2(H_3BTP)]\cdot 1.5H_2O$ (**20**) H_6BTP = 1,3,5-benzenetriphosphonic acid	Triclinic $P\text{-}1$	2D	[17]
$[Cu(2,2'\text{-bipy})]_2V_3O_7(HBTP)$ (**21**)	Orthorhombic $Pbca$	2D	[17]
$[Cu(2,2'\text{-bipy})(H_4BTP)]\cdot H_2O$ (**22**)	Triclinic $P\text{-}1$	1D	[17]
$Zn_{1.5}(pbc)_2(2,2'\text{-bipy})$ (**23**) H_3pbc = 4-phosphonobenzoic acid	Monoclinic $C2/c$	3D	[18]
$Zn_3(pbc)_2(2,2'\text{-bipy})(H_2O)\cdot H_2O$ (**24**)	Triclinic $P\text{-}1$	3D	[19]
$Zn_3(pbc)_2Zn(2,2'\text{-bipy})(H_2O)\cdot 2H_2O$ (**25**)	Monoclinic $P2_1/n$	3D	[19]
$\{[Cu(2,2'\text{-bipy})(H_2O)]MoO_2(hedp)\}\cdot 2H_2O$ (**26**) $hedpH_5$ = 1-hydroxyethylidenediphosphonic acid	Monoclinic $P2_1/c$	1D	[20]
$Zn_2(2,2'\text{-bipy})_2(UO_2)_2(hedp)_2(H_2O)_2$ (**27**)	Orthorhombic $Pna2_1$	2D	[21]
$Cu(2,2'\text{-bipy})(p\text{-}HO_3PC_6H_4PO_3H)$ (**28**)	Triclinic $P\text{-}1$	2D	[22]
$[Cu(2,2'\text{-bipy})(p\text{-xdpaH}_2)]\cdot 2H_2O$ (**29**)	Triclinic $P\text{-}1$	2D	[23]

(Continued)

TABLE 5.1 (*Cont'd*)

Compound	Space group	Dimensionality	Reference
[Cu(2,2′-bipy)(m-xdpaH$_2$)] (**30**)	Triclinic P-1	1D	[23]
m-xdpaH$_4$ = α,α′-m-xylylenediphosphonic acid			
{[Co(phen)$_2$(H$_2$O)$_2$][Co(phen)$_2$(H$_2$O)][(Mo$_5$O$_{15}$) (O$_3$PCH$_2$CH$_2$CH$_2$PO$_3$)]}·6H$_2$O (**31**)	Orthorhombic $Pbca$	1D	[24]
[Cu(phen)]$_2$(V$_2$O$_4$)(O$_3$PCH$_2$CH$_2$CH$_2$PO$_3$H$_2$)(H$_2$O) (**32**)	Monoclinic $C2/c$	1D	[24]
Cu$_2$(phen)$_2$(O$_3$PCH$_2$CH$_2$PO$_3$)(H$_2$O)$_3$]·H$_2$O (**33**)	Monoclinic $P2_1/n$	1D	[25]
Zn(O$_3$PC$_6$H$_5$)(phen) (**34**)	Unknown	2D	[9]
Zn$_2$(O$_3$PCH$_2$C$_6$H$_5$)$_2$(phen) (**35**)	Unknown	2D	[9]
Zn$_2$(O$_3$PC$_2$H$_5$)$_2$(phen) (**36**)	Unknown	Unknown	[9]
Zn$_2$(O$_3$PCH$_3$)$_2$(phen) (**37**)	Unknown	Unknown	[9]
{[Ni(1,10-phen)$_2$(H$_2$O)]$_2$[(Mo$_5$O$_{15}$)(4,4′-dbp)]}·5.75H$_2$O (**38**)	Triclinic P-1	1D	[10]
{[Co(1,10-phen)$_2$(H$_2$O)]$_2$[(Mo$_5$O$_{15}$)(4,4′-dbp)]}·5.5H$_2$O (**39**)	Triclinic P-1	1D	[10]
{[Cu(phen)]VO(O$_3$PC$_6$H$_{5/2}$}·0.5H$_2$O (**40**)	Triclinic P-1	1D	[26]
[Cu$_2$(phen)$_2$(O$_3$PCH$_2$PO$_3$)(V$_2$O$_5$)(H$_2$O)]·H$_2$O (**41**)	Monoclinic $P2_1/n$	2D	[27]
[Cu$_2$(phen)$_2$(O$_3$P(CH$_2$)$_3$PO$_3$)(V$_2$O$_5$)]·C$_3$H$_8$ (**42**)	Monoclinic $C2/c$	2D	[27]
Mn(phen)[HO$_3$P(CH$_2$)$_2$PO$_3$H](H$_2$O)$_2$ (**43**)	Monoclinic $C2/c$	1D	[28]
[Cu(HL)(phen)]·0.5H$_2$O (**44**)	Triclinic P-1	1D	[29]
H$_3$L = m-HO$_3$S–C$_6$H$_4$–PO$_3$H$_2$			
[Y(L)(phen)(H$_2$O)$_2$]·2H$_2$O (**45**)	Monoclinic $P2_1/c$	1D	[29]
H$_3$L = m-HO$_3$S–C$_6$H$_4$–PO$_3$H$_2$			
[Cu(phen)]$_2$V$_2$O$_5$(m-O$_3$PC$_6$H$_4$PO$_3$) (**46**)	Monoclinic $P2/n$	1D	[17]

Compound	Crystal system	Space group	Dimensionality	Reference
Cu(phen)VO$_2$(m-HO$_3$PC$_6$H$_4$PO$_3$) (47)	Monoclinic	$P2_1/c$	2D	[17]
Cu(phen)(H$_2$O)VO$_2$(p-HO$_3$PC$_6$H$_4$PO$_3$) (48)	Monoclinic	$P2_1/n$	2D	[17]
α-[Cu(phen)VO$_2$(p-HO$_3$PC$_6$H$_4$PO$_3$)] (49)	Monoclinic	$P2_1/n$	3D	[17]
β-[Cu(phen)VO$_2$(p-HO$_3$PC$_6$H$_4$PO$_3$)] (50)	Monoclinic	$C2/c$	1D	[17]
Cu(phen)V$_3$O$_6$(H$_2$O)(HBTP) (51)	Monoclinic	$P2_1/c$	2D	[17]
[Cu(phen)(H$_4$BTP)]·2H$_2$O (52)	Monoclinic	$P2_1/n$	3D	[17]
Cu(phen)(m-HO$_3$PC$_6$H$_4$PO$_3$H) (53)	Triclinic	P-1	1D	[16]
Mn(phen)(HL) (54) H$_3$L = m-H$_2$O$_3$PCH$_2$–C$_6$H$_4$–COOH	Triclinic	P-1	1D	[30]
[Zn$_2$(phen)(hedpH$_2$)$_2$]·H$_2$O (55)	Orthorhombic	$Pnma$	1D	[31]
Zn(phen)(hedpH$_2$) (56)	Monoclinic	$P2_1/n$	1D	[31]
Cd$_2$(phen)$_2$(HO$_3$PCH$_2$CH$_2$COO)$_2$ (57)	Triclinic	P-1	1D	[32]
[Co$_2$(phen)$_2$(HO$_3$PCH$_2$CH$_2$COO)$_2$(μ–OH$_2$)]·H$_2$O (58)	Orthorhombic	$Pbcn$	1D	[32]
[Zn$_2$(phen)$_2$(UO$_2$)$_2$(hedp)$_2$(H$_2$O)$_3$]·3H$_2$O (59)	Monoclinic	$P2_1/c$	2D	[21]
[Cu(phen)(p-O$_3$SC$_6$H$_4$PO$_3$H)]·H$_2$O (60)	Monoclinic	$P2_1/n$	2D	[22]
Cu(phen)(H$_2$O)(p-O$_2$CC$_6$H$_4$PO$_3$H) (61)	Monoclinic	$P2_1/c$	1D	[22]

(Continued)

TABLE 5.1 *(Cont'd)*

Compound	Space group	Dimensionality	Reference
{[Cu(phen)]$_2$(p-HO$_3$PC$_6$H$_4$PO$_3$H)(p-HO$_3$PC$_6$H$_4$PO$_3$H$_2$)$_2$}·H$_2$O (**62**)	Triclinic P-1	1D	[22]
Zn$_3$(L)$_2$(phen)$_2$(H$_2$O)$_2$ (**63**) H$_3$L = 2'-carboxybiphenyl-4-ylmethylphosphonic acid	Orthorhombic P2$_1$2$_1$2$_1$	3D	[33]
Cu(phen)Mo$_2$O$_5$(H$_2$O)(hedp) (**64**)	Monoclinic P2$_1$/c	2D	[20]
{[Cu(phen)][Cu(phen)(H$_2$O) MoO$_2$(hedpH)$_2$]}·H$_2$O (**65**)	Monoclinic P2$_1$/c	1D	[20]
[Cu(phen)(p-xdpaH$_2$)]·3H$_2$O (**66**)	Triclinic P-1	2D	[23]
[Cu$_2$(H$_2$O)$_2$(phen)$_2$(p-xdpaH$_2$)]·3H$_2$O (**67**)	Triclinic P-1	1D	[23]
Cu(phen)(o-xdpaH$_2$) (**68**)	Monoclinic P2$_1$/n	2D	[23]
Cu(phen)(m-xdpaH$_2$) (**69**)	Triclinic P-1	1D	[23]
[Cu(H$_4$L)(phen)]·H$_2$O (**70**) H$_6$L = (2,4,6-trimethylbenzene-1,3,5-triyl)tris(methylene) triphosphonic acid	Monoclinic P2$_1$/n	2D	[34]
{[Cu(bpa)$_2$]VO$_2$F(H$_2$O)(m-O$_3$PC$_6$H$_4$PO$_3$)}·H$_2$O (**71**)	Monoclinic P2$_1$/n	1D	[17]
[Cu(bpa)]$_2$-V$_2$O$_4$F$_2$(m-O$_3$PC$_6$H$_4$PO$_3$) (**72**)	Monoclinic C2/c	2D	[17]
[Cu(bpa)]$_2$V$_2$O$_4$F$_2$(p-O$_3$PC$_6$H$_4$PO$_3$) (**73**)	Monoclinic P2$_1$/n	2D	[17]
Cu(bpa)(H$_4$BTP) (**74**)	Triclinic P-1	1D	[17]
[Cu$_2$(bpytrz)(H$_2$O)]VO(O$_3$PC$_6$H$_5$)$_2$(HO$_3$PC$_6$H$_5$) (**75**)	Monoclinic P2$_1$/c	2D	[26]

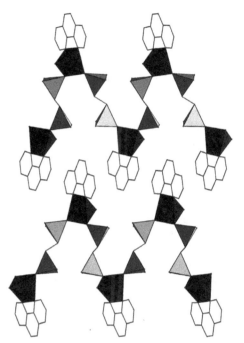

FIGURE 5.1 Polyhedral representation of the structure of compound **33.** PO$_3$C tetrahedra are represented in light grey and metal polyhedra in dark grey.

This kind of ligands has been extensively used for the construction of molecular complexes with practically all the metals of the transition series, because many of these species can act as antennas to harvest the visible light and find application in optical devices and in dye-sensitized solar cells, the most notable being the [Ru(2,2′-bipy)$_3$]$^{3+}$ complex.

When employed as ancillary ligands for the construction of CPs, they do not contribute to extend the dimensionality, and non-polymeric structures are often obtained. Looking at Table 5.1, it can be seen that only 4 CPs out of 75 show 3D structures, whereas 1D and 2D structures are almost equally distributed (37 and 31, respectively): this suggests that 2,2′-bipy and similar ligands are not the best choice if the task is the preparation of MOFs. The majority of the CPs reported in Table 5.1 are based on Cu^{2+} or Zn^{2+}, owing to their good affinity towards both N- and O-donors.

Among the 1D compounds, a very nice example of functional Cu-based CP containing ethylenediphosphonic acid and phen as ligands with formula [Cu$_2$(phen)$_2$(O$_3$PCH$_2$CH$_2$PO$_3$)$_2$(H$_2$O)$_3$]·H$_2$O was reported in 2004 by Lin et al. (compound **33**). Figure 5.1 shows the structure of **33**.

The copper atoms are pentacoordinated by one phen molecule, two oxygen atoms belonging to two different phosphonic groups and one water molecule, giving rise to a zigzag chain; adjacent chains are interdigitated, thanks to the π···π stacking interactions existing between phen molecules belonging to neighbouring units. This arrangement is very simple and resembles that adopted by myosin filaments in the muscles. The material can actually be considered as a supramolecular muscle, stretching in response to an applied potential: upon contacting it with $LiClO_4$, a redox process transforming a fraction of the square pyramidal Cu^{II} ions into square planar Cu^{I} ions is observed, and Li^+ ions are intercalated in the matrix for preserving the electroneutrality; as a consequence, the chain structure is deformed, leading to a 10% increase of the cell volume. The process is reversible, and the original structure is restored if Li^+ ions are de-intercalated, making the material an electrochemical actuator that can undergo several charge–recharge cycles, especially in the range between 3 and 5 V.

Sun et al. reported in 2012 a series of CPs based on molybdenum and copper containing the diphosphonic ligand 1-hydroxyethylidenediphosphonic acid ($hedpH_5$) and 2,2′-bipy or phen as ancillary ligands. Two of them are 1D (compounds **26**, containing 2,2′-bipy, of formula $\{[Cu(2,2′-bipy)(H_2O)]MoO_2(hedp)\}\cdot 2H_2O$ and **65**, containing phen, of formula $\{[Cu(phen)][Cu(phen)(H_2O)MoO_2(hedpH)_2]\}\cdot H_2O$), whereas one is 2D (compound **64**, containing phen, of formula $Cu(phen)Mo_2O_5(H_2O)$ (hedp)), and their structures are shown in Figures 5.2–5.4.

The ancillary ligands coordinate to the Cu atoms, whereas the hedp molecules serve as connection between the Cu and the Mo atoms, behaving in different ways from compound to compound. In all of them, the $\pi\cdots\pi$ stacking interactions between the aromatic rings of 2,2′-bipy or phen play a role in creating supramolecular assemblies of the polymeric units. Although oxomolybdenum organophosphonates are usually studied for their magnetic properties, in this case, the surface photovoltage (SPV) properties were investigated for the first time ever, evidencing that these materials act as p-type semiconductors (Figure 5.5): applying a positive external electric field to the samples, the SPV response increases, whereas if the applied electric field is negative, the SPV response is weakened. This behaviour can be attributed to the positive electric field being beneficial to the separation of photoexcited electron–hole pairs.

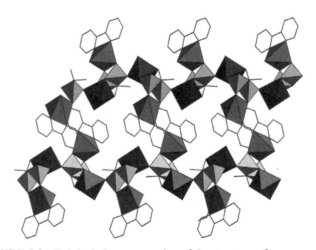

FIGURE 5.2 Polyhedral representation of the structure of compound **26**.

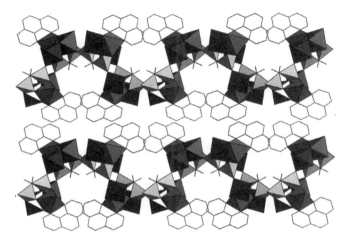

FIGURE 5.3   Polyhedral representation of the structure of compound **65**.

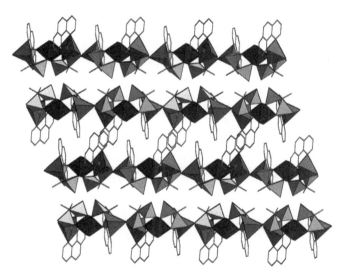

FIGURE 5.4   Polyhedral representation of the structure of compound **64**.

Three out of the four 3D compounds reported to date have been synthesized by Chen et al., using 4-phosphonobenzoic acid (H$_3$pbc) and 2,2′-bipy in combination with zinc (compounds **23**, **24** and **25**). These CPs were obtained in different synthetic conditions and show peculiar features: **23** (Figure 5.6) has formula Zn$_{1.5}$(pbc)(2,2′-bipy), and its structure consists of 1D inorganic chains linked by the pbc ligands to create a framework. Interestingly, reheating the filter liquor of **23**, an open-framework CP with GIS-zeolite topology is obtained, not including 2,2′-bipy as a ligand (Figure 5.7); the authors hypothesize that 2,2′-bipy could play some role in the driving of the structure, because in its absence this open-framework material cannot be obtained.

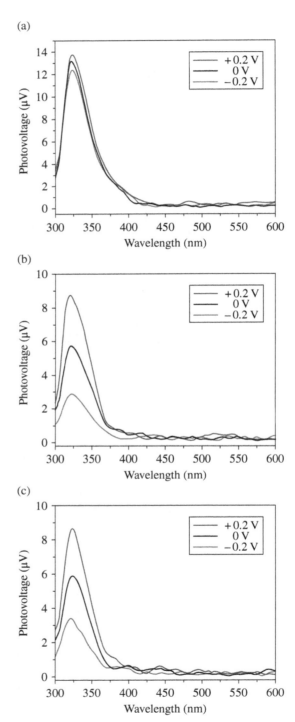

FIGURE 5.5 Surface photovoltage (SPV) curves for compounds **26** (a), **64** (b) and **65** (c).

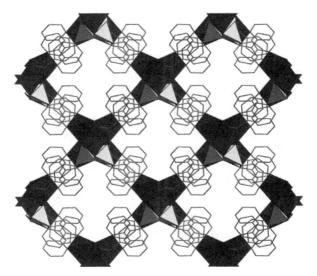

FIGURE 5.6 Polyhedral representation of the structure of compound **23**.

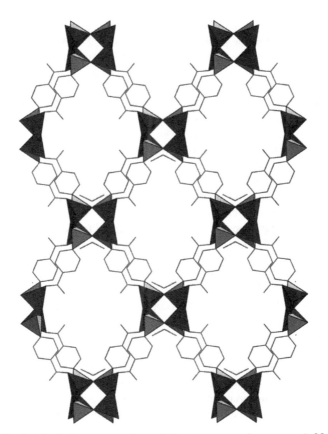

FIGURE 5.7 Polyhedral representation of the structure of compound **23** without the 2,2′-bipy obtained after heating the filtered mother liquors.

This compound shows breathing effect and has a good selectivity towards the adsorption of methanol over water. **24** and **25** (Figures 5.8 and 5.9) of formula $Zn_3(pbc)_2(2,2'-bipy)(H_2O) \cdot H_2O$ and $Zn_2(pbc)_2 \cdot Zn(2,2'-bipy)(H_2O) \cdot 2H_2O$, respectively, show ABW-zeolite topology, even if their structural arrangements are different: in **24**, the ancillary ligand participates in the construction of the framework, whereas in **25** it forms a complex with Zn that acts a template and the framework is built around it from the connection of pbc and Zn. The luminescent properties of **24** and **25** were studied, observing that the emission bands are different and blue shifted a little in contrast with free $Zn(bpy)Cl_2$ due to the different packing of $Zn(bpy)^{2+}$ in the framework, suggesting that the luminescence emissive wavelengths can be modulated by the construction of different crystal structures with various packing fashions.

5.2.2   Terpyridine and Related Molecules

This section deals with the CPs based on terpy and other ancillary ligands structurally related to it, such as bisterpy and mephenterpy. Table 5.2 reports all the compounds found in the literature, along with their space group symmetry and dimensionality.

All of the reported CPs have been synthesized by the Zubieta group and are based on Cu. As already said in the previous section about 2,2'-bipy, also terpy, due to its chelating character, has extensively been used for the construction of molecular complexes having peculiar luminescent properties. Being able to occupy three positions in the coordination sphere of the metal atom, terpy is even less suitable than 2,2'-bipy for the formation of extended structures, and the average dimensionality of the CPs based on it and its analogues is low: 18 out of the 33 compounds reported in Table 5.2 show 1D structure, whereas 10 are 2D and only 4 are 3D.

Compounds **76/107** have basically been studied from the structural point of view, with the aim of rationalizing the relationships between the building block structure and the relative CP structure; for some of them, the magnetic properties were investigated, but they all display the common weak ferromagnetic interactions associated with the presence of weakly coupled Cu^{2+} centres.

Although no peculiar properties have been observed to date for this class of CPs, we decided to include them in this contribution for the sake of completeness.

5.2.3   4,4'-Bipy and Related Molecules

Table 5.3 reports all the phosphonate CPs containing 4,4'-bipy and similar molecules found in the literature.

These ligands, contrary to the chelating ones, display a strong directionality, and they are able to connect different metal ions, therefore, they usually contribute to extend the dimensionality. As a matter of fact, among the 34 compounds found, 17 are 3D, whereas the remaining are 1D and 2D, showing that this kind of ligands can be more suitable for the preparation of MOF-like species. The most part of the reported compounds are based on Cu and Co and, in minor part, on Mn and Ni metal ions owing to their good affinity towards both N- and O-donor ligands.

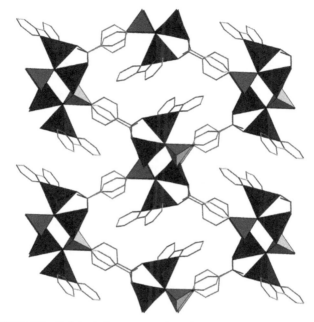

FIGURE 5.8   Polyhedral representation of the structure of compound **24**.

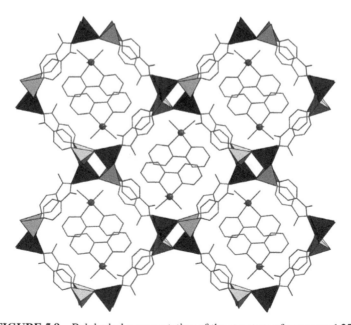

FIGURE 5.9   Polyhedral representation of the structure of compound **25**.

TABLE 5.2 CPs Based on Terpy and Related Ancillary Ligands

Compound	Space group	Dimensionality	Reference
[Cu(terpy)(HO$_3$PCH$_2$CH$_2$PO$_3$H)]·4H$_2$O (**76**)	Triclinic P-1	1D	[35]
[Cu(terpy)](V$_2$O$_4$)(O$_3$PPh)(HO$_3$PPh)$_2$ (**77**)	Triclinic P-1	1D	[36]
[Cu(terpy)]VO(O$_3$PCH$_2$PO$_3$) (**78**)	Triclinic P-1	1D	[36]
[Cu(terpy)]$_2$(V$_4$O$_{10}$)(O$_3$PCH$_2$CH$_2$PO$_3$) (**79**)	Monoclinic $C2/c$	1D	[36]
{[Cu(terpy)](V$_2$O$_4$)[O$_3$P(CH$_2$)$_3$PO$_3$]}·2.5H$_2$O (**80**)	Triclinic P-1	1D	[36]
[Cu(terpy)](V$_2$O$_4$)[O$_3$P(CH$_2$)$_3$PO$_3$] (**81**)	Monoclinic $P2_1/n$	3D	[36]
[Cu(terpy)(H$_2$O)](V$_3$O$_6$)[O$_3$P(CH$_2$)$_4$PO$_3$] (**82**)	Triclinic P-1	2D	[36]
[Cu(terpy)]$_2$V$_2$O$_5$(m-O$_3$PC$_6$H$_4$PO$_3$) (**83**)	Monoclinic $C2/c$	2D	[17]
[Cu(terpy)]$_2$V$_3$O$_6$F(p-O$_3$PC$_6$H$_4$PO$_3$H)$_2$ (**84**)	Monoclinic $P2_1/c$	2D	[17]
[Cu$_2$(bisterpy)]V$_2$O$_4$(O$_3$PCH$_2$PO$_3$H)$_2$ (**85**)	Monoclinic $P2_1/n$	1D	[37]
[Cu$_2$(bisterpy)(H$_2$O)]VO$_2$[O$_3$P(CH$_2$)$_3$PO$_3$][HO$_3$P(CH$_2$)$_3$PO$_3$H$_2$] (**86**)	Triclinic P-1	1D	[37]
{[Cu$_2$(bisterpy)]V$_2$O$_4$[O$_3$P(CH$_2$)$_6$PO$_3$H]}$_2$·2H$_2$O (**87**)	Triclinic P-1	2D	[37]
[Cu$_2$(bisterpy)(H$_2$O)$_2$]V$_2$O$_4$[O$_3$P(CH$_2$)$_2$PO$_3$][HO$_3$P(CH$_2$)$_2$PO$_3$H]$_2$ (**88**)	Triclinic P-1	2D	[37]

Compound	Crystal system / Space group	Dimensionality	Ref.
{[Cu$_2$(bisterpy)]V$_4$O$_8$[O$_3$P(CH$_2$)$_3$PO$_3$]$_2$}·4H$_2$O (**89**)	Triclinic P-1	2D	[37]
{[Cu$_2$(bisterpy)]V$_2$O$_4$(OH)$_2$[O$_3$P(CH$_2$)$_4$PO$_3$]}·4H$_2$O (**90**)	Triclinic P-1	2D	[37]
{[Cu$_2$(bisterpy)]V$_4$O$_4$[O$_3$P(CH$_2$)$_5$PO$_3$H]$_4$}·7.3H$_2$O (**91**)	Triclinic P-1	3D	[37]
Cu2(bisterpy)[HO$_3$P(CH$_2$)$_2$PO$_3$H]$_2$ (**92**)	Triclinic P-1	1D	[38]
{Cu2(bisterpy)[HO$_3$P(CH$_2$)$_2$PO$_3$H]$_2$}·6H$_2$O (**93**)	Triclinic P-1	1D	[38]
{Cu$_2$(bisterpy)(H$_2$O)$_2$[HO$_3$P(CH$_2$)$_2$PO$_3$H]}(NO$_3$)$_2$ (**94**)	Triclinic P-1	1D	[38]
{Cu$_2$(bisterpy)(NO$_3$)$_2$[HO$_3$P(CH$_2$)$_3$PO$_3$H]·2H$_2$O (**95**)	Monoclinic $C2/c$	1D	[38]
[Cu2(bisterpy)]V$_2$F$_2$O$_2$(HO$_3$PCH$_2$PO$_3$)(O$_3$PCH$_2$PO$_3$) (**96**)	Triclinic P-1	1D	[37]
[Cu$_2$(bisterpy)]V$_2$F$_4$O$_4$[HO$_3$P(CH$_2$)$_2$PO$_3$H] (**97**)	Monoclinic $P2_1/n$	1D	[37]
{[Cu$_2$(bisterpy)]V$_2$F$_2$O$_2$(H$_2$O)$_2$[HO$_3$P(CH$_2$)$_2$PO$_3$]$_2$}·2H$_2$O (**98**)	Triclinic P-1	2D	[37]
[Cu$_2$(bisterpy)(H$_2$O)$_2$]V$_2$F$_2$O$_2$[O$_3$P(CH$_2$)$_3$PO$_3$][HO$_3$P(CH$_2$)$_3$PO$_3$H]$_2$ (**99**)	Monoclinic $C2/c$	2D	[37]
{[Cu$_2$(bisterpy)]V$_4$F$_4$O$_4$(OH)(H$_2$O)[HO$_3$P(CH$_2$)$_5$PO$_3$]–[O$_3$P(CH$_2$)$_5$PO$_3$]}·H$_2$O (**100**)	Triclinic P-1	2D	[37]
{[Cu$_2$(bisterpy)]$_3$V$_8$F$_6$O$_{17}$[HO$_3$P(CH$_2$)$_3$PO$_3$]$_4$}·0.8H$_2$O (**101**)	Triclinic P-1	3D	[37]

(Continued)

TABLE 5.2 (*Cont'd*)

Compound	Space group	Dimensionality	Reference
{[Cu$_2$(bisterpy)(H$_2$O)]$_2$V$_8$F$_4$O$_8$(OH)$_4$[HO$_3$P(CH$_2$)$_5$PO$_3$H]$_2$– [O$_3$P(CH$_2$)$_5$PO$_3$]$_3$}·4.8H$_2$O (**102**)	Triclinic P-1	3D	[37]
[Cu(mephenterpy)](VO$_2$)(HO$_3$PCH$_2$PO$_3$) (**103**)	Triclinic P-1	1D	[39]
[Cu(mephenterpy)]$_2$(V$_2$O$_6$)(O$_3$PCH$_2$CH$_2$PO$_3$) (HO$_3$PCH$_2$CH$_2$PO$_3$) (**104**)	Triclinic P-1	1D	[39]
[Cu(mephenterpy)](VO$_2$)(HO$_3$PCH$_2$CH$_2$CH$_2$PO$_3$) (**105**)	Triclinic P-1	1D	[39]
[Cu(mephenterpy)]$_2$(V$_2$O$_{5/2}$)(O$_3$PCH$_2$CH$_2$CH$_2$PO$_3$) (**106**)	Monoclinic C2/c	1D	[39]
{Cu(mephenterpy)[HO$_3$P(CH$_2$)$_4$PO$_3$H]}·H$_2$O (**107**)	Monoclinic C2/c	1D	[39]

TABLE 5.3 CPs Based On 4,4′-bipy and Related Ancillary Ligands

Compound	Space group	Dimensionality	Reference
$[Cu^I_2Cu^{II}(hedpH)_2(4,4'-bipy)_2]\cdot2H_2O$ **(109)**	Monoclinic	2D	[40]
hedpH$_5$ = 1-hydroxyethylidenediphosphonic acid	$P2_1/c$		
$[Co(4,4'-bipy)(HO_3PCH_2CH_2CH_2CH_2PO_3H)(H_2O)_2]\cdot4H_2O$ **(110)**	Monoclinic	2D	[41]
	$C2/c$		
$[Cu_3(hedpH)_2(4,4'-bipy)(H_2O)_2]\cdot2H_2O$ **(111)**	Triclinic	2D	[42]
	$P-1$		
$[CdVO(pmida)-(4,4'-bipy)(H_2O)_2]\cdot(4,4'-bipy)_{0.5}\cdot H_2O$ **(112)**	Monoclinic	3D	[43]
H$_4$pmida = N-(phosphonomethyl)iminodiacetate	$P2_1/c$		
$[CoVO(pmida)-(4,4'-bipy)(H_2O)_2]\cdot(4,4'-bipy)_{0.5}$ **(113)**	Monoclinic	3D	[43]
	$P2_1/c$		
$Cu[(H_2L)(4,4'-bipy)(C_2O_4)_{0.5}]\cdot(ClO_4)\cdot2H_2O$ **(114)**	Monoclinic	3D	[44]
H$_2$L = N-(phosphonomethyl)-aza-18-crown-6	$P2_1/c$		
$\{Cu(4,4'-bipy)\}_{0.5}VO_2(O_3PC_6H_3)$ **(115)**	Orthorhombic	3D	[26]
	$Fdd2$		
$[Cu(4,4'-bipy)]VO_2(o-HO_3PC_6H_4PO_3)$ **(116)**	Monoclinic	3D	[26]
	$P2_1$		
$\{[Cu(4,4'-bipy)]\}[Cu(H_2O)_{0.5}](o-O_3PC_6H_4PO_3)\cdot H_2O$ **(117)**	Orthorhombic	3D	[26]
	$Fdd2$		
$Cu_4(aedp)_2(4,4'-bipy)(H_2O)_4$ **(118)**	Triclinic	3D	[45]
H$_4$aedp = 1-aminoethylidenediphosphonic acid	$P-1$		
$Zn_3(4,4'-bipy)(ppat)_2$ **(119)**	Monoclinic	3D	[46]
ppat = phosphonoacetic acid	Cc		
$(4,4'-bipyH)_2[M(4,4'-bipy)(H_2O)_4][V_2O_2(pmida)_2]\cdot2H_2O$ **(120)**	Monoclinic	1D	[47]
M = Mn, Co	$P2/c$		
$[Ni(HO_3PC_2H_4COO)(4,4'-bipy)(H_2O)]\cdot0.5(4,4'-bipy)$ **(121)**	Monoclinic	2D	[48]
	$P2/n$		

(Continued)

TABLE 5.3 *(Cont'd)*

Compound	Space group	Dimensionality	Reference
[Ni(O$_3$PC$_2$H$_4$COOH)(4,4'-bipy)(H$_2$O)$_3$]·H$_2$O (122)	Monoclinic $P2_1/n$	1D	[48]
Cu(4,4'-bipy)(H$_2$O)$_2$(HO$_3$PCH$_2$CH$_2$CH$_2$PO$_3$H) (123)	Triclinic P-1	2D	[13]
[Zn(H$_2$O)$_6$][Zn$_8$(4-cpp)$_6$(4,4'-bipy)] (124)	Rhombohedral R-3	3D	[49]
4-cppH$_3$ = 4-carboxyphenylphosphonic acid			
M(t-BuPO$_3$H)$_2$(4,4'-bipy)(H$_2$O)$_2$ (125)	Triclinic P-1	1D	[50]
M= Cu^{2+}, Mn^{2+}			
[Co(phen)(4,4'-bipy)(PhPO$_3$H)$_2$]·0.5H$_2$O (126)	Tetragonal $P4_32_12$	1D	[51]
[Mn$_3$(4,4'-bipy)(L)$_2$]·(4,4'-bipy)$_{0.5}$ (127)	Triclinic P-1	3D	[52]
H$_4$L= H$_2$O$_3$PCH(OH)COOH			
[Co$_3$(4,4'-bipy)(H$_2$O)$_2$(L)$_2$]·(4,4'-bipy)$_{0.5}$ (128)	Triclinic P-1	3D	[52]
H$_4$L= H$_2$O$_3$PCH(OH)COOH			
[Cu$_2$(4,4'-bipy)$_{0.5}$(BTP)$_2$]·H$_2$O (129)	Triclinic P-1	2D	[53]
BTP = 4-Br-thienylphosphonate			
[Co(aepa)(4,4'-bipy)(H$_2$O)$_2$]·2H$_2$O (130)	Hexagonal $P6_1$	3D	[54]
H$_2$aepa = 2-aminoethylphosphonic acid			
Ni(4,4'-bipy)(H$_2$L)$_2$(H$_2$O)$_2$ (131)	Orthorhombic $Pccn$	1D	[30]
H$_3$L = p-H$_2$O$_3$PCH$_2$-C$_6$H$_4$-COOH			
Ni$_2$(4,4'-bipy)(L)(OH)(H$_2$O)$_2$ (132)	Triclinic P-1	2D	[30]
H$_3$L= m-H$_2$O$_3$PCH$_2$-C$_6$H$_4$-COOH			
M$_2$(cmdpH$_2$)$_2$(4,4'-bipy)$_{0.5}$(H$_2$O) (133)	Monoclinic $P2_1/c$	3D	[55]
M = Co^{2+}, Mn^{2+}			
cmdpH$_4$ = N-cyclohexylaminomethanediphosphonic acid			
Cu$_2$(H$_4$BTMP)$_2$(H2O)$_2$(4,4'-bipy) (134)	Triclinic P-1	3D	[56]
H$_6$BTMP = benzenetriyl-1,3,5-tris(methylphosphonic acid)	$P1$	2D	

Cu$_4$(hedpH)$_2$(4,4'-bipy)(H$_2$O)$_5$ (**135**)	2D	Triclinic P-1	[57]
Cu$_3$(L)(4,4'-bipy)(OH)$_2$ (**136**) H$_4$L = biphenyl-4,4'-diphosphonic acid	3D	Triclinic P-1	[58]
[Cu(m-xdpaH$_2$)(4,4'-bipy)]·2H$_2$O (**137**) m-xdpaH$_4$ = α,α'-m-xylylenediphosphonic acid	2D	Triclinic P-1	[59]
[Cu(p-xdpaH$_2$)(4,4'-bipy)]·2H$_2$O (**138**) p-xdpaH$_4$ = α,α'-p-xylylenediphosphonic acid	2D	Monoclinic P2$_1$/c	[59]
Cu(H$_4$L)(4,4'-bipy) (**139**) H$_6$L = (2,4,6-trimethylbenzene-1,3,5-triyl)tris(methylene) triphosphonic acid	2D	Triclinic P-1	[59]
[Cu$_3$(H$_4$L)(4,4'-bipy)$_2$]·9H$_2$O (**140**) H$_8$L = N,N,N',N'-tetrakis(phosphonomethyl)-α,α'-p-xylylenediamine	3D	Triclinic P-1	[60]
[Cu$_3$(H$_2$L)(4,4'-bipy)$_2$]·11H$_2$O (**141**) H$_8$L = N,N,N',N'-tetrakis(phosphonomethyl) hexamethylenediamine	3D	Triclinic P-1	[60]
[Cu$_3$(H$_2$L)(etbipy)$_2$]·24H$_2$O (**142**) H$_8$L = N,N,N',N'-tetrakis(phosphonomethyl) hexamethylenediamine	3D	Monoclinic P2$_1$/n	[60]
[Cu$_3$(hedpH)$_2$(4,4'-azpy)(H$_2$O)$_2$]·1.6H$_2$O (**143**)	2D	Triclinic P-1	[42]

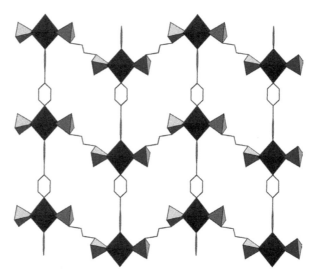

FIGURE 5.10   Polyhedral representation of the structure of compound **110**.

One of the simplest CPs, based on 4,4′-bipy and butylene-diphosphonate (P_2but), was reported in 2003 by Fu and co-workers (compound **110**). **110** has formula $[Co(4,4′-bipy)(P_2but)H_2O]_2 \cdot 4H_2O$, and its structure is composed of CoN_2O_4 distorted octahedra bridged, along the b-axis, by the 4,4′-bipy molecules. The butylene-diphosphonate ligands connect these infinite chains along the a-axis designing a square 2D grid. This structural motif can be found very frequently in many phosphonate CPs with linear diphosphonate groups. These layers are packed along the bc plane by strong H-bonds among the acidic P–O and P–OH groups.

The structure of compound **110** is shown in Figure 5.10.

In 2012, Taddei et al. reported two related isomeric Cu compounds based on p- and m-xylylenediphosphonic acids (m- and p-xdpaH$_4$, respectively) and 4,4′-bipy with formula $Cu(m$-xdpaH$_2)(4,4′$-bipy$)\cdot 2H_2O$ and $Cu(p$-xdpaH$_2)(4,4′$-bipy$)\cdot 2H_2O$ (compounds **137** and **138**). The structural motifs of these two CPs are the same of the aforementioned compound, as the copper atoms, in octahedral coordination, are linked to each other by the 4,4′-bipy groups with the formation of infinite 1D chains internally linked by the diphosphonic groups, thus designing a 2D square grid.

The structures of the two compounds are shown in Figure 5.11.

Both compounds undergo a phase transformation at temperatures above 80°C due to the loss of the apical coordinated water molecule of the Cu coordination sphere. This makes the axial positions of every copper octahedron free, but the metal atoms undergo different processes: half of them retain a distorted octahedral geometry, since the vacancies in their coordination sphere are filled by two oxygen atoms belonging to a chelating bidentate phosphonate group, while the remaining half become square planar. This variation in the coordination geometry is confirmed by EPR measurements that have been performed in order to get information on the copper coordination. The EPR spectra together with the phase transformation of **137** are shown in Figure 5.12.

(a)   (b)

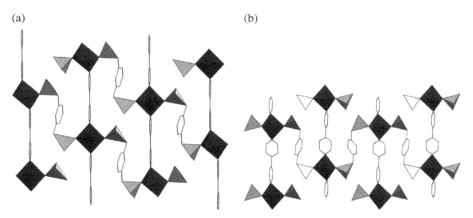

FIGURE 5.11   Polyhedral representation of the structure of compounds **137** (a) **and 138** (b).

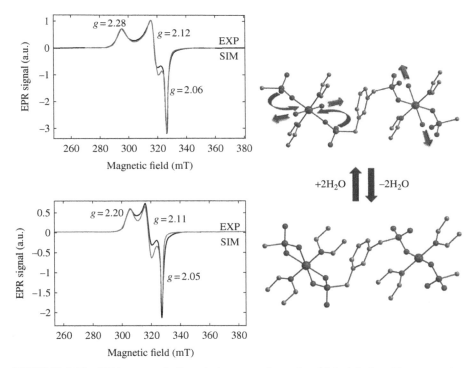

FIGURE 5.12   EPR spectra (left) and phase transformation (rights) induced by one water molecule loss in compound **137**.

The g-values changed from the starting materials, with pure rhombic structure, to new values that merge the contribution of both the octahedral and the square planar Cu environments, with a nice agreement with the observed PXRD structures.

An example of 3D pillared layered Cu-MOF based on the aromatic and rigid biphenyl diphosphonate ligand (P_2biphH_4) and the 4,4′-bipy was reported in 2012

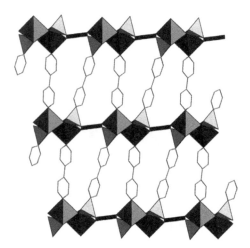

FIGURE 5.13   Polyhedral representation of the structure of compound **136**.

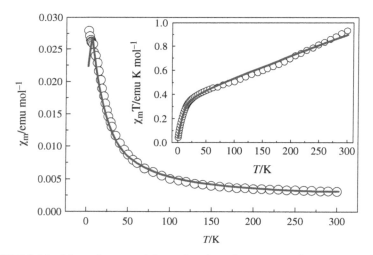

FIGURE 5.14   Magnetic susceptivity as function of temperature for compound **136**.

(compound **136**). **136** has formula $[Cu_3(P_2biph)(4,4\text{-}bipy)(OH)_2]$ with $P_2biphH_4 = H_2O_3P–(C_6H_4)_2–PO_3H_2$, and the structure is composed of 2D inorganic layers $[Cu_3O_6P_2]_n$ connected in the third dimension by the 4,4'-bipy moieties. Figure 5.13 shows the structure of compound **136**.

Despite its 3D structure, the MOF does not possess permanent porosity as the aromatic moieties of both the phosphonate and the 4,4'-bipy occupy the interlayer region in a dense fashion and they are involved in mutual strong $\pi\cdots\pi$ stacking. The magnetic properties of **136** were studied, and the data are reported in Figure 5.14: the χ versus T curve indicated that **136** exhibits antiferromagnetic exchange coupling between the

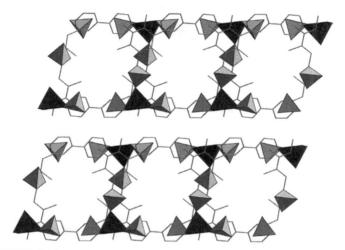

FIGURE 5.15   Polyhedral representation of the structure of compound **139**.

Cu(II) centres. χmT value fell smoothly above 22 K with cooling and then dropped rapidly to a minimum of 0.056 cm^3 mol^{-1}K^{-1} at 2 K, suggesting an antiferromagnetic behaviour.

The use of a tritopic phosphonic ligand, namely, the benzene-1,3,5-triyltris (methylene)triphosphonic (P$_3$tmBTPH$_6$), together with 4,4′-bipy allowed to obtain a copper-based 2D layered CP of formula [Cu(P$_3$tmBTPH$_2$)(4,4′-bipy)] (compound **139**). **139** has a layered structure in which the layers are formed by square pyramidal coordinated copper ion with the oxygen atoms belonging to the phosphonic moieties of the P$_3$tmBTP ligand and with the N atoms of the 4,4′-bipy. The trisphosphonate ligand adopts a *cis–trans–trans* conformation and binds three copper ions with its three phosphonic acid groups in monodentate fashion, forming 1D inorganic units. The 4,4′-bipy co-ligand connects these units along the *b*-axis and designs 1D open channels (3.5 × 7 Å) running along the sheets. The layers are then stacked along the *c*-axis. The structure of **139** is shown in Figure 5.15.

This compound shows a good adsorption selectivity of CO$_2$ over CH$_4$ measured at 273 K and up to 15 bar. Although the absolute amount of CO$_2$ adsorbed is quite low (only 3 wt%), the selectivity over CH$_4$ is pretty good (about 29 times at high pressures).

More recently, Costantino et al. reported the structure and gas sorption properties of three related Cu CPs based on diaminotetraphosphonate groups (viz. the *N,N,N′,N′*-tetrakis(phosphonomethyl)-α,α′-*p*-xylylenediamine (H$_8$P$_4$pxy) and the *N,N,N′,N′*-tetrakis(phosphonomethyl)hexamethylenediamine (H$_8$P$_4$hex)) and the two related N-donor co-ligands 4,4′-bipy and etbipy.

The three CPs (**140/142**) have formulae [Cu$_3$(H$_2$P$_4$pxyl)(4,4′-bipy)$_2$]·9H$_2$O, [Cu$_3$(H$_8$P$_4$hex)(4,4′-bipy)$_2$]·11H$_2$O and [Cu$_3$(H$_8$P$_4$hex)(etbipy)$_2$]·24H$_2$O, respectively. They are 3D open-framework compounds with isoreticular structure constituted by trinuclear composite building units (CBUs), made of two square pyramidal and one

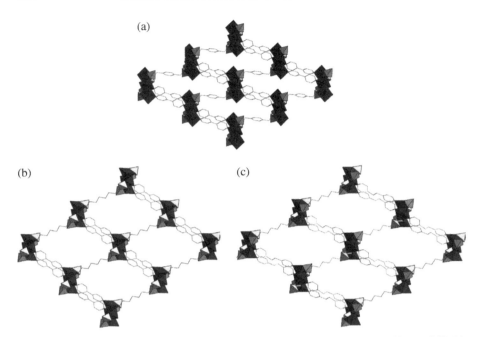

FIGURE 5.16 Polyhedral representation of the structure of compounds **140** (a), **141** (b) and **142** (c).

octahedral coordinated Cu ions and PO_3C groups, that form 1D inorganic chains running along the c-axis. These 1D chains are linked to each other by the N-donor ligands, forming rhombic 1D channels of about 22×8, 21×12 and 24×12 Å for **140**, **141** and **142**, respectively. Their structures are shown in Figure 5.16.

Upon the evacuation of the solvent, the three compounds display different breathing behaviours, due to the partial collapse of the 1D channels with the formation of the so-called *np* phase (*np* for narrow pore). This feature can be ascribed to the different flexibility degrees of both the phosphonic and the N-donor ligands. The CP with the alkyl chains as organic spacer and 4,4′-bipy (compound **141**) displays good absorption selectivity towards CO_2 over N_2, both at 195 K and 1 atm (up to 80 cm^3 g^{-1} CO_2) and 273 K and 10 atm (about 25 cm^3 g^{-1} CO_2) (Figure 5.17).

5.2.4   Imidazole and Related Molecules

Table 5.4 reports CPs based on imidazole and related co-ligands 1,2-bix, 1,3-bix, 1,4-bix, pi and dib. A peculiarity of imidazole is that it can act as an anionic ligand, a feature shared only with bpytrz among all of the N-donors here reported.

Among all the compounds, there is only a 3D structure reported by Zheng and co-workers in 2011. They used a 1,4-bis(imidazol-1-ylmethyl)-benzene (1,4-bix, Scheme 5.2) as a co-ligand for the construction of a Co-based phosphonate CP of formula $[Co_3(pna)_2(1,4\text{-bix})(H_2O)_2]$ (**152**) (pnaH$_3$ = 6-phosphononicotinic acid). This

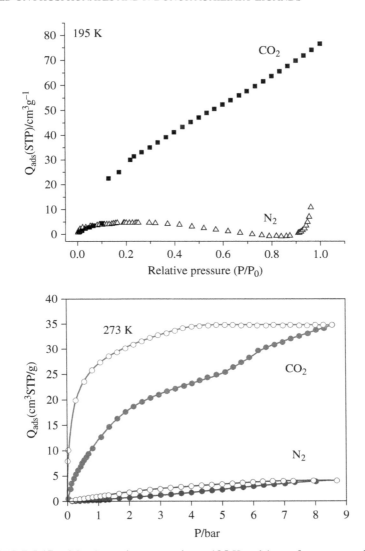

FIGURE 5.17 CO_2 absorption properties at 195 K and 1 atm for compound **141**.

compound possesses a rare $(4 \cdot 6^2)_2(4^2 \cdot 6^{22} \cdot 7 \cdot 8^3)$ topology, and its structure is composed of 1D inorganic units running along the c-axis bridged both by the phosphonocarboxylic and by the N-donor ligands, in a rhombic array. The 1D chains are composed of trinuclear corner-sharing octahedral Co units internally linked by the triply connected PO_3C groups. The structure of **150** is shown in Figure 5.18.

The cobalt-based CBUs are ferromagnetically coupled at low temperature, but, more interestingly, **150** also displays a long-range canted antiferromagnetic coupling mediated by the pna^{3-} ligand with a coercive field of 1130 Oe at 0.5 K. The

TABLE 5.4 CPs Based on Imidazole and Related Ancillary Ligands

Compound	Space group	Dimensionality	Reference
$[Zn_2(AEDP)(im)_3] \cdot H_2O$ **(144)** $AEDPH_4$ = 1-aminoethylidenediphosphonic acid	Monoclinic $P2_1/n$	1D	[14]
$[Cd_3(AEDPH)_2(im)_5] \cdot im$ **(145)**	Triclinic $P\text{-}1$	1D	[14]
$Zn_2(4\text{-pi})(1,2\text{-bix})$ **(146)** 4-piH4 = 4-phosphonoisophthalic acid	Monoclinic $P2_1/c$	2D	[61]
$[Zn_3(4\text{-cpp})_2(1,2\text{-bix})] \cdot H_2O$ **(147)** 4-cppH$_3$ = 4-carboxyphenylphosphonic acid	Triclinic $P\text{-}1$	2D	[62]
$Zn_2(4\text{-pi})(1,3\text{-bix})$ **(148)**	Monoclinic $C2/c$	2D	[61]
$[Zn_3(4\text{-cpp})_2(1,3\text{-bix})_2] \cdot 2.5H_2O$ **(149)**	Triclinic $P\text{-}1$	3D	[62]
$Zn_2(4\text{-cppH})_2(1,4\text{-bix})$ **(150)**	Monoclinic $P2_1/c$	2D	[62]
$[Ni_2(4\text{-cppH})_2(1,4\text{-bix})_3] \cdot 2H_2O$ **(151)**	Triclinic $P\text{-}1$	2D	[62]
$Co_3(pna)_2(1,4\text{-bix})(H_2O)_2$ **(152)** pnaH$_3$ = 6-phosphonic nicotinic acid	Monoclinic $P2_1/c$	3D	[63]
$Zn(pi)_2(UO_2)(PO_3C_6H_5)_2$ **(153)**	Triclinic $P\text{-}1$	1D	[64]
$[Zn(dib)(UO_2)(PO_3C_6H_5)_2] \cdot 2H_2O$ **(154)**	Monoclinic $P2_1/n$	2D	[64]

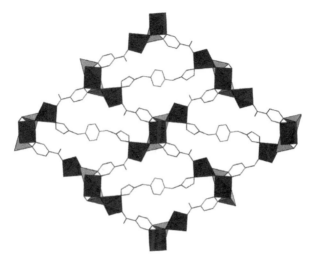

FIGURE 5.18   Polyhedral representation of the structure of compound **150**.

antiferromagnetic behaviour was estimated by using a trinuclear high-spin Co(II) model with a $S = 1/2$ and $S = 3/2$ central and lateral Co ions, respectively. The best fit between 5 and 40 K gave, as values for the anisotropic exchange coupling, $J_z = 5.9$ cm^{-1}, $J_{xy} = 3.5$ cm^{-1}, $g_z = 5.0$, $g_{xy} = 1.4$ and $g = 2.1$. The χT versus T curves together with the field-dependent magnetization are shown in Figure 5.19a and b.

5.2.5   Other Ligands

This section reports on the CPs based on ligands that could not be included as analogues of 2,2′-bipy, terpy, 4,4′-bipy or im because of the lack of straightforward structural relationships. These ligands are pz, tpypz, dabco and pyterpy. Pyterpy appears in this section since it has a substantial difference in the coordination capability if compared to the other terpy-related ligands. Table 5.5 reports the 5 CPs belonging to this heterogeneous class.

The most interesting CP in Table 5.5 is compound **155**, of formula Cu$_4$(hedpH)$_2$(pz)(H$_2$O)$_4$, based on Cu and containing hedp as the phosphonic ligand and pz as the ancillary ligand. The structure of **155** is shown in Figure 5.20.

It shows a 3D framework where layers constituted of the connection of Cu atoms and hedp moieties are pillared by pz. These layers have quite a complex structure: {Cu$_3$(hedpH)$_2$(H$_2$O)$_2$} double chains, containing two different types of Cu atoms organized in interconnected {Cu$_3$(hedp)$_2$} trimer units, are linked by {CuO4} units. The presence of such a variety of distinct metal sites is responsible for the interesting magnetic properties displayed by **155**: it is ferromagnetic above 17.7 K, undergoing a magnetic phase transition around 4.2 K and becoming antiferromagnetic at low field (500 Oe). Although more common in bimetallic systems, ferrimagnetic behaviour has also been observed in homometallic chain

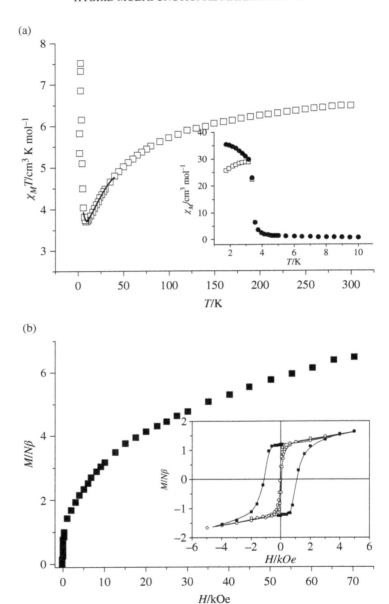

FIGURE 5.19 Magnetic susceptivity (a) and field-dependent magnetization (b) for compound **150**.

systems containing discrete species with an odd number of interacting metal ions or complicated alternating sequences of ferro-/antiferromagnetic interactions. The magnetic behaviour of **155** is mainly attributed to the phosphonate layer, composed of $\{Cu_3(hedpH)_2(H_2O)_2\}$ double chains linked by $\{CuO_4\}$ units.

TABLE 5.5 CPs Based on Pz, Tpypz, Dabco, and Pyterpy Ancillary Ligands

Compound	Space group	Dimensionality	Reference
$Cu_4(hedpH)_2(pz)(H_2O)_4$ (155) hedpH$_5$ = 1-hydroxyethylidenediphosphonic acid	Triclinic P-1	3D	[65]
$[M(pz)(H_2O)_4][M_2(Hpmida)_2(pz)(H_2O)_2]\cdot 2H_2O$ (156) H$_4$pmida = N-(phosphonomethyl)iminodiacetate M = Co^{2+}, Ni^{2+}	Triclinic P-1	2D	[66]
$[Cu_2(H_2O)_2(tpypz)(p$-xdpaH$_2)_2\cdot(p$-xdpaH$_4)_2]\cdot 2H_2O$ (157) p-xdpaH$_4$ = α,α'-p-xylylenediphosphonic acid	Triclinic P-1	1D	[23]
$(dabcoH)2[Zn_8'(OOCC_6H_4PO_3)_6]\cdot 6H_2O$ (158)	Rhombohedral R-3	3D	[49]
$[Cu_3(hedpH)_2(Hpyterpy)_2]\cdot 2H_2O$ (159) hedpH$_5$ = 1-hydroxyethylidenediphosphonic acid	Triclinic P-1	1D	[57]

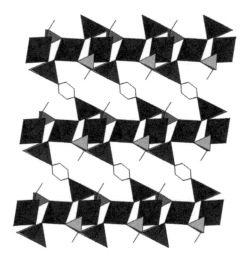

FIGURE 5.20   Polyhedral representation of the structure of compound **155**.

5.3   CPs BASED ON PHOSPHONATES AND O-DONOR AUXILIARY LIGANDS

A large number of mixed-linker phosphonate CPs with O-donor co-ligands (mainly oxalate) have been synthesized in the recent past mainly by the Mao group [67]. A detailed and exhaustive description of these compounds has been already reported in a dedicated chapter on the recent metal phosphonate chemistry book. Therefore, we will not treat about these materials, and the interested readers can find all the information on the aforementioned book.

Herein, we briefly report on some metal phosphonate CPs containing aromatic carboxylates as auxiliary ligands, especially the benzene dicarboxylic acid (hereafter H_2bdc), the 1,2,4,5-benzenetetracarboxylic acid (hereafter H_4btec), the 5-sulfoisophthalic acid (hereafter H_3bts) and the 5-sulfosalicylic acid (hereafter H_3ssc). In all of these compounds, the carboxylic linkers are directly bonded to the metal ions, together with the phosphonic groups. These auxiliary ligands have the capability to extend the connectivity, and thus, the dimensionality of all the reported compounds is usually high: indeed, except two layered compounds, all of the reported CPs have 3D structure.

The molecular formula, symmetry and dimensionality of some recently reported CPs containing these co-ligands are reported in Table 5.6.

In 2009, Liu and co-workers reported on the synthesis, structure and optical properties of four related homochiral Zn-based CPs, containing the enantiomers (S)- or (R)- of (1-phenylethylamino)methylphosphonic acid ($pempH_2$) with bdc and btec co-ligands (**160, 161**).

The four CPs have formulae (R)-$[Zn_4(pempH)_4(bdc)_2]\cdot 2H_2O$ $(R$-**1**$)$, (S)-$[Zn_4(pempH)_4(bdc)_2]\cdot 2H_2O$ $(S$-**1**$)$, (R)-$[Zn_3(pempH)_2(btec)(H_2O)_2]\cdot H_2O$ $(R$-**2**$)$ and (S)-$[Zn_3(pempH)_2(btec)(H_2O)_2]\cdot H_2O$.

TABLE 5.6 CPs Based on O-donor Carboxylic Ancillary Ligands

Compound	Space group	Dimensionality	Reference
(R)-(S)-[Zn$_4$(Hpemp)$_4$(bdc)$_2$]·2H$_2$O (**160**) (R)-, (S)-, pempH$_2$=H$_2$PO$_3$CH$_2$NH$_2$CHCH$_3$C$_6$H$_5$	Orthorhombic $P2_12_12$	3D	[68]
(R)-(S)-[Zn$_3$(pempH)$_2$(btec)(H$_2$O)$_2$]·H$_2$O (**161**)	Monoclinic C2	3D	[68]
[Pb$_4$(Hbts)(bts)(bdc)$_{1.5}$]$_3$ 2H$_2$O (**162**) H$_3$bts=H$_2$O$_3$PCH$_2$–NC$_5$H$_9$–COOH	Triclinic P-1	3D	[69]
[Pb4(Hbts)(bts)(bdc)$_{1.5}$]$_3$ 2H$_2$O (**163**)	Triclinic P-1	3D	[69]
[Pb3(Hbts)2(Hssc)]$_3$ 2H$_2$O (**164**) Hssc=5-sulfoisophthalic acid	Monoclinic $P2_1/c$	3D	[69]
[Cd$_2$(HMorph)(bdc)(Hbdc)] (**165**) H$_2$Morph=O(CH$_2$CH$_2$)$_2$NCH$_2$-PO$_3$H$_2$	Triclinic P-1	3D	[70]
[Zn(HL$_1$)(bdc)$_{0.5}$] (**166**) H$_2$L^1=H$_2$O$_3$PCH(NH$_2$)C$_6$H$_5$	Monoclinic C2/c	2D	[71]
Cd$_{1.5}$(H$_2$bts)(bdc)$_{0.5}$ (**167**)	Triclinic P-1	3D	[71]
Cd$_2$(2-Hcpip)(bdc)] (**168**) 2-H$_3$cpip=H$_2$O$_3$PCH$_2$NC$_5$H$_9$COOH	Monoclinic $P2_1/a$	2D	[72]
[Ln$_3$(4-H$_2$cpip)(4-Hcpip)$_2$(bdc)$_2$(H$_2$O)]·7H$_2$O (**169**) 4-H$_3$cpip=H$_2$O$_3$PCH$_2$NC$_5$H$_9$COOH Ln = La, Ce, Pr, Nd, Sm, Eu, Gd, Tb	Triclinic P-1	3D	[73]

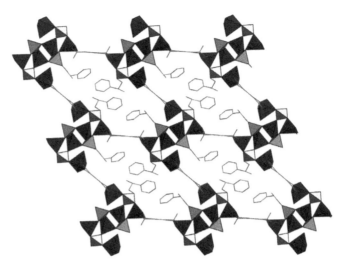

FIGURE 5.21 Polyhedral representation of the structure of compound **160**.

Compounds *R*-**1** and *S*-**1** (**160**) are a pair of optical enantiomers and crystallize in the orthorhombic system, with chiral space group $P2_12_12$, whereas compounds *R*-**2** and *S*-**2** (**161**) crystallize in the monoclinic chiral *C*2 space group. The structure of **160** is 2D and formed by layers of Zn phosphonate chains pillared by the bdc groups. The pempH⁻ serves as a tridentate ligand, coordinating to three different Zn atoms through its three phosphonate oxygen donors. The coordination geometry of Zn atoms is tetrahedral. The homochiral double chains are formed by corner-sharing ZnO_4 and PO_3C tetrahedra. Two adjacent chains are connected by μ_2-bdc groups. The chiral organic moieties of the phosphonate groups are placed on the opposite sides of each hybrid layer. The structure of R-**1** is shown in Figure 5.21.

The homochiral frameworks **161**, constructed by using the tetracarboxylate linkers btec, crystallize in the chiral *C*2 space groups and possess a 3D open-framework structure in which the Zn atoms have again a tetrahedral coordination and are connected by the oxygen atoms belonging to both the phosphonate and the carboxylate moieties. The pempH⁻ acts as a tridentate ligand, with each of its three phosphonate oxygen atoms coordinated to three different Zn atoms (Zn1, Zn2, Zn3). The ZnO_4 and PO_3C tetrahedra are linked in a corner-sharing manner, forming a zigzag homochiral inorganic chain running along the *b*-axis. Each inorganic chain is surrounded by other four equivalent chains through the carboxylic linker with the generation of 1D channels running along the *b*-axis, as shown in Figure 5.22.

The four homochiral phosphonate MOFs display a weak SHG response when irradiated with 560 nm Nd:YAG laser that is 0.8 times faster than the urea, and the circular dichroism spectra confirm the enantiomeric purity of the two MOFs.

Bdc ligand has also been recently used for the attainment of two Zn and Cd 3D CPs containing two different phosphonic acids, namely, $H_2O_3PCH(NH)_2C_6H_5$ (H_2L^1)

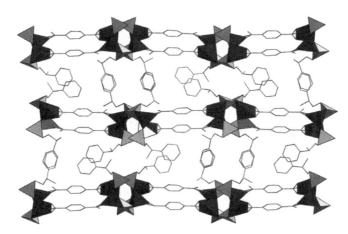

FIGURE 5.22   Polyhedral representation of the structure of compound **161**.

and $H_2O_3PCH_2$–NC_5H_9–COOH (H_3bts). The formula of the two compounds is $[Zn(HL^1)(bdc)_{0.5}]$ **166** and $[Cd_{1.5}(Hbts)(bdc)_{0.5}]$ for **167**, respectively.

In both cases, the phosphonates contribute to the formation of 2D networks that are pillared by the bdc ligand. In compound **166**, Zn has tetrahedral coordination and the ZnO_4 groups are interconnected by phosphonate groups into a 2D layer. The layers are pillared by the bdc^{2-} anions to generate a 3D framework structure with two types of channel system along the c-axis. A notable feature of compound **166** is the presence of alternate left- and right-handed helical chains in the structure. In compound **167**, Cd atoms display two types of coordination modes, CdO_7 and CdO_4, and these coordination polyhedra are assembled into inorganic chains through the connection with the PO_3C groups. These chains are connected by the organic part of the L_2^{2-} ligand to form a double-layer structure in the ab plane. These layers are then linked by the bdc^{2-} anions to form a 3D framework structure with 1D channel systems along the a-axis. The structures of the two Cd and Zn CPs are shown in Figures 5.23 and 5.24.

Another similar 2D CP with Cd, (2-carboxypiperidyl)-N-methylenephosphonic acid (2-H_3cpip) and btc has been reported in 2011 by Sun and others. In this compound, with formula $[Cd_2(2\text{-Hcpip})(bdc)]$ (**168**), Cd atoms have octahedral coordination and are linked by the oxygen atoms belonging to the Hcpip phosphonate ligands forming 1D chains running along the b-axis. The chains are then interconnected by the bdc groups forming hybrid layers stacked along the a-axis. The carboxy-piperidine groups point towards the interlayer region. The structure of **168** is shown in Figure 5.25.

Bdc served as a co-ligand for the attainment of a series of 3D lanthanide CPs containing a carboxyphosphonate group, namely, the (4-carboxypiperidyl)-N-methylenephosphonic acid (4-H_3cpip).

The Ln CPs have the same formula $[Ln_3(4\text{-}H_2cpip)(4\text{-Hcpip})_2(bdc)_2(H_2O)]\cdot7H_2O$ with Ln = La, Ce, Pr, Nd, Sm, Eu, Gd and Tb. They crystallize in the triclinic P-1 space group, designing isostructural 3D frameworks in which Ln(III) polyhedra are

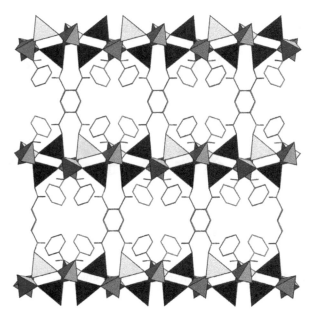

FIGURE 5.23 Polyhedral representation of the structure of compound **166**.

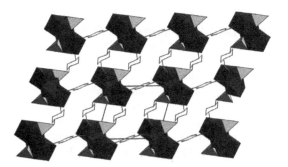

FIGURE 5.24 Polyhedral representation of the structure of compound **167**.

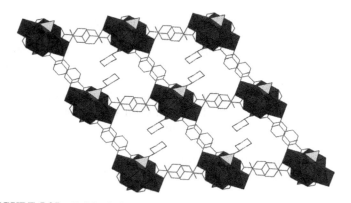

FIGURE 5.25 Polyhedral representation of the structure of compound **168**.

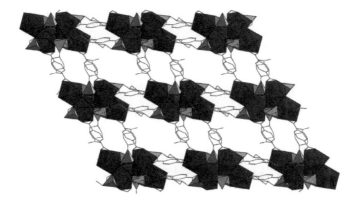

FIGURE 5.26   Polyhedral representation of the structure of compound **169**.

interconnected by bridging PO_3C tetrahedra into 2D inorganic layers parallel to the *ab* plane. In these structures, there are two 9-coordinated and one 8-coordinated Ln ions that are internally connected by the phosphonic groups creating a hexanuclear cluster: the clusters are interconnected via the Hcpip and the bdc groups forming layers. These layers are then pillared by other bdc groups designing a pillared layered structure. The rectangular pores that are formed are occupied by the organic part of the Hcpip ligands.

The structure of the Ln CPs is shown in Figure 5.26.

The Eu- and Tb-based CPs exhibit red and green emissions upon excitation at 360 nm with the typical bands characteristic of metal-centred fluorescence.

5.4   CPs BASED ON PHOSPHINATES AND AUXILIARY LIGANDS

The CPs formed by phosphinate ligands are shown in Table 5.7. The phosphinate ligands used in the materials can be divided in two different categories, namely, the monophosphinic and the diphosphinic acids, where two units are connected by an organic bridge.

As a general trend, phosphinates tend to form network with lower dimensionality than the corresponding phosphonates. In fact, only two 3D structures are reported. Members of the monophosphinic class are hypophosphite, monophenyl and diphenyl phosphinic acids. While the diphenyl phosphinate forms only 1D chains (compounds **173** and **175**), 2D networks were found either for the hypophosphite (**171**) or the monophenyl phosphinate (**174**) as shown in Figure 5.27.

The evident low steric request of the hypophosphite ligand is also testified by the fact that the 2D net and the 1D chain were prepared using phen (**171**) and 2,2′-bipy co-ligands (**172**), while for other phosphinates, only molecular [84] or small polynuclear [85] complexes are obtained. Of particular interest is the Mn CP (**170**) (see Figure 5.28) formed by the porphyrin ancillary ligand together with the monophenyl phosphinate. It can be considered a textbook example of

TABLE 5.7 CPs Based on Phosphinates and N-donor Ancillary Ligands

Compound	Space group	Dimensionality	Reference
[Mn(porphyrine)H$_2$PO$_2$]·H$_2$O (**170**)	Monoclinic C2/c	1D	[74]
Mn(H$_2$PO$_2$)(phen) (**171**)	Monoclinic P2$_1$/c	2D	[75]
Mn(H$_2$PO$_2$)(2,2'-bipy) (**172**)	Monoclinic C2/c	1D	[75]
Hg(Ph$_2$PO$_2$)Py$_2$ (**173**)	Monoclinic P2$_1$/c	1D	[76]
Mn(H(Ph)PO$_2$)(4,4'-bipy) (**174**)	Monoclinic P2/c	2D	[77]
Cu(Ph$_2$PO$_2$)(L) L=6-chloropyridin-2-olato (**175**)	Orthorhombic Pbca	1D	[78]
Co$_2$(pcp)$_2$(4,4'-bipy)(H$_2$O)$_4$ (**176**)	Orthorhombic Pnaa	2D	[79]
Co$_2$(pcp)$_2$(4,4'-bipy)(H$_2$O)$_4$ (**177**)	Monoclinic C2/c	1D	[79]
[Cu$_2$(pcp)$_2$(4,4'-bipy)]·5H$_2$O (**178**)	Orthorhombic I4$_{1/a}$	1D	[80]
[Cu$_2$(pcp)$_2$(4,4'-bpye)]·2.5H$_2$O (**179**)	Orthorhombic I4	1D	[81]
[Cu(pc$_2$p)(4,4'-bipy)(H$_2$O)]·2.5H$_2$O (**180**)	Monoclinic P2$_1$	3D	[82]
[Cu(pc$_2$p)(4,4'-bipy)(H$_2$O)]·3H$_2$O (**181**)	Monoclinic P2$_1$/n	2D	[83]
[Cu(pxylp)(4,4'-bipy)(H$_2$O)$_2$]·2H$_2$O (**182**)	Monoclinic P2/n	2D	[83]
Cu$_2$(pxylp)$_2$(4,4'-bipy) (**183**)	Monoclinic P2$_1$/c	3D	[83]

single-chain magnets. The manganese ion lies in the plane of the porphyrin coordinated by four nitrogen atoms. Two oxygen atoms of two different bridging phosphinates complete the slightly octahedral distorted coordination sphere of the metal atom. The infinite chains elongate along the c-axis. Some characteristics of the material, as the large separation between the chains due to the presence of the porphyrin ligands and the $S = 2$ value for the spin of the Mn(III) ion, have allowed for the first time to simulate a single-chain magnet with a deep level of understanding and to correlate the geometrical parameters with the dynamic behaviour of the system.

The second class is that of diphosphinates, where the two phosphinate units are held together by an organic bridge. The connecting group can be simple as a methylene or ethylene group as for the P,P'-diphenylmethylenediphosphinate (pcp) or P,P'-diphenylethylenediphosphinate (pc$_2$p), respectively, or much longer as in the case of the P,P'-diphenyl-p-xylenediphosphinate (pxylp).

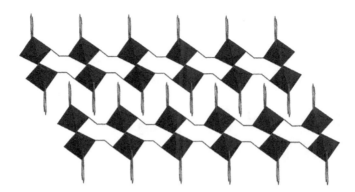

FIGURE 5.27   Polyhedral representation of the structure of compound **174**.

FIGURE 5.28   Polyhedral representation of the structure of compound **170**.

This class of ligands shows a large variety of structures with interesting potential properties. Most of them are synthesized in water under mild conditions. In a typical preparation, a solution of the diphosphinate and of the auxiliary ligand is added to a solution of a divalent metal salt in equimolar ratio at around 90°C. The kind of polymer obtained is critically regulated by the synthetic conditions as pH, temperature and concentration of the reagent as well illustrated, for example, by the Co, pcp and 4,4′-bipy system. In this case, just by varying the counterion of the cobalt metal from acetate to perchlorate, a 2D polymorph was obtained instead of the 1D one. The relationship between the two structures is well shown in Figure 5.29. The 4,4′-bipy is in *trans* position with respect to a water molecule and to an oxygen atom of the pcp ligand in the 2D and 1D structures, respectively.

Another way to induce the change in dimensionality is by using the temperature. The 2D square grid $[Cu(4,4′\text{-bipy})(pxylp)(H_2O)_2]\cdot 2(H_2O)$ polymer (**182**) loses water

(a)   (b)

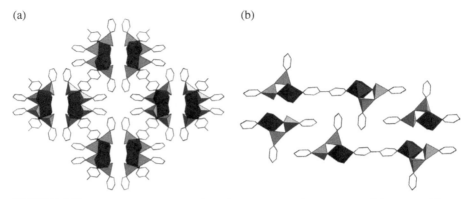

FIGURE 5.29   Polyhedral representation of the structure of compounds **176** (a) and **177** (b).

(a)   (b)

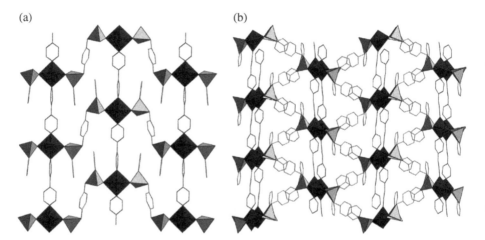

FIGURE 5.30   Polyhedral representation of the structure of compounds **182** (a) **and 183** (b).

molecules at 120°C, and a new crystalline 3D phase is formed (compound **183**). The network is a 4,6-connected binodal net with tcj/hc topology. The copper metal centre is the six-connected node and the pxylp ligand is the four-connected one. A comparison of the 2D and 3D structures (Figure 5.30) shows that the rearrangement occurs, thanks to the slippage of the 2D slabs and to the formation of new bonds between copper and oxygen atoms of neighbour slabs. The presence of $\pi\cdots\pi$ stacking interactions between the xylene and phenyl groups seems to be very important for the formation of a crystalline phase.

A more spectacular change was found for the Cu/pc$_2$p/bipy system. A 3D polymer of formula [Cu(pc$_2$p)(4,4′-bipy)(H$_2$O)]·2.5H$_2$O (compound **180**) was obtained rapidly in water at room temperature. Each copper atom was surrounded by two nitrogen atoms of the bipy, two oxygen atoms of two pc$_2$p ligands and one oxygen from a water molecule. Both the pc$_2$p or the bipy was bridged between two metal atoms.

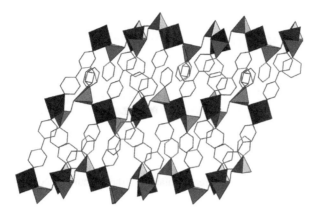

FIGURE 5.31 Polyhedral representation of the structure of compound **181**.

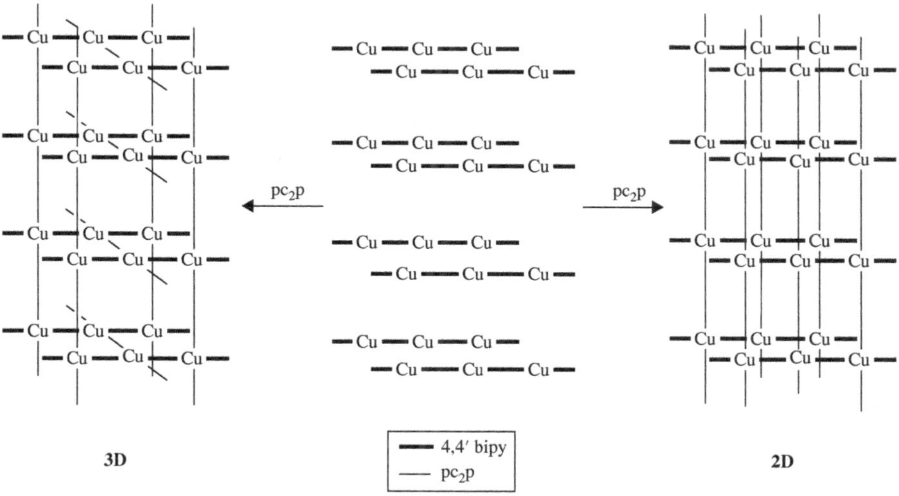

FIGURE 5.32 Schematic representation of the transformation of compound **181** in compound **180**.

The resulting network is a chiral network of cds type, and its architecture can be visualized with the help of Figure 5.31.

When the 3D polymer was placed in water for a few days at room temperature or for few hours at 80–90°C, a new 2D pseudo-polymorph, [Cu(4,4′-bipy)(pc$_2$p) (H$_2$O)]·3H$_2$O (compound **181**), was obtained. The structure of the polymer consists in a 2D square grid with the same type of arrangement found for the [Cu(4,4′-bipy)(pxylp)(H$_2$O)$_2$]·2(H$_2$O) polymer. The relationships between the 2D and 3D polymers are highlighted in Figure 5.32. The common features are the series of the [Cu(4,4′-bipy)] lines represented by the bold lines (bipyridine ligands) and the circles (copper atoms). The lines are aligned in two sets of planes, one below and one

FIGURE 5.33 Schematic representation of the structure of compound **178**.

over the plane of the drawing. The difference between the 3D and the 2D structures is in the way the second linker (pc$_2$p) connects the lines. In the case of the 3D network, the pc$_2$p anions connect the [Cu(4,4'-bipy)] lines either in the planes parallel to the sheet of paper or between the two planes. Instead, in the case of the 2D polymer, square grids are formed and there are no connections between the [Cu(4,4'-bipy)] lines of the planes below and above the paper sheet.

When the 3D polymer is heated up to 120°C, it starts losing some water and the structure changes maintaining a good degree of crystallinity. In an *in situ* temperature-dependent powder diffraction experiment where a sample was heated from 25 to 110°C and successively cooled down to room temperature, it was possible to recognize up to three significant structural changes. In any case, the material was able to recover the starting 3D phase. If the heating temperature was more than 120°C, an amorphous phase was obtained. When the latter is contacted with water for a few minutes, the 2D phase was recovered. The 3D to 2D transformation is very uncommon, since usually upon heating the dimensionality is increased. The 3D phase is a kinetic one, and in some sense, it represents the first stage of the coordination of the ligands around the metal ion, while the 2D one is the thermodynamic product.

The 3D polymer is in principle porous, and its porosity was tested with adsorption and desorption experiments with methanol using the desolvated phase. The total amount of methanol adsorbed by the 3D phase is around 4.0 mol mol^{-1} of material and is consistent with the occupation of all solvent sites, occupied by water molecules. The 3D framework is retained during the absorption measurement as shown by the PXRD analysis after the desorption of methanol.

Finally, it is time to introduce two CPs with a very unusual and rare 1D tube-like structure or metal–organic nanotube (MONT), namely, the [Cu$_2$(pcp)$_2$(4,4'-bipy)]·5H$_2$O (compound **178**) and the corresponding [Cu$_2$(pcp)$_2$(etbipy)]·5H$_2$O (compound **179**). As pointed out by a recent review [86], only few MONTs have been reported in the literature. The schematic representation of the structures is presented in Figure 5.33.

The tubular structure is the result of the connection of four columnar [Cu(pcp)]$_n$ polymeric units using the bipyridine ligands. The tubes are packed together, thanks to C–H⋯π interactions. The two structures are isoreticular. This is the first example of the application of the isoreticular principle for this class of ligand and in general

(a) (b)

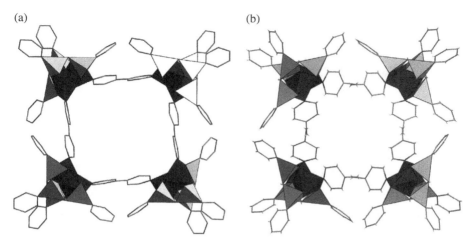

FIGURE 5.34 Polyhedral representation of the structure of compound **178** (a) and **179** (b).

one of the few reported for the field of the phosphinate MOFs. There are also several similarities between the two structures and their properties. For example, the copper is pentacoordinated by one nitrogen atom of the bipyridine and by four oxygen atoms of two different pcp ligands, and the thermal behaviour is very close (see below). From a structural point of view, the main difference is in the shape and the dimension of the internal cavity of the tube. In the case of the **178** (the one with 4,4′-bipy), the cavity is larger and it has a very definite square form with approximate size of 10 Å (Figure 5.34a). On the contrary, in **179**, the cavities have an ellipsoidal shape with a diameter that varies between 4.5 and 15 Å (Figure 5.34b). Also, as well illustrated by Figure 5.34, the planes of the pyridine rings of 4,4′-bipy are parallel to the tube wall for **178**, while in the other case, they are perpendicular. The free van der Waals space of the void per tube (calculated after the removal of all the guest water molecules) is around 20 and 14% of the total volume for **178** and **179**, respectively.

The thermal behaviour of the two materials is comparable. The guest water molecules inside the framework are lost before 80°C without any significant structural change of the tube framework. The materials are stable up to around 260°C, although a significant thermal expansion was noticed. At higher temperature, the bipyridine ligands are lost, but the material forms the polymeric $[Cu(pcp)]_n$ phase [87]. The stability of the dehydrated phases is so high that the single-crystal X-ray structure for anhydrous **179** was also reported.

Nanometric rod-shaped crystals of **178** and **179** were obtained when the synthesis was carried out in ethanol as solvent and at room temperature. A light blue colloidal dispersion formed instantaneously is obtained and a solid could be easily separated by centrifugation. FE-SEM and TEM images have shown a population of nanorods of lengths in the range of 200–1000 nm and with a cross section of around 30 nm. In particular, Figure 5.35 reports a typical single nanorod. Considering that a single tube is 2.7 nm large, it can be estimated as formed by around 50 parallel 1D MONTs.

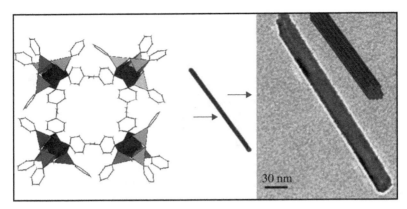

FIGURE 5.35   TEM image of a single tubular nanorod of compound **179**.

This is the first attempt towards the isolation of a single MONT comparable to the carbon nanotube. **178** and **179** either as bulk crystals or nanorods are porous as it was demonstrated by the CO_2 and solvent absorption measurements. For **179**, the nanorod particles have shown an additional porosity and a faster uptake of the CO_2 with respect to the microcrystalline material. The absorption of CH_4 and N_2 is smaller or negligible compared to the CO_2 gas even at 77 K, respectively. A potential use for CO_2 capture and purification in the presence of CH_4 and N_2 was foreseen.

5.5   CONCLUSIONS AND OUTLOOKS

In the context of tailored functional materials, this chapter tries to report an exhaustive survey on the vast class of CPs whose architecture is based on phosphorate building blocks with O- and N-donor auxiliary ligands.

The number of compounds that can be found in the literature is huge, and, as already underlined in the previous paragraphs, the phosphonate and phosphinate ligands display a very high structural variability that is further increased when the co-ligands are employed in the synthesis.

A rough classification of the CPs according to the dimensionality and to the nature of the auxiliary ligand (chelating, directional or both of them) has been done, and it allowed to rationalize some of the numerous structural motifs present.

However, it is clear that the functional properties of these compounds cannot be still rationally tailored in a way similar to the conventional carboxylate MOFs, as there is a lot of work to do especially for as concerns aspects related to the crystal engineering.

As a matter of fact, apart some predictable structural features (i.e. the expected coordination of the metal, the occurrence of $\pi \cdots \pi$ stacking of the aromatic moieties and the coordination capability of the phosphonic ligands), the route towards real isoreticular chemistry is long and still considered a challenge.

On the other hand, these compounds are to be considered very promising owing to their better thermal and chemical stability compared to conventional MOFs, and especially to their resistance to hydrolysis that is unanimously considered as a crucial aspect to be taken into account in designing functional materials for sustainable applications.

REFERENCES

1. (a) S. Ma, H.-C. Zhou, *Chem. Commun.* **2010**, *46*, 44–53; (b) C. Liu, F. Li, L.-P. Ma, H.-M. Cheng, *Adv. Mater.* **2010**, *22*, E28–E62. (c) A. S. Aricò, P. Bruce, B. Scrosati, J.-M. Tarascon, W. van Schalkwijk, *Nat. Mater.* **2005**, *4*, 366–377.

2. (a) B. P. Gomez-Romero, *Adv. Mater.* **2001**, 163–174; (b) P. Judeinstein, C. Sanchez, *J. Mater. Chem.* **1996**, *6*, 511–525; (c) L. Nicole, L. Rozes, C. Sanchez, *Adv. Mater.* **2010**, *22*, 3208–3214. (d) A. L. Mohana Reddy, S. R. Gowda, M. M. Shaijumon, P. M. Ajayan, *Adv. Mater.* **2012**, *24*, 5045–5064.

3. (a) C. Janiak, *Dalton Trans.* **2003**, 2781–2804; (b) A. Y. Robin, K. M. Fromm, *Coord. Chem. Rev.* **2006**, *250*, 2127–2157; (c) S. Kitagawa, R. Kitaura, S. Noro, *Angew. Chem. Int. Ed.* **2004**, *43*, 2334–2375.

4. (a) G. Ferey, *Chem. Soc. Rev.* **2008**, *37*, 191; (b) H.-C. Zhou, J. R. Long, O. M. Yaghi, *Chem. Rev.* **2012**, *112*, 673.

5. (a) AA. VV., in *Metal Phosphonate Chemistry: From Synthesis to Applications*, eds. A. Clearfield and K. Demadis, RSC, Oxford, **2011**; (b) G. Alberti, in *Comprehensive Supramolecular Chemistry*, eds. G. Alberti and T. Bein, Pergamon Press, New York, **1996**, vol. 7.

6. A. Vioux, J. Le Bideau, P. H. Mutin, D. Leclercq, *Top. Curr. Chem.* **2004**, *232*, 145–174.

7. S. Jones, J. Zubieta, in *Metal Phosphonate Chemistry: From Synthesis to Applications*, eds. A. Clearfield and K. Demadis, RSC, Oxford, **2011**.

8. R. C. Finn, J. Zubieta, *J. Chem. Soc. Dalton Trans.* **2000**, 1821–1823.

9. R. C. Clarke, K. Latham, C. J. Rix, M. Hobday, *Chem. Mater.* **2004**, *16*, 2463–2470.

10. H.-S. Zhang, R.-B. Fu, J.-J. Zhang, X.-T. Wu, Y.-M. Li, L.-S. Wang, X.-H. Huang, *J. Solid State Chem.* **2005**, *178*, 1349–1355.

11. G. Yucesan, W. Ouellette, V. Golub, C. J. O'Connor, J. Zubieta, *Solid State Sci.* **2005**, *7*, 445–458.

12. S. Ushak, E. Spodine, D. Venegas-Yazigi, E. Le Fur, J. Y. Pivan, *Microporous Mesoporous Mater.* **2006**, *94*, 50–55.

13. G. Yucesan, W. Ouellette, Y.-H. (Jessica) Chuang, J. Zubieta, *Inorg. Chim. Acta* **2007**, *360*, 1502–1509.

14. S.-P. Chen, Y.-X. Yuan, L.-L. Pan, L.-J. Yuan, *J. Inorg. Organomet. Polym. Mater.* **2008**, *18*, 384–390.

15. B. K. Tripuramallu, R. Kishore, S. K. Das, *Polyhedron* **2010**, *29*, 2985–2990.

16. P. Deburgomaster, J. Zubieta, *Acta Crystallogr.* **2010**, *66*, 1304–1305.

17. P. Deburgomaster, W. Ouellette, H. Liu, J. O. Connor, J. Zubieta, *CrystEngComm* **2010**, 446–469.

18. Z. Chen, Y. Ling, H. Yang, Y. Guo, L. Weng, Y. Zhou, *CrystEngComm* **2011**, *13*, 3378–3382.

19. Z. Chen, H. Yang, M. Deng, Y. Ling, L. Weng, Y. Zhou, *Dalton Trans.* **2012**, *41*, 4079–4083.

20. S.-H. Sun, Z.-G. Sun, Y.-Y. Zhu, D.-P. Dong, C.-Q. Jiao, J. Zhu, J. Li, W. Chu, H. Tian, M.-J. Zheng, et al., *Cryst. Growth Des.* **2013**, *13*, 226–238.

21. H.-Y. Wu, W. Yang, Z.-M. Sun, *Cryst. Growth Des.* **2012**, *12*, 4669–4675.

22. V. Zima, J. Svoboda, Y.-C. Yang, S.-L. Wang, *CrystEngComm* **2012**, *14*, 3469–3477.

23. T. M. Smith, J. Vargas, D. Symester, M. Tichenor, C. J. O'Connor, J. Zubieta, *Inorg. Chim. Acta* **2013**, *403*, 63–77.

24. R.-B. Fu, X.-T. Wu, S.-M. Hu, J.-J. Zhang, Z.-Y. Fu, W.-X. Du, S.-Q. Xia, *Eur. J. Inorg. Chem.* **2003**, 1798–1801.

25. K.-J. Lin, S.-J. Fu, C.-Y. Cheng, W.-H. Chen, H.-M. Kao, *Angew. Chem. Int. Ed.* **2004**, *43*, 4186–4189.

26. G. Yucesan, V. Golub, J. Zubieta, *CrystEngComm* **2005**, 2241–2251.

27. S. Ushak, E. Spodine, E. Le Fur, D. Venegas-yazigi, J. Pivan, W. Schnelle, R. Cardoso-gil, *Inorg. Chem.* **2006**, *45*, 5393–5398.

28. C. Merrill, A. K. Cheetham, *Inorg. Chem.* **2007**, *46*, 278–84.

29. Z.-Y. Du, J.-J. Huang, Y.-R. Xie, H.-R. Wen, *J. Mol. Struct.* **2009**, *919*, 112–116.

30. R. Lei, X. Chai, H. Mei, H. Zhang, Y. Chen, Y. Sun, *J. Solid State Chem.* **2010**, *183*, 1510–1520.

31. W.-N. Wang, Z.-G. Sun, Y.-Y. Zhu, D.-P. Dong, J. Li, F. Tong, C.-Y. Huang, K. Chen, C. Li, C.-Q. Jiao, et al., *CrystEngComm* **2011**, *13*, 6099–6106.

32. E. Fernández-Zapico, J. M. Montejo-Bernardo, R. D'Vries, J. R. García, S. García-Granda, J. Rodríguez Fernández, I. de Pedro, J. Blanco, *J. Solid State Chem.* **2011**, *184*, 3289–3298.

33. W.-Q. Kan, J.-F. Ma, Y.-Y. Liu, J. Yang, B. Liu, *CrystEngComm* **2012**, *14*, 2268–2277.

34. S.-F. Tang, X.-B. Pan, X.-X. Lv, S.-H. Yan, X.-R. Xu, L.-J. Li, X.-B. Zhao, *CrystEngComm* **2013**, *15*, 1860–1873.

35. R. C. Finn, J. Zubieta, *Inorg. Chim. Acta* **2002**, *332*, 191–194.

36. G. Yucesan, V. Golub, J. Zubieta, *Dalton Trans.* **2005**, 2241–2251.

37. W. Ouellette, B.-K. Koo, E. Burkholder, V. Golub, C. J. O'Connor, J. Zubieta, *Dalton Trans.* **2004**, 1527–1538.

38. B.-K. Koo, E. Burkholder, N. G. Armatas, J. Zubieta, *Inorg. Chim. Acta* **2005**, *358*, 3865–3872.

39. G. Yucesan, M. H. Yu, C. J. O'Connor, J. Zubieta, *CrystEngComm* **2005**, *7*, 711–721.

40. L.-M. Zheng, P. Yin, X.-Q. Xin, *Inorg. Chem.* **2002**, *41*, 4084–4086.

41. R.-B. Fu, X.-T. Wu, S.-M. Hu, W.-X. Du, J.-J. Zhang, Z.-Y. Fu, *Inorg. Chem. Commun.* **2003**, *6*, 694–697.

42. P. Yin, Y. Peng, L.-M. Zheng, S. Gao, X.-Q. Xin, *Eur. J. Inorg. Chem.* **2003**, *2003*, 726–730.

43. F. A. Almeida Paz, F.-N. Shi, J. Klinowski, J. Rocha, T. Trindade, *Eur. J. Inorg. Chem.* **2004**, *2004*, 2759–2768.

44. D. Kong, A. Clearfield, *Chem. Commun.* **2005**, 1005–1006.

45. D.-G. Ding, M.-C. Yin, H.-J. Lu, Y.-T. Fan, H.-W. Hou, Y.-T. Wang, *J. Solid State Chem.* **2006**, *179*, 747–752.

46. J. Hou, X. Zhang, *Cryst. Growth Des.* **2006**, 7, 1445–1452.

47. F.-N. Shi, F. Almeida Paz, P. I. Girginova, V. S. Amaral, J. Rocha, J. Klinowski, T. Trindade, *Inorg. Chim. Acta* **2006**, *359*, 1147–1158.

48. Z. Chen, L. Weng, D. Zhao, *Inorg. Chem. Commun.* **2007**, 447–450.

49. J. Li, D. Cao, B. Liu, Y. Li, L. Zheng, *Cryst. Growth Des.* **2008**, *1081*, 6–9.

50. R. Murugavel, S. Shanmugan, *Dalton Trans.* **2008**, 5358–5367.

51. M. Wang, C.-B. Ma, C.-N. Chen, Q.-T. Liu, *J. Mol. Struct.* **2008**, *891*, 292–297.

52. R. Fu, S. Hu, X. Wu, *Dalton Trans.* **2009**, 9843–9848.

53. L.-R. Guo, F. Zhu, Y. Chen, Y.-Z. Li, L.-M. Zheng, *Dalton Trans.* **2009**, *2*, 8548–8554.

54. L.-X. Xie, D.-G. Ding, *Inorg. Chem. Commun.* **2009**, 552–554.

55. Y.-H. Su, D.-K. Cao, Y. Duan, Y.-Z. Li, L.-M. Zheng, *J. Solid State Chem.* **2010**, *183*, 1588–1594.

56. J. Zoń, D. Kong, K. Gagnon, H. Perry, L. Holliness, A. Clearfield, *Dalton Trans.* **2010**, *39*, 11008–11018.

57. K.-R. Ma, F. Ma, Y.-L. Zhu, L.-J. Yu, X.-M. Zhao, Y. Yang, W.-H. Duan, *Dalton Trans.* **2011**, *40*, 9774–9781.

58. X.-M. Zhao, K.-R. Ma, Y. Zhang, X.-J. Yang, M.-H. Cong, *Inorg. Chim. Acta* **2012**, *388*, 33–36.

59. M. Taddei, F. Costantino, R. Vivani, C. Sangregorio, L. Sorace, L. Castelli, *Cryst. Growth Des.* **2012**, *12*, 2327–2335.

60. M. Taddei, F. Costantino, A. Ienco, A. Comotti, P. V Dau, S. M. Cohen, *Chem. Commun.* **2013**, *49*, 1315–1317.

61. H.-J. Jin, P.-F. Wang, C. Yao, L.-M. Zheng, *Inorg. Chem. Commun.* **2011**, *14*, 1677–1680.

62. P.-F. Wang, Y. Duan, D.-K. Cao, Y.-Z. Li, L.-M. Zheng, *Dalton Trans.* **2010**, *39*, 4559–4565.

63. P.-F. Wang, Y. Duan, J. M. Clemente-Juan, Y. Song, K. Qian, S. Gao, L.-M. Zheng, *Chem. Eur. J.* **2011**, *17*, 3579–3583.

64. W. Yang, T. Tian, H.-Y. Wu, Q.-J. Pan, S. Dang, Z.-M. Sun, *Inorg. Chem.* **2013**, *52*, 2736–2743.

65. P. Yin, L.-M. Zheng, S. Gao, X.-Q. Xin, *Chem. Commun.* **2001**, *2*, 2346–2347.

66. F.-N. Shi, F. Almeida Paz, P. I. Girginova, L. Mafra, V. S. Amaral, J. Rocha, A. Makal, K. Wozniak, J. Klinowski, T. Trindade, *J. Mol. Struct.* **2005**, *754*, 51–60.

67. J-.G. Mao, in *Metal Phosphonate Chemistry: From Synthesis to Applications*, eds. A. Clearfield and K. Demadis, RSC, Oxford, **2011**.

68. X.-G. Liu, J. Huang, S.-S. Bao, Y.-Z. Li, L.-M. Zheng, *Dalton Trans.* **2009**, *44*, 9837–9842.

69. K. Chen, Z.-G. Sun, Y.-Y. Zhu, Z.-M. Liu, F. Tong, D.-P. Dong, J. Li, C.-Q. Jiao, C. Li, C.-L. Wang, *Cryst. Growth Des.* **2011**, *11*, 4623–4631.

70. C. Li, J. Li, Z.-G. Sun, K. Chen, C.-L. Wang, C.-Q. Jiao, F. Tong, Y.-Y. Zhu, C.-Y. Huang, *Inorg. Chem. Commun.* **2011**, *14*, 1715–1718.

71. F. Tong, Z.-G. Sun, K. Chen, Y.-Y. Zhu, W.-N. Wang, C.-Q. Jiao, C.-L. Wang, C. Li, *Dalton Trans.* **2011**, *40*, 5059–5065.

72. C.-Q. Jiao, C.-Y. Huang, Z.-G. Sun, K. Chen, C.-L. Wang, C. Li, Y.-Y. Zhu, H. Tian, S.-H. Sun, W. Chu, et al., *Inorg. Chem. Commun.* **2012**, *17*, 64–67.

73. K. Chen, D.-P. Dong, Z.-G. Sun, C.-Q. Jiao, C. Li, C.-L. Wang, Y.-Y. Zhu, Y. Zhao, J. Zhu, S.-H. Sun, et al., *Dalton Trans.* **2012**, *41*, 10948–10956.

74. K. Bernot, J. Luzon, R. Sessoli, A. Vindigni, J. Thion, S. Richeter, D. Leclercq, J. Larionova, A. van der Lee, *J. Am. Chem. Soc.* **2008**, *130*, 1619–1627.

75. T. J. R. Weakley, *Acta Crystallogr.* **1978**, *34*, 3756–3758.

76. M. R. Siqueira, T. C. Tonetto, M. R. Rizzatti, E. S. Lang, J. Ellena, R. A. Burrow, *Inorg. Chem. Commun.* **2006**, *9*, 537–540.

77. J.-H. Liao, P.-L. Chen, C.-C. Hsu, *J. Phys. Chem. Solids* **2001**, *62*, 1629–1642.

78. S. Parsons, D. Robertson, R. Winpenny, S. Harris, P. Wood, Private Communication to the Cambridge Crystallographic Data Bank (Refcode FIGKOG), **2004**.

79. F. Costantino, S. Midollini, A. Orlandini, *Inorg. Chim. Acta* **2008**, *361*, 327–334.

80. T. Bataille, F. Costantino, P. Lorenzo-Luis, S. Midollini, A. Orlandini, *Inorg. Chim. Acta* **2008**, *361*, 9–15.

81. T. Bataille, S. Bracco, A. Comotti, F. Costantino, A. Guerri, A. Ienco, F. Marmottini, *CrystEngComm* **2012**, *14*, 7170–7173.

82. T. Bataille, F. Costantino, A. Ienco, A. Guerri, F. Marmottini, S. Midollini, *Chem. Comm.* **2008**, 6381–6383.

83. F. Costantino, A. Ienco, S. Midollini, *Cryst. Growth Des.* **2010**, *10*, 7–10.

84. S. Midollini, A. Orlandini, *J. Coord. Chem.* **2006**, *56*, 1433–1436.

85. A. Ienco, S. Midollini, A. Orlandini, F. Costantino, *Z. Naturforsch.* **2007**, *62b*, 1476–1478.

86. P. Thanasekaran, T.-T. Luo, C.-H. Lee, K.-L. Lu, *J. Mater. Chem.* **2011**, *21*, 13140–13149.

87. J. Beckman, F. Costantino, D. Dakternieks, A. Duthie, A. Ienco, S. Midollini, C. Mitchell A. Orlandini, L. Sorace, *Inorg. Chem.* **2005**, *44*, 9416–9423.

6

HYBRID AND BIOHYBRID MATERIALS BASED ON LAYERED CLAYS

EDUARDO RUIZ-HITZKY, PILAR ARANDA AND MARGARITA DARDER

Instituto de Ciencia de Materiales de Madrid, ICMM-CSIC, Cantoblanco, Madrid, Spain

6.1 INTRODUCTION: CLAY CONCEPTS AND INTERCALATION BEHAVIOUR OF LAYERED SILICATES

According to the *Handbook of Clay Science* published by Bergaya and Lagaly [1], clays are essentially composed of micro-particulated minerals whose particle size is less than 2 μm, being particularly abundant in the Earth's crust. This family, referred to as the 'clay minerals', is mainly composed of layered silicates (phyllosilicates) of variable composition leading to diverse structural arrangements and properties. Typical layers of clay minerals are built up by cations tetrahedrally and octahedrally coordinated to oxygen atoms designated as tetrahedral (T) and octahedral (O) sheets, respectively. Depending on the sequence of these T and O sheets, these clay minerals can be classified and designed as 1 : 1 or TO and 2 : 1 or TOT phyllosilicates (Figure 6.1). Representative examples of 1 : 1 phyllosilicates are kaolinite (Al silicate) and serpentine (Mg silicate). In the first case, one of the sheets contains silicon atoms in tetrahedral coordination with oxygens condensed with the other sheet composed of aluminium (dioctahedral silicate) or magnesium (trioctahedral silicate) in octahedral coordination to oxygens (i.e. gibbsite-like in kaolinite and brucite-like in serpentine, respectively). The 2 : 1 phyllosilicates contain two sheets of Si tetrahedral sheets sandwiching one octahedral sheet containing Al or Mg ions, giving rise to minerals such as pyrophyllite and talc, respectively. In these 2 : 1 or TOT phyllosilicates, isomorphous substitutions

To Professor José M. Serratosa, in memoriam

Tailored Organic–Inorganic Materials, First Edition. Edited by Ernesto Brunet, Jorge L. Colón and Abraham Clearfield.
© 2015 John Wiley & Sons, Inc. Published 2015 by John Wiley & Sons, Inc.

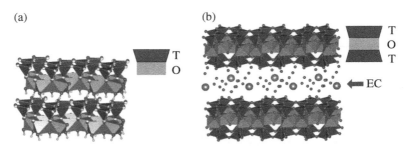

FIGURE 6.1 Schematic representation of the crystalline structure of kaolinite (a) and mont-morillonite (b), which are typical examples of 1 : 1 and 2 : 1 charged phyllosilicates (smec-tites), respectively. They are built up by tetrahedral (T) sheets (mainly silica sheets) and octahedral (O) sheets (aluminium oxyhydroxide substituted by magnesium and other ions in the case of montmorillonite). EC represents the exchangeable cations in its hydrated form. Water molecules are represented in the interlayer space as red small spheres.

of Al or Mg ions, and in some cases of Si atoms in the tetrahedral layers, by ions of similar size but lower charge originate negatively charged layers that are compensated by metal ions (typically Na^+ and Ca^{2+} for natural samples) in the interlayer space, which are denoted as exchangeable cations (EC). Clay minerals belonging to these 2 : 1 charged phyllosilicates comprise the smectite and vermiculite groups with net layer charge per formula unit of 0.2–0.6 and 0.6–0.9, respectively [1, 2]. The most representative example of the smectite group is the layered aluminosilicate named as montmorillonite (*bentonite* clays) where the octahedral layer is mainly occupied by Al^{3+} ions and in minor extent by Mg^{2+}, Fe^{3+} and other divalent or trivalent metal ions (Figure 6.1b). Other examples of smectites are hectorite, saponite, stevensite, beidellite and nontronite. In nature, they result from weathering of precursors based on other silicate rocks, but certain synthetic layered clays can be prepared even at the industrial scale (e.g. synthetic hectorite named Laponite® [3]).

It should be noted that the literature often uses the term 'clays' as an abbreviated term that refers to 'clay minerals', which is considered as a more correct manner to name this class of microcrystalline solids [1]. They exhibit interesting features mainly related to their surface and colloidal properties, driving to a huge variety of applications based on their adsorption, swelling and rheological behaviour. One of the most salient properties of layered clays is their ability to intercalate organic species in the interlayer space leading to an increase of the basal spacing of these phyllosilicates, giving rise to an important class of organic–inorganic hybrid materials [4]. The intercalation, that is, the adsorption of organic species between the layers of electrically neutral phyllosilicates as, for instance, kaolinite, serpentine, talc or pyrophyllite, is difficult or unfeasible to accomplish due to the strong interaction between consecutive layers that avoids penetration of the organic species.

Compared to smectites and vermiculites, relatively few examples of intercalation compounds based on kaolinite and related minerals such as nacrite and halloysite, their hydrated polytype, have been reported [5]. However, smectite- and vermiculite-charged phyllosilicates are able to intercalate neutral molecules or cationic species from low molecular mass to polymers, leading in this last case to *polymer–clay nanocomposites*, a class of improving hybrid materials provided with mechanical, rheological, gas barrier and other interesting properties useful for diverse applications [6–15].

Hybrid materials based on the intercalation of organic compounds in smectite phyllosilicates can be considered the first reported organic–inorganic materials. Nowadays, bottom-up approaches using this type of clay minerals as building blocks for their assembly to a large variety of organic compounds are currently applied [16].

The aim of this chapter is to give a detailed overview of the field of hybrid materials based on the intercalation of organic species into 2D silicates, which is a topic receiving increasing attention in view of the development of materials having both structural and functional properties. Biohybrids derived from clays, an emerging issue dealing with the assembly of smectites and other silicates to organic species from biological origin, will be also addressed.

6.2   INTERCALATION PROCESSES IN 1 : 1 PHYLLOSILICATES

Kaolinite and related clay minerals such as nacrite and dickite are polytypes of $Al_2[Si_2O_5](OH)_2$ formulae in which hydrogen bonds and dipole interactions hold strongly together their elemental aluminosilicate layers (Figure 6.1a). Only strongly polar molecules such as hydrazine, urea, formamide, N-methylformamide (NMF), imidazole and dimethyl sulphoxide (DMSO) are able to weaken the interactions between consecutive layers by formation of strong hydrogen bonds between hydroxyls (aluminol groups) in the interlayer region of the kaolinite-type silicates and the guest molecules, which remain intercalated, forming organic–inorganic hybrid materials of diverse stability [17]. This intercalation can be considered as a reversible and topotactic process as it can be reverted by washing with solvents or by thermal treatments at relatively low temperature, and even by exposure of this type of intercalates to humid atmosphere for long periods of time, driving to de-intercalated materials following a relatively easy removal of the intercalated guest molecules.

Diverse kaolinite intercalation compounds can be prepared in two ways depending on the nature of the guest species: the first one consisting in the intercalation of the host silicate by direct interaction with the guest molecule and the second one consisting in a two-step process in which a replacement of the intercalated molecules by other different organic species takes place (*displacement procedure*) [17–19] (Figure 6.2). Examples of the first type of synthesis are given by the aforementioned polar molecules as, for instance, hydrazine hydrate, NMF and DMSO, which directly intercalate kaolinite from the liquid phase. A particular case belonging to this procedure is related with the ability of certain alkali salts as halides (e.g. CsCl or CsBr) or as salts of organic low molecular weight acids (e.g. potassium and ammonium acetates). These salts can be directly intercalated by grinding with kaolinite in a mortar in presence of high relative humidity or from high concentrated aqueous solutions. In fact, the first intercalation compound of kaolinite was reported in 1961 by Wada [20] who showed that this silicate could be intercalated by repeated

(a)   (b)   (c)

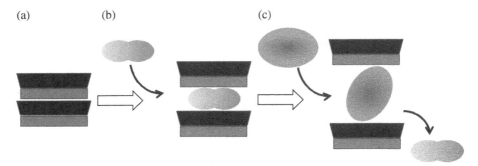

FIGURE 6.2 Schematic representation of the displacement reactions in the preparation of kaolinite (a)-based hybrids by intercalation in a first step of molecules acting as swelling agents giving rise to intermediates (b) allowing further intercalation of bulkier organic species (c).

grinding with potassium acetate and other salts. Practically at the same time, Weiss reported the ability of certain organic polar molecules such as urea to intercalate kaolinite [5]. From this initial pioneering works to nowadays, a great number of contributions have been published dealing with intercalation of kaolinite and related silicates involving a large variety of guest species, from alkylamines to liquid crystals and polymeric materials.

Organic species that are expected to be intercalated in kaolinite and related minerals by direct reaction are (i) compounds with a strong tendency for hydrogen-bond formation such as formamide, NMF, urea, acetamide and imidazole and (ii) molecules having a high dipole moment or able to show mesomeric structures, such as DMSO and pyridine-N-oxide. All these processes need, in general, long time (from hours to days) to achieve a complete molecular coverage of the internal surface of kaolinite by the guest molecules. The access to the intracrystalline region of this 1 : 1 clay mineral and the molecular disposition of the intercalated organic species in its interlayer space are mainly deduced from measurements of the basal spacing $d_{(001)}$ increase measured by X-ray diffraction (XRD) and from the displacement of specific infrared (IR) absorption bands of the aluminol groups on the interlayer surface. Figure 6.3 represents a scheme of the intercalation of NMF, one of the most studied molecules able to accede to the intracrystalline region of kaolinite and related minerals [21–24]. Accommodation of NMF in this interlayer space gives rise to a hybrid material in which the increase of the basal spacing deduced from XRD (from 0.72 to 1.07 nm) agrees with the thickness of the NMF molecule (ca. 0.3 nm) [21–25]. In addition to diffraction methods, spectroscopic techniques such as IR and Raman, as well as solid-state nuclear magnetic resonance (NMR), have been the key to explain the nature of the host–guest interactions between the intercalated species and the hydroxyls in the interlayer region of kaolinite [17, 22, 23, 26–34].

The displacement processes allowing the intracrystalline adsorption of bulkier species into kaolinite were firstly reported by Weiss et al. [35]. In this manner, it is possible to prepare in two steps diverse hybrid materials (Figure 6.4) as reported, for instance, for pyridine-2-methanol (Py2M) molecules by displacement reaction of NMF in the

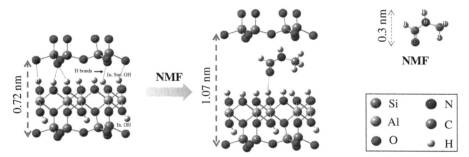

FIGURE 6.3 Schematic representation of the intercalation of *N*-methylformamide (NMF) in kaolinite showing the increase of the basal space deduced from XRD (from 0.72 to 1.07 nm) that corresponds to the thickness of the NMF molecule (0.3 nm). Adapted from Ref. [24].

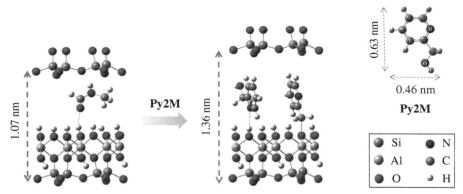

FIGURE 6.4 Schematic representation of the intercalation of pyridine-2-methanol (Py2M) molecules by displacement reaction of the NMF–kaolinite intermediate intercalation compound. Adapted from Ref. [24].

NMF–kaolinite intermediate compound [24]. In some cases, for efficient intercalation of bulkier species, it appears convenient to extend to various steps using two consecutive intermediate compounds. An illustrative example is the use of methanol–kaolinite intermediates, which, in turn, are prepared from kaolinite–NMF [36]. In this way, Kuroda and collaborators reported the assembling of long-chain alkylamines to kaolinite giving rise to intercalation compounds with elevated basal spacing values (e.g. hexylamine = 2.69 nm, octylamine = 3.20 nm and decylamine = 3.71 nm) [30]. These same authors also reported the intercalation of poly(vinylpyrrolidone) (PVP) into kaolinite following a similar approach giving rise to a polymer–kaolinite nanocomposite [37]. In relation to this approach, diverse polymer–clay nanocomposites based on the polymer intercalation in kaolinite using NMF, DMSO, amines, methanol and so on as intermediates were prepared more recently [33] (see following text in more detail; Section 6.4).

Methanol acceding to the intracrystalline region of kaolinite through NMF– or DMSO–kaolinite intermediate compounds could react by thermal activation with the interfacial aluminol groups, giving rise to methoxy-kaolinite nanohybrid materials of $Al_2[Si_2O_5](OH)_{3.13}(OCH_3)_{0.87}$ nominal formula as firstly reported by the Detellier's team [38] and later on studied from both experimental and simulation points of view by Matusik and others [31]. Diverse hydroxyl containing organic species such as ethylene glycol [39] and triethanolamine [40] could be also grafted by the intracrystalline reaction with the Al–OH groups available by previous DMSO intercalation [41]. The formation of Al–O–C covalent bonds resistant to hydrolysis [42] has been confirmed by ^{27}Al solid-state NMR [43]. Amines such as n-hexylamine and n-octadecylamine can be also grafted, using in this case methanol–kaolinite intermediates. These last intercalation compounds show ability towards delamination in toluene, accompanied by the de-intercalation of the amine molecules. The resulting thin kaolinite particles (single kaolinite layers?) rolled up, giving rise to a typical halloysite-like nanotubular morphology [44].

More complex functional nanohybrid materials have been prepared by Detellier and Letaief by grafting of ionic crystals based on pyridinium, imidazolium and pyrrolidinium species [45–49]. The bulky cation counterpart of the ionic crystals grafted into kaolinite acts as pillars between consecutive layers, and the increase of their basal space has given rise to nanoporous and ion-exchangeable materials tested in the development of electrochemical sensors for anion detection [32].

Kaolinite treatment with phenylphosphonic acid results in the interlayer grafting with formation of the silicate derivatives following a reaction with the intracrystalline aluminol groups as claimed by Wypych and co-workers [50] and later by Breen and co-workers [51]. These last authors showed the formation of very stable kaolinite derivatives of $Al_2Si_2O_5(OH)_3(HO_3PPh) \cdot 3H_2O$ formula requiring the reaction long periods of time to accomplish an elevated percentage (ca. 80%) of the reacted kaolinite layers. The same reaction carried out with the halloysite polytype instead of kaolinite shows that a complete reaction takes place with this silicate in a much shorter time [51]. In apparent contradiction with the findings described earlier, Gardolinski et al. reported the lack of evidence of formation of grafted kaolinite derivatives after reaction with phenylphosphonic acid [52]. Probably, the precise adopted experimental conditions are decisive in the course of the reaction and therefore in the nature of the resulting products.

Functional organosilanes have been widely used to prepare organic derivatives of diverse silicates by reaction with silanol groups [9, 17, 53–55]. With the aim to graft kaolinite by reaction of aluminol groups with organochlorosilanes, Ruiz-Hitzky [56] showed the ability of methylvinyldichlorosilane (MVDCS) to accede to the interlayer space in kaolinite previously expanded by intercalation of DMSO. This is a similar procedure that was later successfully applied in the intracrystalline grafting of other types of layer silicates [54, 57], but in the case of kaolinite, a partial elimination of aluminium extracted from its octahedral sheet was observed. The content in this element decreases as the content in organic matter increases. The XRD patterns show a large diffraction peak at values higher than the starting $d_{(001)}$ kaolinite peak, indicating a permanent increase of the basal spacing compatible with the thickness of a silica sheet belonging to de-aluminated kaolinite and the organosilyl-grafted groups. A starting kaolinite (from Zettlitz, Czech Republic) is present as crystals of

(a)   (b)

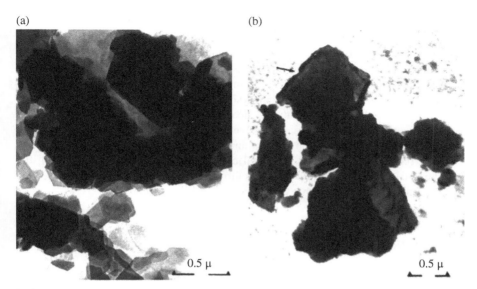

0.5 μ   0.5 μ

FIGURE 6.5 TEM images of (a) starting kaolinite (from Zettlitz, Czech Republic) and (b) kaolinite after reaction with DMSO and MVCS. From Ref. [56].

micrometric size with a regular hexagonal habit that becomes deeply altered after reaction with DMSO and MVCS, as observed in the corresponding transmission electron microscopy (TEM) (Figure 6.5) [56]. These grafting processes with organochlorosilanes in the presence of DMSO follow a complicate mechanism as the kaolinite derivative is formed together with an insoluble compound identified as polymethylsiloxane $[Si(CH_3)O_{3/2}]_x$, which proceeds from a secondary reaction between DMSO and MVDCS in which the vinyl group is eliminated [58]. Actually, in the absence of water, DMSO reacts with chlorosilane functions producing the intermediate DMSO–HCl complex [58], which can act as an extremely corrosive acid system able to attack the kaolinite octahedral sheets, provoking the observed extraction of aluminium. Anyway, the direct treatment of kaolinite with organochlorosilanes fails to produce the grafting of organosilyl groups in the interlayer space of this silicate. However, Kuroda and collaborators [59] reported the grafting of halloysite by reaction with trimethylchlorosilane operating in a mixture of 2-propanol and hydrochloric acid, allowing the aluminium extraction and the consequent generation of silanol reported for other silicates [60–62]. In these experimental conditions, the structural arrangement of halloysite is deeply altered in agreement with the XRD patterns, showing a large diffraction band of similar shape than that observed in kaolinite treated with DMSO and MVDCS [56], in this case centred at around 1.58 nm [59].

As already indicated, halloysite is a polytype of kaolinite showing a better ability than this mineral to intercalate organic compounds. Hydrated halloysite $(Al_2Si_2O_5(OH)_4 \cdot H_2O)$ can be found in nature showing a $d_{(001)}$ basal spacing of about 1 nm instead of 0.715 nm for kaolinite, due to the presence of water molecules in the interlayer space. It has been also indicated that polar molecules can be more easily and rapidly intercalated than into kaolinite [63]. Interestingly, halloysite can exhibit a nanotubular morphology as a result of the rolling of the aluminosilicate single layers giving rise to a multiwalled system, with typical inner diameters in the order of magnitude of 15 nm. This feature represents an additional advantage of these silicates as they can assemble organic species giving rise to hybrid organic–inorganic materials as the lumens of nanotubular halloysite could be loaded with diverse types of organic compounds. In this way, these silicates can be regarded as nanocontainers for molecular encapsulation that can receive diverse applications as drug-releasing agents, as biomimetic reaction vessels and as additives in biocide and protective coatings (corrosion) [64]. The structural characteristics of halloysite allowing the formation of organic–inorganic materials through diverse mechanisms open the way for a variety of applications that will be considered in Section 6.5.

6.3   INTERCALATION IN 2 : 1 CHARGED PHYLLOSILICATES

6.3.1   Intercalation of Neutral Organic Molecules in 2 : 1 Charged Phyllosilicates

Intercalation of neutral molecules into the interlayer space layered clays is possible when the energy released in the adsorption process is sufficient to overcome the attraction between layers. The first report of intercalation of aliphatic amines, alcohols, polyamines and polyglycols was presented by Bradley [65] and MacEwan [66, 67]. In fact, intercalation of ethylene glycol is the most common technique used in clay science to establish the presence of expandable clays in a mineral sample. However, the characteristics of the formed complexes vary with the nature of the mineral, the magnitude and source of the charge on the silicate layers and the number and kind of the EC between the layers [68], parameters that are also relevant in the intercalation of other neutral organic molecules [17].

Exchangeable metal ions and surface oxygens of the tetrahedral sheets are the most probable adsorption sites as they can act as proton acceptors for the formation of H bonds with molecules containing –OH or –NH groups, resulting in coordination compounds. From IR spectroscopic studies, it has been established that hydrogen-bond interactions between surface oxygens of the silicate layers and functional NH or OH groups of the adsorbed organic molecules are only significantly strong when: (i) the layer charge is tetrahedrally located (e.g. in beidellites, saponites, vermiculites); (ii) the clay is saturated with cations of low solvation energy; or (iii) the intercalated organic molecules bear multiple groups capable of forming hydrogen bonds (i.e. O–H, N–H) [17]. For organic compounds of high molecular weight, van der Waals' attractions between molecules and the mineral substrate may also contribute to the adsorption process. A third factor that may influence adsorption in the interlayer region is related to the hydration level as the presence of water molecules weakens interaction between the silicate layers. In fact, organic molecules compete with water for coordination sites around the EC, and depending on the relative values of the hydration and solvation energies of these, they can replace water and become coordinated directly to the inorganic sites, or just occupy sites in a second sphere of coordination around the cation bonded via bridging water molecules, or even accept a proton from the coordinate water molecules or from the cations themselves, for example, H^+ or NH_4^+ [17].

Organic compounds such as alcohols, amines, amides, nitriles, urea and pyridine as well as certain pesticide molecules can be directly coordinated to the EC in smectites and vermiculites. The complexes could be formed with alkaline and alkaline earth cations or with transition metal ions as reviewed by various authors [18, 69–72].

An interesting example to show the influence of the various parameters in the formation of intracrystalline complexes is the systematic study carried out with poly-cyclic ethers of the type of the crown ethers and the cryptands. For non-transition metal ions in the interlayer region, the size and charge of the cations determine the stoichiometry and the arrangement of adsorbed molecules in the interlayer space [73, 74]. When the crown ethers are adsorbed from methanol solution, the interlayer water is practically excluded, and the compounds are coordinated directly to the inter-layer cations in monolayer or bilayer complexes, depending on the size of the cations (Table 6.1). For r_c/r_i ratio (r_c = radius of the macrocyclic cavity and r_i = cationic radius) greater than 1, the cation is occluded into the cavity of the cyclic ligand,

TABLE 6.1 **Schematic models for interlayer complexes of crown ethers, aza crown and cryptand macrocycles intercalated in montmorillonite. Based on Refs [74] and [75]**

r_c/r_i ratio	Interlayer arrangement	Δd_L (nm)	Examples
>1		~0.4	15C6/Na-mont, 18C6/Ba-mont AZA18C6/Li-mont
>1		~0.6	18C6/Na-mont
≤1		~0.4	12C4,15C5,18C6/K- and NH_4-mont
<1		~0.8	12C4/Na-mont; 15C5/Ba-mont
<1		0.6–0.7	12C4/Sr-mont; 12C4/Ba-mont
<1 or >1		0.6–0.8	C(221), C(222)/M-mont (M = Li, Na, K, Cs, Ba, Sr, Cu, Ni,...)

leading in general to the formation of 1 : 1 ligand–cation complexes. When the r_c/r_i ratio is lower than 1, the cation is generally sandwiched between two macrocyclic ligands. In certain cases, for example, K^+ or NH_4^+ interlayer cations and relatively smaller macrocycles, one-layer complexes are formed as the cations are able to coordinate the oxygens of the ditrigonal cavities on the silicate surface in one side and the macrocycle in the other side [75, 76]. Adsorption microcalorimetry measurements determine high exothermic enthalpy values in the formation of intracrystalline complexes of diverse crown ether and cryptand ligands in smectites [77]. The interaction of the intercalated macrocycles and the interlayer cations has been also proved by ^{23}Na solid-state NMR spectroscopy [75, 77]. However, the most conclusive prove that confirms the existence of true interlayer complexes between cations and macrocyclic ligands was obtained from laser microprobe mass spectrometry (LMMS) [78]. This technique has also corroborated the different types of macrocycle–interlayer cation complexation that could take place depending on the nature of the interlayer cation by analysing the fragments removed from the solid in montmorillonites exchanged with Na^+ and Cu^{2+} ions intercalated with the cryptand C222 macrocycle. In the first case, the m/e values correspond to the sum of both the sodium and cryptand atomic mass (i.e. 399 daltons for Na/C222), but in the copper-exchange clay case, the m/e is 377 daltons, corresponding to the removal of a protonated C222 ligand [78].

Other nitrogenated macrocycles based on porphyrin derivatives, such as phenyl- and pyridyl-porphyrins, metalloporphyrins and chlorophyllin, and related macrocycles of synthetic or natural origin have been also intercalated in smectites saturated with different inorganic cations [79, 80]. Part of the interest is related to the fact that the prebiotic synthesis of porphyrins in the primitive Earth, which could have contributed to the evolution of photosynthesis and respiration in living systems, has been postulated [81]. This type of ligand forms very stable complexes with transition metal cations showing a characteristic ability to become protonated in the interlayer space of clays in the same way than the aforementioned intercalation compound of C222 in Cu^{2+}-montmorillonite.

Actually, thermodynamic stability of the complex formed within the interlayer space of 2 : 1 phyllosilicates is high and may be usually higher compared to analogous organic complexes formed in solution, as previously indicated for the case of intercalation of crown ethers into smectites [77]. Cremers and co-workers [82–85] showed that the thermodynamic constants of formation of Cu^{2+}–, Ni^{2+}–, Zn^{2+}–, Cd^{2+}– and Hg^{2+}–montmorillonite complexes with polyamines and Ag^+–montmorillonite with thiourea are two to three orders of magnitude higher than those for the same complexes in solution. Moreover, in certain cases, the combined effect of the silicate-layer electric field and of steric restrictions determines that certain complexes can be formed only in the interlayers. In this sense, Mortland and co-workers [86, 87] showed the stabilization of aromatic molecules in the interlayer region of phyllosilicates. These works reported the formation of complexes of Cu^{2+}-smectites with arenes, which are not formed in solution, via donation of electrons. To reach the formation of the complexes, it is necessary to remove the hydration shell of the interlayer cations, and in the case of benzene, the formation of two types of complexes is possible depending on the extent of dehydration: (1) type I (green colour, formed by exposing the clay to benzene vapour in a desiccator with P_2O_5) in which the benzene is truly coordinated to the cation through the electron system and the aromatic character of the molecule is preserved, and (2) type II complexes (red), formed by further dehydration of the former, in which the benzene ring seems to be a part of a radical cation [17]. This type of arene complexes is typically formed in smectites exchanged with Cu^{2+} and Ag^+ cations, the complex formation of the latter being typically easier because of the lower hydration energy of Ag^+ ions. The location of the clay charge (tetrahedral or octahedral) also influences the formation of these compounds, the complexation being easier in octahedrally charged smectites. This is because water is more firmly bound and more difficult to eliminate in tetrahedrally charged smectites, which indicates that aromatic molecules have greater difficulty entering into direct coordination with the cations [88].

Other aromatic molecules, such as pyridine, become adsorbed on montmorillonite by stabilization in the interlayer region through bridging water molecules to the EC, as proved by IR spectroscopy [89]. IR spectroscopy also provides insides on the type of interactions involving the intercalation of benzonitrile into montmorillonite [90]. In fact, benzonitrile molecules are arranged with the plane of the benzene ring at a large tilt angle to the silicate layers (basal spacing of ~1.5 nm) and with the principal C2 axis parallel to the layers. The IR absorption frequency of the O–H bond depends on the polarizing power of the interlayer cation, varying for alkaline earth cations from 3390 cm^{-1} in the Ca^{2+}-exchanged clay to 3405 cm^{-1} in the one containing Ba^{2+} ions and to 3348 cm^{-1} in the Mg^{2+}-exchanged clay. The C–N stretching vibration band displaces also its frequency due to the existence of hydrogen bonds, although its position is not affected by the nature of the interlayer cation. When the samples are evacuated and heated (80°C), bridged water molecules are lost and benzonitrile is coordinated directly to the cations. This is indicated by the position of the C–N stretching band, which is now sensitive to the metal ion present, appearing at 2240, 2249 and 2261 cm^{-1} for the Ba^{2+}-, Ca^{2+}- and Mg^{2+}-exchanged clays, respectively. In the last sample, some water is still present after prolonged evacuation at 80–100°C, appearing as a relatively sharp IR absorption band at 1650 cm^{-1}, corresponding to the deformation vibration of H_2O. This band is sensitive to the orientation of the clay film with respect to the incident radiation, and this provides relevant information that allows to construct a model in which Mg^{2+} ions are coordinated with six ligands: four benzonitrile molecules arranged as a paddle wheel and two water molecules situated above and below the two remaining corners of the octahedron [17].

The third mechanism that may be implicated on the intercalation of neutral organic molecules refers to the possible protonation of organic bases when adsorbed into layer silicates. Protons present as hydrogen ions in NH_4^+- or short-chain alkylammonium-exchanged clays, and water molecules coordinated to diverse type of EC may be transferred to the adsorbed base. In fact, the exchangeable metal ions coordinated to water molecules are acids of measurable strength whose ionization constants (or pK) have been determined in water, and for certain metal ions, the acidity of that interlayer-coordinated water is even greater than in solution [91]. Molecules such as those containing amino groups can be intercalated by this mechanism as firstly proved by Fripiat and co-workers for the adsorption of short aliphatic amines into montmorillonite [92].

This mechanism of interaction may result very useful to procure the intercalation of diverse organic compounds consisting of amino groups as well as of organic species intentionally modified by incorporation of such functionalities.

6.3.2   Intercalation of Organic Cations in 2 : 1 Charged Phyllosilicates: Organoclays

EC present in the interlayer space of smectite-type silicates can be replaced by positively charged organic molecules following an ion-exchange mechanism. This fact was reported for the first time in the 1930s and 1940s by Smith [93], Gieseking [94] and Hendricks [95]. The penetration of the organic cations into the clay interlayer space was confirmed by means of XRD from the variation in the basal spacing. These exchange reactions were carried out in aqueous solutions using salts of the organic cations, for example, hydrochlorides, and they were mainly driven by electrostatic interactions between the organic molecules and the charged silicate layers, with the contribution of van der Waals forces established between the organic cations and the silicate surface and also between neighbouring adsorbed molecules. As the molecular weight of the organic cations increases, these additional van der Waals' attractions become more important than the electrostatic interaction of the organic molecules with the clay surface [95].

The exchange reactions in layered silicates are applicable to all organic bases that form ionized salts. In this way, the intercalation process of organic cations into the layered solids leads to the development of hybrid organic–inorganic materials, commonly known as organoclays. Since the pioneering works mentioned earlier, a wide variety of organic cations have been intercalated in 2 : 1 charged phyllosilicates in order to study in detail the interactions established between both moieties and also with the aim of using the derived organoclays in diverse applications, from the development of nanocomposites to the removal of pollutants. The studied organic cations include alkyl- and aryl-ammonium ions, nitrogenated heterocycles like pyridinium or imidazolium molecules, organic dyes, amino acids, nitrogenated bases and nucleosides, alkaloids and pesticides [17, 71, 96].

In most cases, the organic bases are aliphatic or aromatic amines, which can interact with the surface oxygens of the silicate layer through H bonding. This interaction is favoured when the groups in the organic cations show trigonal symmetry, like $-NH_3^+$, as oxygens are arranged in ditrigonal six-membered rings in the silicate surface [97]. Electrostatic attraction is added to these H-bond interactions when the charge in the layers is tetrahedrally located, leading to a *keying* effect of the amino groups in the surface cavities [98–102]. The presence of other functional groups acting as electron donors in the organic molecules may hinder the interaction of the amino groups with the surface oxygens, mainly if they are also provided with negative charge. This effect was observed in the adsorption of L-ornithine on vermiculite, where the electrostatic interaction of the $-NH_3^+$ groups with carboxylate groups in adjacent amino acid molecules prevented their coulombic and/or H-bond interactions with the silicate surface [103]. In aliphatic amines, $-CH_2-$ or terminal $-CH_3$ groups can also interact by H bonding with the surface oxygens. The presence of weak interactions between these groups and the silicate oxygen atoms was confirmed from the IR spectra of alkylammonium–vermiculite complexes, which showed a splitting in the symmetric deformation vibration band of $-CH_3$ at 1380 cm^{-1} with the appearance of a new dichroic component at 1395 cm^{-1} [104].

The intercalation of organic bases in the clay interlayer can be also strongly dependent on the pH of the solution and the pK value of the base. The ratio of charged and neutral molecules can be varied as a function of these parameters. At pH values close to the pK value, the amount of neutral molecules is close to that of the positively charged ones, and they can be also adsorbed along with the cation exchange. This was observed for aniline, which was adsorbed as a mixture of monovalent cations and neutral molecules at pH = 3.2 exceeding the cation-exchange capacity (CEC) of the clay [105], suggesting the establishment of intermolecular hydrogen bonds between cations and neutral molecules in the interlayer space. Cations are the predominant species in solution when the pH value is around two units lower than the pK, but the presence of H$^+$ ions or cations released from the silicate framework in such conditions may obstruct the adsorption of the organic cations. The influence of pH on adsorption can be also significant when the solubility of the base in water depends on the pH [106].

The arrangement of organic cations in the interlayer space of phyllosilicates depends not only on the size and charge of the organic cation but also on the charge density and the location of charge (tetrahedral or octahedral) in the silicate layers. For instance, pyridinium ions lie parallel to the silicate layers in clays with low charge density like montmorillonite, while they are organized perpendicularly to the layers in vermiculite [107, 108]. Diverse arrangements were also reported for the intercalation of butylammonium in clays with different charge density. In montmorillonite (half-unit-cell charge of 0.33, octahedrally located), butylammonium molecules were displayed parallel to the silicate layers with an all-*trans* conformation, affording a basal spacing of 1.35 nm [102]. In the case of vermiculites with higher charge density and tetrahedrally located charge, like Beni-Buxera vermiculite (half-unit-cell charge of 0.72) or Llano vermiculite (half-unit-cell charge of 0.95), the NH$_3^+$ end groups of butylammonium were *keyed* into the ditrigonal cavities, and the aliphatic chains adopted different conformations depending on the surface charge density. IR spectra confirmed that the C_3 axes of NH$_3^+$ end groups of butylammonium were perpendicular to the layers and that NH$_3$ groups were hydrogen bonded to the surface oxygens. Intercalation in the clay with the highest charge led to a basal spacing of 1.47 nm, with the molecules tilted 55° to the surface in an all-*trans* conformation, while in the case of the medium-charged Beni-Buxera vermiculite complex, a low basal spacing of 1.32 nm suggested an arrangement of the molecules lying flat on the surface by means of a conformational change via a rotation of 120° around the C_1–C_2 bond [102].

The arrangement of long-chain alkylammonium cations in the interlayer space of 2 : 1 clay minerals has been also studied extensively [17, 109–113]. In this case, the length of the alkyl chain is also considered together with the charge density and location of the silicate layers. Usually, these ions are adsorbed on smectites as a monolayer (Figure 6.6a), and larger adsorbed amounts can be ascribed to the formation of bilayers (Figure 6.6b). Conformational changes through rotations around C–C bonds (*kinks*) are also possible, leading to pseudo-trimolecular arrangements with some

(a)   (b)   (c)   (d)

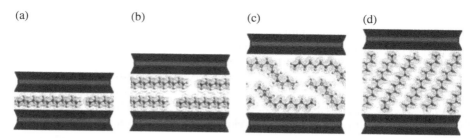

FIGURE 6.6 Schematic representation of alkylammonium ions arranged in the interlayer space of 2 : 1 clay minerals as (a) monolayers, (b) bilayers, (c) pseudo-trimolecular layers of chains lying flat on the silicate surfaces and (d) paraffin-type structure.

chain ends shifted above one another (Figure 6.6c), in which the interlayer spacing corresponds to the thickness of three alkyl chains. In clays with high charge density like vermiculites, long-chain alkylammonium ions tend to adopt paraffin-type structures (Figure 6.6d) [110]. Smectites exchanged with long-chain alkylammonium cations exhibit interesting organophilic characteristics, mainly when the intercalated cations have alkyl chains of 12 carbon atoms or more, and interesting properties that allow their use in many different applications. For instance, they can be used as thickeners and thixotropic agents to improve the rheological properties of dispersions in organic solvents, as adsorbents of organic pollutants, as stationary phases in gas chromatography or as compatibilizers of organophilic polymers, and the clays incorporated can be used as nanofillers in the preparation of polymer–clay nanocomposites, among many other applications that will be widely discussed in Section 6.5 [113].

The arrangement of the intercalated cations can be perturbed due to the adsorption of other organic molecules. This fact was detected in pyridinium–montmorillonite complexes, in which the adsorption of benzene or chlorobenzene changed the disposition of the pyridinium ions from parallel to normal to the silicate layers, increasing the basal spacing from 1.25 to 1.50 nm [114]. The adsorption of long-chain n-alkanols by organoclays incorporating long-chain alkylammonium cations also results in a change of the basal space. Strong van der Waals forces are established between the alkyl chains of these compounds and contribute to its dense packing in ordered bimolecular films between the silicate layers similar to the lipidic biologic membranes. The longitudinal axes of these molecules can be either perpendicular to the silicate layers (montmorillonite) or inclined 54° to the plates (vermiculite). In many cases, the alkyl chains can undergo conformational changes by rotations around the C–C bonds (*kinks*), which were first proposed to occur as defects and structural elements in crystals of the paraffins and polymers [115].

Phosphatidylcholine (PC), the main constituent of cell membranes, has been proposed as an eco-friendly biomodifier of clay minerals that allows the preparation of bio-organoclays [116]. This phospholipid containing a quaternary ammonium as head group can be intercalated in montmorillonite following a cation-exchange mechanism, being arranged as a monolayer and as a bilayer with basal spacings of 1.45 and 5.1 nm, respectively (Figure 6.7). In the last case, the results suggest that

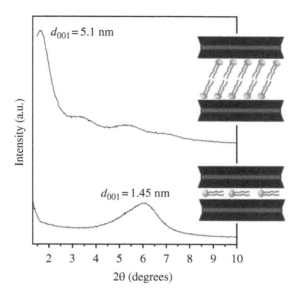

FIGURE 6.7   XRD patterns of the phosphatidylcholine (PC)-modified montmorillonite showing monolayer or bilayer arrangement. Based on Ref. [116].

the biomolecules are tilted 66° with respect to the surface in the lipid bilayer. Due to their hydrophobic character, these bio-organoclays were successfully evaluated for removal of mycotoxins.

Although quaternary alkylammonium salts are the most widely used organic species in the preparation of organoclays, cationic phosphonium derivatives have been also tested for this purpose. Among them, the adsorption of tetraphenylphosphonium on montmorillonite was reported to be up to approximately 90% of its CEC, leading to a complete neutralization of the initial negative charge of the silicate [117]. The resulting complexes were applied as adsorbents for the removal of phenol pollutants from water. Another study reported the enhanced thermal stability of tetrabutylphosphonium– and tetraphenylphosphonium–montmorillonite complexes, which were proposed for the development of polymer–layered clay nanocomposites by melt processing [118]. Similarly, Awad and co-workers reported that organoclays based on alkyl-imidazolium molten salts afforded higher thermal stability than alkyl ammonium-based silicates, showing a decrease in stability as the chain length increased [119].

A recent work reporting the intercalation of stable organic radicals into a fluoromica clay shows the interest in developing organoclays based on piperidine, bipyridinium and imidazole derivatives that can be applied as heterogeneous catalysts [120]. The characterization of the resulting organoclays by XRD suggested several intercalation arrangements, since diverse interlayer distances ranging between 5.1 and 11.9 nm were determined for different adsorption conditions. These results indicated the intercalation of the three organic radicals in parallel, transversal or longitudinal configurations.

Another important group of organic cations used in the preparation of organoclays is the group of positively charged dyes. Complexes based on these organic cations have been prepared with the aim of determining the cation exchange capacity (CEC) of the clays and also to apply them in the environmental field as adsorbents of pollutants or to enhance the photodegradation of adsorbed dyes. Nir and co-workers [121] studied the adsorption of thioflavin T (TFT), methylene blue (MB), crystal violet (CV) and acriflavine (AF) on montmorillonite and determined their strong binding affinity to the clay. The results showed that binding coefficients for the formation of neutral complexes with these organic dyes were more than six orders of magnitude larger than those found for inorganic monovalent cations. The cationic dyes can interact with the clay through non-coulombic interactions, making possible the adsorption of high amounts of dye exceeding the CEC and thus causing a charge reversal. In this case, the adsorbed dyes can adopt a bilayer arrangement in the interlayer space, as determined in the AF–montmorillonite organoclays with 1.2 mmol AF g^{-1} clay that afford a basal spacing value of 1.62 nm [121]. The intercalation compounds involving montmorillonite and two amphiphilic cationic azobenzene derivatives, p-(ω-trimethylammoniopentyloxy)-p'-(dodecyloxy) azobenzene bromide and p-(ω-trimethylammoniodecyloxy)-p'-(octyloxy)azobenzene bromide, showed the formation of the so-called J-like aggregates in the interlayer space of the smectite [122]. Two different models, in mono- and bilayer disposition, were proposed for the resulting intercalation compounds. It was also reported that the intercalated azo dyes can undergo reversible *trans–cis* photoisomerization at room temperature when the complexes are irradiated with UV and visible light.

Some dyes can aggregate in water solution forming dimers and adopting different arrangements in the clay interlayer space, as it has been observed for rhodamine B. For instance, Klika et al. [123] have proposed several models that try to explain the different structural organizations resulting from the intercalation of monomers and dimers of rhodamine B in montmorillonite, giving rise to two phases with basal spacings of 2.3 nm (Figure 6.8a) and 1.8 nm (Figure 6.8b), respectively.

(a)   (b)

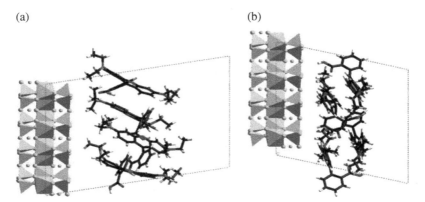

FIGURE 6.8 Side views illustrating the fully exchanged rhodamine B–montmorillonite complex with (a) four [RhB]$^+$ monomers and (b) two [RhB]$_2^{2+}$ H dimers per one 3a × 2b × 1c supercell. Reprinted from Ref. [123] Copyright (2004), with permission from Elsevier.

Dye-based organoclays have been also prepared by synthesizing the smectite while using the dye as a template, as reported by Carrado et al. [124]. In this case, a synthetic hectorite was prepared by refluxing for 2 days aqueous solutions of silica sol, magnesium hydroxide and lithium fluoride in the presence of cationic phthalo-cyanine and metallophthalocyanine dyes. In the resulting intercalation compounds, the phthalocyanine molecules were oriented parallel to the hectorite layers, affording interlayer distances of 0.45–0.65 nm.

Given that the addition of monovalent dyes in small concentrations resulted in a complete displacement of inorganic cations such as Na^+ or Ca^{2+} from the clay mineral, these organic cations were employed to determine the amount of EC and the CEC of smectite clays. For instance, the CEC of montmorillonite was obtained by using MB and CV in combination with inductively coupled plasma emission spectrometry (ICPES) [125]. An incubation time of 3 days was enough, and similar CEC values were obtained with both dyes and using Na–, Ca– and raw montmorillonite.

Dye–clay complexes can be applied to solve environmental problems. An interesting application within this field is related to the enhanced photodegradation of organic dyes adsorbed on certain smectite clays, with the clay particles acting as a catalyst. For instance, this effect has been reported for CV and other triarylmethane dyes adsorbed on Texas vermiculite [126] and for rhodamine B, rhodamine 6G and a stilbazolium derivative (4'-dimethylamino-N-methyl-4-stilbazolium) adsorbed on synthetic sodium saponite [127]. In the first example, no degradation effect was attained when using other vermiculites or montmorillonites. IR spectroscopy confirmed the degradation of the dyes adsorbed on Texas vermiculite, the process being accelerated by a sonication pretreatment. The degradation products remained associated to the clay, as shown by carbon-content analysis [126]. In the case of the saponite-based organoclays, it was also confirmed that the adsorption of the photoactive dyes on the clay surface largely enhanced their photodegradation, even when the amount of adsorbed dyes exceeded the CEC of the clay, around 300% [127]. In clear opposition to this behaviour, other 2 : 1 smectites may contribute to stabilize the adsorbed dyes, as observed in the MB–montmorillonite complex [128]. In this case, the interactions between the oxygen plane of the montmorillonite layer and the aromatic parts of the dye may be responsible for the observed metachromasy, which produces a shift of the λ_{max} of the absorbance spectrum of MB to approximately 570 nm. Although the dye–clay interactions that originate this effect are not still completely understood, metachromasy could be proposed as an indicator of the dye photostabilization in certain clays. Other interesting applications of dye–clay complexes in the environmental field are related to their use as adsorbents of pollutants, as will be detailed in Section 6.5.

Organic cations of biological interest, including amino acids, nitrogenated bases or nucleosides, have been also intercalated in 2 : 1 charged phyllosilicates. The first works on the adsorption of amino acids were focused on determining the interactions established between both moieties and the structural organization of the organoclay [103, 109, 129, 130]. Amino acid–clay complexes exhibit high hydrophilicity due to the adsorbed organic cations, and, for instance, hydrates of L-ornithine–vermiculite complexes show part of the water bound to the silicate surface and another part associated with the functional groups of the intercalated amino acids [103]. Interestingly, the thermal treatment of these complexes at 180°C provokes their dehydration together with the condensation of adjacent L-ornithine molecules into a cyclic peptide of the diketopiperazine type inside the interlayer space.

Due to the interest in the prebiotic processes involved in the origins of life, many studies have centred on the adsorption of purines, pyrimidines and nucleosides on clay surfaces [131–133]. In montmorillonite, the adsorption is stronger for purines, thus being more favourable for adenosine and its derivatives than for pyrimidines like uracil [133]. However, thymine and uracil could be easily adsorbed if adenine is present in the solution, and this co-adsorption is attributed to an H bonding between both nitrogenous bases [132]. It is also remarkable that the extent of the adsorption is maximal at acidic pH values, while at basic pH values, where the molecules are neutral or negatively charged, it is considerably reduced, with the adsorption being attributed to van der Waal forces in this last case [133].

Alkaloids like nicotine and strychnine were employed in one of the pioneering works in the development of organoclays [93]. The high affinity of alkaloids to layered silicates was proven in subsequent studies that confirmed the complete adsorption of alkaloids from even very dilute solutions [109]. It has been also determined that the adsorbed alkaloid cations can significantly block the intracrystalline swelling capacity of the smectite, except in the case of nicotine. The high affinity of smectite clays for alkaloid cations has been exploited for the removal of ergot mycotoxins that contaminate animal feedstuffs, which are efficiently adsorbed on calcium and sodium montmorillonite [134].

The alkaloid berberine, a quaternary ammonium salt extracted from the roots and bark of *Berberis aristata* or *Coptis chinensis* plants, is also adsorbed on montmorillonite through ion exchange accompanied by a secondary organophilic adsorption [135]. The intercalated alkaloid can adopt different arrangements in the interlayer space depending on the amount of adsorbed berberine, from a monolayer of molecules lying parallel to the clay layers to an arrangement in bilayers. As will be mentioned in Section 6.5, this berberine–montmorillonite organoclay has been proposed as platform for the controlled release of herbicides [136].

Several studies have focused on the interactions of pesticides with smectite clays, as, for instance, the divalent cationic herbicides diquat (DQ) and paraquat (PQ) [137]. Both species can be intercalated in montmorillonite originating the loss of interlayer water and a decrease of the basal spacing of the clay from 1.45 nm to around 1.30 nm, which is attributed to the flat orientation of the adsorbed PQ cation and the compact size and keying of the DQ cation into the ditrigonal cavities of the silicate layers. In other cases, the formation of cationic herbicide–clay complexes has been proposed as a method to remove the pollutant compound, as reported for the adsorption of atrazine, whose use is currently prohibited, but it is still present in the environment [138]. In the case of triorganotin species (TOTs), the aim of the study of the TOT–clay complexes was mainly addressed to evaluate their possible desorption from sediments, which could lead to the pollution of marine waters and freshwaters with these biocide compounds [139].

6.4 INTERCALATION OF POLYMERS IN LAYERED CLAYS

6.4.1 Polymer–Clay Nanocomposites

Intercalation of polymers into layered silicates is nowadays a very active field of research mainly focused on the development of polymer–clay nanocomposite materials for diverse applications, as, for instance, plastics with improved mechanical and/or barrier properties [6–15]. Polymer–clay nanocomposites [6, 7, 10, 14], also called clay mineral–polymer nanocomposites by researchers in clay science [9, 11, 140, 141], are nanocomposite materials in which the dispersed phase, frequently named nanofiller, is a clay. Layered clays can be considered as 'one-nanodimensional' nano-materials because the thickness of their individual sheets is in the nanometre range, which is particularly relevant when intercalation of the polymer drives to large separation of the silicate layers and loss of order in their stacking by producing delamination or exfoliation phenomena. Although polymer–clay nanocomposites were firstly reported in 1987 by Toyota's researchers [142], the adsorption of different types of macromolecules on clays was known for many years especially by soil scientists [143]. Those first studies mostly focused on rheological properties of polymer–clay suspensions searching information on the specific interaction mechanisms as a function of the nature of the polymer, the type of solvent and the involved clay mineral. From those investigations, it was deduced that the presence of water molecules, in the solvent or associated with the interlayer cations, was a key point in the process, the gain of entropy being the main driving force to reach the polymer–clay complex formation [17].

Actually, the first polymer–clay intercalation compounds were reported by Blumstein in 1961 [144]. In this example, the presence of the polymer in the interlayer region of smectites was achieved in a two-step reaction because firstly a monomer (e.g. acrylonitrile, vinyl acetate and methyl methacrylate) was intercalated in the presence of an initiator that further provoked the *in situ* polymerization reaction. Other authors found later that layered silicates whose EC were transition metal ions with redox properties could also induce the polymerization of unsaturated monomers, giving intercalation compounds in which the polymer remained strongly associated with the inorganic host [145–148].

Direct intercalation of polymers is possible, but in this case, the affinity between the two counterparts is fundamental to reach the accommodation and stabilization of the polymer in the interlayer region of the clay. This can be easily achieved in smectites if the polymer is a polyelectrolyte having positive sites that could replace the interlayer cations by ion-exchange reactions [149, 150] or if the polymer is highly hydrophilic (for instance, it has oxyethylene functions) and interacts with the hydrated cations located in the interlayer region [151–154]. However, there are many difficulties in intercalating negatively charged or highly hydrophobic polymers in smectites, and routes of synthesis that could favour their compatibility are necessary, for instance, by using organoclays or by incorporation of convenient functionalities in the polymer. In the same way, intercalation of polymers in kaolinite requires specific approaches as previously described for incorporation of discrete organic molecules.

The most part of the polymer–clay nanocomposites are based on natural (montmorillonite, hectorite, saponite, stevensite, etc.) or synthetic (Laponite, fluoro-hectorites) smectites, although more recently vermiculites are also explored as host matrix of polymeric species. Typical methods for reaching intercalation of polymers into layered silicates and other layered solids have been classified in four main synthetic routes (Figure 6.9) [17, 155]:

1. Direct intercalation of the polymer
2. *In situ* intercalative polymerization of monomers
3. Synthesis of the clay in the presence of the polymer (template synthesis)
4. Delamination of intermediate phases followed by restacking in the presence of the organic species (delamination and entrapping–restacking)

It should be noted that it is not always possible to apply all the routes for reaching the formation of the desired polymer–clay nanocomposites because the nature of both the clay and the guest polymer determines that the selected pathway was applicable or not. Furthermore, when various routes may be applicable, the resulting nanocomposite may show differences that can be decisive in the behaviour of the resulting material.

Direct intercalation of the polymer can be in general attained by different synthetic approaches depending on the nature of the polymer and its affinity to the interlayer region of the clay. Adsorption from solutions of the polymer is probably the most common way to intercalate hydrophilic neutral polymers, such as poly(ethylene glycol) (PEG) [151, 152], poly(ethylene oxide) (PEO) [153, 154], poly(vinyl alcohol) [156–158] and so on, which remain in the interlayer region due to ion–dipole interactions of their polar units and the interlayer cations of the clay. Certain interlayer cations, for example, ammonium or ammonium derivative cations, and the presence

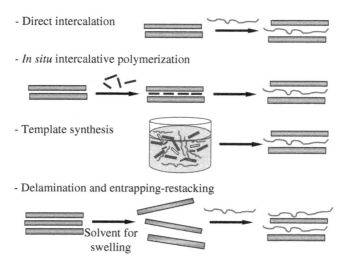

FIGURE 6.9   Most common strategies for the synthesis of polymer–clay nanocomposites.

of water molecules, associated mainly with the EC, favour other interaction forces such as hydrogen bonding or water bridging, respectively, as well as the possibility of hydrogen bonds with hydrogen atoms from aliphatic chains in the polymer, and the oxygen atoms in the silicate surface may also contribute to the stabilization of the intercalated compound [17].

The use of water as solvent may favour the swelling of the clay and the easier adsorption of the polymer within the silicate layers although it may affect the organization of the polymer within the silicate layers and so the final properties of the resulting polymer–clay nanocomposite. This has been clearly corroborated in the case of oxyethylene-based polymers such as PEG and PEO. Adsorption of PEG into Ca^{2+}-exchanged montmorillonite using water as solvent leads to intercalated phases of different basal space increases and different water contents, depending on the polymer length chain [151, 152]. In this case, the polymer adopts a zigzag conformation within the clay layers possibly stabilized by water-bridge interactions with the water molecules of the Ca^{2+} hydration shell. The use of anhydrous solvents, for instance, acetonitrile, and the presence of Na^+ or Li^+ ions in the interlayer region stabilize the helical conformation of PEO in the resulting PEO/montmorillonite intercalation compounds [153, 154, 159]. Although spectroscopy evidences were afforded, for many years, there was a strong controversy regarding the stability of this proposed helicoidal model for the intercalated PEO (Figure 6.10) prepared from acetonitrile solutions and reported as the first organic–inorganic solid polyelectrolyte [153]. In recent years, theoretical calculations have shown that the presence of water molecules in the interlayer region strongly affects the conformation of the intercalated PEO and may drive to the location of the interlayer cation in close interaction with the silicate layers, affecting the ion mobility [160]. In fact, PEO arrangement within the silicate layers depends on the amount of intercalated polymer, location and density of the silicate charge, nature of the EC and water content [155].

In the case of non-polar polymers, it is necessary to employ organoclays to provide an adequate compatibility improving the interactions between the polymer and the clay mineral. In other cases, it is possible to functionalize the polymer introducing

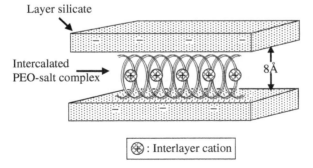

FIGURE 6.10   Schematic representation of the structural arrangement of PEO intercalated into a homoionic 2 : 1 charged phyllosilicate (montmorillonite) with a helix conformation. Reproduced from Ref. [153]. Copyright © 1990 WILEY-VCH Verlag GmbH & Co. KGaA, Weinheim.

polar groups. One of the most popular methods consists in modifying the polymer, for example, polypropylene (PP), by grafting maleic anhydride. PP can be also hydroxylated by grafting acrylic acid or forming random and block copolymers using various polar monomers to favour its intercalation [161].

Ion-exchange reactions can be applied when the polymer to be intercalated contains cationic sites, as, for instance, protonated amino groups. Although in 1952 the intercalation of aminoethylmethacrylate hydroacetate (DMAEM) was already reported [149], there are few examples of this type of polymer–clay intercalation compounds. As reported by Breen [150], certain polycations are able to replace the EC and remain intercalated due to coulombic interactions with the silicate layers. In certain cases, polymers can be functionalized by introducing cationic groups that can facilitate polymer intercalation by ion exchange. Thus, the incorporation of $-NH_2$ functional groups able of being protonated has been used to modify polystyrene (PS) and reach its direct intercalation into smectites [162].

The third way to produce direct intercalation of polymers was firstly reported by Vaia and co-workers [163], and it profits from the fact that certain polymers are able to melt without degradation. Polymer chains exhibit a high mobility in the melt, favouring the mass transport of the polymer within the interlayer space [164]. When this methodology is applied to polymers of hydrophilic character, as it is in the case of PEO, the intercalation of the polymer is easy due to the stabilization of the polymer in the interlayer region via ion–dipole interactions with the interlayer cations. This methodology, using thermal heating [165] or microwave irradiation [166], has been applied to prepare PEO/montmorillonite nanocomposites that here again show remarkable differences with those prepared from PEO solutions. In fact, this route leads to the formation of highly disorganized intercalation compounds as part of the polymer may be located outside the layers connecting different particles and provoking a loss of order in the stacking of layers. This feature becomes advantageous as the resulting nanocomposites show isotropic properties as, for instance, its ionic conductivity. Moreover, the possibility of incorporating lithium salts associated with the non-intercalated PEO may contribute to improve the ionic conductivity of these solid electrolytes [165]. This melt intercalation method has been also applied to PS [163], PP [161] and other thermostable polymers [167], but in these cases, the employed clays need to be conveniently modified to favour the incorporation and/or stabilization of the polymer in the interlayer region. In spite of the difficulties in reaching the appropriate modification of the clay that favour its compatibility with certain polymers, this synthetic approach is one of the major practical steps forwards to reach large-scale nanocomposite preparation especially when it is intended to reach high degree of exfoliation of the clay in a polymeric matrix.

As aforementioned, *in situ* intercalative polymerization was the first reported way to produce intercalation of polymers in smectites [144]. This two-step methodology was also the applied in the preparation of the first so-called polymer–clay nanocomposite reported by Toyota and considered the key point in the development of this type of materials [10, 168]. In this case, a protonated amino acid (e.g., $(NH_3-(CH_2)_n-COOH)$) was intercalated, and then the condensation reaction was thermally activated in the presence of ε-caprolactam to produce the nylon-6/clay nanocomposite [168,

169]. This route of synthesis is also the common way to produce the formation of intercalated conducting polymers, such as polyaniline (PANI) [146, 147, 170, 171], polypyrrole (PPy) [172], polyacrylonitrile [144, 173–175], poly-N-vinylcarbazole [176] or PS. In this last case, it is necessary to use organoclays containing, for instance, an ammonium salt with a styrene group (e.g. $CH_2=CH-C_6H_4CH_2(CH_3)_3N^+Cl^-$), which acts as reactive monomer [177].

In the case of clays, the template synthesis is a synthetic method reduced only to the preparation of nanocomposites based on synthetic clays. By this route, the preparation of diverse polymer–clay nanocomposites incorporating PANI, polyacrylonitrile, poly(N-vinylpyrrolidone), poly(dimethyldiallylammonium) and so on has been reported [178–180]. The smectite clay is a synthetic hectorite formed by *in situ* hydrothermal crystallization of gels (mixture of silica sol, magnesium hydroxide and LiF) and a selected water-soluble polymer, which can be incorporated into the solution in different proportions. Although the limit in the amount of intercalated polymer should be reached when the charged sites on the silicate lattice are neutralized with the cationic sites in the polymer chain, higher content of polymer could be reached as this synthesis method favours the formation of semi-delaminated materials.

In the case of polymer–clay nanocomposites, the delamination and entrapping–restacking synthetic method can be considered as a particular case in diverse synthesis classified within direct intercalation or *in situ* intercalative polymerization. In fact, clays containing Na^+ and other metal ions may swell till practical exfoliation in water, in the same way that organoclays do it in other appropriate organic solvents. In such conditions, it is possible to reach a large separation of their layers, and in that stage, the monomer or the polymer to be intercalated could be added. This methodology has been recently applied not to reach the formation of intercalated solid materials but to incorporate delaminated clay nanoparticles (NPs) in polymeric hydrogels to which the presence of the inorganic phase confers improved mechanical properties [181]. These nanocomposite hydrogels are prepared by *in situ* polymerization of water-soluble monomers containing amide groups, for instance, N-isopropylacrylamide (NIPA), N,N-dimethylacrylamide (DMAA) and acrylamide (AAm), with the delaminated clay acting as inorganic cross-linking agent. In general, small clay particles (about 30 nm diameter) can be more effective than larger ones, which is the reason that, among other smectites, synthetic hectorites and laponites are perhaps the most effective ones in conferring improved mechanical properties to the resulting nanocomposite hydrogels [181]. Other monomers such as those containing carboxyl, sulphonyl or hydroxyl groups can be also used alone or as comonomers in the preparation of this type of nanocomposites [182–184]. Interestingly, nanocomposite hydrogels show swelling–deswelling behaviour in water completely different from those of hydrogels with chemically cross-linked polymer network structures [185]. Certain nanocomposite hydrogel containing hydrophilic and hydrophobic units in the involving polymers may be also stimuli sensitive in response to various external stimuli such as temperature, solvent, pH and salt concentration in aqueous solution [186]. The incorporation of zwitterionic units to the poly(acrylic acid) (PAA) polymer improves the thermosensitive characteristics of the resulting nanocomposite hydrogels [187].

In recent years, the swelling properties of Na$^+$-exchanged clays in water have been also useful for the preparation of highly ordered polymer–clay nanocomposite films by the so-called layer-by-layer (LbL) technique. By this methodology, Kotov and co-workers reported the preparation of the so-called artificial nacre structures using Na$^+$-montmorillonite and polyelectrolytes, such as poly(diallyldimethylammonium) chloride (PDDA) and poly(vinyl alcohol) [188–190]. By this methodology, highly ordered multilayer films can be built up by alternative immersion of a planar substrate into a water suspension of highly expanded smectite clay and a positively charged polyelectrolyte (Figure 6.11). The fact the clay particles are highly exfoliated allows the combination of practically one platelet thickness and a monolayer of polymer, that is, the creation of very thin films of a desired number of polymer–clay intercalated layers. Besides montmorillonite, other smectites such as hectorite and saponite are commonly employed, especially synthetic fluorohectorites and commercial Laponite, due to their small particle size and easy swelling in water. In this way, polymers such as PDDA, sodium polyacrylate or cationic polyacrylamide have been combined with diverse smectites for developing materials with excellent mechanical properties, heat resistance and gas barrier properties [191–193]. Recently, the possibility of applying this technique to the preparation of transparent and gas barrier-resistant films of polyethyleneimine/vermiculite has been also reported [194]. The possibility of using different types of substrates is an advantage, and so it is possible to produce thin coatings, for instance, of branched polyethyleneimine/montmorillonite, on cotton fabrics to provide them with fire-retardant properties [195]. This technique presents other advantages in view of practical applications. For instance, the possibility to tailor the thickness (or spacing) between clay layers by incorporating specific numbers of polymer layers allows to optimize the thickness of the film for the desired gas barrier properties, minimizing deposition steps and so reducing fabrication times [196].

The aforementioned routes work well for smectites, but those methodologies can be hardly applied to reach intercalation of polymers into kaolinite layers. Recently, Detellier and Letaief [33] have reviewed on this topic showing examples

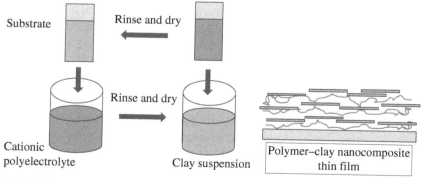

FIGURE 6.11 Schematic representation of the general procedure employed to produce the formation of ordered polymer–clay nanocomposite films by the LbL technique.

of the various specific methodologies used for such purpose. The most classical methodology implies the use of kaolinites intercalated with ammonium acetate, DFM and other polar molecules that are used as intermediates for further incorporation of the appropriate monomer, for instance, acrylonitrile [197, 198] or acrylamide [199]. Once the monomer replaces the intercalated molecule, it is conveniently polymerized to produce the corresponding nanocomposite. Other polymer–kaolinite intercalation compounds prepared by this methodology include the formation of polymers such as poly(N-vinylpyrrolidone) [200], nylon-6 [201], poly(β-alanine) [202], PS [203], PAA [204], poly(methyl methacrylate) [205, 206], poly(vinyl alcohol) [207] and poly(methacrylamide) [208]. Intercalation of melted polymers, such as PEG [209], EVOH [210], PEO and polyhydroxybutyrate [211], has been also achieved in pre-intercalated kaolinites. Intercalation of solved polymers is more complicate and requires finding an appropriated pre-intercalated kaolinite and a suitable solvent. By this methodology, it has been possible to intercalate PVP using a methanol–kaolinite precursor and PVP solved in methanol [37]. In the same way, poly(vinyl chloride) has been intercalated by this method using a DMSO–kaolinite or succinimide–kaolinite precursor [212]. However, exfoliation of kaolinite by the intercalated polymer is not as easier as that for smectites, and few reports on such materials have been reported. These synthetic routes are complex and imply the use of kaolinite grafted with triethanolamine and quaternarized with iodoethane [42] to produce a quaternary ammonium cation that react with sodium polyacrylate [213]. Exfoliation of kaolinite has been also reached in the polymerization of vinyl-modified ionic liquids in the presence of urea–kaolinite [214].

Also, halloysite is being investigated for the development of diverse polymer–clay nanocomposites [215, 216]. In this case, intercalation is not possible, and the nanocomposites intend to disperse the nanotubular silicate within the polymeric matrix. Among others, halloysite-based nanocomposites involving polyamide-6, epoxy resins, ethylene propylene diene monomer (EPDM), PP, ethylene rubber and other polymers have been reported [216]. This is a recent area of research currently in expansion, and the possibility of using halloysite as nanocontainer of different species may open novel opportunities in the development of functional polymer–clay nanocomposites.

6.4.2 Biopolymer Intercalations: Bionanocomposites

Polymers of biogenic origin (biopolymers) can be also intercalated in layered clay minerals by means of ion exchange, hydrogen bonding and other mechanisms. In soils, the interactions of clays with biopolymers such as proteins, polysaccharides, fulvic and humic acids and lignosulphonates are well known since decades ago [143], but the interest in the development of bionanocomposites by intercalation of biopolymers in layered silicates has experienced a huge increase in the last decade [155, 217–219]. This is due to the attractive characteristics of this type of materials, such as their non-toxicity, biodegradability and biocompatibility that are added to the improved mechanical and barrier properties and thermal stability typical of conventional nanocomposites. Biopolymers are abundant and widespread in nature, being directly extracted from the biomass, as, for instance, the agropolymers cellulose or starch, or synthesized from natural sources, such as polylactic acid (PLA) or poly-caprolactone (PCL). A third group is constituted by those biopolymers produced by microorganisms, with xanthan gum and poly-3-hydroxybutyrate (PHB) being the two representative examples. Due to their biodegradability, these naturally occurring polymers contribute to the development of environmentally friendly and renewable nanocomposites, replacing the petroleum-derived polymers and thus helping to reduce plastic waste and dependence on fossil fuel [220].

Biopolymers can be neutral or charged macromolecules, and thus, their interaction with layered clays will differ. The access of neutral species into the interlayer space can be due to the establishment of van der Waals forces, ion–dipole attraction, coordination, hydrogen bonding and water bridges, as well as electron and proton transfer, between both the organic and the inorganic components. When the energy released in the adsorption process is high enough to overcome the attractive forces between the inorganic layers, the intercalation can take place. Other biopolymers are negatively or positively charged, but the intercalation of biopolymers in layered silicates has been mainly reported for polyelectrolytes bearing positive charges, taking place generally by a cation-exchange mechanism with the establishment of strong host–guest electrostatic interactions. In biomacromolecules bearing amino groups, the presence of cationic charges is only achieved at pH values below their acidity constant (pK_a) that allow their protonation. Similarly, in the case of proteins and polypeptides, where the global charge of the macromolecule is pH dependent, the presence of positive or negative charges is controlled by decreasing or increasing the pH value of the solution with respect to the isoelectric point (pI), which is defined as the pH value at which there is an equal number of positive and negative charges in the biomacromolecule. Thus, their intercalation in layered silicates is favoured at pH values below the pI that allow the access of the positively charged species to the interlayer space through cationic exchange [155].

A wide variety of bionanocomposite materials involving layered silicates have been prepared using polysaccharides, proteins and polyesters (Figure 6.12). Among them, the group of polysaccharides is one of the most extensively studied [221], with numerous works addressed to the development of green bionanocomposites involving cellulose or starch.

Cellulose, the most abundant polymer in nature, is a non-charged polymer, and its intercalation by a mechanism of direct exchange is limited due to its low solubility in water. In order to overcome this drawback, water-soluble cellulose derivatives like methyl, hydroxyethyl, hydroxypropyl, methylhydroxyethyl and methylhydroxypropyl cellulose can be prepared; the development of methyl cellulose–montmorillonite bionanocomposite films can be an illustrative example where the silicate layers contributed to a decrease in water adsorption [222]. Other derivatives can be prepared with thermosetting features that allow the application of melting processes, which cannot usually be applied for natural polysaccharides as they decompose below the melting temperature. One of these derivatives is, for instance, cellulose acetate (CA), produced through esterification of cellulose [220]. Usually, the melt extrusion process of CA-based materials requires the addition of plasticizers like triethyl citrate (TEC) and the previous modification of the clay with alkylammonium ions [223], although a

Polysaccharides

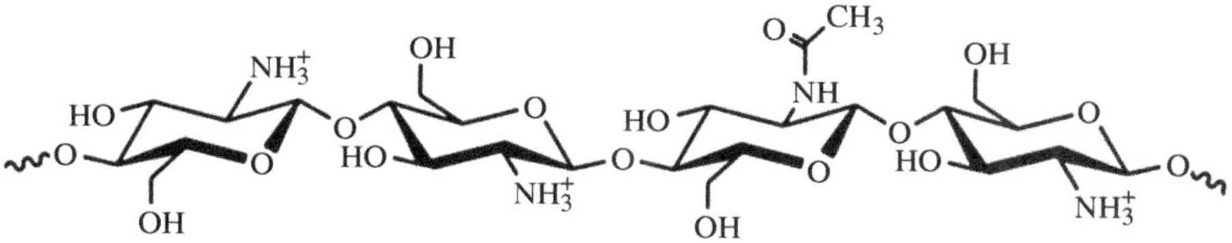

Chitosan

Proteins

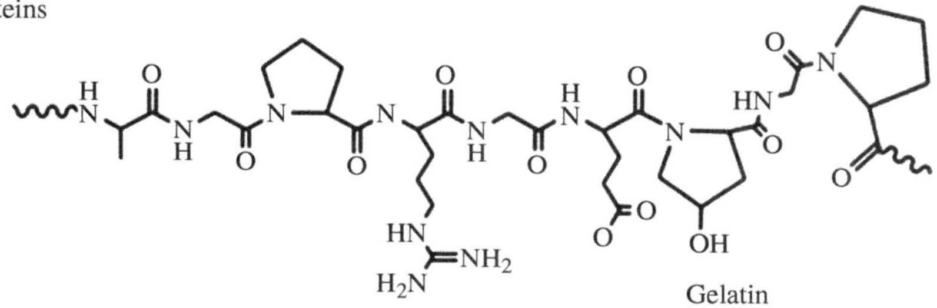

Gelatin

Polyesters

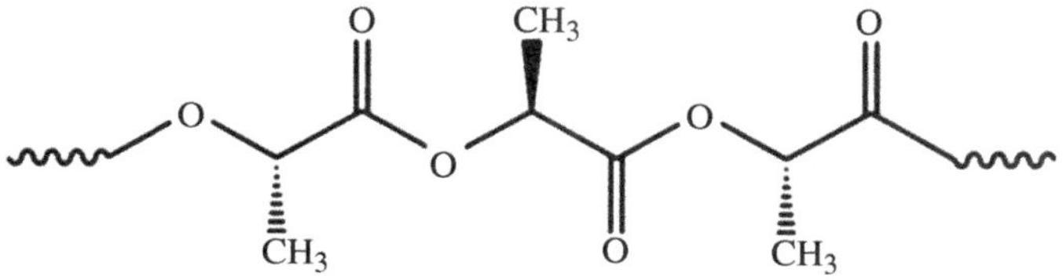

Polylactic acid (PLA)

FIGURE 6.12 Chemical structures of representative biopolymers used in the preparation of bionanocomposites based on layered silicates.

recent work has reported that pristine montmorillonite can afford similar results than organophilic clays, allowing the preparation of exfoliated CA bionanocomposites [224]. In this case, the authors proposed a mechanism consisting in the clay delamination due to the previous intercalation of the plasticizer accompanying the CA matrix.

Recently, the use of appropriate solvents to dissolve cellulose has made possible the preparation of bionanocomposites by the solvent casting method. Thus, N-methylmorpholine-N-oxide (NMMO) was used as solvent, allowing the preparation of bionanocomposite films that involved organoclays as reinforcing fillers [225]. Similarly, ionic liquids like 1-butyl-3-methylimidazolium chloride (BMIMCl) can help to dissolve the cellulose and to obtain cellulose–montmorillonite bionanocomposites by solvent casting, with improved mechanical and thermal properties in comparison to the pristine biopolymer [226].

Another neutral polysaccharide widely used to prepare green nanocomposites is starch, mainly extracted from corn, wheat, rice or potato and composed of the two homopolymers – amylose and amylopectin. Similarly to cellulose, melt processing of starch materials requires the preparation of thermoplastic starch (TPS) by gelatinizing this agropolymer with heat and pressure in the presence of plasticizers, usually glycerol [227, 228]. Urea and formamide were also evaluated as alternative plasticizers, allowing the preparation of starch–montmorillonite nanocomposites by melt extrusion with a high degree of clay exfoliation [229]. The use of formamide–ethanolamine as plasticizers and ethanolamine–montmorillonite as organoclay gave rise to intercalated starch nanocomposites by melt processing, provided with suitable thermal properties and water resistance [230]. Conventional organophilic clays have been also employed to produce starch-based nanocomposites, but several authors have reported that intercalation of starch leading to intercalated structures is favoured in sodium montmorillonite rather than in its organophilic derivatives [227, 231].

The method of solvent casting has been also proposed for the synthesis of starch-layered clay nanocomposites. For instance, Cyras et al. [232] reported the preparation of materials with the intercalated biopolymer rather than exfoliated bionanocomposites, exhibiting significant improvements in the thermal resistance and the Young modulus in comparison to neat starch.

Starch derivatives like carboxymethyl starch (CMS) are also helpful to produce bionanocomposite materials. Thus, exfoliated CMS–kaolinite nanocomposite was synthesized using kaolinite previously modified with DMSO as intermediate [233]. Similarly, a cationic starch (CS) was easily intercalated in montmorillonite, and the resulting material was used as a bio-organoclay to enhance the compatibility of the clay with pristine starch, allowing the production of exfoliated bionanocomposites with better mechanical properties than those based on sodium montmorillonite [234].

Another very abundant polysaccharide on Earth is chitin, a component of the exoskeleton in crustaceans and insects, which is partially deacetylated to produce the chitosan used in the synthesis of bionanocomposites. Chitosan is dissolved in diluted acid solutions, usually acetic acid. It contains primary amino groups that are protonated in these solutions at pH values below their acidity constant ($pK_a = 6.3$) [235]. Thus, chitosan–clay nanocomposites can be prepared by direct cation exchange of the dissolved polysaccharide, which interacts with the negatively charged clay by electrostatic host–guest interactions [236]. Hydrogen bonding and other types of interaction may also contribute to the adsorption mechanism. Varying the synthesis conditions, it is possible to intercalate one, two or more layers of chitosan or even to achieve the exfoliation of the clay in some cases [236–238]. When the equivalents of initial chitosan are below the CEC of the clay, the polysaccharide chains are accommodated as a monolayer in the interlayer space, showing basal spacing values about 1.5 nm [236]. With higher amounts exceeding the CEC, chitosan is intercalated in a bilayer configuration, with basal spacings of 2.1 nm and with an anion-exchange character due to the presence of an excess of protonated amino groups that do not interact with the negatively charged clay sheets. This behaviour as anion exchanger was confirmed by ^{13}C NMR solid-state studies and other techniques [239], and it was profited for diverse applications of these materials in potentiometric sensors for the determination of anions or in drug controlled release, as will be shown in Section 6.5. Other studies have reported the synthesis of partially exfoliated chitosan nanocomposites. Using high amounts of chitosan, the resulting materials showed the coexistence of both intercalated and exfoliated structures [237, 238], which contributed to the improvement in the mechanical and thermal properties. The exfoliation of the clay within the chitosan matrix also contributes to impose a tortuous way to the passage of gases and water vapour that increases the barrier properties of the resulting nanocomposites.

Caramel, derived from the disaccharide sucrose, was intercalated in montmorillonite by a melt intercalation procedure assisted by microwave irradiation [240]. The heating of a sucrose/montmorillonite mixture produces a rapid intercalative polycondensation of the disaccharide forming conventional caramel. These bionanocomposites can be transformed by a subsequent thermal treatment into carbon/clay nanocomposites provided with electrical conductivity, being useful for applications as sensor materials and in energy storage [241, 242].

Other important polymers of natural origin are polypeptides and proteins, which can be intercalated in layered silicates directly from the biopolymer solution, in a similar way to polysaccharides. Due to the fast degradation of most proteins at relatively low temperatures, protein-based nanocomposites usually need to be prepared using solution techniques. Poly-L-lysine, a positively charged polyelectrolyte, was intercalated in Na-montmorillonite by an ion-exchange mechanism, affording bionanocomposite materials provided with good thermomechanical and barrier properties in addition to the biological functionality due to the polypeptide [243]. The polypeptide can be also synthesized inside the clay interlayer space from the previously intercalated monomers. For instance, the kaolinite–poly(β-alanine) bionanocomposite was synthesized by *in situ* polycondensation of β-alanine, which was previously intercalated in kaolinite by means of a displacement method using kaolinite/ammonium acetate as the intermediate [202].

Gelatin is a structural protein obtained by denaturation of collagen, from which it keeps part of its triple helix structure, although at certain temperatures it behaves as a linear polymer with a random coil arrangement [244]. The first attempt to intercalate gelatin in layered silicates was addressed in the 1950s by Talibudeen [245, 246], who proposed that the intercalation mechanism was based on electrostatic interaction between the protein chains and the layers of the montmorillonite and hectorite tested clays, being favoured at pH values below the pI of the gelatin. This led to the complete or partial replacement of the exchangeable sodium ions located in the interlayer space by the gelatin chains. Additional strong van der Waals forces could promote protein uncoiling, making its penetration into the clays interlayer space easier. Up to four gelatin layers can be intercalated in the clay interlayer space depending on the synthesis conditions. The improvement in the thermal stability of the resulting bionanocomposites is attributed to the clay sheets, which can act as physical cross-linking sites able to retard the thermal decomposition of gelatin [247] and enhance its mechanical properties. Due to their excellent properties, gelatin–clay bionanocomposites have been proposed for application in food packaging [244].

Enzymes are complex proteins that act as highly selective catalysts, participating in numerous metabolic reactions. The polypeptide chains in these biomacromolecules are arranged showing specific three-dimensional (3D) structures that determine their specific functionality. Their association with inorganic substrates is usually addressed to provide a protective effect of the inorganic solid, creating a suitable environment that could prevent enzyme denaturation [248]. The intercalation of enzymes in the interlayer space of silicates can help to avoid microbial degradation and to preserve the protein tertiary structure, whereas the open frameworks of these layered solids allow the access of substrates to the immobilized enzymes. Thus, several enzymes like lysozyme, lactoglobulin and chymotrypsin were intercalated in montmorillonite, leading to basal spacing values around 4–5 nm, while lower values about 1.8 nm were achieved in the case of pepsin [249]. Based on these results, which were in accordance with theoretical calculations considering the enzyme dimensions, McLaren and Peterson proposed the use of montmorillonite as a caliper to determine the size of this type of biomacromolecules [249].

Organoclays can be also applied to immobilize enzymes. For instance, myoglobin and haemoglobin were intercalated in the layered sodium silicate magadiite using a tetrabutylammonium–magadiite organoclay as an intermediate, and the resulting materials were proposed for use in biosensor devices [250]. An interesting advantage of this silicate is the possibility of functionalization through the silanol groups present in its surface. Similarly, the organoclay montmorillonite–trimethylpropylammonium has been used to immobilize polyphenol oxidase [251], since it appears that organoclays may provide a favourable environment to certain enzymes, preserving its activity and allowing its use as components in biosensors, as will be shown in Section 6.5.

Within the group of polyesters, PLA is a biodegradable thermoplastic derived from L-lactic acid produced in the fermentation of cornstarch. Green materials based on PLA are attracting great interest as they can help to replace plastics derived from fuel by biodegradable materials obtained from natural resources [220, 252]. Among the numerous inorganic solids used to prepare PLA bionanocomposites with enhanced mechanical and barrier properties, heat resistance and controlled degradability, an important group is that of pristine and organically modified layered clay minerals [253]. Although montmorillonite is the main silicate used in the fabrication of PLA bionanocomposites, other phyllosilicates can be also effective as reinforcing filler due to its high aspect ratio and strong interaction with PLA, leading to materials with improved tensile properties [254].

Melt extrusion processes that do not require solvents are commonly used in the fabrication of PLA-based bionanocomposites [255, 256], but an alternative preparation process based on the *in situ* polymerization of intercalated lactic acid monomers, which can lead to exfoliated nanocomposites, is also feasible [257]. In both cases, the incorporation of layered clay minerals in the fabrication of PLA bionanocomposites allows the production of materials with improved thermomechanical and gas barrier properties as well as better crystallization kinetics. In addition, the incorporated clays can have also influence on both the biodegradation [258] and the hydrolytic degradation [259] of PLA in the bionanocomposites. An additional advantage is the eco-friendly character of the clay minerals, which do not perturb significantly the compostability of the PLA matrix in the bionanocomposites [253, 260].

Another group of biopolymers proposed for the development of green plastics is that of polyhydroxyalkanoates (PHAs), which are linear polyesters produced by

(a)　　　　　　　　　　　　　　　　　(b)

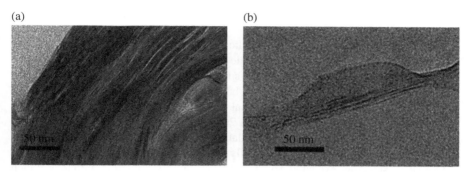

FIGURE 6.13 TEM images showing (a) no intercalation of PHB in Cloisite Na, resulting in well-stacked tactoids of hundreds of platelets, and (b) PHB intercalation in Cloisite 30B showing small tactoids of about 3–10 platelets with a significant increase in the basal spacing. Reprinted from Ref. [264]. Copyright © 2008 WILEY-VCH Verlag GmbH & Co. KGaA, Weinheim.

bacterial fermentation of sugar or lipids. Among them, some studies have reported the use of PHB in the fabrication of bionanocomposites involving layered clays like kaolinite and montmorillonite. Thus, PHB-based materials have been prepared by melt intercalation, using, for instance, a DMSO-modified kaolinite [211], or by means of the solvent casting method as in the case of PHB materials involving an organo-modified montmorillonite [261]. PHB-based copolymers like poly(3-hydroxybutyrate-co-3-hydroxyvalerate) (PHBV) have been also evaluated to fabricate bionanocomposites by melt intercalation or by direct intercalation from solution [262, 263]. A recent study confirmed that both PHB and PHBV gave rise to intercalated materials when an organo-modified montmorillonite was used, due to the good affinity between the clay and these PHAs, while microcomposites were obtained when using sodium montmorillonite (Figure 6.13). The resulting bionanocomposites could be promising materials for biomedical applications, for instance, as implants [264]. However, the low impact resistance and reduced thermal stability of PHB make the use of this polymer difficult, and thus the number of reports on PHB bionanocomposites is still very low in comparison to those other biopolymers mentioned earlier.

6.5 USES OF CLAY–ORGANIC INTERCALATION COMPOUNDS: PERSPECTIVES TOWARDS NEW APPLICATIONS AS ADVANCED MATERIALS

In general, the properties of organic–inorganic materials correspond to the resulting contribution of each one of the two components, but interestingly, the synergistic effect derived from the assembly of the inorganic part and the organic counterpart must be also considered, in which the nature of the corresponding interface between both constituents is a determinant. As in the current case the inorganic component, that is, the clay mineral, could belong to a large variety of silicates of diverse characteristics and, in turn, the organic moiety could be introduced according to a wide choice, the possibilities to prepare hybrids and biohybrids with predetermined properties in view of specific applications are extremely vast.

The properties of the organic–inorganic materials derived from clays offer applications not only for conventional uses such as adsorbents, rheological additives, polymer–clay nanocomposites and so on but also in *advanced applications* based on their electrical, optical or opto-electronical properties, more recently with a focus on biotechnology and biomedicine [17, 113, 265].

So far, the most largely used intercalation compounds with incidence in many industrial applications are the organically modified clays (called as *organoclays*) derived from smectites exchanged with quaternary alkyl-/aryl-ammonium ions, which have been commercialized in the last four decades with many different trademarks (Bentone, Baragel, Nanofil, Cloisite, Smectone, Hectone, Tixogel, Brasgel, etc.). These hybrid materials, also known as organo-bentonites [96], are produced in large amounts (>25,000 tons year^{-1}, data from 1997 [266]) resulting from the intercalation of quaternary alkyl-/aryl-ammonium species of different nature with many diverse of smectite clays, mainly of the montmorillonite type. On the basis of their organophilic and rheological properties, organoclays are receiving intensive application in diverse fields from water treatment as adsorbent of organic pollutants to thixotropic agents for paints and other coatings [267, 268]. Of particular relevance are their applications in polymer–clay nanocomposites [6, 7, 9–11, 13, 14], where the organophilic character inherent to the organoclays is the determinant for the compatibilization of these nanocharges in low-polar polymeric matrixes. Actually, all these applications represent a worldwide and broad sales market, probably the largest known for hybrid materials based on intercalation compounds. Table 6.2 collects some important applications of commercial organoclays based on smectites exchanged with alkyl-/(aryl-) ammonium ions.

New applications of clay-based hybrid materials are possible because the interaction mechanisms between clays and organic compounds, the structural arrangement and organization at the nanometre range of both types of components and the synergistic properties of the resulting materials are reasonably well understood. Nowadays, the controlled modifications of clays by assembling with diverse organic species make possible the design and preparation of clay derivatives with variable functionality and showing suitable properties for potential predetermined applications.

Table 6.3 collects several examples of clay–organic hybrid materials that we have selected to illustrate their interest as potential *advanced* applications.

6.5.1 Selective Adsorption and Separation

Conventional organoclays prepared by intercalation of quaternary alkylammonium species were studied in the last decades with regard to their applications as selective adsorbents and in separation technologies, such as gas chromatography packing materials [304, 305]. In this way, mixtures of light hydrocarbons, that is, from methane to *n*-butane, can be separated using TMA-smectites as packing materials in the gas chromatograph [305].

TABLE 6.2 Current applications of commercial organoclays based on smectites exchanged with alkyl-/(aryl-)ammonium ions

Application field	Main type of application
Agriculture and farming	Carriers and controlled release of agrochemicals (insecticides, herbicides, fungicides, etc.)
	Crop protection formulations
	Mycotoxin sequesters
	Rheological additives in animal feeds
Water purification	Wastewater and sewage treatment
	Treatment of industrial wastewater
Chemical industry	Rheology modifiers: thixotropic, thickening, viscosifier, anti-settling agents
	Oil-based printing inks, adhesives, sealants
	Grease manufacture from mineral and vegetable oils
Petroleum and mineral oils	Oil-based drilling fluids
	Oil spills
	Thicker of lubricating oils
Composite materials	Fillers and additives for reinforcing plastics and rubber
	Components of polymer–clay nanocomposites
Civil engineering and building	Asphalt stabilizers
	Additives in paints
Personal care, cosmetic and pharmacy	Facial masks, skin cleaners, hair shampoos
	Pharmaceutical formulations

The ability of organoclays to absorb pesticides is of interest in the development of 'more ecological' formulations controlling the release of the bioactive compound towards soil and water to prevent indiscriminate pollution [269, 271]. Another important issue in this application field is the stabilization of volatile herbicides, such as alachlor and metolachlor, decreasing its tendency to evaporate when adsorbed on organoclays, which is important in order to prevent air pollution and to increase the effectiveness of those herbicides [270]. Organoclays based on the alkaloid berberine were also effective as platforms for the controlled release of metolachlor [136]. In this case, the use of berberine was proposed as a less hazardous alternative modifier and also in view of the high loadings of herbicide that can be achieved on berberine–montmorillonite organoclays.

Organoclays could be used for removal of pollutants from air, water and soils, particularly low-polar organic molecules such as those based on carcinogenic polycyclic aromatic hydrocarbons [272] and even fullerenes [273]. Interestingly, selective removal of the alkylammonium species in the organoclays-C_{60} intercalation materials can be induced either by thermal treatment at 350°C or by treatment with Al^{3+} solutions, giving rise to clay–fullerene compounds in which C_{60} molecules remain entrapped between the clay silicate layers acting as pillars [273] (Figure 6.14).

Organoclays applied as adsorbents of pollutants can involve not only alkylammonium ions containing aromatic or aliphatic moieties but also monovalent organic dyes with high affinity for the clay mineral. Dye–smectite complexes involving

TABLE 6.3 Development of clay–organic hybrids and biohybrids with properties of interest for potential advanced applications

Potential advanced applications	Selected examples	Authors and references
Selective adsorption and separation	Pesticide adsorbents: stabilizers of formulations and controlled release of pesticides	Rytwo et al. [136] El-Nahhal et al. [269] El-Nahhal et al. [270] Cornejo et al. [271]
	Adsorbents of polycyclic aromatic hydrocarbons and fullerenes	Wefer-Roehl and Czurda [272] Tsoufis et al. [273]
	Selective adsorption of anions (including uptake of anionic radionuclides and other anionic pollutants)	Bors [274] Bors et al. [275] Bors et al. [276] Bors et al. [277] Behnsen et al. [278]
	Modified clays by functionalization with organosilanes for the uptake of heavy metal ions	Mercier and Detellier [279]
Heterogeneous catalysis and supports for organic reactions	Anionic activation reactions	Lin and Pinnavaia [280] Lin et al. [281]
	Acid catalysis	Mortland and Berkheiser [282]
	Transition metal incorporation in the interlayer space of smectite clays as heterogeneous catalysts	Pinnavaia [283] Wang et al. [284]
Membranes, ionic and electronic conductors and sensors	Sensors discriminating hydrocarbons	Yan and Bein [285]
	Kaolinite-based hybrids for anion detection	Letaief et al. [40] Dedzo et al. [32]
	Bionanocomposites based on clays for ion detection	Darder et al. [236] Darder et al. [239]
	Clay/macrocyclic membranes for ion recognition and oxygen sensors	Aranda et al. [286] Ruiz-Hitzky et al. [287] Ceklovsky and Takagi [288]
	Organoclays as components of sensors and biosensors	Peng et al. [250] Mbouguen et al. [251] Colilla et al. [289]
	Ionic and electronic conductors based on clay–organic intercalated materials	Ruiz-Hitzky et al. [287] Aranda et al. [290] Letaïef et al. [291] Rajapakse et al. [292] Aradilla et al. [293]
Photoactive materials	Protection of the photodegradation of labile pesticides	Margulies et al. [294] Rozen et al. [295] Si et al. [296]
	Photodegradation of pollutants	Xiong et al. [297] Meng et al. [298]
Pharmaceutical and biomedical applications	Controlled drug release, tissue engineering and other uses	Ruiz-Hitzky et al. [113] Choy et al. [299] Lin et al. [300] Veerabadran et al. [301] Zhuang et al. [302] Zheng et al. [303]

(a)　　　　　　　　　　(b)　　　　　　　　　　(c)

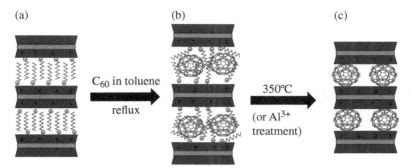

FIGURE 6.14 Scheme representing the interlayer adsorption of fullerene in an organoclay (a) giving rise to an intermediate intercalation compound (b) followed by the removal of alkyl-ammonium species either after heating at 350°C for 24 h or after surfactant exchange with Al^{3+} cations affording a C_{60}-pillared clay (C). Elaborated from data in Ref. [273].

rhodamine B and CV adsorbed on montmorillonite can be also used as adsorbents of non-ionic organic pollutants like naphthalene and phenolic derivatives [117, 306]. These organoclays can be tested in column filters or applied in batch reactors, affording similar efficiencies in the removal of pollutants from aqueous media than high-quality activated carbons, but with the remarkable advantage of exhibiting faster kinetics.

The ability of certain organoclays to adsorb anionic species has been reported, which can be attributed to the inversion of the micelle charge produced by an excess in the adsorption of the long-chain alkylammonium with respect to the CEC of the clay mineral and associating exchangeable anions that remain also located in the interlayer space. These exchangeable anions are useful for exchange with hazardous anions from nuclear waste matrices and contaminated groundwater. So, as reported by Bors and collaborators, long-lived radioiodide ($^{125}I^-$) can be efficiently removed using long-chain alkylammonium organoclays (e.g. hexadecylpyridinium–bentonite) [274–277]. More recently, Behnsen and Riebe showed a clear selectivity of organo-clays such as hexadecylpyridinium-, hexadecyltrimethylammonium- and benzetho-nium-modified bentonites in the uptake of diverse anions. The $ReO_4^- > I^- > NO_3^- Br^- > Cl^- > SO_4^{2-} > SeO_3^{2-}$ experimental affinity sequence for these processes corresponds to the sequence of increasing hydration energies of the considered anions [278].

The modification of the interlayer surface of layered clays (e.g. montmorillonite) by grafting of organosilanes opened the way to new functionalizations in view of introducing groups able to immobilize heavy metals in their cationic form. For instance, the grafting of 3-mercaptopropyltrimethoxysilane (3-MPTMS) affords the ability for complexation of Pb^{2+} and Hg^{2+} ions and, to a lesser extent, Cd^{2+} and Zn^{2+}, through the –SH group [279]. As montmorillonite clay does not contain accessible interlayer hydroxyls for the reaction with alkoxysilanes, it should be considered here that 3-MPTMS can react with $\equiv Si$–OH groups located at the edges of the silicate microcrystals, giving rise to a mercaptopolysiloxane coating on the clay surface with complexing ability towards the aforementioned metal ions [17].

6.5.2   Catalysis and Supports for Organic Reactions

It is well known that anionic activation reactions, for example, the replacement of the halogen atom in alkylbromides by reaction with alkaline salts of different anions, are usually conducted in the presence of organic solvents using phase transfer catalysts. The use of organoclays allows the easy and almost quantitative alkylation of anions such as thiocyanide by reaction with n-alkylbromides in the presence of organoclays, as reported by Lin and Pinnavaia [280]. The alkyl-chain arrangement in the interlayer space of smectites is of great importance in the development of this type of reactions. For instance, the arrangement as bilayers in methyltrioctylammonium–hectorites is determinant for the studied anionic activation reactions [280]. These authors also report on other nucleophilic displacement processes catalysed under triphasic reaction conditions in the presence of organo-hectorites, including alcohol oxidations and C-alkylations of nitriles, among other reactions [281].

The Brönsted acidity introduced in the interlayer space of smectites by the intercalation of certain species such as 1,4-diazabicyclo[2, 2, 2]octane confers the appropriate environment for the catalytic conversion of organic groups such as nitriles to amides [282]. In this way, the 1,4-diazabicyclo [2, 2, 2] octane, which can be considered as the deprotonated triethylene diamine, acts as a pillar separating consecutive layers of the clay silicate facilitating the uptake of small molecules (e.g. acetonitrile) and at the same time procures the suitable acidity for its catalytic transformation (e.g. acetamide) with elevated yield [282]. Transition metal/clay systems with catalytic activity such as $[Rh(PPh_3)_3]^+$-smectite clays have been tested in the hydrogenation of terminal olefins as more efficient heterogeneous catalyst compared with the conventional ones used in homogeneous conditions [283]. These catalysts avoid the competitive isomerization that takes place when the reactions are carried out in homogeneous media, allowing also the easy recovering of the catalyst that can be reused for several cycles without appreciable loss of activity. This represents a significant improvement with respect to the equivalent homogeneous catalysis process. Active rhodium(III) could be also incorporated in the interlayer space by exchange reactions with complexes such as $[Rh(NH_3)_6]^{3+}$. This procedure allows in the same way the incorporation of other diverse complexes such as $[Pd(NH_3)_4]^{2+}$. This type of complexes can be used as precursors for the preparation of Rh and Pd NPs that remain assembled to the clay silicate [307]. Pd–organoclay catalysts, efficient for the aerobic oxidation of benzyl alcohol to benzaldehyde, can be prepared by a different strategy consisting in a first step the incorporation of H_2PdCl_4 in commercial organoclays derived from montmorillonite exchanged with a mixture of octadecyltrimethylammonium (70%) and cetyltrimethylammonium cations (30%) [284]. The further reduction of these intermediate materials with $NaBH_4$ gives rise to Pd NPs that are produced and homogeneously dispersed inside the interlamellar space of the clay, with the quaternary ammonium species playing a role in stabilizing those NPs. This procedure could be extended to the introduction of other types of noble metal NP catalysts (Au, Ru and Pt) [284].

6.5.3 Membranes, Ionic and Electronic Conductors and Sensors

Certain organoclays, in which alkylammonium-intercalated species operate as pillars separating the silicate layers along the c-axis of smectites and therefore creating free galleries in their interlayer spaces, can act as molecular sieves [308]. This is the case of 2 : 1 charged phyllosilicates intercalated, for instance, by tetramethyl- and tetraethylammonium ions (TMA+ and TEA+) resulting in materials able to selectively adsorb small low-polar molecules. Based on this ability, Yan and Bein have developed piezoelectric sensors displaying excellent responses in the discrimination between benzene and cyclohexane [285].

In view of the ability of certain organoclays to adsorb anionic species (vide infra), organoclays based on smectites and alkylammonium-intercalated species such as montmorillonite–cetyltrimethylammonium compounds have been used to prepare modified electrodes, based on the affinity shown by these organoclays to uptake electroactive anions such as $Fe(CN)_6^{4-}$, $Mo(CN)_8^{4-}$ and $Fe(C_2O_4)_3^{3-}$ [309]. Other clay–organic systems can be also used for anion detection. In this way, as previously indicated, kaolinite-based sensors can be developed by interlayer grafting of ionic crystals, which act as pillars between consecutive layers, and their increasing use gives rise to nanoporous and ion-exchangeable materials. The intracrystalline arrangement of those species is useful for applications as electrochemical sensors provided with size selectivity in the detection of anionic species such as cyanide [40], thiocyanate, sulphite and ferricyanide ions [32]. On the other hand, selective detection of anions using intercalated smectites has been also reported for bionanocomposites based on the intercalation of chitosan in montmorillonite, allowing the easy discrimination between mono-, di- and trivalent anions [236]. The anion-exchange ability of these bionanocomposites has been exploited to develop potentiometric sensors based on the intercalation of multilayered chitosan polymer chains that remain positively charged in the interlayer space of smectites. The electrochemical sensors based on these materials show a marked selectivity towards monovalent anions, which can originate from the special arrangement of the biopolymer chains in the clay interlayer space as a nanostructured bidimensional system [236, 239]. This can be also the reason for the observed size discrimination achieved with potentiometric sensors incorporating these bionanocomposites in comparison to those modified with a chitosan film deposited on the electrode surface (Figure 6.15) [310]. Sensors based on the chitosan–montmorillonite nanocomposites were successfully incorporated in an electronic tongue that made use of the artificial intelligence tool known as case-based reasoning (CBR) to evaluate multicomponent solutions used in hydroponic crops [311].

Organoclays based on mercaptopyridinium molecules have been applied as the active phase of carbon paste electrodes for the electrochemical determination of heavy metal ions [289]. The adsorption of 2-mercaptopyridine (2-MPy) and 4-mercaptopyridine (4-MPy) was carried out in acidic methanol/HCl medium that allowed the protonation of the molecules and their intercalation in montmorillonite. Given that the heavy metal ion detection is due to their interaction with the thiol group and that this group was blocked in the protonated 2-MPy due to the formation of a dimer,

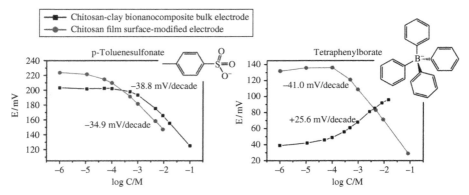

FIGURE 6.15 Potentiometric responses of two sensors, a chitosan–clay bionanocomposite bulk electrode and a chitosan film surface-modified electrode, to *p*-toluenesulphonate showing a similar negative slope in the potential versus log C plot in both cases, and to tetraphenylborate showing only the chitosan film-based electrode the expected response with negative slope. The opposite response obtained with the bionanocomposite-based sensor can be attributed to the difficult access of this anion to chitosan chains confined in the clay interlayer space. From Ref. [310].

only the sensor based on the protonated 4-MPy afforded good results, allowing the amperometric determination of Cd(II), Pb(II), Cu(II) and Hg(II).

Clay–organic materials containing oxyethylene compounds such as crown ethers and PEO as intercalated complexing agents of smectite CE [73] such as Li$^+$ and Na$^+$ (see Section 6.4.1) can modulate their ionic mobility and therefore control the ionic conductivity [287, 290]. The enthalpy of the complexation reaction between cations in the clay interlayer space and diverse $(CH_2O)_n$ crown ethers with different n values ($n = 4,5,6,\ldots$), their nitrogenated and benzo-derivatives, as well as cryptand macrobicyclic compounds, can be correlated with the experimentally found ion conductivity of these systems. Polyoxyethylene (or PEO) compounds of different molecular weight can be also intercalated in smectite clays (vide supra) interacting with the interlayer cations. The ion conductivity of the resulting PEO–clay nanocomposites shows improved characteristics compared to crown ether and cryptand intercalation compounds and also with respect to PEO–salt conventional solid electrolytes. For example, the thermal stability rises in PEO–clay systems to 250–300°C in place of 70–100°C for ion conductors based on PEO–salt complexes [153, 154, 312]. The layer arrangement of these intercalation compounds is clearly anisotropic, presenting ionic in-plane conductivity values in the 10^{-5}–10^{-4} S cm^{-1} range at temperatures higher than 200°C. Such values are several orders of magnitude higher than those showed by the pristine homoionic smectites but lower than those found in PEO–salt complexes (10^{-3}–10^{-4} S cm^{-1} at room temperature). One of the most remarkable characteristics of these solid electrolytes is the values found for the transport number, which is practically equal to unity, deduced from electrical polarization measurements in solid state [166].

Intercalation of electronically conducting polymers in clay smectites using aniline or pyrrole in smectites saturated with transition metal cations results in the corresponding PANI– and PPy–clay nanocomposites [313]. These conducting materials are prepared by intercalative polymerization consisting in the spontaneous polymerization of the adsorbed nitrogenated bases in the interlayer space. For instance, aniline is directly intercalated in Cu-fluorohectorite forming in a first step the non-conducting emeraldine base form, which by exposure to HCl vapours produce a conductive form of PANI showing in-plane electrical conductivities in the order of 10^{-1}–10^{-2} S cm^{-1} [170]. Also, PPy–clay systems display electronic conductivity values that are in the same order of magnitude than those found in PANI–clay nanocomposites [313]. These conductivity values are lower than those corresponding to pure PANI or PPy conducting polymers; however, their encapsulation between the silicate layers avoids the usual polymer degradation when exposed to the ambient [313–315]. Other electronically conducting polymers such as poly(3,4-ethylene-dioxythiophene) known as PEDOT have been later intercalated to montmorillonite. PEDOT can be assembled to the clay by adopting diverse experimental approaches as firstly reported by Letaief et al. [291] and later on by other authors [292, 316]. It appears that the resulting electrical properties of PEDOT–clay nanocomposites mainly depend on the adopted synthetic procedures, the best conductors being the materials prepared by a procedure allowing the production of an intercalated phase together with extra-framework PEDOT [291]. These types of materials, in particular those prepared by *in situ* anodic polymerization as exfoliated PEDOT–montmorillonite nanocomposites, have been used in the fabrication of ultracapacitors using nanometric and micrometric films of PEDOT and the PEDOT nanocomposite [293].

Membranes and sensor devices containing macrocyclic compounds intercalated in smectite clays can be very efficient as active phases of selective sensors. For instance, the intercalation ability of crown ethers and cryptands in 2 : 1 phyllosilicates [73] allows the preparation of composite membranes conformed as sandwich-like materials where the macrocycle–clay material is encapsulated between two thin polybutadiene coatings to improve the mechanical properties of the membrane [286]. The immobilized macrocyclic compounds can modulate the transport properties of cations in aqueous solution, and this can be used for their individual discrimination and recognition. Other types of macrocyclic compounds, such as porphyrins, intercalated in smectite clays can be used for the construction of oxygen sensors [288]. In this way, clay/platinum-porphyrin membranes can serve for the development of effective, reliable and economical optical oxygen sensing devices. The immobilization of phosphorescently active porphyrin molecules that preserve their luminescence activity after intercalation in clays, which is sufficient to reach oxygen sensing processes at aerobic conditions, was positively tested. These results can be extrapolated to analogous systems based on Pd–porphyrin compounds [288].

Organoclays can be also applied as components in biosensors as mentioned earlier. For instance, bionanocomposites based on tetrabutylammonium–magadiite organoclay and two heme proteins, myoglobin and haemoglobin, were proposed for this application, as they show high enzyme-like peroxidase activity after immobilization [250]. In another work, the biosensing device was based on polyphenol oxidase immobilized in montmorillonite–trimethylpropylammonium intercalation compounds, being applied in the electroanalytical determination of catechol chosen as model analyte [251].

6.5.4 Photoactive Materials

Organically modified clays by intercalation of cationic dyes such as MB, methyl green, CV, safranin T, TFT, acridine orange and rhodamine 6G can be used in the preparation of photostable formulations of labile pesticides [113]. The protection towards photodegradation of co-adsorbed labile bioactive species was firstly established by Margulies showing a deactivation mechanism via energy transfer between the two types of organic molecules interacting with the clay surface [294]. This concept of photostabilization can be applied either to insecticides, as, for instance, tetrahydro-2-(nitromethylene)-2H-1,3-thiazine [295], or to herbicides, as, for instance, 2-(4,6-dimethoxypyrimidin-2-carbamoylsulphamoyl)-o-toluic acid methyl ester [296].

Pollutants in water can be removed using organoclay derivatives containing protoactive catalysts. In this context, photosensitized degradation of 2,4,6-trichlorophenol in water was studied using organoclays acting as supports of the palladium(II) phthalocyaninesulphonate photocatalyst under visible light irradiation. A complete dechlorination and total oxidation of TCP have been observed. The efficiency of these photocatalysts can be correlated with the alkyl-chain length of the alkylammonium intercalated from dodecyltrimethylammonium to cetyltrimethylammonium and to octadecyltrimethylammonium [297].

On the other hand, it has been observed that the assembly of photoactive NPs such as SiO_2/TiO_2 to organoclays advantageously combines the high photocatalytic activity of the titanium dioxide NPs with the high adsorption capability of the organoclay, giving rise to improved hybrid materials useful for the removal of pollutants from wastewater by photodegradation [298].

6.5.5 Biomedical Applications

Clays are traditionally applied in biomedicine and particularly in pharmaceutical formulations due to their lack of toxicity together with other properties useful for this area of applications [317–319]. In turn, clay–organic hybrid materials are of great interest in biomedical applications because the organic component assembled to the clay can be a bioactive drug or a biocompatible compound introducing synergistic characteristics helpful for the final uses [113].

Natural and synthetic layered clays can act as reservoir or nanocontainers for diverse type of bioactive molecules that could be intercalated by different mechanisms in the interlayer space of clays for preparation of controlled drug delivery systems (DDS) [299, 320]. This is the case, for instance, of the DDS prepared by intercalation of vitamin B_6 [321] and timolol maleate in montmorillonite [322]. The synthetic smectite known as Laponite has been also employed for drug delivery applications, in this case, for example, of itraconazole, a water-insoluble broad-spectrum drug used for the treatment of fungal infections [323]. The ability of organoclays such as hexadecyltrimethylammonium (HMTA)-montmorillonite to assemble DNA opens the way for future applications of these hybrid materials as non-viral vectors in gene therapy [300]. Similarly, a bionanocomposite based on rectorite, a mica/

smectite superstructure, and quaternized chitosan was evaluated for DNA transfection, showing better results than the pristine polymer [324].

An interesting case of clays for molecular reservoir of drugs is the nanotubular halloysite showing advantages for applications in the pharmaceutical field as well as in medicine, due to its important characteristics such as inner drug entrapment ability, biocompatibility, high stability and high mechanical strength [301, 325–327]. Lvov and co-workers reported the experimental way for the loading of bioactive molecules inside the nanotubules of halloysite allowing the infiltration into the lumen of diverse types of drugs such as furosemide (antihypertensive), dexamethasone (corticosteroid) and nifedipine (antianginal), which can be later released in a controlled manner [301]. Diverse authors showed also the possibility to include into the nanotubules other types of drugs, for instance, Viseras and collaborators who investigated the role of halloysite in the adsorption mechanism of the 5-aminosalicylic acid (anti-inflammatory drug) including the access to the intratubular region [328]. More complex and voluminous bioactive molecules including nicotinamide adenine dinucleotide (NAD) can be also included in the halloysite lumen, observing later a linear release at a rate constant over time, which is very suitable for medical applications [325]. DNA has been assembled to halloysite following a solid-state reaction using a mechanochemical approach that cut the silicate into shorter lengths [329]. These materials could be promising for future uses as non-viral vectors for gene delivery.

Biomimetic and biocompatible clay-based bionanocomposites represent a class of choice of materials for wide applications in biomedicine from DDS to tissue engineering [113]. See, for instance, a new DDS based on the intercalation of procainamide hydrochloride, an antiarrhythmic drug in montmorillonite. To retard the drug release in gastric environments and to get the suitable release of the drug in the intestinal environments in a controlled manner, these intercalation compounds have been assembled firstly to alginate and further coated with chitosan [330]. Chitosan–clay bionanocomposites [236] are good candidates as DDS. As an illustrative example, the antimalarial drug quinine can be loaded in these types of hybrids showing in *in vitro* test a no drug release in the gastric fluid [331]. Similarly, bionanocomposites involving chitosan-*g*-lactic acid and sodium montmorillonite showed a suitable behaviour in the controlled release of ibuprofen in phosphate buffered saline solution, while their biocompatibility and their ability to allow cell proliferation were also confirmed [332]. On the other hand, intercalated chitosan in montmorillonite and more complex bionanocomposite materials such as chitosan–gelatin–clay systems can be considered useful for other biomedical applications, being excellent candidates for tissue engineering as they exhibit good adhesion and proliferation with rat stem cells [302]. For this application, the materials were processed as macroporous scaffolds by means of freeze-drying, showing improved mechanical properties due to the reinforcing effect of the clay filler. In addition, the presence of the clay platelets contributed to reduce the biodegradation rate of the biopolymer matrix [303].

Halloysite-based bionanocomposites prepared by assembling of chitosan and this type of clay and conformed as foams by freeze-drying can form uniform 3D porous scaffolds with enhanced characteristics such as chemical and thermal stabilities, mechanical properties and biological properties for utilization in tissue engineering and drug/gene release applications [333]. Silk fibroin is a protein produced by spiders, the larvae of several moths and other insects. It shows improved mechanical properties, which together its characteristic stiffness and toughness make it a material with applications in several areas, including biomedicine. It has been assembled to montmorillonite following an LbL procedure, resulting in bio-compatible, robust and flexible bionanocomposites with exceptional mechanical properties of great interest for biomedical applications, such as reinforced tissue engineering [334].

REFERENCES

1. F. Bergaya, G. Lagaly (Eds) (**2013**) Handbook of Clay Science. Part A: Fundamentals. Elsevier, Amsterdam.
2. R. T. Martin, S. W. Bailey, D. D. Eberl, D. S. Fanning, S. Guggenheim, H. Kodama, D. R. Pevear, J. Srodon, F. J. Wicks (**1991**) Clays and Clay Minerals *39*, 333–335.
3. B. S. Neumann, K. G. Sansom (**1970**) Clay Minerals *8*, 389–404.
4. E. Ruiz-Hitzky (**2004**) in Functional Hybrid Materials (eds P. Gómez-Romero, C. Sanchez) Wiley-VCH, Weinheim, pp. 15–49.
5. A. Weiss (**1961**) Angewandte Chemie *73*, 736–736.
6. T. J. Pinnavaia, G. Beall (Eds) (**2000**) Polymer-Clay Nanocomposites. John Wiley & Sons, Inc., New York.
7. M. Alexandre, P. Dubois (**2000**) Materials Science & Engineering R-Reports *28*, 1–63.
8. S. S. Ray, M. Okamoto (**2003**) Progress in Polymer Science *28*, 1539–1641.
9. E. Ruiz-Hitzky, A. Van Meerbeek (**2006**) in Handbook of Clay Science (eds F. Bergaya, B. K. G. Theng, G. Lagaly) Elsevier, Amsterdam, pp. 583–621.
10. A. Okada, A. Usuki (**2006**) Macromolecular Materials and Engineering *291*, 1449–1476.
11. K. A. Carrado, F. Bergaya (Eds) (**2007**) Clay-Based Polymer Nanocomposites. The Clay Minerals Society, Chantilly (Virginia, EEUU) Vol. 15.
12. S. Pavlidou, C. D. Papaspyrides (**2008**) Progress in Polymer Science *33*, 1119–1198.
13. D. R. Paul, L. M. Robeson (**2008**) Polymer *49*, 3187–3204.
14. P. Kiliaris, C. D. Papaspyrides (**2010**) Progress in Polymer Science *35*, 902–958.
15. J.-F. Lambert, F. Bergaya (**2013**) in Handbook of Clay Science 2nd Edition (eds F. Bergaya, G. Lagaly) Elsevier Science Ltd, Oxford, pp. 679–706.
16. E. Ruiz-Hitzky, P. Aranda, M. Darder, M. Ogawa (**2011**) Chemical Society Reviews *40*, 801–828.
17. E. Ruiz-Hitzky, P. Aranda, J. M. Serratosa (**2004**) in Handbook of Layered Materials (eds S. M. Auerbach, K. A. Carrado, P. K. Dutta) Marcel Dekker, New York, pp. 91–154.
18. S. Yariv, H. Cross (Eds) (**2002**) Organo-Clay Complexes and Interactions. Marcel Dekker, New York.
19. G. Lagaly, M. Ogawa, I. Dekany (**2013**) in Handbook of Clay Science 2nd Edition (eds F. Bergaya, G. Lagaly) Elsevier Science Ltd, Oxford, pp. 435–506.
20. K. Wada (**1961**) American Mineralogist *46*, 78–91.
21. P. Fenoll Hach-Alí, A. Weiss (**1969**) Anales de Quimica *65*, 769–790.

22. M. Cruz, A. Laycock, J. L. White (**1969**) in Proceedings of the International Clay Conference, Tokyo 1969 (ed L. Heller) Israel University Press, Jerusalem, Vol. 1, pp. 775–789.

23. S. Olejnik, A. M. Posner, J. P. Quirk (**1971**) Clays and Clay Minerals *19*, 83–94.

24. B. Caglar, C. Cirak, A. Tabak, B. Afsin, E. Eren (**2013**) Journal of Molecular Structure *1032*, 12–22.

25. J. M. Adams (**1978**) Clays and Clay Minerals *26*, 169–172.

26. S. Olejnik, L. A. G. Aylmore, A. M. Posner, J. P. Quirk (**1968**) Journal of Physical Chemistry *72*, 241–249.

27. S. Olejnik, A. M. Posner, J. P. Quirk (**1971**) Spectrochimica Acta Part A: Molecular Spectroscopy A *27*, 2005–2009.

28. M. Cruz, H. Jacobs, J. J. Fripiat (**1973**) in Proceedings of the International Clay Conference, Madrid 1972 (ed J. M. Serratosa) Div. Ciencias, C.S.I.C., Madrid, pp. 35–44.

29. S. Hayashi (**1997**) Clays and Clay Minerals *45*, 724–732.

30. Y. Komori, Y. Sugahara, K. Kuroda (**1999**) Applied Clay Science *15*, 241–252.

31. J. Matusik, E. Scholtzova, D. Tunega (**2012**) Clays and Clay Minerals *60*, 227–239.

32. G. K. Dedzo, S. Letaief, C. Detellier (**2012**) Journal of Materials Chemistry *22*, 20593–20601.

33. C. Detellier, S. Letaief (**2013**) in Handbook of Clay Science 2nd Edition. Part A: Fundamentals (eds F. Bergaya, G. Lagaly) Elsevier, Amsterdam, pp. 707–719.

34. S. Yariv (**2002**) in Organo-Clay Complexes and Interactions (eds S. Yariv, H. Cross) Marcel Dekker, New York, pp. 39–111.

35. A. Weiss, W. Thielepape, H. Orth (**1966**) in Proceedings of the International Clay Conference, Jerusalem 1966 (eds L. L. Heller, A. Weiss) Israel University Press, Jerusalem, Vol. 1, pp. 277–293.

36. Y. Komori, Y. Sugahara, K. Kuroda (**1998**) Journal of Materials Research *13*, 930–934.

37. Y. Komori, Y. Sugahara, K. Kuroda (**1999**) Chemistry of Materials *11*, 3–6.

38. J. J. Tunney, C. Detellier (**1996**) Journal of Materials Chemistry *6*, 1679–1685.

39. J. J. Tunney, C. Detellier (**1993**) Chemistry of Materials *5*, 747–748.

40. S. Letaief, I. K. Tonle, T. Diaco, C. Detellier (**2008**) Applied Clay Science *42*, 95–101.

41. J. Gardolinski, G. Lagaly (**2005**) Clay Minerals *40*, 537–546.

42. S. Letaief, C. Detellier (**2007**) Chemical Communications, 2613–2615.

43. D. Hirsemann, T. K. J. Koester, J. Wack, L. van Wuellen, J. Breu, J. Senker (**2011**) Chemistry of Materials *23*, 3152–3158.

44. J. Gardolinski, G. Lagaly (**2005**) Clay Minerals *40*, 547–556.

45. S. Letaief, C. Detellier (**2005**) Journal of Materials Chemistry *15*, 4734–4740.

46. S. Letaief, T. A. Elbokl, C. Detellier (**2006**) Journal of Colloid and Interface Science *302*, 254–258.

47. S. Letaief, C. Detellier (**2007**) Journal of Materials Chemistry *17*, 1476–1484.

48. S. Letaief, C. Detellier (**2008**) Clays and Clay Minerals *56*, 82–89.

49. S. Letaief, T. Diaco, W. Pell, S. I. Gorelsky, C. Detellier (**2008**) Chemistry of Materials *20*, 7136–7142.

50. J. L. Guimaraes, P. Peralta-Zamora, F. Wypych (**1998**) Journal of Colloid and Interface Science *206*, 281–287.

51. C. Breen, N. D'Mello, J. Yarwood (**2002**) Journal of Materials Chemistry *12*, 273–278.

52. J. Gardolinski, G. Lagaly, M. Czank (**2004**) Clay Minerals *39*, 391–404.

53. L. Zapata, A. Van Meerbeek, J. J. Fripiat, M. della Faille, M. van Russelt, J. P. Mercier (**1973**) Journal of Polymer Science Part C-Polymer Symposium, 257–272.

54. E. Ruiz-Hitzky, J. M. Rojo (**1980**) Nature *287*, 28–30.

55. E. Ruiz-Hitzky (**2003**) Chemical Record *3*, 88–100.

56. E. Ruiz-Hitzky, (**1974**) Ph.D. Thesis, Université Catholique de Louvain, Leuven.

57. E. Ruiz-Hitzky, J. M. Rojo, G. Lagaly (**1985**) Colloid and Polymer Science *263*, 1025–1030.

58. E. Ruiz-Hitzky (**1975**) Anales de Quimica *71*, 838–839.

59. K. Kuroda, C. Kato (**1979**) Clays and Clay Minerals *27*, 53–56.

60. J. J. Fripiat, E. Mendelovici (**1968**) Bulletin de la Societe Chimique de France, 483–492.

61. E. Ruiz-Hitzky, J. J. Fripiat (**1976**) Bulletin de la Societe Chimique de France, 1341–1348.

62. A. Van Meerbeek, E. Ruiz-Hitzky (**1979**) Colloid and Polymer Science *257*, 178–181.

63. R. M. Carr, H. Chih (**1971**) Clay Minerals *9*, 153–166.

64. Y. M. Lvov, D. G. Shchukin, H. Mohwald, R. R. Price (**2008**) ACS Nano *2*, 814–820.

65. W. F. Bradley (**1945**) Journal of the American Chemical Society *67*, 975–981.

66. D. M. C. MacEwan (**1946**) Nature *157*, 159–160.

67. D. M. C. MacEwan (**1948**) Transactions of the Faraday Society *44*, 349–367.

68. G. W. Brindley (**1966**) Clay Minerals *6*, 237–259.

69. M. M. Mortland (**1970**) Advances in Agronomy *22*, 75–117.

70. J. A. Rausell-Colom, J. M. Serratosa (**1987**) in Chemistry of Clays and Clay Minerals (ed A. C. D. Newman) Mineralogical Society, London, pp. 371–422.

71. B. K. G. Theng (**1974**). The Chemistry of Clay-Organic Reactions. John Wiley and Sons, Inc., New York.

72. G. Lagaly (**1993**) in Tonminerale und Tone (eds K. Jasmund, G. Lagaly) Steinkopff Verlag, Darmstadt, pp. 89–167.

73. E. Ruiz-Hitzky, B. Casal (**1978**) Nature *276*, 596–597.

74. E. Ruiz-Hitzky, B. Casal, P. Aranda, J. C. Galvan (**2001**) Reviews in Inorganic Chemistry *21*, 125–159.

75. B. Casal, P. Aranda, J. Sanz, E. Ruiz-Hitzky (**1994**) Clay Minerals *29*, 191–203.

76. B. Casal, E. Ruiz-Hitzky, J. M. Serratosa, J. J. Fripiat (**1984**) Journal of the Chemical Society-Faraday Transactions I *80*, 2225–2232.

77. P. Aranda, B. Casal, J. J. Fripiat, E. Ruiz-Hitzky (**1994**) Langmuir *10*, 1207–1212.

78. B. Casal, E. Ruiz-Hitzky, L. Van Vaeck, F. C. Adams (**1988**) Journal of Inclusion Phenomena *6*, 107–118.

79. S. S. Cady, T. J. Pinnavaia (**1978**) Inorganic Chemistry *17*, 1501–1507.

80. H. Van Damme, M. Crespin, F. Obrecht, M. I. Cruz, J. J. Fripiat (**1978**) Journal of Colloid and Interface Science *66*, 43–54.

81. A. G. Cairns-Smith (**1982**). Genetic Takeover and the Mineral Origins of Life. Cambridge University Press, Cambridge.

82. J. Pleysier, A. Cremers (**1975**) Journal of the Chemical Society-Faraday Transactions I *71*, 256–264.

83. A. Maes, P. Peigneur, A. Cremers (**1976**) in Proceedings of the International Clay Conference, Mexico, 1975 (ed S. W. Bailey) Applied Publishing, Wilmette, IL, pp. 319–329.

84. A. Maes, P. Peigneur, A. Cremers (**1978**) Journal of the Chemical Society-Faraday Transactions I *74*, 182–189.

85. P. Peigneur, A. Maes, A. Cremers (**1979**) in Proceedings of the International Clay Conference, Oxford, 1978 (eds M. M. Mortland, V. C. Farmer) Elsevier, Amsterdam, pp. 207–216.

86. H. E. Doner, M. M. Mortland (**1969**) Science *166*, 1406–1407.

87. M. M. Mortland, T. J. Pinnavaia (**1971**) Nature-Physical Science *229*, 75–77.

88. D. M. Clementz, M. M. Mortland (**1972**) Clays and Clay Minerals *20*, 181–186.

89. V. C. Farmer, M. M. Mortland (**1966**) Journal of the Chemical Society, 344–351.

90. J. M. Serratosa (**1968**) American Mineralogist *53*, 1244–1251.

91. J. P. Hunt (**1963**). Metal Ions in Aqueous Solution. Benjamin, New York.

92. J. J. Fripiat, A. Servais, A. Leonard (**1962**) Bulletin de la Societe Chimique de France, *635–644*.

93. C. R. Smith (**1934**) Journal of the American Chemical Society *56*, 1561–1563.

94. J. E. Gieseking (**1939**) Soil Science *47*, 1–13.

95. S. B. Hendricks (**1941**) Journal of Physical Chemistry *45*, 65–81.

96. L. B. de Paiva, A. R. Morales, F. R. Valenzuela Diaz (**2008**) Applied Clay Science *42*, 8–24.

97. A. Weiss, E. Michel, A. L. Weiss (**1958**) in Wasserstoffbruckenbindungen. Ein-und Zweidimensionale Innerkristalline Quellungsvorgänge. Hydrogen Bonding (a Symposium) (ed D. Hazdi) Pergamon Press, London, pp. 495–508.

98. G. F. Walker (**1963**) in Proceedings of the International Clay Conference, Stockholm 1963 (ed I. Rosenqvist) Pergamon Press, Oxford, Vol. 2, pp. 259–261.

99. G. F. Walker (**1967**) Clay Minerals *7*, 129–143.

100. W. D. Johns, P. K. Sen Gupta (**1967**) American Mineralogist *52*, 1706–1724.

101. J. M. Serratosa, W. D. Johns, A. Shimoyama (**1970**) Clays and Clay Minerals *18*, 107–113.

102. J. A. Martin-Rubi, J. A. Rausell-Colom, J. M. Serratosa (**1974**) Clays and Clay Minerals *22*, 87–90.

103. J. A. Rausell-Colom, V. Fornes (**1974**) American Mineralogist *59*, 790–798.

104. T. González-Carreño, J. A. Rausell-Colom, J. M. Serratosa (**1977**) in Proceedings of the III European Clay Conference, Oslo, 1977 (ed I. Rosenqvist) Nordic Society for Clay Research, Oslo, Vol. 1, pp. 73–74.

105. T. Furukawa, G. W. Brindley (**1973**) Clays and Clay Minerals *21*, 279–288.

106. G. W. Brindley, A. Tsunashima (**1972**) Clays and Clay Minerals *20*, 233–240.

107. J. M. Serratosa (**1965**) Nature *208*, 679–681.

108. J. M. Serratosa (**1966**) Clays and Clay Minerals *14*, 385–391.

109. A. Weiss (**1963**) Angewandte Chemie International Edition *2*, 134–144.

110. G. Lagaly (**1986**) Solid State Ionics *22*, 43–51.

111. M. Ogawa, K. Kuroda (**1997**) Bulletin of the Chemical Society of Japan *70*, 2593–2618.

112. G. Lagaly, M. Ogawa, I. Dekany (**2006**) in Handbook of Clay Science (eds F. Bergaya, B. K. G. Theng, G. Lagaly) Elsevier, Amsterdam, Vol. 1, pp. 309–377.

113. E. Ruiz-Hitzky, P. Aranda, M. Darder, G. Rytwo (2010) Journal of Materials Chemistry 20, 9306–9321.
114. J. M. Serratosa (1968) Clays and Clay Minerals 16, 93–97.
115. G. Lagaly (1976) Angewandte Chemie International Edition 15, 575–586.
116. B. Wicklein, M. Darder, P. Aranda, E. Ruiz-Hitzky (2010) Langmuir 26, 5217–5225.
117. G. Rytwo, Y. Kohavi, I. Botnick, Y. Gonen (2007) Applied Clay Science 36, 182–190.
118. H. A. Patel, R. S. Somani, H. C. Bajaj, R. V. Jasra (2007) Applied Clay Science 35, 194–200.
119. W. H. Awad, J. W. Gilman, M. Nyden, R. H. Harris Jr, T. E. Sutto, J. Callahan, P. C. Trulove, H. C. DeLong, D. M. Fox (2004) Thermochimica Acta 409, 3–11.
120. Z. Zeng, D. Matuschek, A. Studer, C. Schwickert, R. Poettgen, H. Eckert (2013) Dalton Transactions 42, 8585–8596.
121. S. Nir, G. Rytwo, U. Yermiyahu, L. Margulies (1994) Colloid and Polymer Science 272, 619–632.
122. M. Ogawa, A. Ishikawa (1998) Journal of Materials Chemistry 8, 463–467.
123. Z. Klika, H. Weissmannova, P. Capkova, M. Pospisil (2004) Journal of Colloid and Interface Science 275, 243–250.
124. K. A. Carrado, J. E. Forman, R. E. Botto, R. E. Winans (1993) Chemistry of Materials 5, 472–478.
125. G. Rytwo, C. Serban, S. Nir, L. Margulies (1991) Clays and Clay Minerals 39, 551–555.
126. G. Rytwo, Y. Gonen, R. Huterer-Shveky (2009) Clays and Clay Minerals 57, 555–565.
127. S. Tani, H. Yamaki, A. Sumiyoshi, Y. Suzuki, S. Hasegawa, S. Yamazaki, J. Kawamata (2009) Journal of Nanoscience and Nanotechnology 9, 658–661.
128. M. Samuels, O. Mor, G. Rytwo (2013) Journal of Photochemistry and Photobiology B-Biology 121, 23–26.
129. G. F. Walker, W. G. Garrett (1961) Nature 191, 1389–1390.
130. C. de la Calle, M. I. Tejedor, C. H. Pons (1996) Clays and Clay Minerals 44, 68–76.
131. G. E. Lailach, T. D. Thompson, G. W. Brindley (1968) Clays and Clay Minerals 16, 285–293.
132. G. E. Lailach, G. W. Brindley (1969) Clays and Clay Minerals 17, 95–100.
133. L. Perezgasga, A. Serrato-Diaz, A. Negron-Mendoza, L. D. Galan, F. G. Mosqueira (2005) Origins of Life and Evolution of Biospheres 35, 91–110.
134. H. J. Huebner, S. L. Lemke, S. E. Ottinger, K. Mayura, T. D. Phillips (1999) Food Additives and Contaminants 16, 159–171.
135. E. Cohen, T. Joseph, I. Lapides, S. Yariv (2005) Clay Minerals 40, 223–232.
136. G. Rytwo, Y. Gonen, S. Afuta (2008) Applied Clay Science 41, 47–60.
137. G. Rytwo, S. Nir, L. Margulies (1996) Soil Science Society of America Journal 60, 601–610.
138. U. Herwig, E. Klumpp, H. D. Narres, M. J. Schwuger (2001) Applied Clay Science 18, 211–222.
139. A. Weidenhaupt, C. Arnold, S. R. Muller, S. B. Haderlein, R. P. Schwarzenbach (1997) Environmental Science & Technology 31, 2603–2609.
140. G. Lagaly (1999) Applied Clay Science 15, 1–9.

141. F. Bergaya, C. Detellier, J.-F. Lambert, G. Lagaly (**2013**) in Handbook of Clay Science 2ⁿᵈ Edition (eds F. Bergaya, G. Lagaly) Elsevier Science Ltd, Oxford, pp. 655–677.

142. Y. Fukushima, S. Inagaki (**1987**) Journal of Inclusion Phenomena *5*, 473–482.

143. B. K. G. Theng (**1979**). Formation and Properties of Clay-Polymer Complexes. Elsevier, New York.

144. A. Blumstein (**1961**) Bulletin de la Societe Chimique de France, *899–905*.

145. H. Z. Friedlander (**1963**) American Chemical Society Division of Polymer Chemistry Reprints *4*, 300–306.

146. P. Cloos, A. Moreale, C. Broers, C. Badot (**1979**) Clay Minerals *14*, 307–321.

147. A. Moreale, P. Cloos, C. Badot (**1985**) Clay Minerals *20*, 29–37.

148. S. Letaief, P. Aranda, E. Ruiz-Hitzky (2005) Applied Clay Science *28*, 183–198.

149. R. A. Ruehrwein, D. W. Ward (**1952**) Soil Science *73*, 485–492.

150. C. Breen (**1999**) Applied Clay Science *15*, 187–219.

151. R. L. Parfitt, D. J. Greenland (**1970**) Clay Minerals *8*, 305–315.

152. R. L. Parfitt, D. J. Greenland (**1970**) Clay Minerals *8*, 317–323.

153. E. Ruiz-Hitzky, P. Aranda (**1990**) Advanced Materials *2*, 545–547.

154. P. Aranda, E. Ruiz-Hitzky (**1992**) Chemistry of Materials *4*, 1395–1403.

155. E. Ruiz-Hitzky, P. Aranda, M. Darder (**2009**) in Bottom-Up Nanofabrication: Supramolecules, Self-Assemblies, and Organized Films (eds K. Ariga, H. S. Nalwa) American Scientific Publisher, Stevenson Ranch, CA, Vol. 3, pp. 39–76.

156. D. J. Greenland (**1963**) Journal of Colloid Science *18*, 647–664.

157. G. Lagaly (**1986**) in Developments in Ionic Polymers (eds A. D. Wilson, H. J. Prosser) Elsevier, London, Vol. 2, pp. 77–140.

158. N. Ogata, S. Kawakage, T. Ogihara (**1997**) Journal of Applied Polymer Science *66*, 573–581.

159. P. Aranda, E. Ruiz-Hitzky (**1994**) Acta Polymerica *45*, 59–67.

160. E. Hackett, E. Manias, E. P. Giannelis (**2000**) Chemistry of Materials *12*, 2161–2167.

161. E. Manias, A. Touny, L. Wu, K. Strawhecker, B. Lu, T. C. Chung (**2001**) Chemistry of Materials *13*, 3516–3523.

162. B. Hoffmann, C. Dietrich, R. Thomann, C. Friedrich, R. Mulhaupt (**2000**) Macromolecular Rapid Communications *21*, 57–61.

163. R. A. Vaia, H. Ishii, E. P. Giannelis (**1993**) Chemistry of Materials *5*, 1694–1696.

164. E. P. Giannelis, R. Krishnamoorti, E. Manias (**1999**) in Advances in Polymer Science (ed S. Granick) Springer, Vol. 138, pp. 107–147.

165. R. A. Vaia, S. Vasudevan, W. Krawiec, L. G. Scanlon, E. P. Giannelis (**1995**) Advanced Materials *7*, 154–156.

166. P. Aranda, Y. Mosqueda, E. Perez-Cappe, E. Ruiz-Hitzky (**2003**) Journal of Polymer Science Part B-Polymer Physics *41*, 3249–3263.

167. J. W. Cho, D. R. Paul (**2001**) Polymer *42*, 1083–1094.

168. Y. Fukushima, A. Okada, M. Kawasumi, T. Kurauchi, O. Kamigaito (**1988**) Clay Minerals *23*, 27–34.

169. A. Usuki, Y. Kojima, M. Kawasumi, A. Okada, Y. Fukushima, T. Kurauchi, O. Kamigaito (**1993**) Journal of Materials Research *8*, 1179–1184.

170. V. Mehrotra, E. P. Giannelis (**1991**) Solid State Communications *77*, 155–158.

171. T. C. Chang, S. Y. Ho, K. J. Chao (**1992**) Journal of the Chinese Chemical Society *39*, 209–212.

172. V. Mehrotra, E. P. Giannelis (**1992**) Solid State Ionics *51*, 115–122.

173. F. Bergaya, F. Kooli (**1991**) Clay Minerals *26*, 33–41.

174. L. Duclaux, E. Frackowiak, T. Gibinski, R. Benoit, F. Beguin (**2000**) Molecular Crystals and Liquid Crystals *340*, 449–454.

175. R. Blumstein, A. Blumstein, K. K. Parikh (**1973**) Abstracts of Papers of the American Chemical Society, 7–7.

176. M. Biswas, S. S. Ray (**1998**) Polymer *39*, 6423–6428.

177. A. S. Moet, A. Akelah (**1993**) Materials Letters *18*, 97–102.

178. K. A. Carrado, L. Q. Xu (**1998**) Chemistry of Materials *10*, 1440–1445.

179. K. A. Carrado (**2000**) Applied Clay Science *17*, 1–23.

180. K. A. Carrado, L. Xu, S. Seifert, R. Csencsits (**2000**) in Polymer–Clay Nanocomposites (eds T. J. Pinnavaia, G. W. Beall) Wiley, West Sussex, pp. 47–93.

181. K. Haraguchi (**2007**) Current Opinion in Solid State and Materials Science *11*, 47–54.

182. S. H. Nair, K. C. Pawar, J. P. Jog, M. V. Badiger (**2007**) Journal of Applied Polymer Science *103*, 2896–2903.

183. S. Kundakci, O. B. Uezuem, E. Karadag (**2008**) Reactive & Functional Polymers *68*, 458–473.

184. K. Xu, J. Wang, S. Xiang, Q. Chen, W. Zhan, P. Wang (**2007**) Applied Clay Science *38*, 139–145.

185. H.-Y. Ren, M. Zhu, K. Haraguchi (**2011**) Macromolecules *44*, 8516–8526.

186. K. Haraguchi, K. Murata, T. Takehisa (**2012**) Macromolecules *45*, 385–391.

187. J. Ning, G. Li, K. Haraguchi (**2013**) Macromolecules *46*, 5317–5328.

188. Z. Y. Tang, N. A. Kotov, S. Magonov, B. Ozturk (**2003**) Nature Materials *2*, 413–418.

189. P. Podsiadlo, S. Paternel, J. M. Rouillard, Z. F. Zhang, J. Lee, J. W. Lee, L. Gulari, N. A. Kotov (**2005**) Langmuir *21*, 11915–11921.

190. P. Podsiadlo, A. K. Kaushik, B. S. Shim, A. Agarwal, Z. Tang, A. M. Waas, E. M. Arruda, N. A. Kotov (**2008**) Journal of Physical Chemistry B *112*, 14359–14363.

191. T. Ebina, F. Mizukami (**2007**) Advanced Materials *19*, 2450–2453.

192. V. Vertlib, M. Dietiker, M. Ploetze, L. Yezek, R. Spolenak, A. M. Puzrin (**2008**) Journal of Materials Research *23*, 1026–1035.

193. W.-S. Jang, I. Rawson, J. C. Grunlan (**2008**) Thin Solid Films *516*, 4819–4825.

194. M. A. Priolo, K. M. Holder, S. M. Greenlee, J. C. Grunlan (**2012**) ACS Applied Materials & Interfaces *4*, 5529–5533.

195. Y.-C. Li, J. Schulz, S. Mannen, C. Delhom, B. Condon, S. Chang, M. Zammarano, J. C. Grunlan (**2010**) ACS Nano *4*, 3325–3337.

196. M. A. Priolo, K. M. Holder, S. M. Greenlee, B. E. Stevens, J. C. Grunlan (**2013**) Chemistry of Materials *25*, 1649–1655.

197. Y. Sugahara, S. Satokawa, K. Kuroda, C. Kato (**1988**) Clays and Clay Minerals *36*, 343–348.

198. D. Sun, Y. Li, B. Zhang, X. Pan (**2010**) Composites Science and Technology *70*, 981–988.

199. Y. Sugahara, S. Satokawa, K. Kuroda, C. Kato (**1990**) Clays and Clay Minerals *38*, 137–143.

200. Y. Sugahara, S. Satokawa, K. Kuroda, C. Kato (**1992**) Nippon Kyokay Seramikksu Kyokai *100*, 413–416.

201. A. Matsumura, Y. Komori, T. Itagaki, Y. Sugahara, K. Kuroda (**2001**) Bulletin of the Chemical Society of Japan *74*, 1153–1158.

202. T. Itagaki, Y. Komori, Y. Sugahara, K. Kuroda (**2001**) Journal of Materials Chemistry *11*, 3291–3295.

203. T. A. Elbokl, C. Detellier (**2006**) Journal of Physics and Chemistry of Solids *67*, 950–955.

204. B. Zhang, Y. Li, X. Pan, X. Jia, X. Wang (**2007**) Journal of Physics and Chemistry of Solids *68*, 135–142.

205. H. A. Essawy (**2008**) Colloid and Polymer Science *286*, 795–803.

206. Y. Li, B. Zhang, X. Pan (**2008**) Composites Science and Technology *68*, 1954–1961.

207. Z. Jia, Q. Li, J. Liu, Y. Yang, L. Wang, Z. Guan (**2008**) Journal of Polymer Engineering *28*, 87–100.

208. T. A. Elbokl, C. Detellier (**2009**) Canadian Journal of Chemistry-Revue Canadienne De Chimie *87*, 272–279.

209. J. J. Tunney, C. Detellier (**1996**) Chemistry of Materials *8*, 927–935.

210. L. Cabedo, E. Gimenez, J. M. Lagaron, R. Gavara, J. J. Saura (**2004**) Polymer *45*, 5233–5238.

211. J. E. Gardolinski, L. C. M. Carrera, M. P. Cantao, F. Wypych (**2000**) Journal of Materials Science *35*, 3113–3119.

212. T. A. Elbokl, C. Detellier (**2008**) Journal of Colloid and Interface Science *323*, 338–348.

213. S. Letaief, C. Detellier (**2009**) Langmuir *25*, 10975–10979.

214. S. Letaief, J. Leclercq, Y. Liu, C. Detellier (**2011**) Langmuir *27*, 15248–15254.

215. M. Du, B. Guo, D. Jia (**2010**) Polymer International *59*, 574–582.

216. R. Deepak, Y. K. Agrawal (**2012**) Reviews on Advanced Materials Science *32*, 149–158.

217. E. Ruiz-Hitzky, M. Darder, P. Aranda (**2010**) in Annual Review of Nanoresearch (eds G. Cao, Q. Zhang, C. J. Brinker) World Scientific Publishing, Singapore, Vol. 3, pp. 149–189.

218. L. Avérous, E. Pollet (Eds) (**2012**) Environmental Silicate Nano-Biocomposites. Springer-Verlag, London.

219. E. Ruiz-Hitzky, P. Aranda, M. Darder (**2008**) in Kirk-Othmer Encyclopedia of Chemical Technology (ed Kirk-Othmer) John Wiley & Sons, Inc., Hoboken, NJ, pp. 1–28.

220. S. S. Ray, M. Bousmina (**2005**) Progress in Materials Science *50*, 962–1079.

221. F. Chivrac, E. Pollet, L. Averous (**2009**) Materials Science & Engineering R-Reports *67*, 1–17.

222. S. Tunc, O. Duman (**2010**) Applied Clay Science *48*, 414–424.

223. H. M. Park, M. Misra, L. T. Drzal, A. K. Mohanty (**2004**) Biomacromolecules *5*, 2281–2288.

224. R. B. Romero, M. M. Favaro Ferrarezi, C. A. Paula Leite, R. M. Vercelino Alves, M. D. C. Goncalves (**2013**) Cellulose *20*, 675–686.

225. S.-W. Jang, J.-C. Kim, J.-H. Chang (**2009**) Cellulose *16*, 445–454.

226. S. Mahmoudian, M. U. Wahit, A. F. Ismail, A. A. Yussuf (**2012**) Carbohydrate Polymers *88*, 1251–1257.

227. H. M. Park, W. K. Lee, C. Y. Park, W. J. Cho, C. S. Ha (**2003**) Journal of Materials Science *38*, 909–915.

228. J. K. Pandey, R. P. Singh (**2005**) Starch-Starke *57*, 8–15.

229. X. Tang, S. Alavi, T. J. Herald (**2008**) Carbohydrate Polymers *74*, 552–558.

230. M. F. Huang, J. G. Yu, X. F. Ma, P. Jin (**2005**) Polymer *46*, 3157–3162.

231. B. S. Chiou, E. Yee, G. M. Glenn, W. J. Orts (**2005**) Carbohydrate Polymers *59*, 467–475.

232. V. P. Cyras, L. B. Manfredi, M.-T. Ton-That, A. Vazquez (**2008**) Carbohydrate Polymers *73*, 55–63.

233. X. Zhao, B. Wang, J. Li (**2008**) Journal of Applied Polymer Science *108*, 2833–2839.

234. F. Chivrac, E. Pollett, M. Schmutz, L. Averous (**2008**) Biomacromolecules *9*, 896–900.

235. R. Muzzarelli (**1978**) in Proceedings of the First International Conference on Chitin/Chitosan (eds R. Muzzarelli, Pariser ER) MIT, Boston, MA, pp. 335–354.

236. M. Darder, M. Colilla, E. Ruiz-Hitzky (**2003**) Chemistry of Materials *15*, 3774–3780.

237. S. F. Wang, L. Shen, Y. J. Tong, L. Chen, I. Y. Phang, P. Q. Lim, T. X. Liu (**2005**) Polymer Degradation and Stability *90*, 123–131.

238. S. F. Wang, L. Chen, Y. J. Tong (**2006**) Journal of Polymer Science Part A-Polymer Chemistry *44*, 686–696.

239. M. Darder, M. Colilla, E. Ruiz-Hitzky (**2005**) Applied Clay Science *28*, 199–208.

240. M. Darder, E. R. Ruiz-Hitzky (**2005**) Journal of Materials Chemistry *15*, 3913–3918.

241. E. Ruiz-Hitzky, M. Darder, F. M. Fernandes, E. Zatile, F. J. Palomares, P. Aranda (**2011**) Advanced Materials *23*, 5250–5255.

242. C. Ruiz-García, J. Pérez-Carvajal, A. Berenguer-Murcia, M. Darder, P. Aranda, D. Cazorla-Amorós, E. Ruiz-Hitzky (**2013**) Physical Chemistry Chemical Physics 15, 18635–18641.

243. V. Krikorian, M. Kurian, M. E. Galvin, A. P. Nowak, T. J. Deming, D. J. Pochan (**2002**) Journal of Polymer Science Part B-Polymer Physics *40*, 2579–2586.

244. F. M. Fernandes, M. Darder, A. I. Ruiz, P. Aranda, E. Ruiz-Hitzky (**2011**) in Nanocomposites with Biodegradable Polymers. Synthesis, Properties, and Future Perspectives (ed V. Mittal) Oxford University Press, New York, pp. 209–233.

245. O. Talibudeen (**1950**) Nature *166*, 236–236.

246. O. Talibudeen (**1955**) Transactions of the Faraday Society *51*, 582–590.

247. J. P. Zheng, P. Li, Y. L. Ma, K. D. Yao (**2002**) Journal of Applied Polymer Science *86*, 1189–1194.

248. C. Mousty (**2004**) Applied Clay Science *27*, 159–177.

249. A. D. McLaren, G. H. Peterson (**1961**) Nature *192*, 960–961.

250. S. Peng, Q. M. Gao, Q. G. Wang, J. L. Shi (**2004**) Chemistry of Materials *16*, 2675–2684.

251. J. K. Mbouguen, E. Ngameni, A. Walcarius (**2006**) Analytica Chimica Acta *578*, 145–155.

252. S. S. Ray (**2012**) Accounts of Chemical Research *45*, 1710–1720.

253. P. Bordes, E. Pollet, L. Averous (**2009**) Progress in Polymer Science *34*, 125–155.

254. J. H. Chang, Y. U. An, D. H. Cho, E. P. Giannelis (**2003**) Polymer *44*, 3715–3720.

255. M. Pluta, A. Galeski, M. Alexandre, M. A. Paul, P. Dubois (**2002**) Journal of Applied Polymer Science *86*, 1497–1506.

256. M. A. Paul, M. Alexandre, P. Degee, C. Henrist, A. Rulmont, P. Dubois (**2003**) Polymer *44*, 443–450.

257. M. A. Paul, C. Delcourt, M. Alexandre, P. Degee, F. Monteverde, A. Rulmont, P. Dubois (**2005**) Macromolecular Chemistry and Physics *206*, 484–498.

258. S. S. Ray, K. Yamada, M. Okamoto, K. Ueda (**2003**) Macromolecular Materials and Engineering *288*, 203–208.

259. M. A. Paul, C. Delcourt, M. Alexandre, P. Degee, F. Monteverde, P. Dubois (**2005**) Polymer Degradation and Stability *87*, 535–542.

260. S. S. Ray, K. Yamada, M. Okamoto, A. Ogami, K. Ueda (**2003**) Chemistry of Materials *15*, 1456–1465.

261. S. T. Lim, Y. H. Hyun, C. H. Lee, H. J. Choi (**2002**) Journal of Materials Science Letters *22*, 299–302.

262. G. X. Chen, G. J. Hao, T. Y. Guo, M. D. Song, B. H. Zhang (**2004**) Journal of Applied Polymer Science *93*, 655–661.

263. S. F. Wang, C. J. Song, G. X. Chen, T. Y. Guo, J. Liu, B. H. Zhang, S. Takeuchi (**2005**) Polymer Degradation and Stability *87*, 69–76.

264. P. Bordes, E. Pollet, S. Bourbigot, L. Averous (**2008**) Macromolecular Chemistry and Physics *209*, 1473–1484.

265. C. Sanchez, B. Julian, P. Belleville, M. Popall (**2005**) Journal of Materials Chemistry *15*, 3559–3592.

266. H. H. Murray (**1997**) in Proceedings of the 11th International Clay Conference, Ottawa 1997 (eds H. Kodama, A. R. Mermut, J. K. Torrance) ICC97 Organizing Committee, Ottawa, pp. 3–11.

267. Elementis Specialties. http://www.elementis.com/esweb/esweb.nsf/pages/adhesivesseal ants?opendocument. Accessed 26 September 2013.

268. Organoclays Get Steamed. http://www.chemicalprocessing.com/articles/2003/287/. Accessed 26 September 2013.

269. Y. El-Nahhal, S. Nir, T. Polubesova, L. Margulies, B. Rubin (**1998**) Journal of Agricultural and Food Chemistry *46*, 3305–3313.

270. Y. El-Nahhal, S. Nir, C. Serban, O. Rabinovitch, B. Rubin (**2000**) Journal of Agricultural and Food Chemistry *48*, 4791–4801.

271. L. Cornejo, R. Celis, C. Dominguez, M. C. Hermosin, J. Cornejo (**2008**) Applied Clay Science *42*, 284–291.

272. A. Wefer-Roehl, K. A. Czurda (**1997**) in Proceedings of the 11th International Clay Conference, Ottawa (eds H. Kodama, A. R. Mermut, J. K. Torrance) ICC97 Organizing Committee, Ottawa, pp. 123–128.

273. T. Tsoufis, V. Georgakilas, X. Ke, G. Van Tendeloo, P. Rudolf, D. Gournis (**2013**) Chemistry – A European Journal *19*, 7937–7943.

274. J. Bors (**1990**) Radiochimica Acta *51*, 139–143.

275. J. Bors, A. Gorny, S. Dultz (**1994**) Radiochimica Acta *66–67*, 309–313.

276. J. Bors, A. Gorny, S. Dultz (**1997**) Clay Minerals *32*, 21–28.

277. J. Bors, S. Dultz, B. Riebe (**2000**) Applied Clay Science *16*, 1–13.

278. J. Behnsen, B. Riebe (**2008**) Applied Geochemistry *23*, 2746–2752.

279. L. Mercier, C. Detellier (**1995**) Environmental Science & Technology *29*, 1318–1323.

280. C. L. Lin, T. J. Pinnavaia (**1991**) Chemistry of Materials *3*, 213–215.

281. C. L. Lin, T. Lee, T. J. Pinnavaia (**1992**) ACS Symposium Series *499*, 145–154.

282. M. M. Mortland, V. Berkheiser (**1976**) Clays and Clay Minerals *24*, 60–63.

283. T. J. Pinnavaia (**1983**) Science *220*, 365–371.

284. H. Wang, S.-X. Deng, Z.-R. Shen, J.-G. Wang, D.-T. Ding, T.-H. Chen (**2009**) Green Chemistry *11*, 1499–1502.

285. Y. G. Yan, T. Bein (**1993**) Chemistry of Materials *5*, 905–907.

286. P. Aranda, J. C. Galvan, B. Casal, E. Ruiz-Hitzky (**1994**) Colloid and Polymer Science *272*, 712–720.

287. E. Ruiz-Hitzky, P. Aranda, B. Casal, J. C. Galvan (**1995**) Advanced Materials *7*, 180–184.

288. A. Ceklovsky, S. Takagi (**2013**) Central European Journal of Chemistry *11*, 1132–1136.

289. M. Colilla, M. Darder, P. Aranda, E. Ruiz-Hitzky (**2005**) Chemistry of Materials *17*, 708–715.

290. P. Aranda, J. C. Galvan, B. Casal, E. Ruiz-Hitzky (**1992**) Electrochimica Acta *37*, 1573–1577.

291. S. Letaief, P. Aranda, R. Fernandez-Saavedra, J. C. Margeson, C. Detellier, E. Ruiz-Hitzky (**2008**) Journal of Materials Chemistry *18*, 2227–2233.

292. R. M. G. Rajapakse, S. Higgins, K. Velauthamurty, H. M. N. Bandara, S. Wijeratne, R. M. M. Y. Rajapakse (**2011**) Journal of Composite Materials *45*, 597–608.

293. D. Aradilla, D. Azambuja, F. Estrany, M. T. Casas, C. A. Ferreira, C. Aleman (**2012**) Journal of Materials Chemistry *22*, 13110–13122.

294. L. Margulies, H. Rozen, E. Cohen (**1985**) Nature *315*, 658–659.

295. H. Rozen, L. Margulies (**1991**) Journal of Agricultural and Food Chemistry *39*, 1320–1325.

296. Y. B. Si, J. Zhou, H. M. Chen, D. M. Zhou (**2004**) Chemosphere *54*, 943–950.

297. Z. G. Xiong, Y. M. Xu, L. Z. Zhu, J. C. Zhao (**2005**) Langmuir *21*, 10602–10607.

298. X. Meng, Z. Qian, H. Wang, X. Gao, S. Zhang, M. Yang (**2008**) Journal of Sol-Gel Science and Technology *46*, 195–200.

299. J. H. Choy, S. J. Choi, J. M. Oh, T. Park (**2007**) Applied Clay Science *36*, 122–132.

300. F. H. Lin, C. H. Chen, W. T. K. Cheng, T. F. Kuo (**2006**) Biomaterials *27*, 3333–3338.

301. N. G. Veerabadran, R. R. Price, Y. M. Lvov (**2007**) Nano *2*, 115–120.

302. H. Zhuang, J. Zheng, H. Gao, K. De Yao (**2007**) Journal of Materials Science: Materials in Medicine *18*, 951–957.

303. J. P. Zheng, C. Z. Wang, X. X. Wang, H. Y. Wang, H. Zhuang, K. De Yao (**2007**) Reactive & Functional Polymers *67*, 780–788.

304. T. Gonzalezcarreno, J. A. Martinrubi (**1977**) Journal of Chromatography *133*, 184–189.

305. H. B. Lao, C. Detellier (**1994**) Clays and Clay Minerals *42*, 477–481.

306. M. Borisover, E. R. Graber, F. Bercovich, Z. Gerstl (**2001**) Chemosphere *44*, 1033–1040.

307. H. A. Patel, H. C. Bajaj, R. V. Jasra (**2008**) Journal of Nanoparticle Research *10*, 625–632.

308. R. M. Barrer (**1989**) Clays and Clay Minerals *37*, 385–395.

309. P. Falaras, D. Petridis (**1992**) Journal of Electroanalytical Chemistry *337*, 229–239.

310. M. Darder, M. Colilla, E. Ruiz-Hitzky (unpublished results).

311. M. Darder, A. Valera, E. Nieto, M. Colilla, C. J. Fernandez, R. Romero-Aranda, J. Cuartero, E. Ruiz-Hitzky (**2009**) Sensors and Actuators B-Chemical *135*, 530–536.

312. E. Ruiz-Hitzky, P. Aranda, E. Pérez-Cappe, A. Villanueva, Y. Mosqueda Laffita (**2000**) Revista Cubana de Química *12*, 58–63.

313. E. Ruiz-Hitzky, P. Aranda (**2000**) in Polymer-Clay Nanocomposites (eds T. J. Pinnavaia, G. W. Beall) Wiley, West Sussex, pp. 19–46.

314. E. Ruiz-Hitzky (**1993**) Advanced Materials *5*, 334–340.

315. E. Ruiz-Hitzky, P. Aranda (**1997**) Anales de Quimica *93*, 197–212.

316. I. Ahmad, M. Hussain, K.-S. Seo, Y.-H. Choa (**2010**) Journal of Applied Polymer Science *116*, 314–319.

317. M. I. Carretero, M. Pozo (**2010**) Applied Clay Science *47*, 171–181.

318. M. I. Carretero, M. Pozo (**2009**) Applied Clay Science *46*, 73–80.

319. C. Viseras, C. Aguzzi, P. Cerezo, M. C. Bedmar (**2008**) Materials Science & Technology *24*, 1020–1026.

320. J.-M. Oh, T. T. Biswick, J.-H. Choy (**2009**) Journal of Materials Chemistry *19*, 2553–2563.

321. G. V. Joshi, H. A. Patel, H. C. Bajaj, R. V. Jasra (**2009**) Colloid and Polymer Science *287*, 1071–1076.

322. G. V. Joshi, B. D. Kevadiya, H. A. Patel, H. C. Bajaj, R. V. Jasra (**2009**) International Journal of Pharmaceutics *374*, 53–57.

323. H. Jung, H.-M. Kim, Y. Bin Choy, S.-J. Hwang, J.-H. Choy (**2008**) International Journal of Pharmaceutics *349*, 283–290.

324. X. Y. Wang, X. F. Pei, Y. M. Du, Y. Li (**2008**) Nanotechnology *19*, art. #375102.

325. Y. M. Lvov, R. R. Price (**2008**) in Bio-inorganic Hybrid Nanomaterials: Strategies, Syntheses, Characterization and Applications (eds E. Ruiz-Hitzky, K. Ariga, Y. M. Lvov) Wiley-VCH, Weinheim, pp. 419–441.

326. K. Ariga, Q. Ji, M. J. McShane, Y. M. Lvov, A. Vinu, J. P. Hill (**2012**) Chemistry of Materials *24*, 728–737.

327. D. Rawtani, Y. K. Agrawal (**2012**) Reviews on Advanced Materials Science *30*, 282–295.

328. M. T. Viseras, C. Aguzzi, P. Cerezo, C. Viseras, C. Valenzuela (**2008**) Microporous and Mesoporous Materials *108*, 112–116.

329. M. H. Shamsi, K. E. Geckeler (**2008**) Nanotechnology *19*, #075604.

330. B. D. Kevadiya, G. V. Joshi, H. C. Bajaj (**2010**) International Journal of Pharmaceutics *388*, 280–286.

331. G. V. Joshi, B. D. Kevadiya, H. M. Mody, H. C. Bajaj (**2012**) Journal of Polymer Science Part A-Polymer Chemistry *50*, 423–430.

332. D. Depan, A. P. Kumar, R. P. Singh (**2009**) Acta Biomaterialia *5*, 93–100.

333. M. Liu, C. Wu, Y. Jiao, S. Xiong, C. Zhou (**2013**) Journal of Materials Chemistry B *1*, 2078–2089.

334. E. Kharlampieva, V. Kozlovskaya, R. Gunawidjaja, V. V. Shevchenko, R. Vaia, R. R. Naik, D. L. Kaplan, V. V. Tsukruk (**2010**) Advanced Functional Materials *20*, 840–846.

7

FINE-TUNING THE FUNCTIONALITY OF INORGANIC SURFACES USING PHOSPHONATE CHEMISTRY

BRUNO BUJOLI AND CLÉMENCE QUEFFELEC

CNRS, CEISAM UMR 6230, University of Nantes, Nantes Cedex, France

7.1 PHOSPHONATE-BASED MODIFIED SURFACES: A BRIEF OVERVIEW

One attractive route towards functional materials consists in the surface modification of metals and metal oxides, which can be achieved by covalent binding of organic groups. For that purpose, silicon-based (SiH_3, $Si(OR)_3$ or $SiCl_3$) and phosphorus-based ($P(O)(OR)_2$ with R = alkyl or H) functional molecules are the most popular.

Interestingly, while silane chemistry has been used for decades for the covalent coating of inorganic surfaces, especially glass, silicon and silica, the potential of phosphonate chemistry was only recently developed. This is particularly surprising since phosphonic acids have proved to bind strongly to surface-exposed metal ions on a wide variety of metals and metal oxides, driven by the formation of PO–metal bonds, which are usually very stable, in particular for metal ions of high oxidation state (Scheme 7.1).

In this context, a recent review [1] has highlighted the current state of the art related to the modification of inorganic surfaces using phosphonic acids and their esters; this chapter gives evidence of the richness of applications, which can be considered using the resulting functional materials. These include protective layers (e.g. anti-corrosion coatings), dye-sensitized solar cells, photocatalytic systems, supported catalysts, devices for the detection or trapping of soluble chemical species, etc. In addition, the field of biotechnologies is certainly one of the area for which

Tailored Organic–Inorganic Materials, First Edition. Edited by Ernesto Brunet, Jorge L. Colón and Abraham Clearfield.
© 2015 John Wiley & Sons, Inc. Published 2015 by John Wiley & Sons, Inc.

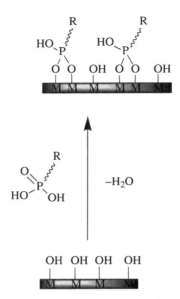

SCHEME 7.1   Covalent binding of phosphonic acid coatings on metal oxide substrates.

phosphonate-derivatized inorganic surfaces have received a rapidly increasing interest, playing a key role in the design of original concepts and novel products.

This chapter proposes to outline this aspect, with a particular focus (i) on the surface chemistry of biological microarrays and (ii) bone regeneration using calcium phosphate composites.

7.2   BIOLOGICAL APPLICATIONS OF PHOSPHONATE-DERIVATIZED INORGANIC SURFACES

7.2.1   Phosphonate Coatings as Bioactive Surfaces

7.2.1.1   Supported Lipid Bilayer   The lipid bilayer is one of the most important self-assembled structures in nature. It hosts much of the machinery for cellular communication and transport across the cell membrane and is characterized by structural heterogeneity and fluidity. Solid-supported lipid bilayers provide an excellent biomimetic model for studying many characteristics of cell membranes. Different routes have been developed for forming biomimetic bilayer films, including the sequential transfer of phospholipid layers from an air/water interface or fusion of vesicles onto an appropriate solid substrate. McConnell and co-workers [2–4] pioneered the discovery of supported phospholipid bilayers (SPBs) by vesicle fusion.

A concern for many applications is that supported bilayers are generally unstable to air. Several strategies have been explored to generate air-stable supported bilayers, such as adding stabilizing components, for example, cholesterol to the lipid mixture

[5] or modified lipids such as poly(2-oxazoline) lipopolymers [6]. Very recently, Budvytyte et al. [7] prepared more robust tethered bilayer lipid membranes (tBLMs) with fewer defects by combining rapid solvent exchange and vesicle fusion.

Interestingly, phosphonate-based materials have been demonstrated to be well suited for preparing supported artificial biological membranes. Oberts *et al.* [8] used zirconium phosphate chemistry to immobilize lipid monolayers or bilayers, drawing on the affinity of polar phospholipid head groups such as choline, ethanolamine, glycerol or serine for the zirconium surface. For example, they treated gold substrates modified with 6-mercapto-1-hexanol with $POCl_3$, followed by Zr^{4+} to form zirconium phosphonate-modified gold surfaces, and then deposited the lipids. Cyclic voltammetry, optical ellipsometry and water contact angle measurements confirmed the presence of lipid layers, and the complexation of 1,2-dimyristoyl-*sn*-glycero-3-phosphatidylcholine (DMPC) to zirconated surfaces was confirmed by ^{31}P NMR [8]. They found that phosphocholine and especially phosphatidic acid lipids bind strongly to Zr^{4+} ions, compared to other lipids with H-bonding head groups [9]. They also studied the strength of interaction of phospholipids with other metal ions relative to Zr^{4+} [10] by depositing 1,2-dimyristoyl-*sn*-glycero-3-phosphatidic acid (DMPA) vesicles on different metal phosphate surfaces using Cu^+, Cu^{2+}, Fe^{3+}, Zn^{2+}, Ni^{2+}, Ca^{2+} and Mg^{2+}. The conclusion was that surfaces with Fe^{3+} and Zr^{4+} ions, with high ionic charge and small radius, lead to relatively well-organized monolayer structures upon complexation with DMPA, while with Cu^{2+}, a bilayer structure forms, and with other metal ions, a partial adlayer forms with limited organization.

Talham and co-workers have also developed stable supported lipid bilayers on zirconium phosphonate surfaces. Zirconium phosphonate-modified substrates were prepared via the Langmuir–Blodgett (LB) method [11, 12]. The lipid bilayer could be formed either directly by vesicle fusion or by first transferring the inner layer onto the zirconium phosphonate surface followed by vesicle fusion of a different lipid composition to form asymmetric bilayers. They noticed for the latter method that the addition of a small percentage of phosphatidic acid into the inner layer increased the stability of the bilayer (Figure 7.1). X-ray photoelectron spectroscopy (XPS) and surface plasmon resonance-enhanced ellipsometry (SPREE) were used to characterize the layers and to investigate the vesicle fusion and stability of the lipid bilayer. The inner layer was found to possess low fluidity, but the outer layer was found to be highly fluid, comparable to other systems [13].

As a proof of concept, they studied the binding of melittin to the lipid bilayer. Melittin is a peptide contained in honeybee venom, and its interaction with lipid membranes has already been well described [14–16]. The membrane composition (zwitterionic or anionic lipids) had a strong influence on the binding of melittin, and the binding was five times higher with the anionic lipid, proving the viability of those membrane models.

In a subsequent paper, the same group developed skeletonized zirconium phosphonate-modified surfaces as supports for lipid bilayers in order to accommodate transmembrane proteins [17]. To form the skeletonized supports, octadecanol was mixed with octadecylphosphonic acid in an LB monolayer, and after treatment with zirconium ions, the monolayer was rinsed with ethanol to remove the alcohol

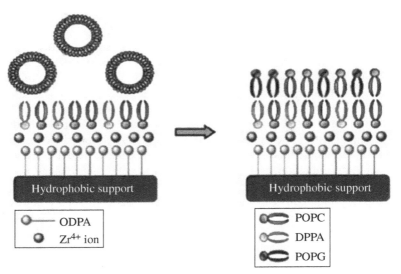

FIGURE 7.1 Asymmetric lipid bilayer on Langmuir–Blodgett substrates. Adapted with permission from Ref. [13]. Copyright 2013 American Chemical Society.

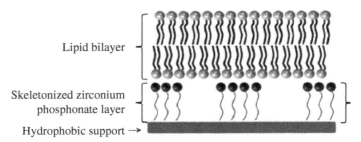

FIGURE 7.2 Lipid bilayer on skeletonized zirconium phosphonate films. Adapted with permission from Ref. [17]. Copyright 2013 American Chemical Society.

molecules, leaving voids in the film. The lipid bilayer was built on the skeletonized zirconium phosphonate as described before, either by direct vesicle fusion or by the LB/Langmuir–Schaefer method, providing available space between the support and the membrane to accommodate proteins (Figure 7.2).

SPREE analyses were conducted to study the interaction mechanism of two different proteins, integrin $\alpha_5\beta_1$, the primary receptor for fibronectin, or a modified BK channel, which is a calcium-activated potassium channel. The BK channel interacted strongly and was inserted into the lipid bilayers while supported on a skeletonized zirconium phosphonate, in contrast to nonskeletonized surfaces where the association was weak. As for integrin, it maintained its ability to recognize fibronectin while inserted into the lipid bilayer on skeletonized supports.

This strategy is general because the surface treatment can be applied to any surface materials and those supports can be used without the need to extensively modify the biomolecules.

7.2.1.2 Surface-Modified Nanoparticles

The development of novel materials for different applications in nanoscience and nanotechnology continues to be a very active field of chemical research, and metal phosphonate chemistry is increasingly used for the modification of inorganic nanoparticles (NPs) [18–24].

In a recent review [1], Queffélec et al. described the phosphonate-mediated modification of metal NPs based on superparamagnetic iron oxide (SPIO) (Figure 7.3) and their bioconjugation. Derivatization of NPs with functional groups such as primary amine or carboxylic acid groups allowed bioconjugation via carboxylic acid or isothiocyanate groups present on the biomolecules (NH_2 platform) or via primary amine groups on the biomolecules (CO_2H platform).

SPIO, along with other inorganic particles, are being studied for use with magnetic resonance imaging (MRI) and optical imaging techniques. MRI is a non-invasive medical imaging technique that allows visualization of the internal structure of the body and discrimination between normal and pathological tissues. MRI uses the fact that body tissues contain a large amount of water and a MR image is generated from the nuclear magnetic resonance of water protons. The signal intensity and contrast in MRI depend largely on the following factors: the proton density, the longitudinal relaxation rate (T_1) and transverse relaxation time (T_2) of water protons. In some cases, contrast agents are needed to increase the MR contrast, by changing both T_1 and T_2 of water molecules.

Gadolinium phosphate NPs [25] have recently been shown to be effective MRI contrast agents. Gadolinium is a lanthanide with seven unpaired electrons, and the unique magnetic properties of the Gd(III) ion have made it a compound of choice for T_1 or positive contrast agents. In that regard, different types of gadolinium-based hybrid nanomaterials have been developed, such as GdF_3 NPs [26–28], layered gadolinium hydroxide (LGdH) [29] or delaminated LGdH doped with europium [30] and ultrasmall gadolinium oxide particles [31]. Moreover, gadolinium compounds such as gadolinium phosphate ($GdPO_4$) have also been proven to be good candidates for MRI applications because they are poorly soluble and form small paramagnetic NPs [32–34].

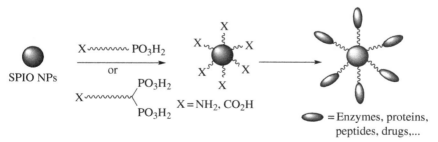

FIGURE 7.3 Example of derivatization of SPIO NPs via phosphonic acid or bisphosphonic acid anchors.

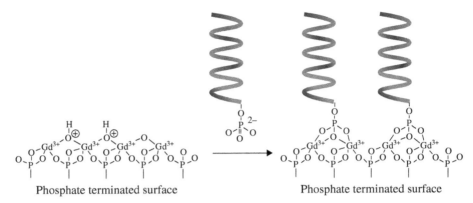

FIGURE 7.4 Covalent binding of phosphate-terminated oligonucleotides on the surface of gadolinium phosphate nanoparticles. Adapted with permission from Ref. [35]. Copyright 2013 American Chemical Society.

Talham and co-workers have functionalized gadolinium phosphate NPs (GdPO$_4$ NPs) with oligonucleotides and measured the magnetic resonance relaxivity [35]. The ability of phosphate-terminated biomolecules to bind to the gadolinium phosphate surface was demonstrated in a similar manner to what was reported previously in the case of zirconium phosphonate surfaces used to prepare DNA microarrays (see Section 7.2.2) [36, 37]. They proved that the same type of chemistry could be applied to gadolinium surfaces, since Gd^{3+} and Zr^{4+} have similar acid dissociation constant (K_a) values. From a microemulsion containing Gd(NO$_3$)$_3$ and NaH$_2$PO$_4$ and using IGEPAL CO-520 as a surfactant, monodisperse NPs of 50 nm in length and 10 nm in width were obtained. A final treatment with Gd^{3+} was applied to assure a Gd^{3+}-rich surface. A phosphorylated oligonucleotide sequence, usable as a cancer cell marker, was selected, and a strong binding of the probe to the NPs occurred via its terminal 5′-phosphate group under biological conditions. Moreover, the immobilized oligonucleotides bound to the surface were able to hybridize with their complementary sequence (Figure 7.4). The DNA-functionalized GdPO$_4$ NPs proved also to generate negative MR contrast, with relaxivities comparable to the commercial contrast agents. In addition, a high stability of the NPs was observed since the release of Gd^{3+} was found to be low, even after 18 months.

7.2.2 Specific Binding of Biological Species onto Phosphonate Surfaces for the Design of Microarrays

7.2.2.1 Single- and Double-Stranded Oligonucleotides
Major challenges in chemistry have to be addressed for the design of new concepts in biomaterials engineering. For example, the construction of biosensors and biochips for biological applications requires efficient attachment of biologically active molecules on the surface of solid inorganic substrates while minimizing non-specific adsorption.

The choice of the substrate and its surface modification is always the primary critical step in the design of a bio-microarray technology. To date, silanized glass substrates are the most commonly used. Monolayers deposited on gold slides are also widely studied for biomolecule immobilization [38–41] but one of the problems for these self-assembled films is the lability of the Au–S bond under stringent conditions. For metal oxides other than silica, the metal–Si–O linkage is often of low stability, limiting the utility of silanizing agents for surface modifications [42, 43]. Recently, zirconium phosphonate or zirconium phosphate-modified surfaces have emerged as substrates for binding biomolecules and have proven to be especially useful for selectively binding phosphate-containing biomolecules.

Although there are a number of different methods to prepare zirconium phosphonate coatings [44–50], zirconium phosphonate surfaces prepared using the LB technique were thus proposed as reactive surfaces able to provide covalent attachment of phosphate-terminated biological probes. Indeed, free phosphate groups are expected to bind strongly to the surface-exposed zirconium ions present on the surface of these monolayers. An additional interest of this original approach lies in the fact that phosphorylation of many biomolecules can be achieved using enzymes without affecting the functionality of the modified biological probes. This concept was first demonstrated by Nonglaton *et al.* [51] in the case of 5′-phosphate-terminated oligonucleotides for which specific binding to the zirconium phosphonate surface occurred. Probes modified with a terminal phosphate group have been found to bind strongly to the zirconium surface compared to the non-phosphorylated analogues. This method thus proved to be effective for forming oligonucleotide microarrays.

To further rationalize these results, XPS studies were conducted to assess the surface coverage of oligonucleotides grafted on the zirconium phosphonate monolayer. The intensity of the N 1s signal related to the ss-DNA probes was compared with the Zr 3d signal, only present in the phosphonate coating [36]. The surface coverage was found to be lower compared to thiol-modified DNA immobilized on gold (2.8×10^{11} molecules/cm^2 vs. $12–37 \times 10^{12}$ molecules/cm^2). An explanation for the lower surface coverage was that in addition to the terminal phosphate covalently bound to the surface, there may be some weak non-specific adsorption along the ss-DNA backbone, hindering further covalent binding of more oligonucleotides.

Target capture was enhanced by introducing a poly-guanine spacer between the probe and the terminal phosphate. The insertion of a spacer between the phosphate end group of the oligonucleotide and the surface increased the surface coverage [52]. Poly(dG) and poly(dA) spacers were studied, and the poly(dA)-containing probe showed a higher surface density by XPS when using mild rinsing conditions (12.7×10^{11} molecules/cm^2 vs. 6.35×10^{11} molecules/cm^2), while for more stringent rinsing conditions, ss-DNA with the poly(dG) spacer had a surface coverage two times higher than the poly(dA)-containing probe (5.01×10^{11} molecules/cm^2 vs. 2.53×10^{11} molecules/cm^2). The conclusion was that the probes with the poly(dG) spacer produce a higher target capture because of the higher probe density, which might suggest a stronger affinity for the zirconium phosphonate surface.

This methodology was then successfully extended to the immobilization of phosphate-terminated ds-DNA and used to investigate ds-DNA/protein interactions

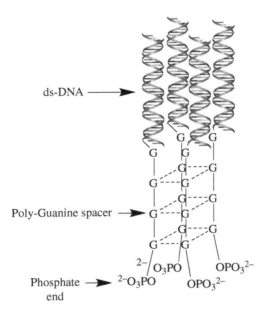

FIGURE 7.5 Structuration of ds-DNA functionalized with $(G)_n$–OPO_3 end groups via the formation of G-quadruplex structures.

[37]. Different features of the probes were shown to influence their capture efficiency when immobilized. For instance, the location and numbers of phosphate anchors in the duplex influenced the sensitivity of the biosensor and the specificity of the probe binding onto the surface. The best results were observed when two phosphate groups were introduced on the two ends (3′ and 5′) of one of the two strands of the DNA complex.

As observed in the case of ss-oligonucleotides (*vide supra*), the sensitivity of the biosensor was shown to be enhanced by the use of a poly-guanine segment $((G)n, n > 5)$ as a spacer between the phosphate linker and the protein interaction domain. This was explained by the formation of G-quadruplex structures (Figure 7.5), which were experimentally evidenced by circular dichroism (CD) with the observation of a positive band close to 260 nm typical of parallel G-quadruplex structures (Figure 7.6). It was assumed that this association of polyG segments as a tetraplex led to the formation of multidentate aggregates that raise the avidity for the surface relative to individual probes.

7.2.2.2 Proteins and Other Biomolecules

7.2.2.2 Proteins and Other Biomolecules The development of protein microarrays is another area under intense investigation worldwide. This is a truly challenging issue needing original approaches to allow stable surface attachment of proteins while controlling their orientation. The use of amino or carboxylic acid residues present on the protein to bind surfaces can be considered, but this strategy does not allow the control of the orientation of the immobilized probes. For that purpose, an innovative solution

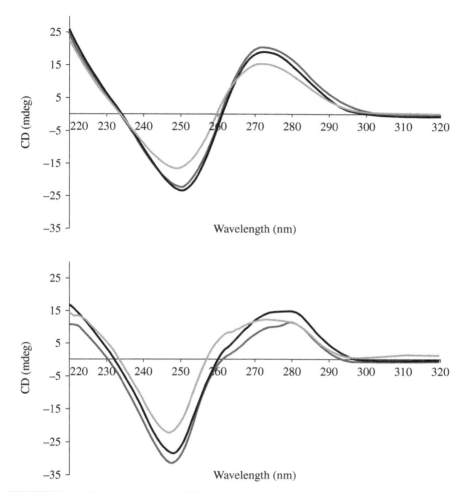

FIGURE 7.6   Circular dichroism (CD) spectra of phosphate-terminated ds-DNA with different spacers (ds-DNA-(spacer)-OPO_3H_2) – ☆, spacer = 9-mer random sequence; □, spacer = G_1; △, spacer = G_9. Left part, CD spectra recorded in water. Right part, CD spectra recorded in 1X SCC (saline sodium citrate, pH 6) showing a positive band at 260 nm, indicating the presence of parallel G-quadruplex for a G_9 spacer. Adapted with permission from Ref. [37]. Copyright 2013 American Chemical Society.

is to fuse a suitable short peptide sequence at the N- or C-terminus of the proteins that would make it able to bind to the active surface. This method has great potential since fusion of such peptide tags should in principle be feasible for any type of proteins expressed by genetic engineering, thus preserving the activity of the protein probes.

In that context, interaction between a polyhistidine-tagged protein and nickel chelated to a nitrilotriacetic acid (NTA) moiety has been reported. Ni-NTA has primarily been used for immobilized metal ion affinity chromatography (IMAC), as a simple and practical method for protein purification. Then, studies have been developed on direct methods for forming SAMs presenting the Ni-NTA group on gold [53] or silicon [54–56] for immobilizing a wide variety of proteins onto the surfaces without altering their activities. Similarly, Cinier *et al.* [57] have demonstrated that a zirconium phosphonate surface engineered by the LB technique can be functionalized with a bifunctional adaptor, containing a bisphosphonic acid group at one end to provide strong and stable binding with the zirconium interface and one or two Ni-NTA groups at the other end in order to bind His-tagged proteins. This method provided stable, uniform surfaces presenting a high and controlled density of Ni-NTA groups, allowing highly efficient protein binding (Figure 7.7, left). The capture of a labelled target protein in a microarray format was reported with a very good sensitivity, and the performance of these Ni-NTA-coated slides compared favourably to commercially available substrates.

More recently, Cinier *et al.* [58] have developed a phosphorylatable tag, fused at the C-terminus of proteins, to allow efficient and oriented direct binding of these proteins on zirconium phosphonate-coated glass slides. This concept was applied to a specific class of proteins, nanofitins, which can be expressed in high amounts in *Escherichia coli* and are highly stable to pH and temperature. The peptide tag contained four serine units, which were suitably positioned in the peptide tag to allow its *in vitro* phosphorylation by casein kinase II. This provided a cluster of four phosphorylated serine moieties allowing a multivalent binding, thus strengthening the interaction with the zirconium phosphonate surface (Figure 7.7, right). This system was prepared in a protein microarray format and led to very high signal-to-noise ratios and very high sensitivity and specificity.

Very recently, Han *et al.* [59] have successfully prepared phosphonate self-assembled monolayers (SAMs) on Zr^{4+}-doped glass substrates. This consists in the treatment of glass slides with zirconium oxychloride after a preliminary etching step, as reported by Hong *et al.* [45]. Then, a 10-mercaptodecanylphosphonic acid (MDPA) monolayer can be deposited upon binding of the phosphonic groups to exposed zirconium ions. Photolithographic patterning of the substrates was then achieved by exposing the supports to 254 nm UV light through a mask, which results in the conversion of the thiol groups to sulphonates. The non-oxidized thiol groups were then reacted with 3-maleimidopropionic acid *N*-hydroxysuccinimide ester (MPS) used as a cross-linker for the coupling of proteins and antibodies (Figure 7.8).

Proteins such as goat anti-mouse IgG or amino-modified biomolecules such as NH_2-biotin were thus immobilized specifically into the unexposed areas. The efficiency of this patterning procedure was confirmed by fluorescence imaging (Figure 7.9).

7.2.3   Calcium Phosphate/Bisphosphonate Combination as a Route to Implantable Biomedical Devices

It is worth noting that the use of gem-bisphosphonate anchors for surface modification is rapidly expanding and is more and more preferred to monophosphonic acids, since they bind more strongly to many types of metal oxides and thus provide higher

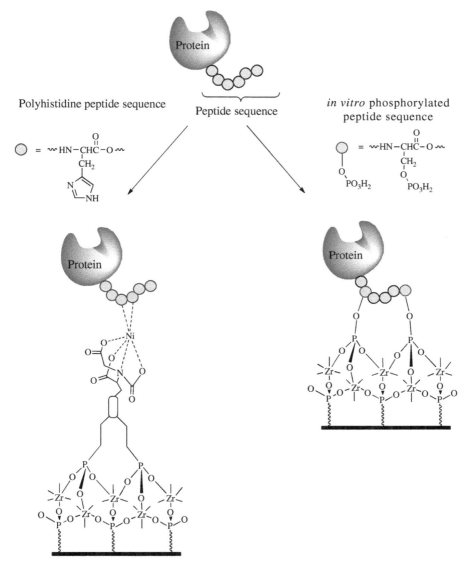

FIGURE 7.7 Left: binding of his-tagged-proteins on Ni-NTA zirconium phosphonate surfaces. Right: binding of proteins bearing a phosphorylated peptide tag on zirconium phosphonate surfaces. Adapted with permission from Ref. [1]. Copyright 2013 American Chemical Society.

stability of the resulting coatings, due to the higher avidity of the bifunctional chelating moiety [60–62]. Moreover, another interest of gem-bisphosphonates lies in the fact that some of them, in particular those substituted with both a hydroxyl and an aminoalkyl group on the bridging P–C–P carbon atom, are approved drugs for

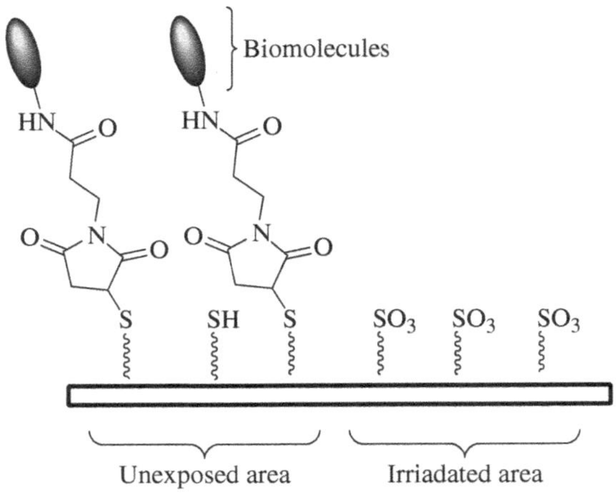

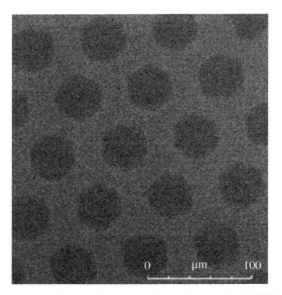

FIGURE 7.8 Schematic representation of a biomolecule patterned glass surface using the photolithography procedure reported by Han *et al.* [59].

FIGURE 7.9 Fluorescence image of a patterned sample treated with MPS and amino-biotin and subsequent exposure to a solution of FITC-labelled avidin. Adapted with permission from Ref. [59]. Copyright 2013 Elsevier.

the treatment of osteoporosis and bone metastasis [63–65]. These drugs are known for their high affinity for bones due to their strong ability to bind to the mineral component of bone tissues, made of calcium-deficient apatite (CDA, $Ca_{10-x}[\ ]_x(HPO_4)_y(PO_4)_{6-y}(OH)_{2-z}[\ ]_z$).

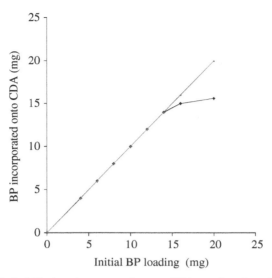

FIGURE 7.10 Typical bisphosphonate uptake onto CDA as a function of the initial amount of BP introduced in the reaction medium.

The binding mechanism of BPs onto CDA was investigated by Bujoli and co-workers [66–68], by mixing the calcium phosphate and the bisphosphonate in water for different liquid/solid ratios and bisphosphonate concentrations and then measuring the amount of bisphosphonate chemisorbed onto CDA. It was thus found that a phosphate release takes place concomitantly to the bisphosphonate incorporation onto the CDA surface, corresponding to a 1 : 1 molar ratio. The interaction process consists of a 'ligand exchange' between one phosphonate function of the bisphosphonate and a phosphate group located on the surface of the calcium phosphate matrix (Eq. 7.1), in full agreement with ^{31}P 2D through space DQ–SQ MAS NMR correlation spectra:

$$BP + P - X \equiv \rightarrow P - X \equiv + BP \qquad (7.1)$$

where $X \equiv$ corresponds to the surface binding sites of the CDA that must be in interaction with either a bisphosphonate (BP) or a phosphate (P) moiety.

Thus, when suspending powdered samples of CDA in bisphosphonate aqueous solutions (batch experiments), a quantitative bisphosphonate uptake by CDA was observed, until a plateau was reached corresponding to the saturation of the surface phosphate exchangeable sites (Figure 7.10). In addition, the amount of bisphosphonate binding sites was found to be quite similar whatever the nature of the bisphosphonate (i.e. ca. $0.25\,mmol.g^{-1}$).

The same binding mechanism of BPs was found to be present in human bone tissues, as confirmed by Oldfield et al. using NMR spectroscopy investigations [69, 70].

Calcium hydroxyapatite enters the composition of various medical devices used in dental and bone surgery. They can be used, for example, as coatings for orthopaedic implants (i.e. hip prostheses) or be part of biphasic calcium phosphate (BCP) ceramics applied for bone void filling. BCPs are made of a mixture of apatite, which provides good biocompatibility of the ceramic, and tricalcium phosphate (β-TCP), which improves the *in vivo* degradability of the implant. On the other hand, in the case of apatitic calcium phosphate cements (CPCs) [71–76], which are increasingly used in bone repair in orthopaedics and trauma, the final product at the end of the setting process is mainly a CDA. Quite naturally, many attempts to bind bisphosphonates to these implants have been reported in the literature, with the purpose of designing bisphosphonate delivery systems having the ability to release the drug locally in osteoporotic pathological sites.

In this context, the treatment of BCPs ceramics with bisphosphonates led to the precipitation of a bisphosphonate calcium complex onto the surface of these materials, due to the partial dissolution of the calcium phosphate matrix and subsequent trapping of the released calcium ions by the bisphosphonate, as shown in the case of β-TCP [67].

On the contrary, BPs were found to bind apatitic layers coated on titanium alloys, following a mechanism similar to that of Equation 7.1. The resulting materials were implanted in healthy and osteoporotic animal models, showing improvement of the mechanical fixation in the case of BP-doped implants when compared to undoped analogues, due to significantly higher new bone formation around the implant (Figure 7.11) [77–83]. Similar *in vivo* results were obtained in large animal osteoporotic models when performing bone void filling using bisphosphonate-loaded CDA granules.

Finally, in the particular case of apatitic CPCs, the main issue to address when incorporating gem-bisphosphonates in the composition is related to their use as setting retardants (especially for Portland cement), due to their ability to strongly inhibit the hardening process of cements. Bisphosphonic acids are excellent calcium scavengers, which rapidly trap the calcium ions released during the setting reaction, which first consists in the dissolution of the main component of the CPC (most often α-TCP). This first step results in a medium supersaturated with Ca^{2+} and PO_4^{3-} ions, which then favours the precipitation of a less soluble calcium phosphate, namely, CDA. Many different strategies were thus investigated to minimize inhibition of the setting, since different options are possible for adding the BP in the CPC formulation, including:

1. Dissolution of the BP in the liquid phase. This is the most unfavourable situation for which the amount of solubilized bisphosphonate is maximum, thus leading to a significant increase of setting times in direct proportion to the BP loading in the CPC formulation [84, 85].

2. Addition of the solid form of the BP to the ground solid phase. Here again, this mode of introduction leads to a significant amount of BP dissolved in the cement paste and a quite pronounced retarding effect [85].

3. 'Chemical combination' of the BP with one of the calcium phosphate components of the solid phase, in particular CDA. As mentioned earlier, in the case of CDA, the bisphosphonate can be grafted on the surface of the calcium phosphate according to Equation 7.1, and the release of the BP proceeds via the

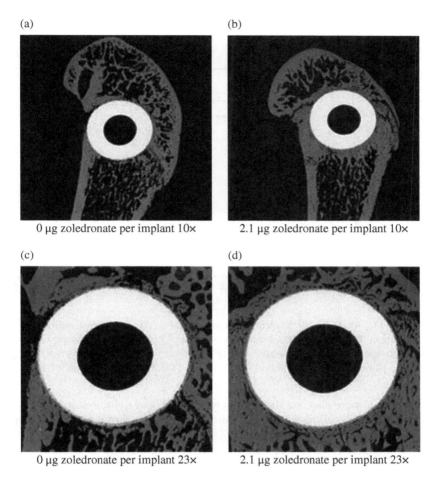

(a)

(b)

0 µg zoledronate per implant 10×   2.1 µg zoledronate per implant 10×

(c)

(d)

0 µg zoledronate per implant 23×   2.1 µg zoledronate per implant 23×

FIGURE 7.11   SEM pictures of two implanted condyles at magnification of 10X and 23X. Panel (a) shows the bone structure of a condyle implanted with a coated implant containing no zoledronate, and panel (b) shows the bone structure of the condyle containing an implant coated with HA grafted with 2.1 µg of zoledronate. The same implants and their peri-implant bone are shown in panels (c) and (d) for the coatings loaded with 0 and 2.1 µg of zoledronate, respectively, at a magnification of 23X. Adapted with permission from Ref. [81]. Copyright 2013 Elsevier.

reverse reaction, driven by the phosphate concentration in the medium. This option is very attractive since it was demonstrated that the amount of BP present in the cement paste is minimized, and therefore, its influence on the setting properties of the cement is limited [85]. Moreover, investigation of the BP release under simulated *in vivo* conditions showed no flash release phenomenon. The drug release was found to be directly driven by the bisphosphonate/cement interaction, leading to a constant and low delivery of the BP (Figure 7.12).

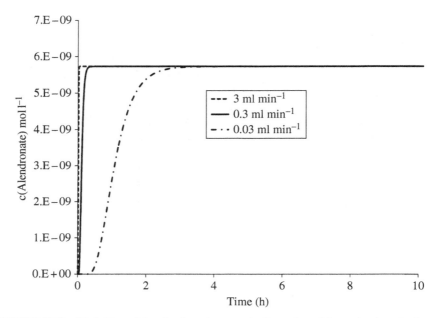

FIGURE 7.12 Modelling of the alendronate concentration released from alendronate-doped apatitic cement block versus time, for a 0.03–3 ml.min^{-1} percolation flux range. Elution = 1 mmol.l^{-1} phosphate buffer; alendronate loading in the cement block = 0.057 wt.% with respect to the solid phase. Adapted with permission from Ref. [85]. Copyright 2013 Elsevier.

Other methods were also investigated, showing rather satisfying results, which consists in the decrease of the liquid/solid ratio, in addition to insertion of gelatin in the cement composition [86].

More importantly, it is very likely that such combinations of bisphosphonates with calcium phosphate biomaterials will lead in the near future to groundbreaking technologies for bone reinforcement and fracture prevention in osteoporotic sites.

7.3 CONCLUSION

The field of organically modified functional surfaces is rapidly expanding in materials science for the design of novel materials applicable in diverse areas. These applications include the derivatization of substrates with reactive end groups for further modification (in particular bioconjugation), protective layers, analytical or biological sensors, catalysis, biomedical devices, solar batteries and so on. The recent advances have focused on the development of easy and reproducible synthetic strategies for the construction of these surfaces by using the appropriate functional molecules for the modification/grafting step, in order to provide a good control of the density and orientation of the organic component on the surface. For this purpose, the organic backbone to be bound onto the surface requires the presence of anchoring groups compatible with the chemical nature of the substrate to be modified. Phosphonic acids RPO_3H_2 have proven to be good candidates because they react easily with metals and metal oxides, leading to a large variety of metal organic hybrid frameworks. Importantly, they are stable, water compatible and easy to handle, all these features being highly desirable for upscaling chemical processes. Phosphonic acids are currently being intensively explored for modifying metal NPs and biomedical devices or preparing supported bilayers, while their use for the design of biological sensors and immobilized catalysts has also been topics of investigations. In conclusion, it appears that phosphonic acids promise to play a leading role in the ever-growing world of surface modifications, although this field still needs fundamental research to make it blossom in a close and exciting future.

REFERENCES

1. Queffelec, C.; Petit, M.; Janvier, P.; Knight, D. A.; Bujoli, B. *Chem. Rev.* **2012**, *112*, 3777.
2. Tamm, L. K.; McConnell, H. M. *Biophys. J.* **1985**, *47*, 105.
3. Watts, T. H.; Brian, A. A.; Kappler, J. W.; Marrack, P.; McConnell, H. M. *Proc. Natl. Acad. Sci. U. S. A.* **1984**, *81*, 7564.
4. Watts, T. H.; Gaub, H. E.; McConnell, H. M. *Nature* **1986**, *320*, 179.
5. Deng, Y.; Wang, Y.; Holtz, B.; Li, J. Y.; Traaseth, N.; Veglia, G.; Stottrup, B. J.; Elde, R.; Pei, D. Q.; Guo, A.; et al. *J. Am. Chem. Soc.* **2008**, *130*, 6267.
6. Purrucker, O.; Fortig, A.; Jordan, R.; Tanaka, M. *Chemphyschem* **2004**, *5*, 327.
7. Budvytyte, R.; Mickevicius, M.; Vanderah, D. J.; Heinrich, F.; Valincius, G. *Langmuir* **2013**, *29*, 4320.
8. Oberts, B. P.; Blanchard, G. J. *Langmuir* **2009**, *25*, 2962.
9. Oberts, B. P.; Blanchard, G. J. *Langmuir* **2009**, *25*, 13918.
10. Oberts, B. P.; Blanchard, G. J. *Langmuir* **2009**, *25*, 13025.
11. Byrd, H.; Pike, J. K.; Talham, D. R. *Chem. Mater.* **1993**, *5*, 709.
12. Byrd, H.; Pike, J. K.; Talham, D. R. *J. Am. Chem. Soc.* **1994**, *116*, 7903.
13. Fabre, R. M.; Talham, D. R. *Langmuir* **2009**, *25*, 12644.
14. Chen, X. Y.; Wang, J.; Boughton, A. P.; Kristalyn, C. B.; Chen, Z. *J. Am. Chem. Soc.* **2007**, *129*, 1420.
15. Papo, N.; Shai, Y. *Biochemistry* **2003**, *42*, 458.
16. Wessman, P.; Stromstedt, A. A.; Malmsten, M.; Edwards, K. *Biophys. J.* **2008**, *95*, 4324.
17. Fabre, R. M.; Okeyo, G. O.; Talham, D. R. *Langmuir* **2012**, *28*, 2835.
18. Basly, B.; Felder-Flesch, D.; Perriat, P.; Billotey, C.; Taleb, J.; Pourroy, G.; Begin-Colin, S. *Chem. Comm.* **2010**, *46*, 985.
19. Cushing, B. L.; Kolesnichenko, V. L.; O'Connor, C. J. *Chem. Rev.* **2004**, *104*, 3893.
20. Daniel, M. C.; Astruc, D. *Chem. Rev.* **2004**, *104*, 293.
21. Daou, T. J.; Greneche, J. M.; Pourroy, G.; Buathong, S.; Derory, A.; Ulhaq-Bouillet, C.; Donnio, B.; Guillon, D.; Begin-Colin, S. *Chem. Mater.* **2008**, *20*, 5869.
22. Das, M.; Mishra, D.; Maiti, T. K.; Basak, A.; Pramanik, P. *Nanotechnology* **2008**, *19*, 415101.

23. Lalatonne, Y.; Paris, C.; Serfaty, J. M.; Weinmann, P.; Lecouvey, M.; Motte, L. *Chem. Comm.* **2008**, 2553.

24. Laurent, S.; Forge, D.; Port, M.; Roch, A.; Robic, C.; Elst, L. V.; Muller, R. N. *Chem. Rev.* **2008**, *108*, 2064.

25. Caravan, P.; Ellison, J. J.; McMurry, T. J.; Lauffer, R. B. *Chem. Rev.* **1999**, *99*, 2293.

26. Cheung, E. N. M.; Alvares, R. D. A.; Oakden, W.; Chaudhary, R.; Hill, M. L.; Pichaandi, J.; Mo, G. C. H.; Yip, C.; Macdonald, P. M.; Stanisz, G. J.; van Veggel, F.; Prosser, R. S. *Chem. Mater.* **2010**, *22*, 4728.

27. Evanics, F.; Diamente, P. R.; van Veggel, F.; Stanisz, G. J.; Prosser, R. S. *Chem. Mater.* **2006**, *18*, 2499.

28. Rodriguez-Liviano, S.; Nunez, N. O.; Rivera-Fernandez, S.; de la Fuente, J. M.; Ocana, M. *Langmuir* **2013**, *29*, 3411.

29. Lee, B. I.; Lee, K. S.; Lee, J. H.; Lee, I. S.; Byeon, S. H. *Dalton Trans.* **2009**, 2490.

30. Yoon, Y. S.; Lee, B. I.; Lee, K. S.; Heo, H.; Lee, J. H.; Byeon, S. H.; Lee, I. S. *Chem. Comm.* **2010**, *46*, 3654.

31. Guay-Begin, A. A.; Chevallier, P.; Faucher, L.; Turgeon, S.; Fortin, M. A. *Langmuir* **2012**, *28*, 774.

32. Hifumi, H.; Yamaoka, S.; Tanimoto, A.; Akatsu, T.; Shindo, Y.; Honda, A.; Citterio, D.; Oka, K.; Kuribayashi, S.; Suzuki, K. *J. Mater. Chem.* **2009**, *19*, 6393.

33. Hifumi, H.; Yamaoka, S.; Tanimoto, A.; Citterio, D.; Suzuki, K. *J. Am. Chem. Soc.* **2006**, *128*, 15090.

34. Rodriguez-Liviano, S.; Becerro, A. I.; Alcantara, D.; Grazu, V.; de la Fuente, J. M.; Ocana, M. *Inorg. Chem.* **2013**, *52*, 647.

35. Dumont, M. F.; Baligand, C.; Li, Y. C.; Knowles, E. S.; Meisel, M. W.; Walter, G. A.; Talham, D. R. *Bioconjug. Chem.* **2012**, *23*, 951.

36. Lane, S. M.; Monot, J.; Petit, M.; Bujoli, B.; Talham, D. R. *Colloids Surf. B Biointerfaces* **2007**, *58*, 34.

37. Monot, J.; Petit, M.; Lane, S. M.; Guisle, I.; Leger, J.; Tellier, C.; Talham, D. R.; Bujoli, B. *J. Am. Chem. Soc.* **2008**, *130*, 6243.

38. Blawas, A. S.; Reichert, W. M. *Biomaterials* **1998**, *19*, 595.

39. Briand, E.; Humblot, V.; Landoulsi, J.; Petronis, S.; Pradier, C. M.; Kasemo, B.; Svedhem, S. *Langmuir* **2011**, *27*, 678.

40. Castelino, K.; Kannan, B.; Majumdar, A. *Langmuir* **2005**, *21*, 1956.

41. Viel, P.; Walter, J.; Bellon, S.; Berthelot, T. *Langmuir* **2013**, *29*, 2075.

42. Howarter, J. A.; Youngblood, J. P. *Langmuir* **2006**, *22*, 11142.

43. Krasnoslobodtsev, A. V.; Smirnov, S. N. *Langmuir* **2002**, *18*, 3181.

44. Benitez, I. O.; Bujoli, B.; Camus, L. J.; Lee, C. M.; Odobel, F.; Talham, D. R. *J. Am. Chem. Soc.* **2002**, *124*, 4363.

45. Hong, H. G.; Sackett, D. D.; Mallouk, T. E. *Chem. Mater.* **1991**, *3*, 521.

46. Katz, H. E.; Schilling, M. L. *Chem. Mater.* **1993**, *5*, 1162.

47. Katz, H. E.; Schilling, M. L.; Chidsey, C. E. D.; Putvinski, T. M.; Hutton, R. S. *Chem. Mater.* **1991**, *3*, 699.

48. Lee, H.; Kepley, L. J.; Hong, H. G.; Akhter, S.; Mallouk, T. E. *J. Phys. Chem.* **1988**, *92*, 2597.

49. Lee, H.; Kepley, L. J.; Hong, H. G.; Mallouk, T. E. *J. Am. Chem. Soc.* **1988**, *110*, 618.

50. Wu, A. P.; Talham, D. R. *Langmuir* **2000**, *16*, 7449.

51. Nonglaton, G.; Benitez, I. O.; Guisle, I.; Pipelier, M.; Leger, J.; Dubreuil, D.; Tellier, C.; Talham, D. R.; Bujoli, B. *J. Am. Chem. Soc.* **2004**, *126*, 1497.

52. Lane, S. M.; Monot, J.; Petit, M.; Tellier, C.; Bujoli, B.; Talham, D. R. *Langmuir* **2008**, *24*, 7394.

53. Lee, J. K.; Kim, Y. G.; Chi, Y. S.; Yun, W. S.; Choi, I. S. *J. Phys. Chem. B* **2004**, *108*, 7665.

54. Alonso, J. M.; Reichel, A.; Piehler, J.; del Campo, A. *Langmuir* **2008**, *24*, 448.

55. Han, H. M.; Li, H. F.; Xiao, S. J. *Thin Solid Films* **2011**, *519*, 3325.

56. Li, H. F.; Han, H. M.; Wu, Y. G.; Xiao, S. J. *Appl. Surf. Sci.* **2010**, *256*, 4048.

57. Cinier, M.; Petit, M.; Williams, M. N.; Fabre, R. M.; Pecorari, F.; Talham, D. R.; Bujoli, B.; Tellier, C. *Bioconjug. Chem.* **2009**, *20*, 2270.

58. Cinier, M.; Petit, M.; Pecorari, F.; Talham, D. R.; Bujoli, B.; Tellier, C. *J. Biol. Inorg. Chem.* **2012**, *17*, 399.

59. Han, X. S.; He, T. *Colloids Surf. B Biointerfaces* **2013**, *108*, 66.

60. Denizot, B.; Hindre, F.; Portet, D. Patent WO2006/053910, **2006**.

61. Karimi, A.; Denizot, B.; Hindre, F.; Filmon, R.; Greneche, J. M.; Laurent, S.; Daou, T. J.; Begin-Colin, S.; Le Jeune, J. J. *J. Nanopart. Res.* **2010**, *12*, 1239.

62. Portet, D.; Denizot, B.; Rump, E.; Lejeune, J. J.; Jallet, P. J. *Colloid Interface Sci.* **2001**, *238*, 37.

63. Russell, R. G. G. *Bone* **2007**, *40*, S21.

64. Russell, R. G. G.; Watts, N. B.; Ebetino, F. H.; Rogers, M. J. *Osteoporos. Int.* **2008**, *19*, 733.

65. Shane, E. *N. Engl. J. Med.* **2010**, *362*, 1825.

66. Josse, S.; Faucheux, C.; Soueidan, A.; Grimandi, G.; Massiot, D.; Alonso, B.; Janvier, P.; Laib, S.; Gauthier, O.; Daculsi, G.; et al. *Adv. Mater.* **2004**, *16*, 1423.

67. Josse, S.; Faucheux, C.; Soueidan, A.; Grimandi, G.; Massiot, D.; Alonso, B.; Janvier, P.; Laib, S.; Pilet, P.; Gauthier, O.; et al. *Biomaterials* **2005**, *26*, 2073.

68. Roussiere, H.; Montavon, G.; Laïb, S.; Janvier, P.; Alonso, B.; Fayon, F.; Petit, M.; Massiot, D.; Bouler, J. M.; Bujoli, B. *J. Mater. Chem.* **2005**, *15*, 3869.

69. Mukherjee, S.; Huang, C.; Guerra, F.; Wang, K.; Oldfield, E. *J. Am. Chem. Soc.* **2009**, *131*, 8374.

70. Mukherjee, S.; Song, Y. C.; Oldfield, E. *J. Am. Chem. Soc.* **2008**, *130*, 1264.

71. Ambrosio, L.; Guarino, V.; Sanginario, V.; Torricelli, P.; Fini, M.; Ginebra, M. P.; Planell, J. A.; Giardino, R. *Biomed. Mater.* **2012**, *7* (SI), 024113.

72. Bohner, M.; Gbureck, U.; Barralet, J. E. *Biomaterials* **2005**, *26*, 6423.

73. Brown, W. E.; Chow, L. C. *J. Dent. Res.* **1983**, *62*, 672.

74. Dorozhkin, S. V. *J. Mater. Sci.* **2008**, *43*, 3028.

75. Dorozhkin, S. V. *Materials* **2009**, *2*, 221.

76. LeGeros, R.; Chohayeb, A.; Shulman, A. *J. Dent. Res.* **1982**, *61 (Special Issue)*, 343.

77. Bobyn, J. D.; McKenzie, K.; Karabasz, D.; Krygier, J. J.; Tanzer, M. *J. Bone Joint Surg.* **2009**, *91*, 23.

78. Denissen, H.; Martinetti, R.; van Lingen, A.; van den Hooff, A. *J. Periodontol.* **2000**, *71*, 272.

79. Gao, Y.; Zou, S. J.; Liu, X. G.; Bao, C. Y.; Hu, J. *Biomaterials* **2009**, *30*, 1790.

80. Peter, B.; Gauthier, O.; Laib, S.; Bujoli, B.; Guicheux, J.; Janvier, P.; van Lenthe, G. H.; Muller, R.; Zambelli, P. Y.; Bouler, J. M.; Pioletti, D. P. *J. Biomed. Mater. Res. A* **2006**, *76A*, 133.

81. Peter, B.; Pioletti, D. P.; Laib, S.; Bujoli, B.; Pilet, P.; Janvier, P.; Guicheux, J.; Zambelli, P. Y.; Bouler, J. M.; Gauthier, O. *Bone* **2005**, *36*, 52.

82. Tanzer, M.; Karabasz, D.; Krygier, J. J.; Cohen, R.; Bobyn, J. D. *Clin. Orthop. Rel. Res.* **2005**, *441*, 30.

83. Yoshinari, M.; Oda, Y.; Inoue, T.; Matsuzaka, K.; Shimono, M. *Biomaterials* **2002**, *23*, 2879.

84. Panzavolta, S.; Torricelli, P.; Bracci, B.; Fini, M.; Bigi, A. *J. Inorg. Biochem.* **2009**, *103*, 101.

85. Schnitzler, V.; Fayon, F.; Despas, C.; Khairoun, I.; Mellier, C.; Rouillon, T.; Massiot, D.; Walcarius, A.; Janvier, P.; Gauthier, O.; et al. *Acta Biomat.* **2011**, *7*, 759.

86. Panzavolta, S.; Torricelli, P.; Bracci, B.; Fini, M.; Bigi, A. *J. Inorg. Biochem.* **2010**, *104*, 1099.

8

PHOTOFUNCTIONAL POLYMER/ LAYERED SILICATE HYBRIDS BY INTERCALATION AND POLYMERIZATION CHEMISTRY

GIUSEPPE LEONE AND GIOVANNI RICCI

CNR-ISMAC, Istituto per lo studio delle Macromolecole, Milano, Italy

8.1 INTRODUCTION

The second half of the twentieth century, with the development of composites based on micrometre-sized reinforcing particles, witnessed an enormous transformation in the chemical design, engineering and performance of structured materials. Actually, the emergence of nanometre-sized particles (such as platelets, fibres and tubes) is leading to a second revolution in composite materials.

Polymer nanocomposites (PNs) are currently the subject of extensive worldwide research [1]. PNs are two-phase hybrid systems consisting of a polymer filled with high-surface-area reinforcing nanoparticles, for which at least one dimension of the dispersed particles is in the nanometre range. Three types of nanocomposites can be distinguished, depending on how many dimensions of the dispersed particles are in the order of nanometres. Thus, we are dealing with isodimensional nanoparticles, such as spherical silica nanoparticles and semiconductor nanoclusters, when all the three dimensions are in the order of nanometres; with nanotubes or whiskers, such as carbon nanotubes or cellulose whiskers, if two dimensions are in the nanometre scale; or with layered silicates when only one dimension is in the nanometre range.

Nowadays, PNs have displaced a lot of traditional composite materials in a variety of applications because the intimate interactions between components can provide enhancement of the bulk polymer properties (i.e. mechanical and barrier properties, thermal stability, flame retardancy and abrasion resistance). Since PNs are

Tailored Organic–Inorganic Materials, First Edition. Edited by Ernesto Brunet, Jorge L. Colón and Abraham Clearfield.
© 2015 John Wiley & Sons, Inc. Published 2015 by John Wiley & Sons, Inc.

used as engineering materials, layered silicates (e.g. dioctahedral 2 : 1 phyllosilicates) have attracted much academic and industrial attention: the interest is mainly due to their abundance, relative cheapness and large contact area [2, 3]. Researches dealing with the fabrication of polymer/layered silicate hybrids mostly focused on thermoplastic, elastomers and thermosetting polymers of different polarities including polyamide, polystyrene, polycaprolactone, poly(ethylene oxide), epoxy resin, polysiloxane, polyurethane, poly(ethylene terephthalate), polypropylene, and so on [4]. For instance, improvement of the polyolefin technologies, joined to the synthesis of targeted nanocomposites, allows the substitution of well-known homo- and copolymers such as acrylonitrile butadiene styrene resins, poly(vinyl chloride) and polyurethane in unexpected applications.

The demand for materials with superior chemical and physical properties has motivated vigorous research, and layered silicates have provided an accessible and low-cost additive to enhance the properties of polymers for fabricating hybrid materials for applications in the area of fibres, clothes, soft drink bottling and automotive and for the developments of polymer nanotechnologies [5, 6]. The research opportunities offered by these materials have not ended. Research efforts are now focused into the possibility of functionalizing layered silicates in order to develop host–guest materials with specific functional features for application in advanced fields. This strategy is leading to breakthrough results in sensors, drug release, functional coatings and other applications in biomedicine, electronics and energy sectors.

In this chapter, the research area about the preparation of photofunctional layered silicate-based hybrids has been mainly considered. The chapter themes are not fully covered: the applications of such hybrids range from 'traditional' uses, that is, decorative and transparent coatings, protective hard coatings for transparent plastics, coatings for protection of glass optical fibres (which improve mechanical strength, water corrosion, mechanical and chemical damage, abrasion, high-temperature degradation), hybrids for photonics, and so on, but the reader can refer to other chapters of this book and to recent reviews [7].

Specifically, this chapter aims to give an overview on the recent advances in the use of layered silicates as hosts to fabricate functional organic–inorganic structures by intercalation of π-conjugated molecules and polymers for application in light-emitting devices (LEDs) and polymer light-emitting devices (PLEDs).

8.2   LIGHTING IS CHANGING

The evolution of light bulbs lives up to Darwin's scale: from the arc lamp to incandescent light bulbs and from gas-discharge lamps to LEDs. Over the years, the technology in this field has changed its objectives accordingly. If at the beginning the aim was to improve the lighting and, subsequently, to prolong the life of the light bulbs, today, the focus is to consume less energy. To date, fluorescent lamps and incandescent lamps convert only 20 and 5%, respectively, of the incoming electric current in visible light. This means that more than 80% of electric energy used worldwide for lighting is lost and, therefore, the development of more efficient lighting systems is urgent. In this regard, solid-state lighting (SSL) has the potential to be a 100% efficient technology, although it is still far from large-scale application and marketing.

Traditionally, this field of application is dominated by inorganic materials, silicon and gallium arsenide being the most common ones. However, over the past decade, an increasing attention has been paid to organic materials, both π-conjugated molecules and polymers. This resulted in the development of organic light-emitting diodes (OLEDs) that now offer a commercially viable alternative to conventional inorganic semiconductor-based LED technology. The introduction of OLEDs represents a great advance in SSL due to their peculiar properties (i.e. low costs, flexibility and conformability, high brightness, fast response time and low driven voltage (working voltage)), and power efficiencies exceeding conventional SSL sources are reported for OLEDs [8]. Nonetheless, OLEDs exhibit some drawbacks, that is, the lack of long-term chemical and thermal stability and poor mechanical strength. The need to maximize the OLED performances, joined with the demand for smaller devices and components requiring less resources and energy, has been rapidly pushing the academic and industry research towards nanometre-scale hybrids. Indeed, the combination of organic compounds with inorganic nanoparticles in a single material may overcome the above drawbacks. An ideal SSL hybrid technology would be characterized by high quantum efficiency and robustness coupled with low-cost and solution processability of the materials. Layered silicate-based hybrids are promising candidates for this paradigm shift because they combine the chemical and thermal stability of the inorganic counterparts with the organic molecule/polymer functions while optimizing and maximizing complementary properties (e.g. density, permeability, mechanics) [9, 10].

8.3 GENERALITIES

8.3.1 Layered Silicates

Layered nanoparticles of interest for functional hybrid technology have platelets from ca. 0.7 to 2.5 nm thick [5]. A partial and non-exhaustive list of candidates is given in Table 8.1.

TABLE 8.1   Some layered nanoparticles for the potential use in polymer hybrid materials

Chemical nature	Example
Smectite clays	Montmorillonite, bentonite, volkonskoite, saponite, laponite sepiolite
Synthetic clays	Hectorite, $MgO(SiO_2)_s(Al_2O_3)_a(AB)_b(H_2O)_x$ (where AB is a ion pair, viz. NaF)
Other clays	Micas, vermiculite, illite, tubular attapulgite layered double
Hydroxides	$M_6Al_2(OH)_{16}CO_3 \cdot nH_2O$; M = Mg, Zn
Metal chalcogenides	TiS_2, MoS_2, MoS_3, $(PbS)_{1.18}(TiS_2)_2$
Metal oxides	V_2O_5, MoO_3
Others	Graphite, graphite oxide, etc.

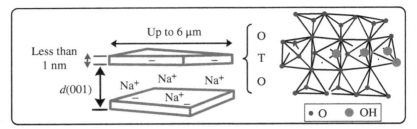

FIGURE 8.1 Schematic structure of a sodium-exchanged layered silicate mineral.

The silicates commonly used in polymer (nano)composites are essentially aluminosilicate (2 : 1 phyllosilicates), constituted of stacks of hydrated layers whose crystal structure is composed of an octahedral alumina $Al(O,OH)_6$ sheet (O-network) sandwiched between two silicon–oxygen tetrahedral sheets (T-network) (Figure 8.1). The layers are separated by galleries where cations (e.g. Na^+, K^+) balance the negative charge of the aluminosilicate sheets arising from the isomorphic substitution of Al or Si with other metals [2]. The layer thickness is around 1 nm, and the lateral dimensions of these layers may vary from 30 nm to several microns.

The layers organize themselves to form stacks with a regular van der Walls gap in between them, named the interlayer region. Isomorphic substitution within the layers (e.g. Al^{3+} replaced by Mg^{2+} or by Fe^{2+}, Mg^{2+} replaced by Li^+) generates negative charges that are counterbalanced by alkali or alkaline earth cations situated in between the lamellae. Based on the extent of the substitutions, a term called layer charge density is defined. The layer surface has 0.25–1.2 negative charges per unit cell and a commensurate number of exchangeable cations in the interlamellar galleries. For the anionic clays (e.g. smectites), the ion concentration is usually expressed as the cation-exchange capacity (CEC), which ranges from about 0.5 to $2 \, meq \, g^{-1}$. The distance from the T-network to its analogue in one of the neighbouring layers is defined as interlayer d-spacing. This spacing mainly depends on the size of the exchangeable cations and the amount of interlayer water. In order to make these hydrophilic phyllosilicates more organophilic, the hydrated cations of the interlayer can be exchanged with cationic surfactants, such as alkylammonium or alkylphosphonium salts, to give organically modified silicates. The intercalation of small molecules in between the layers is easy, being the forces holding the stacks together relatively weak [11]. When the alkali cations are exchanged with organic cations, a larger d-interlayer spacing usually results. Studies on these topics have been extensively reviewed elsewhere [12, 13].

8.3.2 Polymer/Layered Silicate Hybrid Structures

Depending on the nature of the components and the preparation method, three types of PNs can be obtained [6]. Phase-separated microcomposites (conventional composites) are obtained when the polymer chains are unable to intercalate within the inorganic sheets: silicate lamellae remain stacked in structures marked as tactoids as in the pristine mineral. Otherwise, when the polymer chains penetrate in between the silicate galleries, an intercalated system is obtained. In this case, the nanocomposite shows, at least in principle, a well-ordered multilayer morphology built up with alternating polymeric and silicate layers. When the inorganic platelets are randomly dispersed in the polymer matrix and the lamellae are far apart from each other, so that the periodicity of this platelet arrangement is totally lost, an exfoliated structure is achieved.

PNs are usually characterized by means of X-ray diffraction (XRD), transmission electron microscopy (TEM) and scanning electron microscopy (SEM). These tools indeed provide information about the nanocomposite structures and morphologies. Specifically, XRD is used to identify intercalated structures: in such nanocomposites, the repetitive multilayer structure is well preserved, allowing the interlayer spacing to be determined. The intercalation of the polymer chains increases the interlayer spacing compared to the spacing of pristine silicate, leading to a shift of the diffraction peak towards lower angle values (angle and layer spacing values being related through Bragg's equation: $2d_{hkl} \sin \theta = \lambda$, where λ corresponds to the wavelength of the X-ray radiation used in the diffraction experiment, d is the spacing between diffractional lattice planes and θ is the measured diffraction angle).

In the case of exfoliated structures, no more diffraction peaks are detectable by XRD, either because of a too much large spacing between the layers (i.e. the polymer separates the silicate platelets by 8–10 nm or more) or because the nanocomposites do not present ordering anymore. Moreover, it must be pointed out that very often the lost of the XRD signal of the inorganic component may be due to the lower silicate amount (≤ 1 wt. %) in the nanocomposites. Hence, for a better check of the nanocomposite structure, TEM investigation is the powerful analytical method, allowing a qualitative understanding of the internal structure and directly providing information on morphology and defect structures [14]. However, the classification of the composite structures as exfoliated or intercalated is not very realistic, since a mixture of different morphologies can exist, presenting both intercalation and exfoliation features. In this case, a broadening of the XRD peak is observed, and TEM observation too does not allow to assign the exact structure to the investigated nanocomposites: only a qualitative classification of the morphology as more or less intercalated or exfoliated can be made.

8.3.3 Methods of Preparation of PNs

There are two approaches to operate at the nanometre level. One refers to the approach known as 'top-down', which means reducing, with physical methods, the dimensions of the structures to nano levels. The other one is the so-called bottom-up, which indicates the approach in which small components, usually molecules, are used as building blocks to fabricate hybrids nanoassemblies [7].

Polymer/layered silicate hybrids are currently prepared by (i) mixing in the molten state (melt intercalation), (ii) intercalation of polymer or prepolymer (in the case of insoluble polymers) from solution, (iii) latex compounding, (iv) layer-by-layer assembly and (v) *in situ* intercalative polymerization [15–17].

Among them, one of the most intriguing approaches is the direct formation by *in situ* polymerization ('bottom-up' approach). Unlike the latex and melt compounding, which are conducted directly with the polymer, the *in situ* method involves the direct addition of the silicate, intercalated by a catalyst or an initiator, to a polymerizable (co)monomer mixture [4, 18–21]. According to this method, the intimate mixing of the 'soft' polymer matrix with the 'hard' filler is promoted by the functionalized silicate-mediated monomer polymerization. Various different polymerization methods have been used in the production of well-dispersed silicate layer hybrids, including atom transfer radical polymerization (ATRP), surface-initiated nitroxide-mediated polymerization (SI-NMP), reversible addition–fragmentation chain transfer (RAFT) polymerization, ring-opening polymerization (ROP), ring-opening metathesis polymerization (ROMP), metal-catalysed polymerization, living cationic polymerization and living anionic polymerization [22].

8.4   FUNCTIONAL INTERCALATED COMPOUNDS

Structured hybrid materials are widely observed in nature, for example, in plant cell walls, bones and skeletons, where specific functionalities are obtained by structuring the matter across a range of length scales. For instance, the supramolecular organization of chlorophyll molecules in polypeptide cages, involved in photosynthesis, explains why plants are so efficient optoelectronic systems. This nature's peculiarity has inspired scientists to design hybrids that are potentially useful in the optimization of LED performance. In this sense, the intercalation of π-conjugated molecules (dyes) and polymers in between layered silicates allows the construction of hybrids that self-organize through chemical or physical processes, the structure and functions of which are easily tunable and controlled. Indeed, layered silicates are particularly appealing for fabricating functional hybrids due to their adsorptive properties, ion-exchange ability and high specific surface area, which permit to easily tune the interaction between the emitting centres by surface chemistry, and a sandwich-type intercalation [23, 24].

8.4.1   Dyes Intercalated Hybrids and (Co)intercalated PNs

Layered silicates offer a unique two-dimensional expandable interlayer space for constructing hybrid materials with unique intercalated guest dye arrangement. They can accommodate, at least in principle, a large variety of molecules by simple ion-exchange reactions while maintaining the structural features of the host. Among them, rhodamine, oxazine and cyanine laser dyes (Table 8.2) have been extensively studied and well reviewed by Ogawa and Kuroda [23].

Two mainly aspects are expected to be covered by intercalation of organic laser dyes within the silicate interlayer region:

1. To improve the dye photo-, thermo- and chemical stability, which is generally insufficient for use in practical optoelectronic devices
2. To control the accommodation of the guest species for organizing efficient dye assemblies, thus allowing tuning their photofunctions

TABLE 8.2 Some examples of cationic dyes studied for the intercalation within layered silicates

Name/abbreviation	Structural formula	λ^{MAX} (nm)	λ^{MAX} (nm)	PL-QY (%)
Oxazine 1 (Ox1)		646	663	14
Oxazine 4 (Ox4)		614	651	62
Oxazine 170 (Ox170)		627	645	58
Rhodamine 123 (R123)		505	560	90
Rhodamine 6G (R6G)		530	566	95
Rhodamine 101 (R101)		564	587	91
Cryptocyanine (Cc)		710	720	1.2

Counterions = perchlorate; PL-QY = photoluminescence quantum yield.

Since the forces holding the stacks together are relatively weak, the intercalation of these cations by ion exchange is almost easy. However, the inclusion of photoactive molecules in the restricted geometry comes with the formation of molecular aggregates, which could results, if ordered, in cooperative and coherent phenomena.

Basically, there are two types of ordered molecular assemblies. The first, named H-aggregate, is based on a sandwich-type association, and it is characterized by a light absorption at higher energies if compared to single molecules and a poor

TABLE 8.3 Dimers with different geometries – a schematic view

Geometrical arrangement	Aggregate type	θ	σ
	H-type dimers	Any value	90
		0	$0 < \theta < 90$
	J-type dimers	0	0
	J-type dimers	Any value	90

photoluminescence efficiency. The second one, called J-aggregate (head-to-tail alignment of dyes), is an assembly of molecules with coherently coupled transition dipole moments [25, 26]. This configuration shows a red-shifted narrow absorption band with respect to isolated molecules, a negligible Stokes shift between absorption and emission and a very intense emission, making it an appealing arrangement for various technological applications such as OLEDs, fluorescent sensors, organic photoconductors, artificial light harvesting and organic transistors [27–31].

Depending on the specific geometrical arrangement of the monomer units in the aggregates, dimers with different spectroscopic characteristics can be described, as reviewed by Arbeloa et al. [24]. Table 8.3 illustrates some dimer geometries, defined by the angle between the transition moments of the monomers in the dimer (σ) and the angle between the direction of the dipolar moments and the line linking the molecular centres (θ). Some cases in Table 8.3 are extreme and idealistic geometries, being the aggregate that can adopt structures with intermediate θ and σ angles.

Research focuses mainly on partially crystalline synthetic smectite-type silicates with a general structural formula $Na^{+0.7}[(Si_8)(Mg_{5.3}Li_{0.7})O_{20}(OH)_4]^{0.7-}$ of the unit cell and small particle size (<30 nm) [32]. The light scattering in the visible region from those particles is very low as they form transparent aqueous suspensions and solid films, making smectites ideal for hosting cationic dyes by simple immersion of the silicate film in a dye's solution. Nonetheless, these materials are still far from being used in LEDs. The reason is that smectite silicates have high charge unit density (≈ 0.60 nm^2) so that dyes progressively arrange in a more perpendicular way with respect to the layer surface. Such dye's tail-to-tail association (H-coupling) arrangement allows for a more compact state of the dyes and a maximal host layer cover, exhibiting light absorption at higher energies when compared to single molecules and poor photoluminescence efficiency [33–36]. Thus, there has been considerable debate as to limiting H-type packing formation, and many strategies have been pursued for fabricating hybrids of interest for practical use. López Arbeloa et al. [37, 38] and Iyi et al. [39] claimed the convenience of modifying laponite silicate by (co)intercalation of the dye with non-emissive ammonium salts that acted as dye aggregation

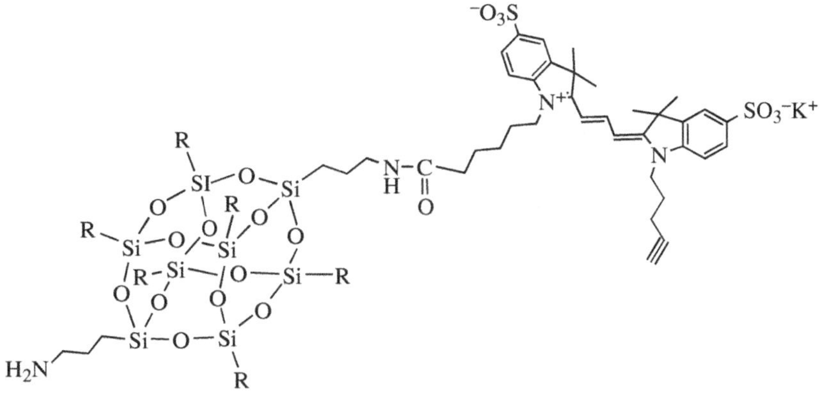

FIGURE 8.2 The luminescent POSS was obtained by linking a cyanine dye on the POSS cage functionalized with two amino groups. One of the amino groups served for the intercalation within the silicate interlayer space.

inhibitors. Marchese et al. intercalated bifunctional polyhedral oligomeric silsesquioxanes (POSS) bearing a cyanine moiety and an amino group, reducing both the dye's mobility and aggregation (Figure 8.2) [40]. Bujdak and Komadel investigated reduced charge Li-silicate embedding methylene blue (MB) [36]. They reported that the lower layer charge density leads to a larger distance between the neighbouring intercalated MB cations, inhibiting the MB aggregation. In addition, spin coating on glass substrates [41–43], layer-by-layer assembly [44], the Langmuir–Blodgett technique [45, 46] and the search for new host matrices are the subject of intense studies for fabricating hybrids of interest for practical use.

While smectite silicates are in the focus due to their easy solution processing, crystalline fluoromica-type silicates with a structural formula $Na_{0.66}Mg_{2.68}(Si_{3.98}Al_{0.02})O_{10.02}F_{1.96}$ are favoured due to the lower charge density ($\approx 0.77\,nm^2$), and higher CEC with respect to smectites. However, widespread development of dye-intercalated fluoromica hybrids is still a challenge. Indeed, mica platelets are larger than smectite (micas have an aspect ratio of 6000), thus increasing the interplate crystalline welding. In addition, a partial replacement of –OH groups by –F reduces the hydrophilic character. These two factors limit solution processability.

In this framework, our research group pioneered the use of fluoromica for hosting oxazine-1 dye and by dispersing the resultant host–guest compound within a polymer matrix [47]. The aim of the polymer hybrid synthesis is to uniformly disperse and distribute the dye-doped fluoromica inorganic component, initially composed of aggregates of stacks of parallel layers, within the polymer, while fabricating transparent films, and avoid the formation of traditional filled microcomposites, less processable by spin coating. Decreasing the particle size allows a much more homogeneous distribution of a material and leads to a drastic increase of the polymer/inorganic particle interfacial area. For instance, if the particle size is reduced by a factor of 10, the number of particles has to be increased by a factor of 1000 to achieve

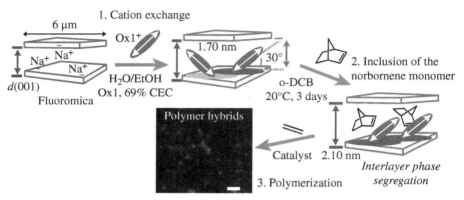

FIGURE 8.3 Schematic and simplified illustration of the poly(norbornene) hybrids' synthesis by means of cation exchange and metal-catalysed polymerization. Scale bar = 10 μm.

the same inorganic content and the interfacial area increase by a factor of 10 at the same time. A homogeneous dispersion of the inorganic counterpart on the nanoscale is of primary importance for fabricating transparent thin films for LED applications. In this respect, the development of synthetic methods is the key challenge. In the *in situ* polymerization technique, the monomer, together with an initiator and/or catalyst, is (co)intercalated within the silicate layers, and the polymerization is initiated by external stimulation such as thermal, photochemical or chemical activation. The chain growth in between the silicate galleries triggers the intercalation/exfoliation process and hence the nanocomposite formation.

In this way, our first attempt to synthesize efficient multicomponent, integrated hybrids mainly consisted of two steps (Figure 8.3).

In the first one, fluoromica was modified by cation exchange with oxazine-1 dye (Table 8.2) in a solution of water/ethanol. The dye arrangement was easily tuned by modulating the dye loading: increasing the dye loading up to 69%CEC, the oxazine took a 30° tilted distribution of its long molecular axis with respect to the fluoromica interlayer surface. We found that this arrangement creates interlayer voids that facilitate the entry of polymerizable monomers, that is, norbornene and ethylene, in between the silicate interlayer region. The low viscosity of the monomer with respect to the preformed polymer makes it more easy to intercalate and to break up silicate aggregates in the second reaction step by adding the proper catalytic system, that is, $(PCyp_3)_2CoCl_2/MAO$ (MAO = methylaluminoxane). The metal-catalysed polymerization leads to the growth of the polymer chains that wrapped the fluoromica stacks, as also observed by AFM (Figure 8.4), thus reducing the activity of the surface atoms and the sedimentation commonly observed in simple blends. In fact, along with scanning of the sample surface in contact mode configuration, the scanner executes a vertical periodic motion. The cantilever 'feels out' the sample surface. In agreement with the local elasticity of the sample, force modulation imaging was obtained, and sharper details were observable with respect to the corresponding height image: bright areas with size in the range 30–70 nm correspond to stiffer domains (silicates) surrounded by a softer matrix (polymer) (Figure 8.4).

(a)

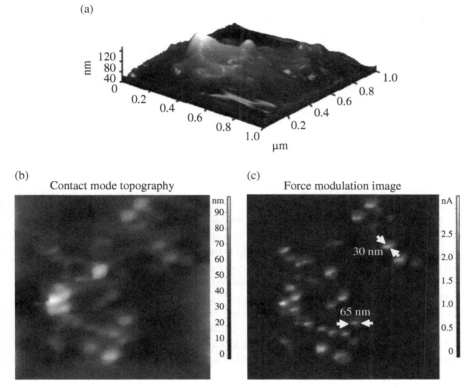

(b) Contact mode topography

(c) Force modulation image

FIGURE 8.4 AFM 1×1 mm^2 contact mode topography (a and b) and force modulation images (c) of the polymer hybrid film.

Besides, we observed that the intercalated poorly polar polymer chains determined an interlayer phase segregation driving the oxazine-1 molecules to adopt less conformational degrees of freedom up to the point of increasing the head-to-tail interaction (J-aggregates) between them by aligning their electronic transition dipole moments. Optical and photophysical evidences of effective J-band formation were given by means of low-temperature photoluminescence and ultrafast pump–probe measurements [48]. The evolution of the photoluminescence spectra, showing an increase in the J-aggregate emission upon the norbornene monomer addition, demonstrated, for the first time, that the driving force for J-aggregation was the interpenetration of the hydrophobic norbornene monomers, producing a such surrounding environment for the dye, up to the point of aligning their electronic transition dipole moments and producing J-aggregates of different sizes and uniformity with respect to angle and distance (see also Table 8.3).

It is worth to note that, so far, techniques to assemble J-aggregate are still quite rare. While in solution the formation of such assemblies is quite simple, its control in the solid state remains a challenge. J-aggregation, if not self-induced, can be promoted by template agents such as proteins, dendrimers, polysaccharides, helical polyacetylene, micelle structure and inorganic solids [49–52]. Several recent investigations have shown that cyanine J-aggregates can also form in restrictive environments such as DNA grooves, within the cavities of mesoporous structures and on the surface of colloidal particles [53–55].

Nonetheless, we found that the hybrids obtained with this procedure were poorly efficient once plugged into a device. Indeed, the approach becomes detrimental likely due to the introduction of the cobalt metal-based catalyst, which acted as impurities, and therefore luminescence quenchers.

To overcome this issue, we next explored the SI-NMP approach [56]. Compared to the transition metal complex-mediated polymerization, SI-NMP represents a valuable expedient for the controlled fabrication of polymer brushes and hybrid nanoassemblies [57–60]. An advantage of SI-NMP is that no further catalysts are required. This obviates the need for additional purification steps and reduces the chance to introduce impurities, which is advantageous for applications in LEDs.

To apply the SI-NMP technique to fluoromica, firstly, we synthesized a 2,2, 6,6-tetramethyl-1-piperidyloxy (TEMPO)-based alkoxyamine initiator bearing an ammonium functional moiety for anchoring to the layer inorganic surface. Successively, the synthetic route can be divided into two steps, as sketched in Figure 8.5 [56].

In the first, the TEMPO initiator was (co)intercalated with the dye by cation exchange; in the second one, the polymerization of styrene was exploited. By applying this 'grafting-from' approach, we successfully fabricated poly(styrene)-graft–/rhodamine 6G (R6G)–fluoromica hybrids. The materials exhibited (i) higher solution processability than that of both R6G–fluoromica and pristine fluoromica, allowing us to process the polymer hybrids as solid films, and (ii) a strongly improved dye thermostability, thus improving the materials' operative temperature up to 435 °C. XRD data and photophysical investigation by absorption and continuous-wave and time-resolved photoluminescence measurements allowed us to shed light on the rhodamine supramolecular organization, that is, the different spatial configuration of the dye in between the fluoromica layers, and to find a relationship among the dye's arrangement and photophysical features: the R6G monomer emission progressively red shifts towards J-like coupling by increasing the dye loading, which in turn involves a higher R6G tilt angle with respect to the host layer surface. As a consequence, films showing different emission colours were obtained.

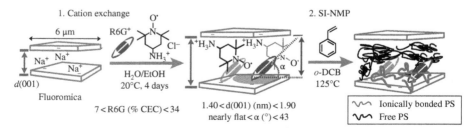

FIGURE 8.5 Schematic and simplified illustration of the poly(styrene) hybrids' synthesis by means of cation exchange and surface-initiated nitroxide-mediated polymerization (SI-NMP).

The synthesis of these new three component hybrids, still in progress, is easy to scale up and applicable to a variety of cationic dyes, nitroxide initiators, catalysts and monomers. The increased solution processability of the materials strongly encourages technological applications of fluoromica that, otherwise, would not have been considered interesting because of very poor swelling capability.

8.4.2 Light-Emitting Polymer Hybrids

While the intercalation of small conjugated molecules, as well as the study of the interactions between the inorganic host and the organic guest, has been widely studied, the intercalation of photoactive polymers in between layered silicates is still in an embryonic stage, and only few examples are reported. This paragraph aims to give an overview on the advances in the use of layered silicates as starting units to fabricate hybrid organic–inorganic structures by intercalation of light-emitting polymers (LEPs) for PLED technology application.

The driving force of polymer electronics research in lighting is the fact that LEPs are suitable for miniaturization of devices likely because their optical features stem from molecules and aggregates rather than the device itself structure. Among p-type polymers, poly(p-phenylene vinylene), poly(p-phenylene)s, poly(9,9-dialkylfluorene)s and poly(thiophene)s have shown interesting chemical and physical features, making them the most investigated ones for application in PLEDs [61, 62]. However, among others, one key issue regarding the commercialization of PLEDs with this class of polymers is overcoming their poor stability against oxygen and moisture. To date, one general way to enhance the stability is to prevent oxygen and moisture from penetrating into the device emissive layer by using a poly(aniline) buffer layer on the top of an indium tin oxide (ITO) anode or an environmentally stable cathode such as Al. In the same manner, the intercalation of LEPs within the silicate lamellae would combine the advantages of layered silicate, that is, high chemical and thermal stability, high aspect ratio and layered nanoparticle form, improving the barrier to the oxygen and moisture and thus preserving the emitter against degradation. Indeed, the silicate stacks or single layers involve a strongly tortuous diffusion pathway of permanent gas molecules, greatly improving the environmental material stability.

8.4.2.1 Poly(p-Phenylene Vinylene)-Based Polymer Hybrids Since poly(p-phenylene vinylene)-based PLEDs were reported by Friend in 1990 [63], LEPs have been considered promising candidates for flat, large-area displays. However, in addition to the poor stability, commercial use of such polymers in device is limited by their low quantum efficiency. One of the major reasons is that the electron injection current of single-layer LEDs is too low. Thus, to make more efficient electroluminescent device, it is necessary to enhance the injection and confinement of the charge carrier. Different approaches have been tried, such as (i) use of an electron-injecting or transporting layer [64], (ii) use of a cathode with a low work function [65], (iii) use of an insulating layer such as Al_2O_3, [66], and so on.

Otherwise, Park et al. first reported the fabrication of layered silicate hybrids including intercalated emitting poly[2-methoxy-5-(2′-ethylhexyloxy)-1,4-phenylenevinylene]

(MEH-PPV) (Figure 8.6) with the aim to simultaneously improve the electroluminescence efficiency and the environmental stability [67]. The hybrids were prepared by solution methods, mixing solutions in 1,2-dichloroethane of both a mica-type silicate and MEH-PPV. The PLEDs of the MEH-PPV/silicate were fabricated by spin coating, from a 1,2-dichloroethane solution, on ITO-coated glass substrates.

The material exhibited (i) high photostability due to its good gas barrier property, (ii) higher lifetime (indeed, the photoluminescence from the hybrid film decayed with time more slowly than that of the pure MEH-PPV film) and (iii) highly enhanced photoluminescence intensity.

PLED fabricated from such composite exhibited a high external quantum efficiency, 0.38% photons/electrons, with an Al cathode, which was enhanced by 100 times compared with that of the pure MEH-PPV device. In addition, the authors found that the hole mobility of the device was reduced by using MEH-PPV/silicate as active layer. The hole mobility of the ITO/MEH-PPV/silicate/Al device architecture was three times lower than by using ITO/MEH-PPV/Al device, confirming that the charge carrier was effectively confined.

Karasz et al. reported the synthesis and hybridization/intercalation of a poly[2',5'-dihexyloxy-p-terphenyl-4,4''-ylenevinylene-alt-2,7-(9,9'-di-n-hexylfluorene) diylvinylene] (Figure 8.7) [68]. The material was obtained by dissolving a mixture of the polymer and an organo-modified silicate in chloroform. LED fabrication was performed from a chloroform solution of the composite spin coated onto PEDOT-/ITO-coated glass substrates.

With respect to the first study by Park [67], who observed a change in the photoluminescence emission maximum over the polymer intercalation, Karasz and co-workers observed that no change occurred, likely due to the only partially conjugated structure of the polymer with respect, on the contrary, to the fully conjugated MEH-PPV. Both the photo-oxidation and photodegradation processes, measured with monochromatic 400 nm light from a xenon lamp in air, were retarded in the hybrid films in which the silicate reduced oxygen and moisture penetration into the emissive layer.

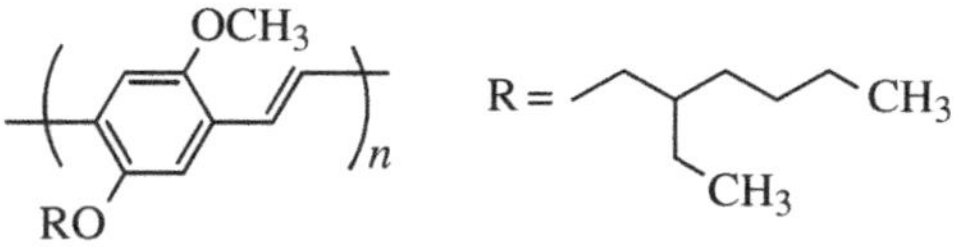

FIGURE 8.6   Poly[2-methoxy-5-(2'-ethylhexyloxy)-1,4-phenylenevinylene] (MEH-PPV).

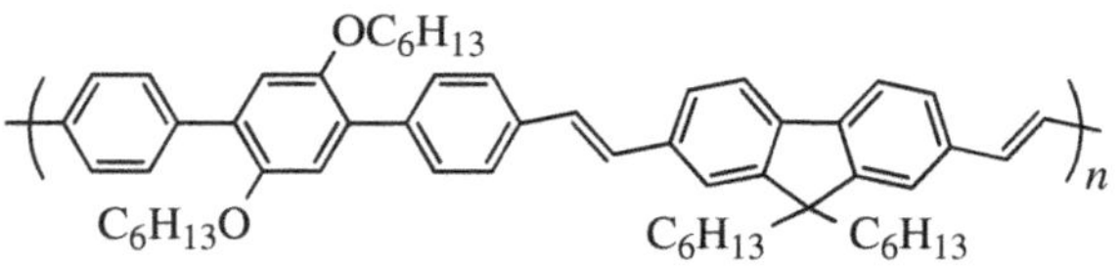

FIGURE 8.7   Poly[2',5'-dihexyloxy-p-terphenyl-4,4''-ylenevinylene-alt-2,7-(9,9'-di-n-hexylfluorene)diylvinylene].

8.4.2.2 Poly(fluorene)-Based Polymer Hybrids

Poly(fluorene)s (PFs) and its derivates have shown interesting chemical and physical features, making them the most investigated blue LEPs for application in PLEDs. PFs have very high luminescence efficiencies and superior thermal stability with respect, for instance, to PPV-based polymers. However, there are some problems to successfully employ PF-type polymers in PLED. One key issue regarding the use of fluorene polymers in optical devices is overcoming the low-energy emission bands at 2.2–2.3 eV, which may be caused by either aggregates/excimers or ketone defects, being the other one the faster photodegradation of the polymer by the oxygen residual in the polymer or released from the ITO electrode [69–71]. Indeed, amorphous polymer films have sufficient free volume to allow diffusion of oxygen atoms that acts as luminescence quenching centres.

In order to circumvent both the practical issues, Kim et al. prepared poly(9,9'-dioctylfluorene) (PDOF)/layered silicate intercalated hybrids by trichloroethylene solution method (Figure 8.8) [72]. The materials exhibited (i) reduced low-energy emission with smaller long-wavelength tails, and hence improved emission colour purity, likely due to the reduced inter-chain interactions between the conjugated macromolecular chains nanoconstrained within the silicate lamellae, and (ii) increased thermal stability.

Comparing the performance of the neat ITO/PVK/PDOF/Li : Al device with that of ITO/PVK/PDOF–C20A/Li : Al [PVK=poly(vinylcarbazole)], the authors observed that the 'hybrid' electroluminescence device showed much higher optical power. For the device fabricated from the hybrid material, the current decreased with increasing the silicate amount, meaning that the layered nanoparticles blocked the flow of electrons and holes. The real confinement led to an improvement of the quantum efficiency, while the external quantum efficiency of the ITO/PVK/PDOF–C20A/Li : Al (PDOF–C20A = 1 : 2 weight ratio) LED reached 1.0%, being a fivefold increase with respect to pristine PDOF device (Figure 8.9). Note that excellent results were obtained by Frey and co-workers by intercalation of PDOF within the interlayer space of layered metal dichalcogenide materials, metallic MoS_2 and semiconducting SnS_2 [73]. Such hybrids were synthesized by Li intercalation, exfoliation and restacking in the presence of the polymers.

Before introducing the following examples, it is important to make a note. Indeed, in the examples we have seen so far (Sections 4.2.1. and 4.2.2), the silicate

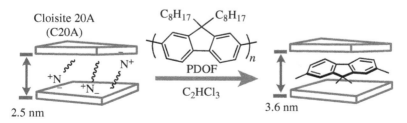

FIGURE 8.8 Schematic diagram of the PDOF intercalation.

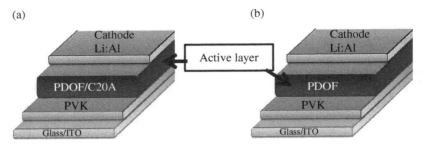

FIGURE 8.9 Hybrid optimized (a) and neat polymer (b) device architecture.

was effectively an organo-modified silicate, and the polymers were non-ionic polymers. This means that the intercalation mainly occurred by diffusion of the polymer chains within the inorganic galleries, being then physisorbed to the layer surface. In this respect, the replacement of the native silicate cations by ammonium surfactants was fundamental not only to match the polarity of the silicate surface with that of the polymer but, and especially, to enlarge the layer spacing, facilitating the polymer intercalation.

On the contrary, the study by Malik, Dana, et al. is different because native silicates were used without any pre-organic modification [74]. Rather than modifying the inorganic filler, they modified the polymer to cationic form. Specifically, they reported the synthesis and intercalation of poly[(9,9′-bis(6″-(N,N,N,-trimethylammonium)hexyl)fluorene-co-alt-1,4-phenylene)dibromide] (PFNBr) (Figure 8.10) within kaolinite silicate hybrids. Two solution intercalation/exfoliation blending methods were explored. In the first one, the materials were prepared by mixing both the kaolinite and PFNBr in a solution of N,N-dimethylformamide, followed by alternating heating and sonication process. In the second, kaolinite pre-intercalated by dimethyl sulphoxide dipolar molecules has been successfully used as precursor for the polymer intercalation.

In both cases, the intercalated/exfoliated nanocomposites exhibited superior thermal stability with respect to the pristine polymer, as demonstrated by thermogravimetric studies. The keto defect sites of the fluorene polymer unit were strongly reduced likely due to the intercalation of single PFNBr chains. The kaolinite sheets acted as barrier to exciton and oxygen diffusion, and consequently, the hybrids exhibited enhanced photostability in air compared to neat PF.

Very recently, in a continuation of the work of our research group on the *in situ* synthesis of such functional materials, we reported a facile and robust approach for fabricating hierarchically structured, blue-emitting polymer hybrids by grafting poly(styrene) incorporating π-conjugated oligo(fluorene) side chains through SI-NMP and water-templated assembly (Figure 8.11) [75].

Besides the advantages already described, control over the electronic structure and the interlayer geometrical conformation of the polymer was expected to be achieved by judicious choice of (co)monomers and by tuning the initiator grafting density and polymerization conditions.

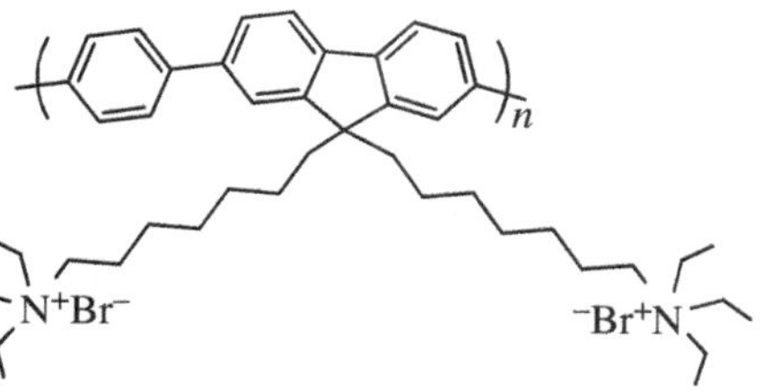

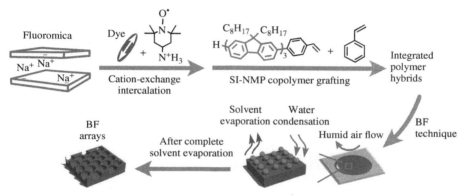

FIGURE 8.10 Cationic fluorene-based polymer.

FIGURE 8.11 Schematic and simplified illustration of the fluorene-type polymer hybrid synthesis by means of SI-NMP and breath-figure (BF) templating conditions.

The experimental results indicated that the strategy was rational and efficacious: the hybrids exhibited high solution processability, improved thermal stability and blue photoluminescence quantum yield (PL-QY) as high as 0.90, even in the solid state, which makes the materials appealing for PLEDs. Notably, we found that, under optimized conditions, the hybrids spontaneously assemble into highly ordered microporous films, where an organization of the matter at different length scales was obtained.

Indeed, the adjustment of overall hydrophilicity of the material through a proper balancing of the amounts of the components (i.e. dye, fluorene-type (co)monomer, fluoromica) led to hierarchically structured hybrids in the form of patterned films organized at two different length scales, that is, intercalated polymer chains and micrometric periodic structures. Specifically, the polymer hybrids were let self-assemble into microporous films, by applying the breath-figure (BF) templating conditions. This technique in fact allows for the realization of microporous films with cavities arranged in a packed hexagonal fashion. Films prepared with such (co)polymer hybrid materials showed extremely regular cavities having external diameter of 1.5 μm and centre-to-centre distance of 2.3 μm (Figure 8.12a and b). The hexagonal arrangement of cavities was homogeneous on the whole film, and the ordered areas without defects, except for the local discontinuities due to the presence

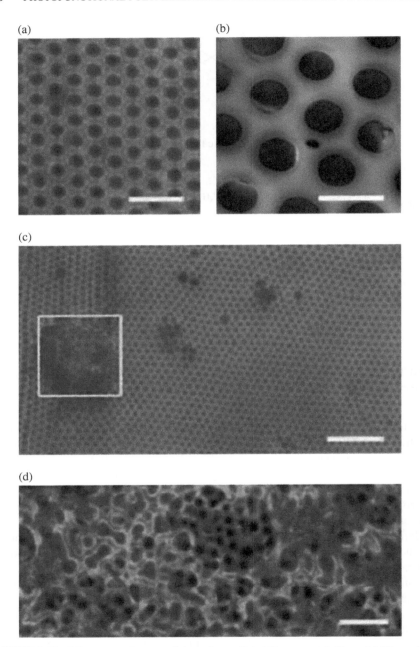

FIGURE 8.12 Microscopy images of the polymer hybrid honeycomb films. (a) Close view by fluorescence microscopy (UV excitation, scale bar 4 μm). (b) SEM micrograph (scale bar 2 μm). (c) Wider fluorescence view showing long-range order (scale bar 15 μm). (d) Fluorescence image of the bottom of film by UV excitation (scale bar 5 μm).

of fluoromica tactoids, were as wide as 1 mm^2 (Figure 8.4c). However, to a closer view (Figure 8.12c), it was possible to notice that these silicate aggregates, which exceed the thickness of the film, were coated by a thin layer of the copolymer and that this layer was patterned as well, confirming that the inorganic nanoparticles were wrapped by the polymer, as we previously observed by AFM in similar composites (Figure 8.4). This multilevel organization is certainly appealing for an optimization of polymer hybrid-based LEDs. In particular in PLEDs, since most of the light generated in the active layer of a device with conventional architecture remains entrapped as waveguide and substrate modes, the organization into patterned films is functional to introducing a perturbation in the waveguide modes, which may enhance the light outcoupling [76–78].

Remarkably, as showed in Figure 8.12d (i.e. patterned film viewed at the fluorescence microscope from the bottom, i.e. through the glass cover slip used as substrate), underneath the regularly arranged micrometric network, a second level of interconnected and disordered voids was visible. Such foam-like structure suggests that these films are, at least partially, opened at the bottom. This observation, quite uncommon for BFs, opens the way to the possible application of these films as functional membranes.

8.5 CONCLUSIONS AND PERSPECTIVES

Artificial lighting is accounting for the consumption of 20% of the global electricity. In 2008, the associated CO_2 emission was estimated to about 2000 millions of tons. The implication is that the international policies and development strategies are emphasizing the importance of energy saving and reduction of gas emission. In this context, the removal of traditional light bulbs in favour of efficient SSL able of emitting 75% or more light for the same adsorbed watts is developing quickly. Compared to incandescent lighting, SSL creates visible light with reduced heat generation. Thus, there is a tremendous interest in making available a new generation of efficient materials, and some of the polymer/layered silicate hybrid materials described in this chapter are appealing and challenging candidates for SSL technologies. However, to date, the application of such hybrids in device is still limited to few examples. To make available hybrids for commercialization, some key technical issues should be properly addressed, and a deeper understanding of the interaction between the inorganic and the organic components is fundamental.

From our point of view, the chemical approach is the key point to fabricate efficient multicomponent hybrids, and the bottom-up one may be the best practice to synthesize integrated LEP/silicate hybrids, with respect to solution and melt compounding, which are conducted directly with preformed polymers.

In situ fabrication methods have been progressing day by day along with advances in polymerization chemistry. The key advantages of such methods are the superior diffusion/intercalation ability of the (co)monomers, their controlled polymerization propagation behaviour and hence, the possibility to tune the (co)polymer microstructure and morphology by copolymerization chemistry or simply by changing the reaction conditions. Thus, it is expected that the *in situ* fabrication of such functional hybrids will continue to be used to exploit different potentialities. We believe that in the near future, much effort will be directed to the *in situ* preparation of integrated layered silicate-based hybrids comprising of intercalated donors and acceptors that provide the transfer of excited-state energy. The assembling/(co)intercalation of LEPs in between silicate layers, pre-intercalated with red- and green-emitting dyes, should provide the fabrication of integrated hybrids with hierarchical control over interfaces, structure and morphology, possibly taking advantage of a cascade energy transfer from the high band-gap polymer to the intercalated dyes, thus leading to PL-QY increase and emission colour tunability.

In conclusion, the growing interest in the synthesis of new materials for SSL technologies could offer new opportunities for the development of nano-manufacturing processes to fabricate polymer hybrids, including 'white' layered silicate host/filler.

REFERENCES

1. Pinnavaia, T. J., Beale, G. W., Polymer-Clay Nanocomposites; John Wiley & Sons, Inc.: New York, **2000**.
2. Brindley, G. W., Brown, G., Crystal Structure of Clay Minerals and Their X-Ray Identification; Mineralogical Society: London, **1980**.
3. Bailey, S. W., Reviews in Mineralogy; Virginia Polytechnic Institute and State University: Blacksburg, VA, **1984**.
4. Mittal, V., ed., In situ Synthesis of Polymer Nanocomposites; Wiley-VCH Verlag GmbH: Weinheim, **2012**.
5. Utracki, L. A., Sepehr, M., Boccaleri, E., Polym. Adv. Technol. **2007**, *18*, 1.
6. Alexandre, M., Dubois, P., Mater. Sci. Eng. **2000**, *28*, 1.
7. Ruiz-Hitzky, E., Aranda, P., Darder, M., Ogawa, M., Chem. Soc. Rev. **2011**, *40*, 801.
8. Sasabe, H., Kido, J., J. Mater. Chem. C. **2013**, *1*, 1699.
9. Sanchez, C., Lebeau, B., Chaput, F., Boilot, J.-P., Adv. Mater. **2003**, *15*, 1969.
10. Holder, E., Tessler, N., Rogach, A. L., J. Mater. Chem. **2008**, *18*, 1064.
11. Theng, B. K. G., The Chemistry of Clay-Organic Reactions; John Wiley & Sons,Inc.: New York, **1974**.
12. Pinnavaia, T. J., Science **1983**, *220*, 365.
13. Vaia, R. A., Teukolsky, R. K., Giannelis, E. P., Chem. Mater. **1994**, *6*, 1017.
14. Morgan A. B., Gilman J. W., J. Appl. Polym. Sci. **2003**, *87*, 1329.
15. Pavlidou, S., Papaspyrides, C. D., Prog. Polym. Sci. **2008**, *33*, 1119.
16. Ray, S. S., Okamoto, M., Prog. Polym. Sci. **2003**, *28*, 1539.
17. Podsiadlo, P., Kaushik, A. K., Arruda, E. M., Wass, A. M., Shim, B. S., Xu, J., Nandivada, H., Pumplin, B. G., Lahann, J., Ramamoorthy, A., et al. Science **2007**, *318*, 80.
18. Heinemann, J., Reichert, P., Thomann, R., Mülhaupt, R., Macromol. Rapid Commun. **1999**, *20*, 423.
19. Leone, G., Bertini, F., Canetti, M., Boggioni, L., Conzatti, L., Tritto, I., J. Polym. Sci. Part A Polym. Chem. **2009**, *47*, 548.

20. Leone, G., Boglia, A., Bertini, F., Canetti, M., Ricci, G., Macromol. Chem. Phys. **2009**, *210*, 279.

21. Leone, G., Boglia, A., Bertini, F., Canetti, M., Ricci, G., J. Polym. Sci. Part A Polym. Chem. **2010**, *48*, 4473.

22. Tasdelen, M. A., Kreutzer, J., Yagci, Y., Macromol. Chem. Phys. **2009**, *210*, 1867.

23. Ogawa, M., Kuroda, K., Chem. Rev. **1995**, *95*, 399.

24. López Arbeloa, F., Martínez Martínez, V., Arbeloa, T., López Arbeloa, I., J. Photochem. Photobiol. C **2007**, *8*, 85.

25. Iyi, N., Sasai, R., Fujita, T., Deguchi, T., Sota, T., López Arbeloa, F., Kitamura, K., Appl. Clay. Sci. **2002**, *22*, 125.

26. Kobayashi, T., ed., J-Aggregates; World Scientific Publishing: Singapore, **1996**.

27. Mal'tsev, E., Lypenko, D. A., Shapiro, B. I., Brusentseva, M. A., Milburn, G. H. W., Wright, J., Hendriksen, A., Berendyaev, V. I., Kotov, B. V., Vannikov, A. V., Appl. Phys. Lett. **1999**, *75*, 1896.

28. Hannah, K. C., Armitage, B. A., Acc. Chem. Res. **2004**, *37*, 845.

29. Walker, B. J., Dorn, A., Bulović, V., Bawendi, M. G., Nano Lett. **2011**, *11*, 2655.

30. Yamamoto, Y., Fukushima, T., Suna, Y., Ishii, N., Saeki, A., Seki, S., Tagawa, S., Taniguchi, M., Kawai, T., Aida, T., Science **2006**, *314*, 1761.

31. Kim, K. H., Bae, S. Y., Kim, Y. S., Hur, J. A., Hoang, M. H., Lee, T. W., Cho, M. J., Kim, Y., Kim, M., Jin, J.-I., et al., Adv. Mater. **2011**, *23*, 3095.

32. Mering, J. J., Smectites; Springer-Verlag: New York, **1975**.

33. Kaneko, Y., Iyi, N., Bujdák, J., Sasai, R., Fujita, T., J. Colloid Interface Sci. **2004**, *269*, 22.

34. Bujdák, J., Martínez Martínez, V., López Arbeloa, F., Iyi, N., Langmuir **2007**, *23*, 1851.

35. Bujdák, J., Iyi, N., Kaneko, Y., Czímerová, A., Sasai, R., Phys. Chem. Chem. Phys. **2003**, *5*, 4680.

36. Bujdák, J., Komadel, P., J. Phys. Chem. B. **1997**, *101*, 9065.

37. Sasai, R., Iyi, N., Fujita, T., López Arbeloa, F., Martínez Martínez, V., Takagi, K., Itoh, H., Langmuir **2004**, *20*, 4715.

38. Salleres, S., López Arbeloa, F., Martínez Martínez, V., Arbeloa, T., López Arbeloa, I., Langmuir **2010**, *26*, 930.

39. Bujdák, J., Iyi, N., Chem. Mater. **2006**, *18*, 2618.

40. Olivero, F., Carniato, F., Bisio C., Marchese, L., J. Mater. Chem. **2012**, *22*, 25254.

41. Lotsch, B. V., Ozin, G. A., ACS Nano **2008**, *2*, 2065.

42. López Arbeloa, F., Martínez Martínez, V., Chem. Mater. **2006**, *18*, 1407.

43. Martínez Martínez, V., López Arbeloa, F., Bãnuelos Prieto, J., Lópe Arbeloa, I., Chem. Mater. **2005**, *17*, 4134.

44. Glinel, K., Laschewski, A., Jonas, A. M., J. Phys. Chem. B **2002**, *106*, 11246.

45. Syed Arshad, S. A. Hussain, Schoonheydt, R. A., Langmuir **2010**, *26*, 11870.

46. Ras, R. H. A., Umemura, Y., Yamagishi, A., Schoonheydt, R. A., Phys. Chem. Chem. Phys. **2007**, *9*, 918.

47. Leone, G., Giovanella, U., Porzio, W., Botta, C., Ricci, G., J. Mater. Chem. **2011**, *21*, 12901.

48. Giovanella, U., Leone, G., Ricci, G., Virgili, T., Suarez Lopez, I., Rajendran, S. K., Botta, C., Phys. Chem. Chem. Phys. **2012**, *14*, 13646.

49. Rainó, G., Stöferle, T., Park, C., Kim, H. C., Chin, I. J., Miller, R. D., Mahrt, R. F., Adv. Mater. **2010**, *22*, 3681.

50. Busby, M., Blum, C., Tibben, M., Fibikar, S., Calzaferri, G., Subramaniam, V., De Cola, L., J. Am. Chem. Soc. **2008**, *130*, 10970.

51. Calzaferri, G., Langmuir **2012**, *28*, 6216.

52. Xu, W., Guo, H., Akins, D. L., J. Phys. Chem. B **2001**, *105*, 7686.

53. Wang, M., Silva, G. L., Armitage, B. A., J. Am. Chem. Soc. **2000**, *122*, 9977.

54. Kometani, N., Tsubonishi, M., Fujita, T., Asami, K., Yonezawa, Y., Langmuir **2001**, *17*, 578.

55. Lu, L., Jones, R. M., McBranch, D., Whitten, D., Langmuir **2002**, *18*, 7706.

56. Leone, G., Giovanella, U., Bertini, F., Porzio, W., Meinardi, F., Botta, C., Ricci, G., J. Mater. Chem. C **2013**, *1*, 1450.

57. Brinks, M. K., Studer, A., Macromol. Rapid Commun. **2009**, *30*, 1043.

58. Fan, X., Xia, C., Advincula, R. C., Langmuir **2003**, *19*, 4381.

59. Pyun, J., Matyjaszewski, K., Chem. Mater. **2001**, *13*, 3436.

60. Tebben, L., Studer, A., Angew. Chem. Int. Ed. **2011**, *50*, 5034.

61. Scherf, U., Neher, D., Advances in Polymer Science, Vol. 212, Springer: Berlin, **2008**.

62. Chen, S. A., Lu, H. H., Huang, C. W., Adv. Polym. Sci. **2008**, *212*, 49.

63. Burroughes, J. H., Bradley, D. D. C., Brown, A. R., Marks, R. N., Mackay, K., Friend, R. H., Burns, P. L., Holmes, A. B., Nature **1990**, *347*, 539.

64. Lee, T.-W., Park, O. O., Appl. Phys. Lett. **2000**, *76*, 3161.

65. Parker, I. D., J. Appl. Phys. **1994**, *75*, 1659.

66. Li, F., Tang, H., Anдеregg, J., Shinar, J., J. Appl. Phys. Lett. **1997**, *70*, 1233.

67. Lee, T.-W., Park, O. O., Yoon, J., Kim, J. J., Adv. Mater. **2001**, *13*, 211.

68. Zheng, M., Ding, L., Lin, Z., Karasz, E., Macromolecules **2002**, *35*, 9939.

69. Xie, L. H., Yin, C. R., Lai, W. Y., Fan, Q. L., Huang, W., Prog. Polym. Sci. **2012**, *37*, 1192.

70. Kappaun, S., Slugovc, C., List, E. J. W., Adv. Polym. Sci. **2008**, *212*, 273.

71. Monkman, A., Rothe, C., King, S., Dias, F., Adv. Polym. Sci. **2008**, *212*, 187.

72. Park, H., Lim, Y. T., Park, O. O., Kim, J. K., Yu, J.-W., Kim, Y. C., Adv. Funct. Mater. **2004**, *14*, 377.

73. Aharon, E., Albo, A., Kalina, M., Frey, G., Adv. Funct. Mater. **2006**, *16*, 980.

74. Chakraborty, C., Dana, K., Malik, S., J. Colloid Interf. Sci. **2012**, *368*, 172.

75. Leone, G., Giovanella, U., Bertini, F., Hoseinkhani, S., Porzio, W., Ricci, G., Botta, C., Galeotti, F., J. Mater. Chem. C. 2013,1,6585:10.1039/C3TC31122H.

76. Bai, Y., Feng, J., Liu, Y. F., Song, J. F., Simonen, J., Jin, Y., Chen, Q. D., Zi, J., Sun, H. B., Org. Electron. **2011**, *12*, 1927.

77. Koo, W. H., Jeong, S. M., Araoka, F., Ishikawa, K., Nishimura, S., Toyooka, T., Takezoe, H., Nat. Photonics **2010**, *4*, 222.

78. Bocksrocker, T., Hoffmann, J., Eschenbaum, C., Pargner, A., Preinfalk, J., Maier-Flaig, F., Lemmer, U., Org. Electron. **2013**, *14*, 396.

9

RIGID PHOSPHONIC ACIDS AS BUILDING BLOCKS FOR CRYSTALLINE HYBRID MATERIALS

JEAN-MICHEL RUEFF[1], GARY B. HIX[2] AND PAUL-ALAIN JAFFRÈS[3]

[1] ENSICAEN, CNRS UMR 6508, Laboratoire CRISMAT, Caen, France
[2] School of Science and Technology, Nottingham Trent University, Nottingham, United Kingdom
[3] CEMCA, CNRS UMR 6521, Université Européenne de Bretagne, Université de Brest, Brest, France

9.1 INTRODUCTION

Metal–organic framework (MOF) or coordination polymers [1] are constructed by the assembly of a selected metallic precursor (usually from a salt) and an organic molecule possessing suitable functional groups to produce coordination complexes (pyridine, nitrile, etc.) or iono-covalent bonds as occurring with carboxylate, phosphonic or sulphonic acid functional groups. While a significant number of the hybrid materials reported so far were synthesized from carboxylate and polycarboxylate derivatives, likely due to the commercial availability of many of the precursors, this chapter is dedicated to the use of synthetic non-commercial phosphonic acid derivatives as organic precursors. The main difference between carboxylate functional groups and phosphonic acids is that the latter possess a tetrahedral geometry around the central atom (phosphorus atom) and the higher number of oxygen atoms in phosphonic acid functional group leads to an increase in the number of possible coordination modes as schematically illustrated in Figure 9.1.

Which of the possible coordination modes is adopted is deeply influenced by the nature of the metallic salt, but other parameters (synthetic conditions, thermal rearrangement [2]) can also influence the formation of the coordination network. Since many previous review articles or book chapters have been dedicated to the use of phosphonic

Tailored Organic–Inorganic Materials, First Edition. Edited by Ernesto Brunet, Jorge L. Colón and Abraham Clearfield.

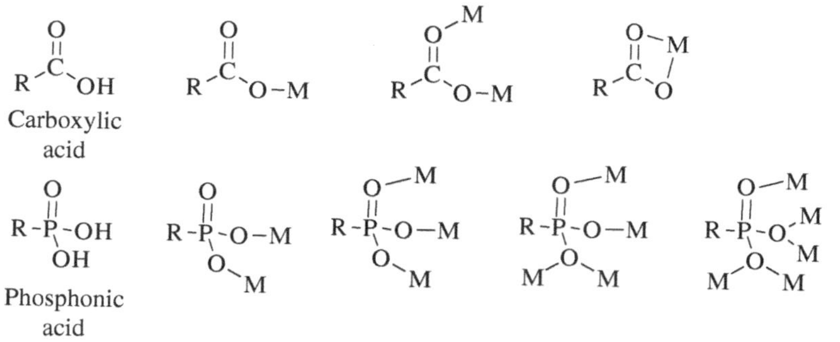

FIGURE 9.1 Selection of few representations of coordination modes for carboxylic acid and phosphonic acid functional groups.

acid derivatives for the production of hybrid structures [3] or focused on a specific application including luminescence [4], porous materials [5] or chirality [6], we will not further detail these points in this chapter. The selected focus for this chapter is the use of rigid phosphonic acids as precursors of hybrid materials. If monofunctional rigid precursors (e.g. phenylphosphonic acid) usually produce low-dimensional structures, the incorporation of additional functional groups, having the capacity to form iono-covalent bond with the inorganic network, can increase the dimensionality of the materials and can produce original structures as illustrated in the following. A distinction must be made within the scope of rigid polyfunctional organic precursors since the general term covers both homo-polyfunctional (all the functional groups are identical, herein polyphosphonic acid derivatives) and hetero-polyfunctional molecules (one or several phosphonic acid groups and one or many additional groups such as carboxylic acid, sulphonic acid, amine, nitro groups, and so on are present on a rigid scaffold).

The concept of the rigidity of an organic building block must be first defined. Basically, the degree of rigidity of an organic precursor will be correlated with the number of conformations that this compound can adopt. Accordingly, the presence of an aromatic ring and the limiting of sp^3-hybridized carbon atoms will reduce the number of conformations and therefore produce what we can call rigid organic precursors. As shown schematically in Figure 9.2, the direct connection of the phosphonic acid functional group (or any other functional groups) to a sp^2-hybridized atom of an aromatic ring will produce the most rigid organic precursors (compound A). The use of fused aromatic rings also keeps the rigidity roughly unchanged, while for biphenyl derivatives (compound B), the incorporation of sp^3-hybridized carbon atom (compounds C and D) or the absence of aromatic ring (compound E) increases the number of conformations and finally (as in the case of compound E) defines, what we call herein, non-rigid organic precursors.

In this chapter, we will discuss only hybrid structures synthesized from organic compounds in which the functional groups are bonded directly to a (hetero)aromatic ring. As will be illustrated in this chapter, rigid polyfunctional precursors can

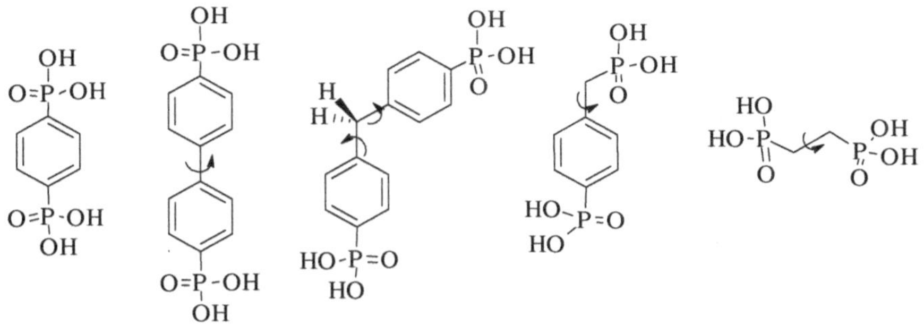

FIGURE 9.2 Illustration of diphosphonic acid derivative with a decreased degree of rigidity (from a to e). The arrow indicates the conformation involving C–C bond rotation.

significantly influence the topology of the hybrid, and original topologies including non-centrosymmetric materials have been reported. Moreover, hybrids built from rigid organic building blocks often possess an enhanced thermal stability that can be of a great interest for some applications (e.g. luminescence or catalysis).

This chapter will first summarize the usual methods employed for the synthesis of rigid phosphonic acid (phosphonic acid group directly bonded to an aromatic ring), which is a key step in designing original organic precursors and consequently original hybrids. Then, the method and experimental procedure to produce hybrid materials will be summarized. In the next two sections, we will illustrate hybrids obtained from polyphosphonic acid derivatives (homo-polyfunctional precursors) and, subsequently, hetero-polyfunctional precursors. In these two sections, we will compare the structure of the hybrids obtained from different region isomers, and we will address the question of chemoselectivity when hetero-polyfunctional precursors are employed as organic substrates. For some organic precursors, the consequences of changing the metallic precursor upon the structure of the hybrid will be also reported. A special section will focus on the use of heteroaromatic unit as rigid platform, and the final section will illustrate some recent applications.

Despite this chapter being focused on hybrids constructed from rigid precursors possessing at least one phosphonic acid functional group, it is difficult to cover all the studies reported so far. As such, we have selected the most representative results to illustrate the interest and the original topology but also to point out the applications and the perspectives opened by this family of hybrid materials.

9.2 OVERVIEW OF THE SYNTHESIS OF RIGID FUNCTIONAL AROMATIC AND HETEROAROMATIC PHOSPHONIC ACIDS

Phosphonic acids are generally synthesized in a two-step sequence in which the first step is to synthesize a phosphonate and the second to hydrolyse the alkyl phosphonate functional group to yield the phosphonic acid derivative. Different methodologies can be applied for the synthesis of Csp^2–P bond including Friedel–Crafts reaction [7], Grignard reagent or organolithium precursors (the reader interested with an overview of these methods can refer to a specialized book [8]). Nevertheless, methods that can be applied at a large scale (more than 10 g of phosphonate) are more appropriate for the synthesis of precursors of hybrid materials. Most of these methods involve the phosphonation of an aromatic halide in the presence of a catalyst as shown in Figure 9.3. The first method (method 1), which was reported initially by Tavs [9], is based on the use of nickel chloride and trialkyl phosphite at high temperature. It must be noted that this reaction, also identified as nickel-catalysed Arbuzov reaction, must be carried out with care when applied on a large scale. Indeed, the reaction proceeds only when the reaction mixture is heated at a temperature greater than 160 °C, but the reaction itself is highly exothermic. Accordingly, the addition of trialkyl phosphite to a suspension of hot aryl halide and nickel bromide must be slow. We found it beneficial (in terms of reproducibility and yield) to replace nickel chloride with nickel bromide. Moreover, the addition of a high boiling point solvent (e.g. mesitylene) was helpful when the starting aryl halide is a solid. Accordingly, phosphonation or polyphosphonation of aromatic (e.g. synthesis of 1,3,5-benzenetriphosphonate [10]) was achieved in high yield (usually >80%). Several optimization procedures of Tavs's method were proposed including microwave heating [11]. Hirao and co-workers [12] proposed an alternative to the Tavs's method that was based on the use of dialkyl phosphite (instead of trialkyl phosphite), in which a palladium coordination complex as catalyst ($Pd(PPh_3)_4$ and triethylamine was used (method 2, Figure 9.3). This method was further improved by the use of $Pd(OAc)_2$ as pre-catalyst and a diphosphine as ligand (1,1'-bis(diphenylphosphino) ferrocene, dppf) [13] or by the use of microwave activation [14]. Finally, the most recent modification of this methodology involved the replacement of the catalyst (Ni, Pd) by a copper salt (method 3). This approach was first reported by Buchwald et al. [15] and subsequently applied to aryl iodides [16] and boronic acids [17].

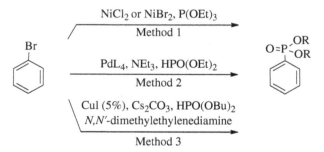

FIGURE 9.3 Common methodologies for the synthesis of aromatic phosphonate from aryl bromide.

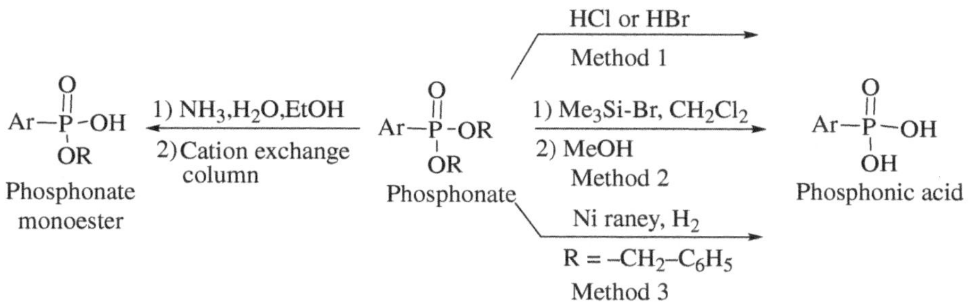

FIGURE 9.4   Usual methods for the synthesis of phosphonic acids from phosphonates.

The synthesis of phosphonic acids from phosphonates can be achieved according to several different methods (Figure 9.4). The most useful method, which is widely employed, is based on refluxing the selected dialkyl phosphonate in aqueous hydrochloric or hydrobromic acid solution (method 1, Figure 9.4). According to this procedure, the final phosphonic acid can be recovered after removing the excess of acid and water solution. Recrystallization can be also achieved (e.g. 3-phosphonobenzoic acid **12-H₃**) [18]. However, in many cases, it is difficult to produce phosphonic acid without any trace of water. Indeed, it was shown that phosphonic acids can be very hygroscopic and can even crystallize with water [19]. When this situation occurs, elemental analysis indicates the number of water molecules present within the phosphonic acid derivative. These hydrated materials can then be used to produce hybrids with controlled stoichiometry. When using hydrochloric acid as reagent to hydrolyse phosphonate, degradation of the organic molecule is sometimes observed [20]. As an example, we found that the hydrolysis of diethyl 4-hydroxybenzenephosphonate produced simultaneously phenol and a phosphate due to phosphorus–carbon bond breaking. With such acid-sensitive substrates, a softer method (method 2, Figure 9.4), originally proposed by McKenna et al. [21], involving the addition of bromotrimethylsilane followed by a methanolysis step, is recommended. This method, which is also quantitative, can be viewed as a trans-esterification reaction, thus producing a bis-trimethylsilylphosphonate as intermediate that is subsequently transformed to a phosphonic acid after methanolysis; the methanolysis also produces methoxytrimethylsilane that is easily removed by evaporation. Other methods such as the debenzylation of bis-benzylphosphonate can be also used [22] (method 3, Figure 9.4) but are much less commonly employed than the two previous methods mentioned earlier. To conclude on the transformation of phosphonate to phosphonic acid, it must be noted that *in situ* hydrolysis, during the synthesis of the hybrid, constitutes an alternative. Nevertheless, the crystallinity of the produced materials depends on the nature of the reaction media as exemplified with zirconium methylphosphonate materials [23]. Finally, it is worth noting that mono-hydrolysis of dialkylarylphosphonate can be achieved with LiN₃ in DMF [24] or with ammonia [25] in water/ethanol mixture of solvent as shown in Figure 9.4.

9.3   SYNTHETIC METHODS TO PRODUCE PHOSPHONIC-BASED HYBRIDS

A vast number of studies realized to date on hybrid materials (36,887 results for the topic *hybrid materials* and 16,024 for the topic *MOFs* in Web of Knowledge – consultation October 2013) have allowed us to establish and to understand the influence of the synthetic conditions on the architectures of the hybrids (i.e. their dimensionalities: one-dimensional (1D), two-dimensional (2D) or three-dimensional (3D)) and on their properties in order (i) to control and optimize the link between structures and properties and (ii) to offer the chemist the opportunity to create 'à façon' an unlimited number of materials. The different synthetic methods (soft chemistry, sol–gel, solvo- or hydrothermal syntheses, etc.) are extremely dependent on the nature (shape, rigidity, flexibility) and of the reactivity (number of reactive functional groups, type of reactive functional groups, etc.) of the organic precursors in the reaction media [26]. Naturally, similar arguments can also be considered regarding the inorganic precursors involved in this synthesis. Thus, the organic building block will form the organic subnetwork and act as a linker or connector between the inorganic subnetworks made, for instance, of planes, columns, isolated cations or small inorganic building blocks. A classification of these hybrid materials has been established by Sanchez and Ribot [27], which is based on the nature and the strength of the binding interactions between the inorganic and organic subnetworks. For their part, Cheetham, Rao and Feller proposed a description of the global dimensionality of these materials considering the specific dimensionality of each subnetwork component of these materials (1D chain or column, 2D layer, 3D network). Finally, each subnetwork, organic or inorganic, can be chosen as possessing an intrinsic physical property that can be combined with the other one in order to obtain a final multifunctional material.

Keeping all these points in mind, polyfunctional rigid phosphonic and/or carboxylic acids are usually engaged in hydro- or solvothermal synthesis. This technique of synthesis is preferentially employed for the formation of zeolites [28], mesoporous materials [29], MOFs [30] or hybrid materials [31] because it exhibits the clear advantage of leading to well-crystallized final materials, obtained as powder or mono-crystals. This last point is a desirable target in order to correlate the observed physical properties to the molecular structure of the material. Moreover, due to the large number of studies involving hydro- or solvothermal synthesis, several complementary strategies, based on the study of the formation of structural building units (SBUs), have been envisaged to predict and to understand the architecture of the final materials such as computational modelling and structure prediction [32] or *in situ* NMR [33]. Different equipment can be used for the synthesis depending on the nature of the synthesized material (inorganic, organic or hybrid materials), the range of working pressure (mild or high pressure) or the heating mode (classical heating, microwave heating). These last points are generally closely related. The general procedure for this type of synthesis consists of mixing all the precursors in the chosen solvent, water for hydrothermal synthesis or organic solvent (alcohol, DMF, DMSO, etc.) for solvothermal conditions, and adjusting several parameters including concentration, temperature, time of synthesis, addition of mineralizing agent, nature of the inorganic precursor or value of the pH, in order to promote crystal growing. For this last parameter, NaOH or KOH solution, amine or urea is frequently used to adjust the initial pH or modify it during the synthesis. In order to increase the crystallinity of the final product, some authors have [34] described the use of a layered-solvothermal method that consists of deposing a layer of a solution, which contains one of the precursors, on a layer of a solution with the other precursor(s) avoiding, in a first step, the rapid mixing of all the reactants in the mother solution. To conclude this part and as already mentioned, further different methods of synthesis have been described and used in order to obtain hybrid materials from phosphonic and/or carboxylic acid building blocks such as soft chemistry, organometallic routes, ionothermal synthesis (where an ionic liquid is used as solvent) and others, but they will not be developed furthermore here since they were already extensively described in the literature [35] and would merit to be developed in a more substantial chapter.

9.4 HYBRID MATERIALS FROM RIGID DI- AND POLYPHOSPHONIC ACIDS

The initial works dedicated to the use of aromatic phosphonic acid **1** (Figure 9.5) were reported by Alberti [36] and Clearfield [37].

The first hybrid structure $Zr(O_3P–C_6H_5)_2$ featured inorganic layers (ZrO_6) and phenyl groups located in the interlayer space (Figure 9.6).

Many other studies reported the structure of hybrid materials synthesized from phenylphosphonic acid or alkylphosphonic acids. These works, which were previously reviewed [38], are not detailed in this section. Indeed, we have selected to report herein only the hybrid materials synthesized from di- or polyphosphonic acid functional groups possessing a rigid structure. Accordingly, these functional groups are directly bonded to an aromatic ring (hybrid materials synthesized from heteroaromatic derivatives are reported in Section 9.6). All the di- and polyphosphonic acid precursors of hybrid materials reviewed herein are summarized in Figure 9.6. Benzene-1,4-diphosphonic **2** (Figure 9.6) was first employed with the aim of connecting two planar inorganic layers (such as those observed in materials constructed from the monophosphonic acid benzenephosphonic acid **1** [39]) by an organic bridge. The incorporation of this organic pillar was first investigated by Clearfield [40] who intercalated this compound into a preformed inorganic 2D material $ZrCu(PO_4)_4$. This exchange strategy was then adapted to introduce diphosphonic acid **2** in γ-$ZrPO_4H_2PO_4 \cdot 2H_2O$ acting as a preformed inorganic starting material [41]. Accordingly, with this topotactic exchange, one phosphate group from the inorganic structure was replaced by one phosphonic acid functional group leading to the formation of a hybrid structure in which the pillar is randomly distributed within the structure. Hybrid materials can also be produced from a metallic salt and a phosphonic acid derivative in water solution. The synthetic procedure was first reported by Clearfield et al. [42] who produced a zirconium phosphonate hybrid by reaction of a zirconium salt and two phosphorus-based precursors (diphosphonic

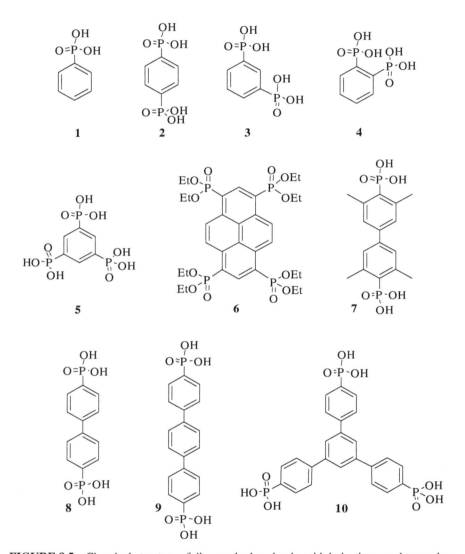

FIGURE 9.5 Chemical structure of di- or polyphosphonic acid derivatives used to produce hybrid materials.

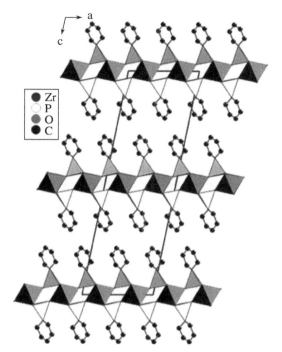

FIGURE 9.6 Crystal structure of Zr(O$_3$P–C$_6$H$_5$)$_2$, monoclinic, *C2/c,* a = 9.0985(5) Å, b = 5.4154(3) Å, c = 30.235(2) Å, β = 101.233(1)°. Adapted from Ref. [37].

acid **2** and phosphate), which were randomly distributed within the structure. This material possessed a layered structure, but its crystal structure was not resolved from X-ray diffraction due to the low crystallinity of the sample. The crystallinity can be improved when zirconium salt was used by the addition of HF in the reaction media. Indeed, the stable ZrF$_6^{2-}$ anion is then slowly dissociated in the presence of phosphonic acid. This slow crystallization process explains the better crystallinity observed. The synthesis of hybrids, based on the reaction of both diphosphonic acid and phosphorous acid with a zirconium salt, was further studied by Alberti et al. [43] with the aim of producing mesoporous materials. Accordingly, structures featuring the general formula Zr(O$_3$P–H)$_x$(O$_3$P–C$_6$H$_4$–PO$_3$)$_y$ were produced from benzene-1,4-diphosphonic acid **2**. This rigid diphosphonic precursor acts as a template, thus creating pores within the structure of the hybrid material (volume of the pores: 0.3–0.7 cm^3 g^{-1}). Another strategy for the production of microporous materials, investigated by Alberti et al. [44], involved mixing rigid diphosphonic acid with a bis-phosphate possessing the same length. The resulting hybrid was placed under hydrolytic conditions to selectively hydrolyse the phosphate functional groups (P–O bonds are more readily cleaved than the P–C bond present in phosphonic acid), thus creating pores within the structure. A further strategy studied by Alberti et al. used the biphenyl-bisphosphonic acid **7** possessing methyl groups in ortho-position relative to the phosphonic acid functional group; these methyl groups aim to produce, due to steric hindrance, a homogeneous distribution of the organic precursors in the final structure of the hybrid. After hydrolysis of the phosphate groups, porous materials were characterized [45]. These two strategies are however not frequently used since both require additional synthetic efforts to produce either tetramethyl-bis-phenyl derivatives or bisphosphonic acid and bis-phosphate of similar lengths. A highly porous zirconium phosphonate was also characterized by the use of benzene-1,4-diphosphonic acid and zirconium (IV) salt (used in excess) in an organic solvent. This material was then post-functionalized with sulphonic acid functions to produce a hybrid structure with acidic properties that was tested as an acid catalyst [46]. The production of hybrid structures possessing significant long-range crystal-linity allowed structural resolution from X-ray diffraction data, which were reported when other metallic salts were mixed with benzene-1,4-diphosphonic acid **2**. Accordingly, copper- [47] and zinc-based [48] hybrids were fully characterized, as shown in Figure 9.7, from X-ray powder diffraction (XRPD) patterns. The structure of these materials was quite similar to the layered materials $Zn(O_3P-C_6H_5) \cdot H_2O$ [49] or $Cu(O_3P-C_6H_5) \cdot H_2O$ [50] (zinc atoms were octahedrally coordinated, while copper

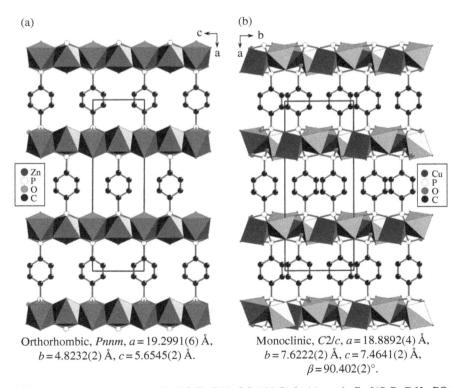

(a)

(b)

Orthorhombic, *Pnnm*, $a = 19.2991(6)$ Å, $b = 4.8232(2)$ Å, $c = 5.6545(2)$ Å.

Monoclinic, *C2/c*, $a = 18.8892(4)$ Å, $b = 7.6222(2)$ Å, $c = 7.4641(2)$ Å, $\beta = 90.402(2)°$.

FIGURE 9.7 Structure of $Zn_2[(O_3P-C_6H_4-PO_3)(H_2O)_2]$ (a) and $Cu_2[(O_3P-C_6H_4-PO_3)(H_2O)_2]$ (b). Adapted from Refs (a) [48] and (b) [47].

atoms adopted a distorted square pyramidal geometry); the main differences arise from the connection of the two inorganic planes by the bifunctional molecules when the benzene-1,4-diphosphonate anion is employed.

An isostructural compound to $Zn_2[(O_3P-C_6H_4-PO_3)(H_2O)_2]$ was characterized when benzenebisphosphonic acid 2 was replaced with 4,4′-biphenylenebis(phosphonic acid) 8 [51]. This change leads to an increase in the distance between the two inorganic planes from 19.299(6) to 27.904(1) Å. The increase of the size of the rigid aromatic scaffold, which was aimed at producing porous materials, was studied by the use of 4,4′-terphenylbis(phosphonic acid) 9 as organic precursor. Compound 9 was successfully reacted with zirconium salt (used in excess) in a mixture of organic solvents (DMSO/EtOH) to produce a microporous hybrid [52]. This compound was subsequently sulphonated with SO_3 to yield a strong Brönsted acid material. γ-Zirconium phosphate including the same terphenylbisphosphonic acid 9 also produced porous materials that possess the capacity to absorb to $74 \, cm^3 g^{-1} \, H_2$ at 650 torr and 77 K [53].

Besides the use of zirconium salt to produce zirconium phosphonic acid materials, a variety of hybrids were produced using different metallic salts other than those already mentioned earlier (Cu, Zn). Tin-based materials were reported by Clearfield et al. These materials include either Sn(II) [54] or Sn(IV) [55] species. In both cases, 3D materials were produced in which the rigid organic molecules linked two inorganic planes. In the case of Sn(IV)-based materials, the synthesis was achieved under solvothermal conditions that make use of either a mixture of water/ethanol or water/DMSO. In both cases, porous materials were characterized (pore size 8–20 Å), but with the water/ethanol solvent mixture, the distribution of the pore size was larger. These materials were also tested as catalysts for the oxidation of aldehyde (anisaldehyde) with hydroperoxide. These experiments indirectly demonstrate the stability of rigid benzene-1,4-diphosphonique 2 even under severe oxidative conditions.

Transition metal-based hybrids, which contain benzene-1,4-diphosphonic acid, were also reported. Zubieta et al. [56] first reported the use of this ligand 2 and its regioisomers 3 and 4 (respectively, 1,3 and 1,2 substituted) under hydrothermal conditions, producing oxovanadium-arylphosphonate having different structures from the lamellar structures reported by these authors for the V_xO_y/alkyldiphosphonate series [57]. Using the 1,4-regioisomer (compound 2) resulted in the formation of a layered structure made of unconnected VO_6 polyhedra, while the 1,3-regioisomers (precursors 3) produce a complex 3D structure in which the VO_6 polyhedra are connected by the corners to form chains. In this work, Zubieta et al. also reported hybrid structures synthesized from di- and triphosphonic acids possessing biphenyl-type structures (bis-phenyl 8 or benzene-1,3,5-tris(phenyl)-4,4′,4″triphosphonic acid 10) (Figure 9.8). The structure of the material obtained from 8 featured an inorganic network similar to that found when 2 was used as organic precursor. Indeed, in both of these cases, a layered structure was observed, while in the material made from 10, trinuclear VO_6 clusters were connected by the ligands to form a 2D structure. These results clearly demonstrate that the oxovanadium-arylphosphonate materials possess a structure distinct from those reported for oxovanadium-alkyldiphosphonate. As already mentioned, V_xO_y/alkyldiphosphonate compounds present a lamellar structure in which the interlayer spacing depends on the length of the alkyl chain. Thus, three

(a)   (b)   (c)

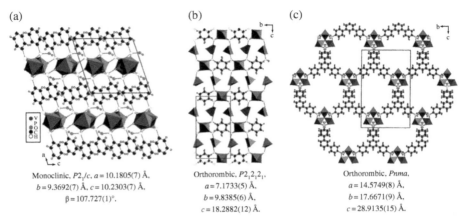

Monoclinic, $P2_1/c$, $a = 10.1805(7)$ Å,   Orthorombic, $P2_12_12_1$,   Orthorombic, $Pnma$,
$b = 9.3692(7)$ Å, $c = 10.2303(7)$ Å, $a = 7.1733(5)$ Å, $a = 14.5749(8)$ Å,
$\beta = 107.727(1)°$. $b = 9.8385(6)$ Å, $b = 17.6671(9)$ Å,
    $c = 18.2882(12)$ Å. $c = 28.9135(15)$ Å.

FIGURE 9.8 Structure of oxovanadium-based hybrids synthesized from benzene-1,4-diphosphonic acid **2** ($VO(HO_3P-C_6H_4-PO_3H)$): (a) benzene-1,3-diphosphonic acid **3** ($[V_2O_2(H_2O)_2(O_3P-C_6H_4-PO_3)]\cdot1.5H_2O$ (b) benzene-1,3,5-tris(phenyl)-4,4',4''triphosphonic acid **10** ($[V_3O_3(OH)((HO_3P-C_6H_4-)C_6H_3)_2]\cdot7.5H_2O$ (c). Adapted from Ref. [56].

types of structures have been describe: a layer pillar (type 1) for short chains (2–5 carbons), a 2D structure (type 2) for medium chains (6–8 carbons) and a 3D structure (type 3) for long chains (9–12 carbons) [57]. This observation clearly illustrates the consequences of the use of rigid organic precursors on the topology of the produced materials that is likely governed by a reduced number of conformations when rigid organic precursors are employed. Zubieta et al. also reported other oxovanadium hybrids based on benzene-1,4-diphosphonic acid that also incorporated copper coordination complexes. It was further shown that the HF/vanadium ratio used in the synthesis directly impacted upon the composition of the hybrids. Indeed, at high concentration of HF, the fluorine atoms were incorporated within the structure [58]. In all these materials, the copper complexes with either phenanthroline or terpyridine were constituent building blocks of the extended networks. The incorporation of bis-amine in oxovanadium-aryl-bisphosphonic acid was also reported by Zubieta et al. [59] The diamine was introduced in the reaction media in the presence of V_2O_5, benzene-1,4-diphosphonic acid **2** and HF. The structures produced were based on the anionic species $\{V_2O_2(O_3PC_6H_4PO_3H)_2\}^{2-}$, while neutrality was reached by the inclusion of diprotonated diamine cations. Other hybrid structures based on vanadium oxide were reported in which the rigid bisphosphonic acid (benzene-1,4-diphosphonic acid **2** or biphenyl-4,4'-bisphonic acid **8**) acts as an organic tether between vanadium-based polyoxymetalates [60].

Benzene-1,4-benzenephosphonic acid **2** was also used to prepare lanthanide-based hybrids under hydrothermal conditions using conventional or microwave heating [61]. In this study, 15 lanthanide elements were used producing either isostructural anhydrous materials (Ln = La, Ce, Pr, Nd, Sm, Eu, Gd, Tb, Dy and Ho) having the general composition $Ln[O_3P(C_6H_4)POH_3]$ or hydrated materials whose structures were not solved (Er, Tm, Yb and Lu). The anhydrous materials, which were characterized by X-ray diffraction including single-crystal X-ray diffraction (Ln = Pr), possessed a high thermal stability (the onset of degradation occurs at 600 °C). The structure is formed from inorganic layers containing Pr^{3+} cations bonded to eight oxygen atoms coming from phosphonic acid functional groups. The geometry of the coordination sphere of Pr^{3+} was a distorted bicapped trigonal prism (PrO_8). The organic molecules link the $\{PrO_8\}_n$ inorganic planes.

More recently, benzenebisphosphonic acids **2** and **4** were used to prepare a series of actinide hybrids. The group of Albrecht-Schmitt has reported several uranyl hybrids that include benzene-1,4-diphosphonic acid **2**. The structure of the hybrids depends upon the experimental conditions. Indeed, with HF, fluorine atoms can be included in the structure (e.g. $[H_3O]_4\{(UO_2)_4[C_6H_4(PO_3)_2]_2F_4\} \cdot H_2O$) [62] or absent (e.g. [phen]-$\{(UO_2)_3[p\text{-}O_3P\text{-}C_6H_4\text{-}PO_3]_2\} \cdot H_2O$) [63]. The addition of an auxiliary ligand for the second material and the different hydrothermal conditions (200 °C vs. 180 °C) likely explained these results. Benzene-1,4-diphosphonic acid **2** was also engaged in hydrothermal synthesis in the presence of trimethylammonium hydroxide or tetraethylammonium hydroxide acting as a base and a source of organic cation. As an example, $\{[(CH_3CH_2)_4N]\{(UO_2)[p\text{-}HO_3P\text{-}C_6H_4\text{-}PO_3H_{1.5}]_2(H_2O)\}$ features inorganic chains connected by the organic ligand, thus producing a 3D framework with cavities filled with the organic cation [64]. The replacement of organic cations with Cs^+ as inorganic templating agent produced nanotubular structure $Cs_{3.62}H_{0.38}$ $[(UO_2)_4\{p\text{-}HO_3P\text{-}C_6H_4\text{-}PO_3H\}_3\{p\text{-}O_3P\text{-}C_6H_4\text{-}PO_3\}F_2]$, which features nanotubes of approximately 1×2 nm. Cs^+ cations were located in the tube and also between the tubes [65]. The authors have shown that Cs^+ cations can be exchanged by monovalent cations including Ag^+, Tl^+ and, to a lesser extent, K^+ and Na^+ [65, 66]. The use of crown ether (18-crown-6) as template produced, under mild hydrothermal conditions (150 °C, 2 days), materials characterized by anionic layers separated by H_3O^+–18-crown-6 inclusion complexes [67].

Since benzene-1,4-diphopshonic acid **2** and benzene-1,2-diphopshonic acid **4** were both employed by Albrecht-Schmitt et al. to produce uranium-based hybrids, it is interesting to point out the consequences of a selected diphosphonic acid on the structure of the hybrid. With the para-diphosphonic acid **2**, a layered structure was formed in which the fully deprotonated ligand **2** connects the inorganic sheets as shown in Figure 9.9a [62]. With the ortho-regioisomer **4**, and despite the synthetic conditions not being strictly identical, the topology of the materials is completely different (Figure 9.9b) [68]. In that case, a 2D topology was formed with the benzene ring pointing towards the interlayer space. Interestingly, the kinetics of reaction between dioxouranium(VI) and benzene-1,2-diphosphonic acid **4** revealed that the rate of reaction decreases when organic solvents were added to (or replaced) the water media [69].

Benzene-1,3,5-tris(phosphonic acid) **5**, which was synthesized from 1,3,5-tribromobenzene, is a rigid and planar derivative that in the solid state produced a supramolecular ladder-like structure [10]. Benzene-1,3,5-tris(phosphonic acid) **5** was first used by Clearfield et al. [70] in association with copper salts ($CuCl_2 \cdot 6H_2$

(a) (b)

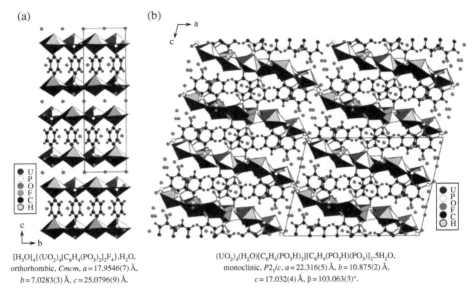

$[H_3O]_4\{(UO_2)_4[C_6H_4(PO_3)_2]_2F_4\}.H_2O$,
orthorhombic, $Cmcm$, $a = 17.9546(7)$ Å,
$b = 7.0283(3)$ Å, $c = 25.0796(9)$ Å.

$(UO_2)_4(H_2O)[C_6H_4(PO_3H)_2][C_6H_4(PO_3H)(PO_3)]_2.5H_2O$,
monoclinic, $P2_1/c$, $a = 22.316(5)$ Å, $b = 10.875(2)$ Å,
$c = 17.032(4)$ Å, $\beta = 103.063(3)°$.

FIGURE 9.9 Structure of uranium-based materials synthesized from 1,4-phosphonoben-zoic acid 2 (a) and 1,2-phosphonobenzoic acid 4 (b). Adapted from Refs (a) [62] and (b) [68].

O or $Cu(ClO_4)_2 \cdot 6H_2O$) and a mixture of bis-pyridines (4,4'-bipyridine and 4,4'-tri-methylenedipyridine) acting as a base. After the hydro-/solvothermal treatment (water/DMF mixture), three types of crystals were isolated and characterized. Two of them incorporated 4,4'-bipyridine in their structure that acted as an organic template, while the third was composed only of copper and the fully deprotonated tris-phos-phonic acid **5** ($\{Cu_6[(O_3P)_3C_6H_3]_2 \cdot 8H_2O\} \cdot 5.5H_2O$). The structure of this last compound (Figure 9.10a) involved three copper atoms surrounded by the oxygen atoms of the phosphonate groups and the water molecules leading to two different geometries: square planar CuO_4 for one copper and square pyramidal CuO_5 for the other two. The copper atoms are connected to the organic part by the PO_3 groups leading to the formation of a porous 3D materials made of channel delimited by the benzene rings and filled by water molecules.

A second structure based on ligand **5** that did not include any additional organic molecules in the hybrid material produced was reported by Shimizu et al. This material, $Zn_3[(O_3P)_3C_6H_3 \cdot (H_2O)_2] \cdot 2.5H_2O$, was isolated as single crystals by dif-fusion of acetone into an aqueous solution of $Zn(ClO_4) \cdot 6H_2O$ and benzene-tris(phosphonic acid) **5** [71]. This material features a layered structure composed of compound **5** connected to a tetrahedral zinc atom that is fully coordinated by oxygen atoms from phosphonic acids (Figure 9.10b). In this structure, compound **5** formed π-stacked dimers with 3.72 Å between two adjacent aromatic rings. These layers are connected via a third tetrahedral zinc atom possessing in its coordination sphere two oxygen atoms from phosphonic acid groups and two others from two

(a)   (b)

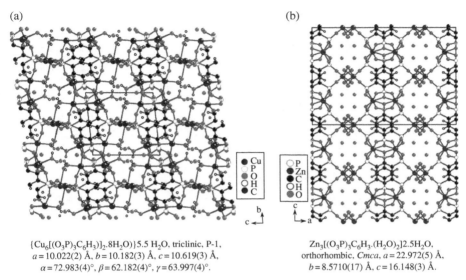

{Cu$_6$[(O$_3$P)$_3$C$_6$H$_3$)]$_2$.8H$_2$O)}5.5 H$_2$O, triclinic, P-1,   Zn$_3$[(O$_3$P)$_3$C$_6$H$_3$.(H$_2$O)$_2$]2.5H$_2$O,
a = 10.022(2) Å, b = 10.182(3) Å, c = 10.619(3) Å,   orthorhombic, $Cmca$, a = 22.972(5) Å,
α = 72.983(4)°, β = 62.182(4)°, γ = 63.997(4)°. b = 8.5710(17) Å, c = 16.148(3) Å.

FIGURE 9.10   Structure of (a) {Cu$_6$[(O$_3$P)$_3$C$_6$H$_3$)]$_2$ · 8H$_2$O)} · 5.5H$_2$O and (b) Zn$_3$[(O$_3$P)$_3$ C$_6$H$_3$ · (H$_2$O)$_2$] · 2.5H$_2$O synthesized from benzene-1,3,5-tris(phosphonic acid) **5**. Adapted from Ref. (a) [70] and (b) [71].

water molecules. This material exhibited proton conductivity (3.5×10^{-5} S cm^{-1} at 25 °C and 98% relative humidity) with, interestingly, a low activation energy for the proton transfer (0.17 eV). These proton conduction properties were further improved by the synthesis of a hybrid structure that incorporated both benzene-1,3,5-triphosphonic **5** and another C$_3$-symmetric ligand (2,4,6-trihydroxy-1,3,5-trisulphonate benzene) [72].

Benzene-1,3,5-trisphosphonic acid was also employed by Zubieta et al. [58] to produce copper/vanadium bimetallic hybrid materials. The synthetic strategy involved the addition of a bidentate ligand known to form coordination complexes with copper (e.g. 2,2′-bipyridine, 2,2′ : 6,2″-terpyridine or 1,10-phenanthroline). As a consequence, the (CuL)$^{2+}$ coordination complexes, produced *in situ*, acted as a charge-compensating group, which was incorporated into the structure of the hybrid possessing a general formula of {CuL}/V$_x$O$_z$/aromatic phosphonic acid. The authors have shown the crucial role of the quantities of HF added in the reaction media. Indeed, [{Cu(bpy)}$_2$V$_3$O$_7${(O$_3$P)$_2$C$_6$H$_3$–PO$_3$H}] and [Cu(bpy)VO$_2${(HO$_3$P)$_3$C$_6$H$_3$}] · 1.5H$_2$O were isolated by using HF/benzene-1,3,5-triphosphonic acid **5** in ratios of 14 and 53, respectively. These two materials possessed 2D structures composed of hybrid layers of respective formula {V$_3$O$_7$[(O$_3$P)$_2$C$_6$H$_3$–PO$_3$H]}$_n^{4n-}$ and {VO$_2$[(HO$_3$P)$_3$C$_6$H$_3$]}$_n^{2n-}$. In this study, benzene-1,4-diphosphonic acid **2** and benzene-1,3-diphosphonic acid **3** were also considered to produce heterobimetallic structures.

La[(HO$_3$P)$_3$C$_6$H$_3$] · 2H$_2$O, recently synthesized from benzene-1,3,5-triphosphonic acid **5** and La(NO$_3$)$_3$ under hydrothermal conditions (120 °C, 72 h) [73],

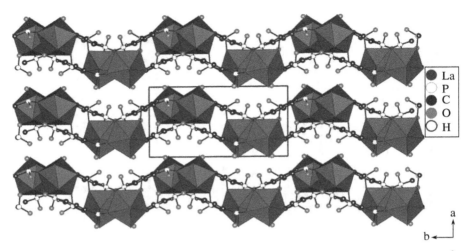

FIGURE 9.11   Structure of La[(HO$_3$P)$_3$C$_6$H$_3$]·2H$_2$O, monoclinic, *P2$_1$/c*, *a* = 9.6708(2) Å, *b* = 16.3872(3) Å, *c* = 9.2210(2) Å, *β* = 117.8512(6)°. Adapted from Ref. [73].

possesses a layered structure (each layer being linked only by hydrogen bonds). This material was exfoliated by polar solvent (DMF) to produce nano-sheets with 1.3 nm thickness. Europium- and terbium-doped exfoliated structures were also reported, and their photoluminescence (red and green, respectively) was reported (Figure 9.11).

Finally, Bhaumic et al. [74] have recently reported the synthesis of titanium-based nanomaterials by the formation of a gel (created from benzene-1,3,5-triphosphonic acid and titanium(IV) isopropoxide) that was subsequently heated under hydro-thermal conditions (180 °C, 24 h). The structure of this material was not solved from X-ray diffraction but was characterized by ^{31}P and ^{13}C MAS NMR. These character-izations suggested the presence of P–O–Ti bonds (111 type of bonding), and it is also noteworthy that the material exhibits a high thermal stability (up to 450 °C) and mesoporosity (255 m^2 g^{-1}).

Besides the use of a benzene ring as a rigid scaffold to produce rigid polyphos-phonic acid derivatives, fused aromatic scaffolds were also explored recently. Pyrene functionalized with four diethylphosphonate functional groups (compound **6**) was directly engaged with a barium salt (BaBr$_2$) under solvothermal conditions (ethanol, 120 °C) to produce a porous hybrid possessing a 3D structure containing rectangular 1D pores (3.9×4.6 Å). During the synthesis, a monodealkylation of each diethylphosphonate group was observed. The remaining ethyl ester groups are located at the periphery of these pores, making them hydrophobic [75] (Figure 9.12). In this structure, it is the rigidity of this tetraphosphonate mono-ester that prevents folding of the organic moiety, in association with coordination behaviour of Ba^{2+} ions, which accounts for the formation of the porous framework.

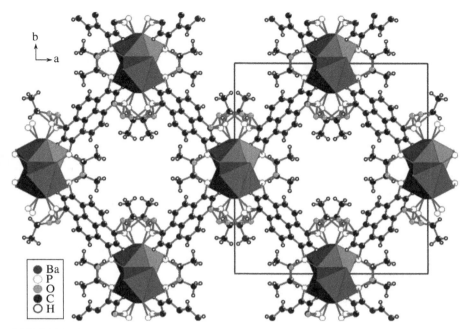

FIGURE 9.12   Structure of barium-pyrene-tetrakis(phosphonic ethyl ester) **6** (Ba[(O$_3$P)$_4$C$_{24}$ H$_{26}$·H$_2$O]), monoclinic, C2/c, $a = 19.9200(6)$ Å, $b = 21.4700(8)$ Å, $c = 8.2480(3)$ Å, $\beta = 107.6080(19)°$. Adapted from Ref. [75].

9.5   HYBRID MATERIALS FROM RIGID HETERO-POLYFUNCTIONAL PRECURSORS

9.5.1   Phosphonic–Carboxylic Acids

The association of phosphonic acid and carboxylic acid functional groups, which are both independently employed to produce crystalline hybrid materials, has been widely studied to produce original materials. In this section, we have selected only organic precursors in which these two functional groups are attached directly to a benzene ring as shown in Figure 9.13. As discussed in the following, a large variety of hybrid structures were characterized in which the topology was directly influenced by the nature of the metallic salt, the synthetic conditions and the structure of the rigid di- or polyfunctional organic precursor employed.

Precursor **11-H$_3$** (p-(HO)$_2$OP–C$_6$H$_4$–CO$_2$H or 4-(HO)$_2$OP–C$_6$H$_4$–CO$_2$H) was the first rigid phospho-carboxy-bifunctional compound used to design hybrid materials. The structure of Ca(p-HO$_3$P–C$_6$H$_4$–CO$_2$H)$_2$ (Ca(**11-H$_2$**)$_2$) was solved from XRPD [76]. This material, which was synthesized at room pressure by mixing compound **11-H$_3$** with CaCl$_2$ in a boiling water/ethanol mixture, possesses a 1D structure with [Ca(p-HO$_3$P–C$_6$H$_4$–CO$_2$H)$_2$] ribbons that are held together by hydrogen bonds that involve a

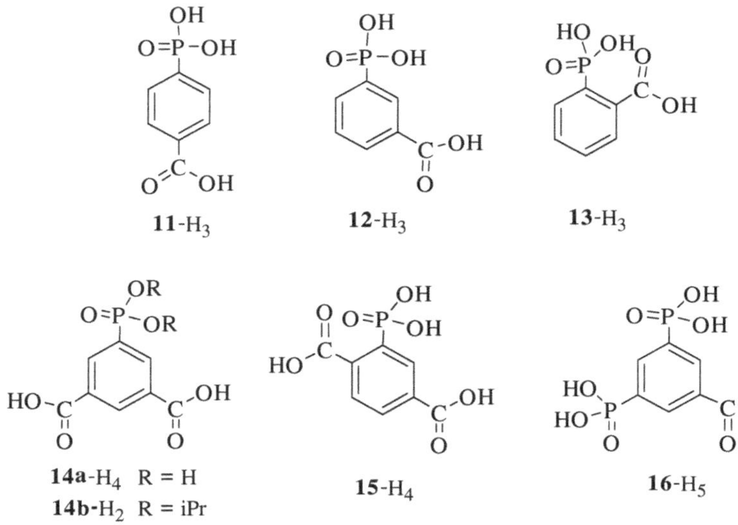

FIGURE 9.13 Chemical structure of rigid bifunctional organic precursors based on phosphonic acid and carboxylic acid functions.

protonated carboxylic acid. It must be noted that in that case only the most acidic function (phosphonic acid vs. carboxylic acid) was linked to the calcium-based framework.

Interestingly, the addition of NH_4OH to a suspension of $Ca(p-HO_3P-C_6H_4-CO_2H)_2$ in $CaCl_2$ containing water/ethanol solution produced $Ca_3(p-O_3P-C_6H_4-CO_2)\cdot 6(H_2O)$ ($Ca_3(11)$) as confirmed by XRPD and thermogravimetric measurements. We note that the material $Ca(11-H_2)_2$ can be recovered from $Ca_3(11)$ by the addition of the organic precursor $11-H_3$. This result suggested an equilibrium between these two structures, which shifts towards one of the structures depending on the pH and/or the addition of reagents ($CaCl_2$, compound $11-H_3$). A similar equilibrium, governed by the pH of the reaction media, was observed when a strontium salt was reacted with precursor $11-H_3$ [77]. Accordingly, $Sr(p-HO_3P-C_6H_4-COOH)_2$ ($Sr(11-H_2)_2$) and $Sr_3(p-O_3P-C_6H_4-COO)_2\cdot 5.7(H_2O)$ ($Sr_3(11)_2$) were characterized by XRPD. $Sr(11)_2$ was formed by mixing 11 with $Sr(NO_3)_2$ in water/ethanol mixture, while $Sr_3(p-O_3P-C_6H_4-COO)_2\cdot 5.7(H_2O)$ was formed in the presence of NH_4OH. The same authors reported similar behaviour with a barium-based hybrid. Indeed, depending on the pH of the reaction media, either $Ba_3(p-O_3P-C_6H_4-CO_2)_2\cdot H_2O$ ($Ba_3(11)_2$) or $Ba(HO_3P-C_6H_4-COOH)_2$ ($Ba(11-H_2)_2$) was formed [78]. It can be shown that $Ba(11-H_2)_2$ is converted to $Ba_3(11)_2$ by increasing the pH of the surrounding media, and vice versa. Finally, $Ba(11-H_2)_2$ and $Sr(11-H_2)_2$ as dehydrated forms were evaluated as anti-corrosion coating for magnesium alloys [79]. p-Phosphonobenzoic acid $11-H_3$ was also associated with different transition metallic salts. With copper, a first study reported the structure of $Cu(p-O_3P-C_6H_4-CO_2H)\cdot 2H_2O$ ($Cu(11-H)\cdot 2H_2O$) and $Cu(p-O_3P-C_6H_4-CO_2H)$. The hydrated material was isolated by simply mixing compound $11-H_3$ with $CuCl_2$ in water/ethanol mixture, while the dehydrated compound was synthesized from CuO under hydrothermal conditions (160 °C, 20 h). Finally, a pH increase resulting from the addition of NH_3 produced at room pressure $Cu_3(p$-O_3P-C_6H_4-$CO_2)_2 \cdot 3H_2O$ ($Cu(11) \cdot 3H_2O$) [80]. The hybrid $Cu(p$-O_3P-C_6H_4-$CO_2H)$ possesses a layered structure formed by inorganic planes (CuO_6 octahedra) linked together in a supramolecular way via hydrogen bonds involving the protonated carboxylic acid functional groups. Access to copper-based hybrid constructed from precursor 11-H_3 was also addressed by adding a nitrogen-based auxiliary ligand to the reaction media [81]. Accordingly, 1,10-phenanthroline was mixed in the reaction media before hydrothermal treatment to produce $Cu(Phen)(p$-HO_3P-C_6H_4-$COO)(H_2O)$ that featured helical inorganic chains linked together to produce a centrosymmetric 3D structure (right- and left-handed helices alternate in the crystal). After copper-based hybrids, the association of zinc salts with precursor 11-H_3 was reported in several studies. First, 11-H_3 was mixed with $Zn(Ac)_2 \cdot 2H_2O$ and heated in a mixture of DMF and water under hydrothermal conditions. Three different structures were characterized including $Zn(p$-O_3P-C_6H_4-$COO)(Me_2NH_2)$ [34]. This layered material features hybrid planes held together by protonated dimethylamines that are present in the interlayer space and form hydrogen bonds. Dimethylamine that acts as a structure-directing agent is produced *in situ* by degradation of DMF. Other amine or polyamines were also considered to produce hybrids [34]. Due to the basicity of the additional amine derivative, both functional groups (phosphonic acid and carboxylic acid of compound 11-H_3) were connected to the inorganic framework. Different topologies were formed, including interconnected inorganic chains in $Zn_2(p$-O_3P-C_6H_4-$COO) \cdot (H_2teta) \cdot H_2O$ (teta: triethylenetetramine) (Figure 9.14a) or a 3D open framework (sodalite-like topology) as observed in $Zn(p$-O_3P-C_6H_4-$COO) \cdot H_2O$ (Figure 9.14b). The synthetic conditions of these two materials were almost identical except that a different co-solvent (respectively, butane-1,3-diol and ethylene glycol) was used in association with water.

DABCO (1,4-diazabicyclo[2.2.2]octane) or 4,4′-bipyridine was also used to produce the pillared layered framework materials $(DABCOH)_2[Zn_8(p$-O_3P-C_6H_4-$COO)_6] \cdot 6H_2O$ and $[Zn(H_2O)_6][Zn_8(p$-O_3P-C_6H_4-$COO)_6(4,4′$-$bipy)]$ [82]. In these isotype structures, protonated DABCO or hexaaqua-zinc cations, which compensated the negative charge of the inorganic framework, are localized in the cavity of these materials. The addition of 2,2′-bipyridine as templating agent produced $Zn_{1.5}(p$-O_3P-C_6H_4-$COO)(2,2′$-$bipy)$ (solvothermal synthesis). Interestingly, the heating of this material under solvothermal conditions (water/DMF 1/1 mixture) produced $Zn(p$-O_3P-C_6H_4-$COO)(NH_2Me_2)_{0.5} \cdot 0.5H_2O$ that presented an open-framework topology [83]. This material exhibited molecular exchange capacity (amine and solvent). These exchanges induced a variation of the size of the pore that can be seen as a breathing effect of the framework. Hydrated crystalline zinc-based hybrids were also produced (hydrothermal condition 180 °C for 3 days) when 2,2′-bipyridine was mixed into the reaction media. $Zn_3(p$-O_3P-C_6H_4-$COO)_2(2,2′$-$bipy)(H_2O) \cdot H_2O$ and $Zn(bipy)(H_2O)_4Zn_2(p$-O_3P-C_6H_4-$COO)_2 \cdot 2H_2O$, which possessed interconnected inorganic chains (the organic molecules act as a tether between these chains), also exhibited luminescent properties [84]. Three bis-imidazole-based regioisomers (1,2-, 1,3- or 1,4-bis(imidazol-1-ylmethyl)benzene, respectively, 1,2-bimb, 1,3-bimb and

(a)

(b)

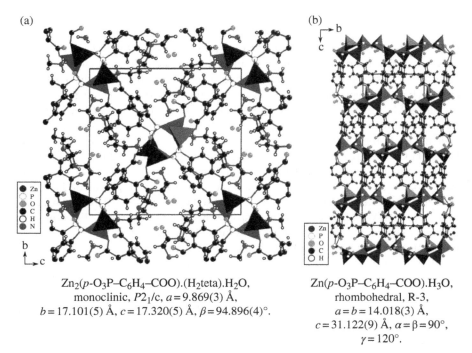

$Zn_2(p\text{-}O_3P\text{-}C_6H_4\text{-}COO).(H_2\text{teta}).H_2O$,
monoclinic, $P2_1/c$, $a = 9.869(3)$ Å,
$b = 17.101(5)$ Å, $c = 17.320(5)$ Å, $\beta = 94.896(4)°$.

$Zn(p\text{-}O_3P\text{-}C_6H_4\text{-}COO).H_3O$,
rhombohedral, R-3,
$a = b = 14.018(3)$ Å,
$c = 31.122(9)$ Å, $\alpha = \beta = 90°$,
$\gamma = 120°$.

FIGURE 9.14   Structure of $Zn_2(p\text{-}O_3P\text{-}C_6H_4\text{-}COO) \cdot (H_2\text{teta}) \cdot H_2O$ (a) and $Zn(p\text{-}O_3P\text{-}C_6H_4\text{-}COO) \cdot H_3O$ (b). Adapted from Ref. [34].

1,4-bimb) were also studied as more flexible nitrogen-based auxiliary ligands [85]. As an example, $Zn_2(p\text{-}O_3P\text{-}C_6H_4\text{-}COOH)_2(1,4\text{-bimb})$ featured hybrid layers formed by zinc atoms and 1,4-bimb in a *trans* conformation. These layers are interconnected with 4-phosphonobenzoic acid in its monoprotonated form (**11-H**). Besides the use of bipyridine or bis-imidazole as auxiliary ligand, other nitrogen polydentate ligands were used (e.g. terpyridine) to produce oxomolybdenum-based hybrids [86].

A polydentate nitrogen-based auxiliary ligand can therefore favour the formation of crystalline materials; it was also shown that an inorganic salt (LiF) can have the same effect producing crystalline hybrids. $Zn_2Li_2(p\text{-}O_3P\text{-}C_6H_4\text{-}COO)_2$ was synthesized from $ZnSO_4$, compound **11-H$_3$**, LiF in water at pH 3.55 (pH adjusted with NaOH) under hydrothermal treatment (160 °C, 72 h), and $Zn_2(p\text{-}O_3P\text{-}C_6H_4\text{-}COO)F$ (reduced quantities of compound **11-H$_3$** were introduced in the reaction media and the temperature of reaction was 140 °C) possessed a layered structure [87]. For the lithium-containing material, the lithium cation was connected to four oxygen atoms coming from three phosphonic acid groups to form LiO_4 tetrahedra. The organic molecules, which occupied the interlayer space in both materials, are connected to a selected inorganic plane by the phosphonic and carboxylic acid function in an alternating fashion (Figure 9.15a). In the second material, the fluorine atom served as a linking atom between three zinc atoms.

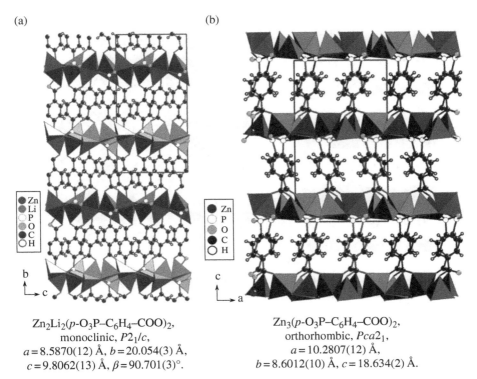

(a)

(b)

Zn$_2$Li$_2$(p-O$_3$P–C$_6$H$_4$–COO)$_2$,
monoclinic, $P2_1/c$,
$a = 8.5870(12)$ Å, $b = 20.054(3)$ Å,
$c = 9.8062(13)$ Å, $\beta = 90.701(3)°$.

Zn$_3$(p-O$_3$P–C$_6$H$_4$–COO)$_2$,
orthorhombic, $Pca2_1$,
$a = 10.2807(12)$ Å,
$b = 8.6012(10)$ Å, $c = 18.634(2)$ Å.

FIGURE 9.15 Structure of Zn$_2$Li$_2$(p-O$_3$P–C$_6$H$_4$–COO)$_2$ (a) and Zn$_3$(p-O$_3$P–C$_6$H$_4$–COO)$_2$ (b). Adapted from Refs (a) [87] and (b) [88].

The production of the pillared layered material Zn$_3$(p-O$_3$P–C$_6$H$_4$–COO)$_2$ in which molecule **11** connects the inorganic planes was carried out under hydrothermal conditions (160 °C, 96 h) from ZnSO$_4$ and **11**-H$_3$ in water solution at pH that was adjusted to between 4 and 5 with pyridine. Interestingly, the 4-phosphonobenzoic acid **11** was uni-oriented in the structure (Figure 9.15b), leading to the production of a material with dielectric properties that exhibited second harmonic generation activity [88].

Other metallic salts were associated with 4-phosphonobenzoic acid **11** including europium (Eu(p-O$_3$P–C$_6$H$_4$–COO)) [89], thorium (ThF$_2$(p-O$_3$P–C$_6$H$_4$–COO)) [90] or uranium (Cs$_2${(UO$_2$)$_2$(p-HO$_3$P–C$_6$H$_4$–COOH)$_3$(p-O$_3$P–C$_6$H$_4$–COOH)F}) [91]. In these hybrids, the rigidity and the bifunctionality of the organic precursor produced a 3D structure in which the rigid organic molecule acts as a linker between inorganic planes (e.g. thorium oxyfluoride (ThO$_4$F$_4$)) or chains (e.g. uranium oxide (UO$_6$)). Interestingly, with europium, high thermal stability was observed (up to 500 °C), which is explained by both the absence of water in the structure and the stability of the organic molecules likely due to the stable aromatic ring.

Aside from the use of the rigid organic precursor **11**-H$_3$ possessing the phosphonic and carboxylic acid groups in para position on the benzene ring, its regioisomer 3-phosphonobenzoic **12**-H$_3$ was also used to produce hybrids. Firstly, it is interesting

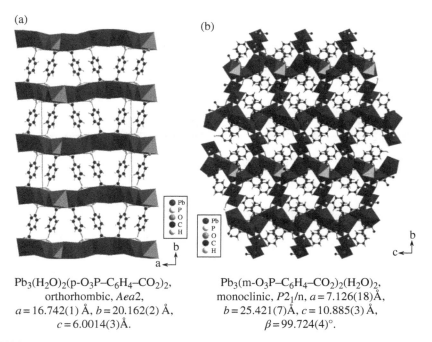

(a)   (b)

$Pb_3(H_2O)_2(p\text{-}O_3P\text{-}C_6H_4\text{-}CO_2)_2$, $Pb_3(m\text{-}O_3P\text{-}C_6H_4\text{-}CO_2)_2(H_2O)_2$,
orthorhombic, $Aea2$,   monoclinic, $P2_1/n$, $a = 7.126(18)$Å,
$a = 16.742(1)$ Å, $b = 20.162(2)$ Å, $b = 25.421(7)$Å, $c = 10.885(3)$ Å,
$c = 6.0014(3)$Å. $\beta = 99.724(4)°$.

FIGURE 9.16 Structure of lead-based hybrid synthesized from either 4-phosphonobenzoic acid **11** (a) or from 3-phosphonobenzoic acid **12** (b). Adapted from Refs (a) [92] and (b) [93].

to compare typical examples of lead-hybrid structures obtained from either **11**-H_3 or **12**-H_3. With the para regioisomer **11**-H_3, the structure $Pb_3(H_2O)_2(p\text{-}O_3P\text{-}C_6H_4\text{-}CO_2)_2$ was characterized by a layered 3D structure with the rigid organic compound **11** connecting inorganic planes formed by PbO_n ($n = 6$ or 7) polyhedra [92]. With the 1,3-regioisomer **12**-H_3, the structure $[Pb_3(m\text{-}O_3P\text{-}C_6H_4\text{-}CO_2)_2(H_2O)_2]$ (Figure 9.16b) features a 3D structure [93]. These two materials point out the consequences of the structure of the organic molecules, which, due to its rigidity, seems to compel the topology of inorganic framework to adapt accordingly.

2-Phosphonobenzoic acid **13**-H_3 (ortho-regioisomer) is the third hetero-difunctional organic precursor. The two functional groups of this compound **11**-H_3 are sufficiently close, and the limited number of conformations due to the rigidity of the scaffold leads these groups to be connected to the same side of an inorganic plane as observed in $Co_3(o\text{-}O_3P\text{-}C_6H_4\text{-}CO_2)(H_2O)_3 \cdot H_2O$ (Figure 9.17a) [94]. In this 2D layered material, the interlayer space was filled by the aromatic ring. Similar 2D layered topology was observed for $Mn(o\text{-}HO_3P\text{-}C_6H_4\text{-}CO_2)(H_2O)$ [95] or $(VO)_3(o\text{-}HO_3P\text{-}C_6H_4\text{-}CO_2)_2(H_2O)_6 \cdot H_2O$. Of note, this vanadium oxide-based material crystallized in a non-centrosymmetric space group, but after a partial release of coordinated water molecules, a centrosymmetric crystal was isolated [96]. Finally, heterobimetallic materials were also produced from the precursors **12**-H_3 or **13**-H_3.

(a)   (b)

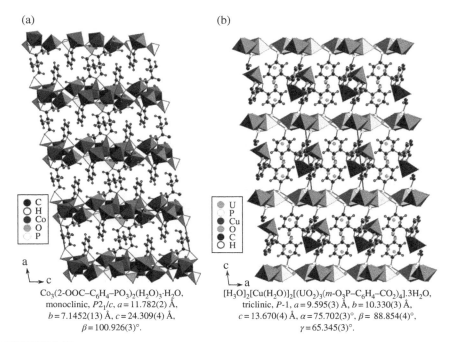

Co$_3$(2-OOC–C$_6$H$_4$–PO$_3$)$_2$(H$_2$O)$_3$·H$_2$O,
monoclinic, $P2_1/c$, $a = 11.782(2)$ Å,
$b = 7.1452(13)$ Å, $c = 24.309(4)$ Å,
$\beta = 100.926(3)°$.

[H$_3$O]$_2$[Cu(H$_2$O)]$_2$[(UO$_2$)$_3$(m-O$_3$P–C$_6$H$_4$–CO$_2$)$_4$].3H$_2$O,
triclinic, P-1, $a = 9.595(3)$ Å, $b = 10.330(3)$ Å,
$c = 13.670(4)$ Å, $\alpha = 75.702(3)°$, $\beta = 88.854(4)°$,
$\gamma = 65.345(3)°$.

FIGURE 9.17   Two-dimensional (a) and heterobimetallic (b) hybrid materials synthesized from 2-phosphonobenzoic **13** and 3-phosphonobenzoic **12**. Adapted from Refs (a) [94] and (b) [97].

Interestingly, one of the hybrid materials ([H$_3$O]$_2$[Cu(H$_2$O)]$_2$[(UO$_2$)$_3$(m-O$_3$P–C$_6$H$_4$–CO$_2$)$_4$]·3H$_2$O) (Figure 9.17b) exhibited an original structure since the phosphonic acids are exclusively connected to uranium atoms, while the carboxylate groups are connected to copper atoms [97]. Therefore, this material features a chemoselective connection involving two types of functional groups and two distinct metallic ions.

In the previous study reported in the preceding text, the position of the functional groups on the rigid benzene ring induced the formation of hybrids with different topologies. The pH of the reaction media is another factor that can also influence the topology of the hybrids. Indeed, with phosphonobenzoic derivatives, the phosphonic acid group is the most acidic function (pK_a of phenylphosphonic acid is 1.29 and 7.1 [98], while the pK_a of benzoic acid is 4.2 in water). Consequently, the pH of the reaction media influences the protonation state of these two functional groups. At low pH (pH < 3), only the phosphonic acid is partially deprotonated, leading to the production of hybrids in which only the phosphonic acid was bonded with the inorganic framework as illustrated by Cu(H$_2$O)(m-O$_3$P–C$_6$H$_4$–COOH) or the isostructural materials (Mn, Co, Zn) [18] (Figure 9.18). In this material, the carboxylic acid functional groups are forming hydrogen bonds in the interlayer space. On the other hand, a higher pH obtained by the addition of a base in the reaction media or a precursor that will form a base *in situ* by degradation under hydrothermal condition (degradation of urea) produces a hybrid with a completely different topology due

(a)   (b)

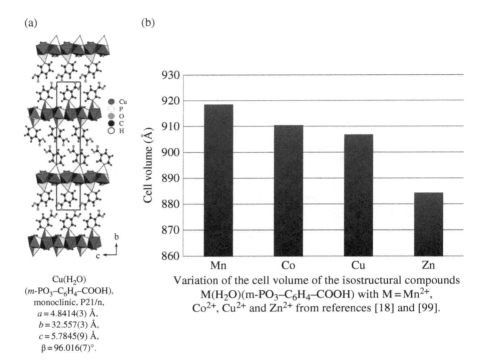

Cu(H$_2$O)
(m-PO$_3$–C$_6$H$_4$–COOH),
monoclinic, P2$_1$/n,
a = 4.8414(3) Å,
b = 32.557(3) Å,
c = 5.7845(9) Å,
β = 96.016(7)°.

Variation of the cell volume of the isostructural compounds M(H$_2$O)(m-PO$_3$–C$_6$H$_4$–COOH) with M = Mn^{2+}, Co^{2+}, Cu^{2+} and Zn^{2+} from references [18] and [99].

FIGURE 9.18   Structure of copper 1,3-phenylphosphonic acid Cu(H$_2$O)(m-PO$_3^-$)$_6$H$_4^-$) (OH). Adapted from Ref. [99].

to the interaction of the two functional groups with the inorganic framework. With copper, an original helicoidal structure (Cu$_6$(H$_2$O)$_7$(m-O$_3$P–C$_6$H$_4$–COO)$_4$) was produced (Figure 9.19) [99]. Interestingly, the single crystals were homochiral. Once more, this last structure demonstrates that the rigidity of the hetero-bifunctional organic precursor forces the inorganic framework to adopt a special topology (chiral in that case), thus illustrating the complexity of the hybrid structure obtained that can be produced from quite simple rigid difunctional precursor.

Finally, the complexity of the hybrid can be even further increased by adding further functional groups (carboxylic and/or phosphonic acid) as illustrated with the precursors 4-phosphonoisophthalate **14a**-H$_4$, 5-phosphonoisophthalate **15**-H$_4$ and 3,5-diphosphonobenzoic **14a**-H$_5$. With the precursor **14a**-H$_4$, 3D materials featuring inorganic chains (e.g. Cu$_3$(H$_2$O)[O$_3$P–C$_6$H$_3$(COOH)(COO)]$_2$·2H$_2$O) [100] or inorganic planes (e.g. Zn$_2$(H$_2$O)(O$_3$P–C$_6$H$_3$(COO)$_2$)·H$_2$O) [101] were reported. Interestingly, it was observed that a proton shift was induced when Zn$_3$(O$_3$P–C$_6$H$_3$(COO)$_2$)$_2$·2H$_2$O, synthesized through layered-solvothermal synthesis, was heated to 200 °C [34, 102]. Still with the organic precursor **14a**-H$_4$, the use of a structure-directing agents (piperidine NHC$_5$H$_{10}$) was beneficial to produce a 3D rutile-type zinc-based material (Zn$_3$[O$_3$P–C$_6$H$_3$(COO)$_2$]$_2$·(NH$_2$C$_5$H$_{10}$)·H$_3$O·5H$_2$O) exhibiting enhanced CO$_2$ adsorption due to a partition of the pore space [103]. Of note, this rutile-like porous material,

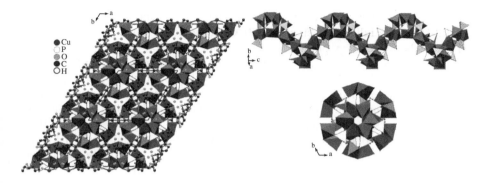

Cu$_6$(H$_2$O)$_7$(*m*-O$_3$P–C$_6$H$_4$–COO)$_4$, hexagonal,
*P*6$_5$22, *a* = *b* = 14.646(3) Å, *c* = 16.905(2) Å, α = β = 90°, γ = 120°.

FIGURE 9.19 Structure of helical copper chain material Cu$_6$(H$_2$O)$_7$(*m*-PO$_3$–C$_6$H$_4$–COO)$_4$ obtained from 1,3-phenylphosphonic acid [99]. Adapted from Ref. [99].

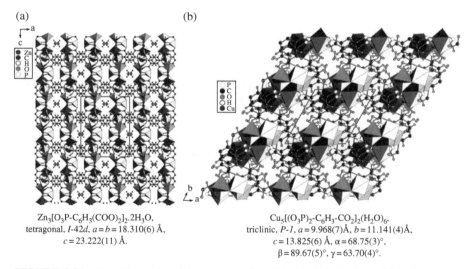

(a)

(b)

Zn$_3$[O$_3$P-C$_6$H$_3$(COO)$_2$]$_2$·2H$_3$O,
tetragonal, *I*-42*d*, *a* = *b* = 18.310(6) Å,
c = 23.222(11) Å.

Cu$_5$[(O$_3$P)$_2$-C$_6$H$_3$-CO$_2$]$_2$(H$_2$O)$_6$,
triclinic, *P-1*, *a* = 9.968(7)Å, *b* = 11.141(4)Å,
c = 13.825(6) Å, α = 68.75(3)°,
β = 89.67(5)°, γ = 63.70(4)°.

FIGURE 9.20 Hybrids produced from 5-phosphonoisophthalate 15 (a) and 3,5-diphospho-nobenzoic 16 (b). Adapted from Refs (a) [104] and (b) [109].

Zn$_3$[O$_3$P–C$_6$H$_3$(COO)$_2$]$_2$·2H$_3$O (Figure 9.20a), exhibited excellent catalytic properties for Friedel–Crafts benzylation reactions [104].

The synthesis of porous materials, which represents a great challenge, was attempted following an original strategy based on the use of the phosphonic acid protected as its ester form (compound **14b**-H, isopropyl ester). It was shown that after hydrothermal treatment, the ester group was still present if the reaction was carried out at low temperature ($<60\,^{\circ}C$) but at higher temperature a monodeprotection (at $120\,^{\circ}C$) and a full deprotection (at $180\,^{\circ}C$) occurred in the reaction media [105]. The dicarboxylate monophosphonic regioisomer **15**-H_4 (4-phosphonoisopthalic acid) was also employed to produce porous 3D structure with gismondine zeolite topology (e.g. $[Mn_2(4-(O_3P)-C_6H_3(COO)_2)(H_2O)_4]\cdot 1\cdot 5H_2O)$ [106]. The pores of this material were filled with water molecules. The association of bidentate ligand with nitrogen donor atoms (e.g. 1,10-phenthroline or bis-imidazole [107]) was also employed to produce, for instance, manganese-based layered 3D material [108]. The incorporation of two phosphonic acid and one carboxylic acid functional group on a benzene ring (compound **16**-H_5) was used in one study to produce the copper-based hybrid $Cu_5[(O_3P)_2-C_6H_3-CO_2]_2(H_2O)_6$ [109]. This material features a 3D topology (Figure 9.20b) with both phosphonic and carboxylic acid functions bonded to inorganic columns.

9.5.2 Phosphonic–Sulphonic Acids

Early in the development of hybrid materials synthesized from phosphonic acids as organic precursors, the incorporation of an additional sulphonic acid functional group was investigated with the aim to produce materials possessing cationic exchange properties, proton conductivity or acid catalytic properties (Figure 9.21).

The first investigations of this type were based on the synthesis of zirconium sulphophenylphosphonate in which the sulphonophenylphosphonate moieties were spaced within the structure by phosphate [110], methylphosphonate [111], ethylphosphonate [112], hydroxymethylphosphonate [112, 113] or phenylphosphonic acid groups [114]. The presence of this additional phosphonic acid in the structure aimed to space out the sulphonic acid functional group. One consequence of using two precursors during the synthesis of the hybrid was the production of materials with an amorphous structure or with a low crystalline structure due to a random distribution of sulphophenylphosphonic acid and the spacer in the structure. However, the crystallinity was increased by the addition of fluoride ions in the reaction media. The *in situ* formation of the stable ZrF_6^{2-} ion slowed down the formation of the

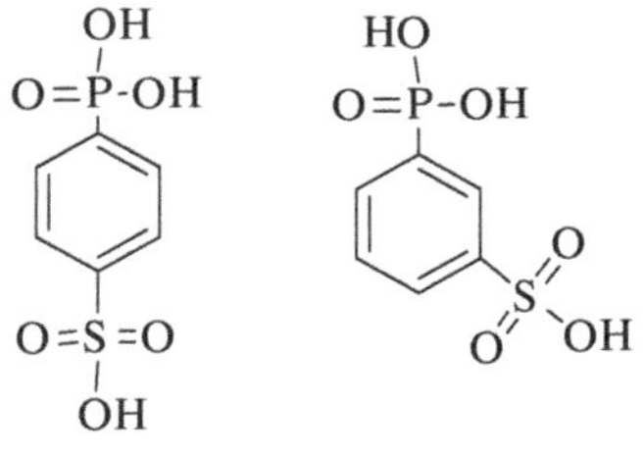

FIGURE 9.21   Chemical structure of 4-phosphonophenylsulphonic acid **17**-H_3 and 3-phosphonophenylsulphonic acid **18**-H_3.

hybrid and consequently improved the crystallinity. With this experimental procedure, it was confirmed that the hybrid possessed a layered structure [115]. These materials were studied for alkali and alkaline earth metal ion-exchange capacity [110], exchange with cationic coordination complexes exhibiting catalytic properties $(Ru(Bipy)_3^{2+})$ [116], catalytic properties (hydrolysis of oximes, semicarbazone or tosylhydrazone [111]; trimethylsilylation of alcohol [117] or phenol [118]; formation of heterocyclic compounds [119]; Ferrier rearrangement to produce 2-desoxy sugar [120]) and proton conductivity [112, 113, 121, 122]. These materials were also incorporated into polymers to produce nanocomposites exhibiting proton conductivity [122,123,124]. A recent study [125] compared the proton conduction of $Zr(HPO_4)_{0.7}(O_3P-C_6H_4-SO_3H)_{1.3}$ with $Zr(O_3P-C_6H_4-SO_3H)_2$. It was shown that the former material (phosphate-/phosphonate-based material) was more conductive ($\sigma=0.063\,S\,cm^{-1}$ at $100\,^{\circ}C$ and 90% relative humidity). Titanium-based hybrid material $(Ti[HPO_4]_1[O_3P-C_6H_4-SO_3H]_{0.85}[OH]_{0.30}\cdot nH_2O)$ was also studied for its proton conduction properties [126].

A second strategy, which was less studied, consisted of post-functionalizing aryl phosphonic hybrid by reaction with SO_3. Accordingly, Clearfield et al. [46] reported the sulphonation of the porous $Zr(O_3P-C_6H_4-C_6H_4-PO_3)$, which was synthesized from 4,4′-biphenylbis(phosphonic acid) under solvothermal conditions. It was observed that the use of an excess of zirconium salt during the synthesis produced large surface area ($400\,m^2\,g^{-1}$) with the presence of pores having an estimated diameter between 10 and 20 Å. The treatment of this material with SO_3 produced very strong Brönsted acid catalyst. It was shown that this material was more acidic than $Zr(O_3P-C_6H_4-SO_3H)_2$ [127].

Surprisingly, the first crystalline metal – sulphophenylphosphonate – was reported quite recently by Gao et al. [128]. Crystalline samples were prepared by hydrothermal reaction of Zn(II) carbonate with 3-sulphonophenylphosphonic acid and 1,10-phentholine or 4,4′-bipyridine. When phenanthroline was used, Zn atoms were connected to a phosphonic acid by either two or four bonds in a tetranuclear Zn(II) cluster $(Zn_4(O_3P-C_6H_4-SO_3)_4(Phen)_4)$. The four negative charges of this cluster arising from the deprotonated sulphonic acid groups are compensated by two $[Zn(Phenanthroline)_3]^{2+}$ cations (Figure 9.22a). In this material, all the organic anions are rigid, and the sulphonic acid functional group, which is deprotonated, contributes to form the anionic cluster. With the use of 4,4′-bipyridine, a 3D network was formed, $[Zn_6(m-O_3S-C_6H_4-PO_3)_4-(bipy)_6(H_2O)_4]\cdot 18H_2O$. In this material, the 4,4′-bipyridine molecules act as a linker between the inorganic columns, and the deprotonated sulphonic acid functions point into a rectangle cavity of 4.082 by 5.276 Å (Figure 9.22b).

The use of 1,10-phenanthroline or 4,4′-bipyridine was subsequently systematically added during hydrothermal or solvothermal syntheses to produce hybrids. A large number of metallic salts were used including lanthanide (Eu, Er, Nd, La) [129], cadmium [130], manganese [131], lead [132], copper [81, 133], yttrium [132] and tin [134]. Some of these materials (Cu [135], Pb [136]) were identified after a systematic high-throughput screening process. The geometry of 1,10-phenanthroline that favours κ^2 coordination produced material with a low dimensionality (e.g. a 1D inorganic framework illustrated by $Cu(1,10-Phen)(H_2O)(O_2C-C_6H_4-PO_3H)$ [81]. On the other hand, the use of 4,4′-bipyrine (rigid linear μ^2 coordination ligand) produced

(a)   (b)

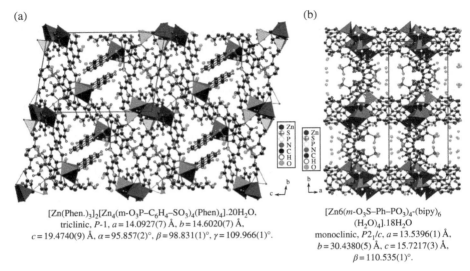

[Zn(Phen.)$_3$]$_2$[Zn$_4$(m-O$_3$P–C$_6$H$_4$–SO$_3$)$_4$(Phen)$_4$].20H$_2$O,
triclinic, *P*-1, *a* = 14.0927(7) Å, *b* = 14.6020(7) Å,
c = 19.4740(9) Å, *α* = 95.857(2)°, *β* = 98.831(1)°, *γ* = 109.966(1)°.

[Zn6(*m*-O$_3$S–Ph–PO$_3$)$_4$-(bipy)$_6$
(H$_2$O)$_4$].18H$_2$O
monoclinic, *P*2$_1$/*c*, *a* = 13.5396(1) Å,
b = 30.4380(5) Å, *c* = 15.7217(3) Å,
β = 110.535(1)°.

FIGURE 9.22   Crystal structure of [Zn$_4$(*m*-O$_3$P–C$_6$H$_4$–SO$_3$)$_4$(Phen)$_4$] · 20H$_2$O (a) and [Zn$_6$(*m*-O$_3$S–Ph–PO$_3$)$_4$–(bipy)$_6$(H$_2$O)$_4$] · 18H$_2$O (b). Adapted from Ref. [128].

3D networks in which this organic ligand acts as an organic cross-linking agent between chains (e.g. [Cd$_3$(O$_3$P–C$_6$H$_4$–SO$_3$)$_2$(bipy)$_3$(H$_2$O)$_6$] · 4H$_2$O) [130] formed by the cluster of metal ions and sulphophenylphosphonic acid.

9.5.3   Other Functional Groups

The group of Lin et al. has explored the use of chiral and functional rigid bisphosphonic acid from 1,1′-binaphthyl derivatives as organic building blocks for the construction of hybrid materials. The additional functional group was either an ether, a crown ether, a phenol or a triarylphosphine group. First, the bis ethyl ether **19a**-H$_4$ was added to Ln(NO$_3$)$_3$ (Ln = La, Ce, Pr, Sm, Gd, Tb) in methanol to produce homochiral lanthanide–bisphosphonic acid [Ln(**19a**-H$_2$)$_2$(**19a**-H$_3$)$_2$ (H$_2$O)$_n$]. The Gd-based hybrid, which crystallized in a chiral space group (chiral space group P2$_1$2$_1$2$_1$), was characterized by single-crystal X-ray diffraction [137]. Only binaphthyl units of (*R*) configuration were present in the crystal. This material adopts a layered structure – each layer being formed by a rhombohedral grid constituted by the bisphosphonic acid **19a**-H$_2$ and Gd atoms placed at each corner. Attempts to use this hybrid chiral material as separating agent of racemic mixture of trans-1,2-diaminocyclohexane were not fully convincing since an enantio-enrichment of only 10% was obtained. These first results were extended towards the incorporation of a crown ether group in a hybrid structure. Accordingly, reaction of 2,2′-pentaethylene glycol-1,1′-binaphthyl-6,6′-bis(phosphonic acid) (compound **20**-H$_4$, Figure 9.23) with lanthanide (Nd, Sm) produced single crystal ([Nd$_2$(**20**-H)$_2$(MeOH)$_8$] · (**20**-H$_4$) · (HCl)$_3$ · (H$_2$O)$_6$) by slow evaporation of an acidic mixture of Nd(NO$_3$)$_3$ and diphosphonic acid **20**-H$_4$ in methanol [138]. This material

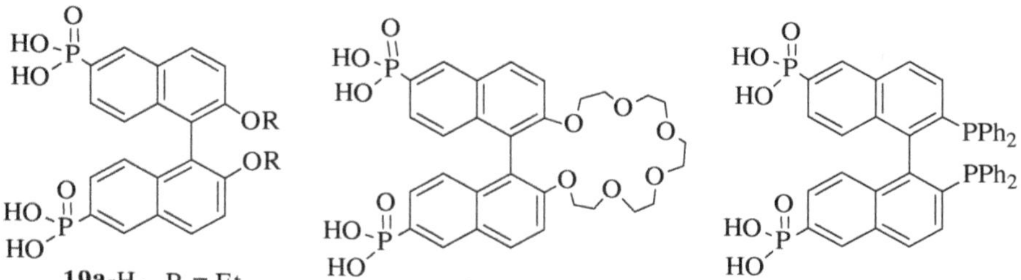

FIGURE 9.23 Chemical structures of bisphosphonic-binaphthyl derivatives used to design hybrid materials (H_n indicates the number of acidic protons present in the structure of the ligand).

crystallized in the chiral space group C2. In this homochiral structure, the rigid binaphthyl units of (R) configuration were connected to lanthanide to produce a lanthanide phosphonate layer. The rather flexible crown ether groups were located between these lanthanide phosphonate layers. Other materials synthesized from the binaphthol-diphosphonic acid $19b$-H_4 and BINAP-diphosphonic 21-H_4 and zirconium salts were also reported. These materials, which possess a porous and amorphous structure, were characterized by microscopy, solid-state NMR and circular dichroism spectroscopy. The addition of Ti(OiPr)$_4$ to the binaphthol-based hybrid produced an active catalyst for the enantioselective addition of diethylzinc on aromatic aldehyde to give enantiomeric excess (ee) up to 72% [139]. With the BINAP-containing hybrid (compound 21-H_4, Figure 9.23), ruthenium coordination complexes were formed, and this material was used as a heterogeneous catalyst for asymmetric hydrogenation of β-keto esters [140] or aromatic ketones [141].

9.6 HYBRID MATERIALS FROM PHOSPHONIC ACIDS LINKED TO A HETEROCYCLIC COMPOUND

9.6.1 Aza-heterocyclic

Many aza-heterocyclic derivatives have been reported, but in this section, we will focus on pyridine phosphonic acids, which are certainly the most studied rigid aza-heterocyclic compounds. Examples of materials synthesized from the phosphonic acids 22–27 (Figure 9.24) are briefly discussed. First, it must be stressed that the replacement of the benzene aromatic ring by the pyridine moiety corresponds to the addition of a coordination or protonation site. Hence, in both cases, this heterocyclic moiety will impact and contribute to the topology of the hybrid framework.

4-Pyridylphosphonic acid 22 and 3-pyridylphosphonic acid 23 were used by Lin et al. to form divalent transition metal hybrids (Zn, Co, Cu, Cd) under hydrothermal conditions using mild conditions (70–130 °C, 1–7 days) in aqueous-organic solvents [142]. With zinc and compound 23, a 1D structure was formed with the repetition of

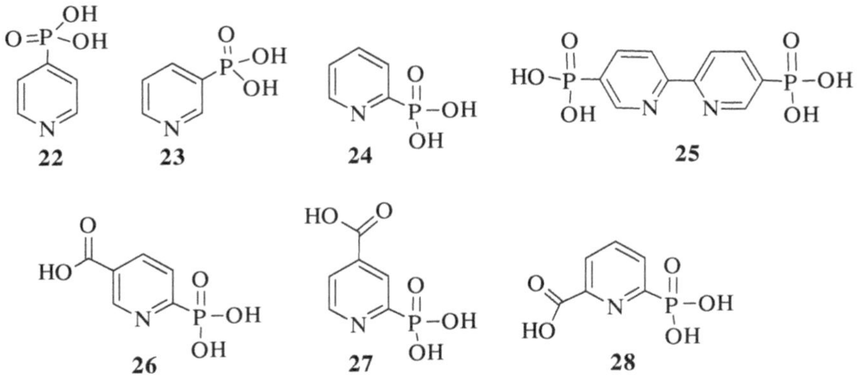

FIGURE 9.24 Structure of pyridine phosphonic acids used to produce hybrid materials.

Zn–O–P–O–Zn motifs to form a ladder-like structure and with the protonated pyridinyl moiety placed between these chains jointly with bromide ions to assure the electroneutrality (Figure 9.25a). This type of ladder-like structure was also observed with tin-based hybrid (e.g. Sn(3-O$_3$P–C$_5$H$_5$N)) [143]. With cobalt and compound **22**, a 2D grid structure Co(4-O$_2$P–C$_5$H$_5$N)(H$_2$O)$_3$ was observed. In that case, the nitrogen atom of the pyridyl moiety was connected to a cobalt atom to form hybrid layers. Two interlayer phosphonic acid oxygen atoms were involved in hydrogen bonds with water molecules placed in the interlayer space (Figure 9.25b). Finally, Cd(4-O$_3$P–C$_5$H$_5$N)$_2$ illustrates the formation of a 3D topology. The phosphonic acid function, which was fully deprotonated, and the nitrogen atom were bonded to cadmium atom (Figure 9.25c).

Cu$_4$(4-O$_3$–C$_5$H$_5$N)$_4$·nH$_2$O was directly produced from diethyl 4-pyridylphosphonate [141] or 4-pyridylphosphonic acid [144]; in this material that featured open channels occupied by water molecules, both phosphonic acid and the nitrogen atom from pyridine are connected to copper atoms (Figure 9.26a). Interestingly, with the use of the regioisomer 3-pyridylphosphonic acid, a layered 2D material was produced likely due to the rigidity of this bifunctional organic precursor [145]. In that case, the aromatic rings are pointing towards the interlayer spaces (Figure 9.26b). These last two examples are illustrative of the consequences of the rigidity of hetero-bifunctional precursors on the topology of the hybrids.

2-Pyridylphosphonic acid **24** was also considered for the production of hybrids. With aluminium and gallium salts and by slow crystallization at room temperature, a 0-dimensional material was produced featuring cage topology [146]. The association of compound **24** with other metallic salts including zinc, Cd, Hg and Ag was also reported [147].

The diphosphonic acid 2,2′-bipyridine-5,5′-bisphosphonic acid **25** produced zirconium-based hybrid in association with phosphite (H$_3$PO$_3$) to produce a non-crystalline material hybrid Zr(HPO$_3$)$_{0.8}$(O$_3$P–C$_5$H$_3$N–C$_5$H$_3$N–PO$_3$)$_{0.6}$·5.3H$_2$O [148]. The likely layered structure of this material was suggested on the basis of solid-state NMR experiments and other spectroscopic methods.

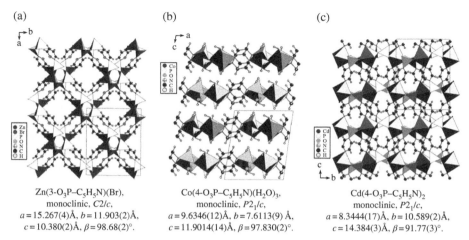

FIGURE 9.25 Examples of 1D (a), 2D (b) and 3D (c) topologies obtained with 4-pyridylphosphonic acid **22** or 3-pyridylphosphonic acid **23** and divalent metallic salts. Adapted from Ref. [142].

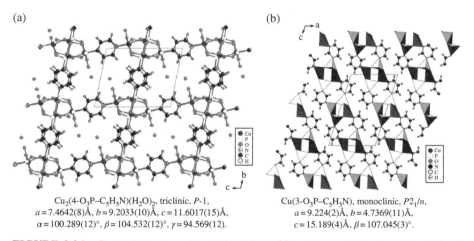

FIGURE 9.26 Channel structure (a) obtained from **22** and layered structure (b) obtained from **23** with copper(II). Adapted from Refs (a) [142] and (b) [145].

Pyridyl derivatives functionalized with one phosphonic and one carboxylic acid functions were used in different studies to produce crystalline hybrids. Three regioisomers were considered: 2-phosphononicotinic acid **26**, 2-phosphono-isonicotinic acid **27** and 6-phosphonopyridine-2-carboxylic acid **28**. The reaction of **26** with copper or cobalt salts under hydrothermal conditions (140 or 180 °C) produced the isostructural materials $M_3(O_3P-C_5H_3-CO_2)_2(H_2O)_2$ with M=Cu or Co [149]. These materials possess a pillared layered structure in which the organic compounds are connected to two successive inorganic layers (Figure 9.27a). The synthesis at room temperature (simple crystallization) produced material $Co(O_3P-C_5H_3-CO_2H)_2(H_2O)_3$ that featured a lower dimensionality (1D material). In this material, only the phosphonic acid and nitrogen atom from the pyridyl moiety are bonded to cobalt atoms. The disposition of the organic molecules produces helicoidal chains. The packing of this material was assumed by hydrogen bonds involving carboxylic acid, water molecules and deprotonated phosphonic acid functions [150]. In another set of experiments, a bis-imidazole derivative was associated with compound **26** and cobalt salt to produce an open 3D structure in which the pores were filled with the bis-imidazole derivative (Figure 9.27b). This material exhibited ferromagnetic inorganic chains [151]. Next, the phosphono-isonicotinic derivative **27** was reacted with copper [152], zinc and cadmium [153]. Accordingly, pillared layered materials were formed. Finally, the last regioisomer **28** was reacted with copper, cobalt and nickel to produce hybrid materials under hydrothermal conditions (150 °C, 24 h) [154]. In each case, compound **28** acts as a pincer ligand towards the metallic salt with a κ^3 coordination mode involving one nitrogen and two oxygen atoms. For the copper-based material, $[Cu(HO_2C-C_5H_3N-PO_3H)(HO_2C-C_5H_3N-PO_3H)\cdot 3H_2O]$, the copper atom is present in a distorted octahedral environment constituted by two tridentate ligands. The packing of this coordination complex involves hydrogen bonds to produce 2D hydrogen bond network. The cobalt- and nickel-based hybrids feature metallic atoms bonded to only one tridentate ligand and three water molecules. The packing of the cobalt-based material

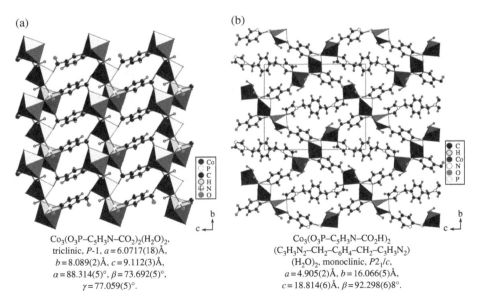

(a)

$Co_3(O_3P-C_5H_3N-CO_2)_2(H_2O)_2$,
triclinic, P-1, $a = 6.0717(18)$Å,
$b = 8.089(2)$Å, $c = 9.112(3)$Å,
$\alpha = 88.314(5)°$, $\beta = 73.692(5)°$,
$\gamma = 77.059(5)°$.

(b)

$Co_3(O_3P-C_5H_3N-CO_2H)_2$
$(C_3H_3N_2-CH_2-C_6H_4-CH_2-C_3H_3N_2)$
$(H_2O)_2$, monoclinic, $P2_1/c$,
$a = 4.905(2)$Å, $b = 16.066(5)$Å,
$c = 18.814(6)$Å, $\beta = 92.298(6)8°$.

FIGURE 9.27   Pillared structure (a) obtained from **26** and open 3D structure (b) obtained from **26** and bis-imidazole derivative with cobalt(II). Adapted from Refs (a) [149] and (b) [151].

$[Co(O_2C–C_5H_3N–PO_3H)(H_2O)_3 \cdot 3H_2O]$ results from a hydrogen bond network that produces interconnected channels (8×7Å). Finally, the nickel-based hybrid $[Ni(O_2C–C_5H_3N–PO_3) \cdot 0.5(Ni(H_2O)_6 \cdot 2H_2O]$ is formed by the assembly of two types of nickel coordination complexes: one being positively charged $[Ni(H_2O)_6^{2+}]$ and the other negatively charged $[Ni(O_2C–C_5H_3N–PO_3)]$. The packing involves a complex 3D hydrogen-bonding network.

9.6.2 Thio-heterocycles

Even though the synthesis of thiophene ring functionalized with one phosphonic acid was described some time ago [155, 156] and thiophene bisphosphonic acid was reported more recently [157], very few structures of hybrid materials including these organic moieties have been reported in the literature (Figure 9.28). In the first case, which concerns more precisely the use of 2- or 3-thienylphosphonic acid building blocks (29 and 30), the synthesis of $Zn[O_3P(2-C_4H_3S)]$ and $Zn[O_3P(3-C_4H_3S)]$ was achieved by using organometallic precursors. These materials were obtained as white powders in solution by reaction between $Zn(CH_3)_2$ and the corresponding thienylphosphonic acid [155]. Even though no crystal structure was reported, the X-ray diffraction study carried out on powder samples confirmed the layered structure of these materials, which are characterized by an interlayer distance of 13.4 and 14.2Å for materials based on 2- and 3-substituted thiophene rings, respectively. These two materials can be hydrated in a second step when suspended in water, leading to the formation $Zn[O_3P(2-C_4H_3S)] \cdot H_2O$ and $Zn[O_3P(3-C_4H_3S)] \cdot H_2O$ exhibiting an interlayer distance of 13.6 and 14.3Å. In another study, the impact of the synthetic conditions was clearly underlined in the case of the metal pyrophosphonate compound families [158] such as Ag(I) pyrophosphonates [159] with the general formula $[Ag_n(RPO_2(O)O_2PR)_m](CN)$ [160], where R is either a benzene, thiophene or naphthalene ring. These last materials were obtained from the corresponding arylphosphonate precursors and silver nitrate ($AgNO_3$) in acetonitrile as solvent under solvothermal conditions. A three-step mechanism was suggested by Zheng et al. to explain the formation of these silver pyrophosphonate materials during the solvothermal process including a condensation process between two different hydroxyl groups of two ligands.

The structure of the thiophene derivative $[Ag_3(3\text{-pyrothienylphosphonate})](CN)$ featured positively charged columns of silver ions and pyrothienylphosphonate

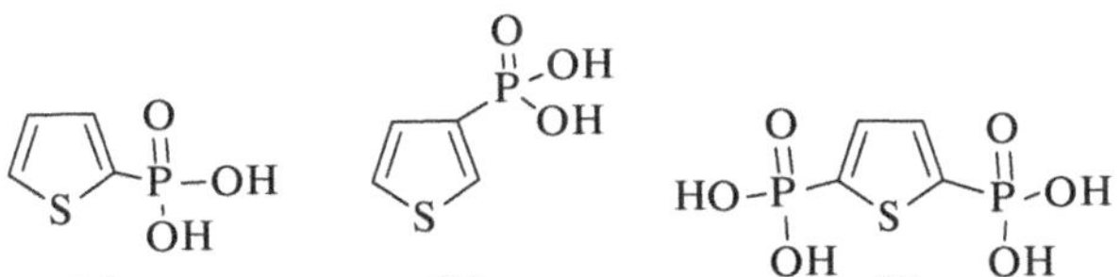

FIGURE 9.28 Structure of thiophene phosphonic acids used to produce hybrid materials.

connected together via the negatively charged cyano groups, leading to the formation of layers (Figure 9.29).

The last example involving thiophene phosphonate precursor concerns the hydrothermal synthesis of $Mn_2(O_3P-C_4H_2S-PO_3)\cdot 2H_2O$ obtained from thiophene-2,5-diphosphonic acid $(OH)_2OP-C_4H_2S-PO_3(OH)$ (31) and manganese nitrate $Mn(NO_3)_2\cdot 6H_2O$, in the presence of urea $CO(NH_2)_2$ to control the pH of the reaction media. To the best of our knowledge, $Mn_2(O_3P-C_4H_2S-PO_3)\cdot 2H_2O$ [157] was the first material synthesized from thiophene phosphonate derivates for which the structure was fully resolved from X-ray diffraction on single crystals. The pristine compound crystallizes in a layer architecture made of alternating organic and inorganic subnetworks connected together by the phosphonate function present on the thiophene ring. The inorganic layer $[MnO_6H_2]_n$ is constituted of PO_3C tetrahedron and MnO_6 octahedron. Each tetrahedron shares one edge and one corner with two adjacent octahedra, leading to the formation of an inorganic layer of the same type as that found for the isostructural material $M(H_2O)(m\text{-}PO_3-C_6H_4-COOH)$ with $M=Mn^{2+}$, Co^{2+}, Cu^{2+} and Zn^{2+} [18, 99]. This material presents a surprising and drastically different orientation of the thiophene rings present in two consecutive organic layers. Thus, in a layer, all the thiophene rings are pointing along the b direction, whereas in the next layer, they are oriented head to tail along the a direction (Figure 9.30a). A polymorphic compound was isolated after a dehydration–rehydration process, and its structure was solved by single-crystal X-ray diffraction. This compound also presents a layered architecture,

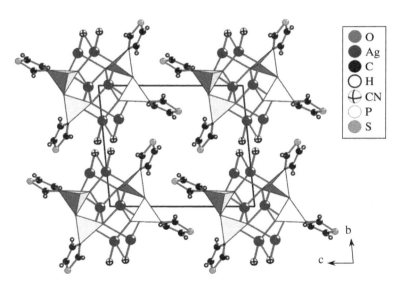

[Ag$_3$(3-pyrothienylphosphonate)](CN), triclinic, $P-1$,
$a = 5.8048(9)$ Å, $b = 10.7814(16)$ Å, $c = 12.4020(18)$ Å,
$\alpha = 83.785(1)°$, $\beta = 85.432(3)°$, $\gamma = 81.368(2)°$.

FIGURE 9.29 Crystal structure of a silver pyrophosphonate obtained from **29**. Adapted from Ref. [160].

(a)

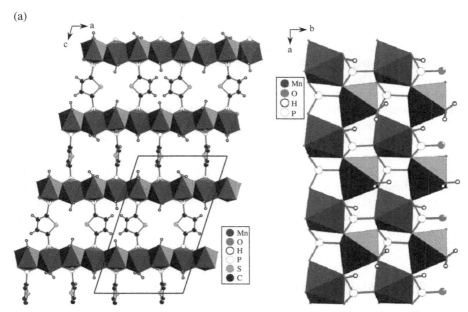

Mn$_2$(O$_3$P–C$_4$H$_2$S–PO$_3$).2H$_2$O, monoclinic, P2, a = 11.60(1) Å, b = 4.943(5) Å, c = 19.614(13) Å, β = 107.22°.

(b)

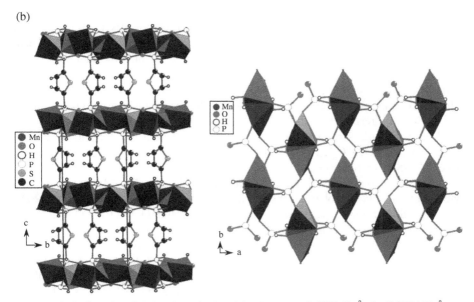

Mn$_2$(O$_3$P–C$_4$H$_2$S–PO$_3$).2H$_2$O, orthorhombic, *Pnam*, a = 7.5359(3) Å, b = 7.5524(3) Å, c = 18.3050(9) Å.

FIGURE 9.30 Crystal structure of the two polymorphs Mn$_2$(O$_3$P–C$_4$H$_2$S–PO$_3$)·2H$_2$O *P2* space group (a) and *Pnam* space group (b). Adapted from Ref. [157].

but the structures of the two subnetworks are totally different from the pristine. Indeed, the inorganic layer is composed of isolated ($Mn_2O_8H_4$) dimers connected together by the PO_3C tetrahedron. The projection of two adjacent organic layers along the b direction revealed a fishbone arrangement of the thiophene heterocycles (Figure 9.30b). The magnetic properties of these two materials have shown antiferromagnetic behaviour for the pristine material and a weak ferromagnetic component for the second polymorph, in good agreement with the structural study. A combination of thermogravimetric analysis (TGA) and thermodiffraction study (XRDT) performed on the pristine compound has determined that the compound obtained after dehydration presents a layered organization (300 °C, XRDT), with a thermal stability up to 500 °C (TGA) and paramagnetic behaviour. These two results confirm the thermal stability of the thiophene-2,5-diphosphonic acid $(HO)_2OP-C_4H_2S-PO(OH)_2$ **31** building block, when involved in the framework of hybrid materials, and the key role played in general by homo- or hetero-rigid aryl polyfunctional building blocks in the design of such multifunctional materials.

9.7 PHYSICAL PROPERTIES AND APPLICATIONS

9.7.1 Magnetism

The effect of the phosphonate anions upon the magnetic properties and behaviour of metal phosphonate materials is rooted in the influence the anion has over the structure of the inorganic subnetwork. Many magnetic properties rely upon coupling of magnetically active ions, and hence, it would be expected that the number of near neighbours, their proximity and the identity of the atoms linking them are the key criteria. Materials with a wide variety of different magnetic behaviours have been reported in the literature, more than can realistically be covered within a general chapter covering properties, so a selection of materials containing rigid phosphonate anions will be presented.

An isostructural series of layered compounds containing the 1,2-phosphonobenzoate anion $[M(OOCC_6H_4PO_3H)(H_2O)]$ (M(II) = Mn, Co, Ni) have been reported by Li et al. [95] (Figure 9.31).

The layers contain di-μ_3-O(P)-bridged $M^{II}_2O_2$ dimers. The Mn derivative is a weak antiferromagnet obeying the Curie–Weiss law at temperatures above 20 K, with the antiferromagnetic interaction occurring through O–P–O and O–C–O path, rather than via paths comprised of O–P–O paths alone. The Co derivative displays a room temperature effective magnetic moment of 4.96 μ_B per Co atom, somewhat larger than the spin-only value of 3.86 μ_B per Co atom. $\chi_M T$ decreases to a minimum value at 26 K, whereupon the values increase to a maximum of 2.11 cm^3 K mol^{-1} at 4 K, which is indicative of ferromagnetic ordering, which is enabled by a decrease in the M–O–M bond angle in the dimer. Below 4 K, a subsequent decrease in $\chi_M T$ is attributed to interlayer antiferromagnetic interactions. The Ni derivative again exhibits a larger room temperature magnetic moment (3.36 μ_B/Ni) than expected. The material obeys the Curie–Weiss law, with a Weiss constant of +6.20 K,

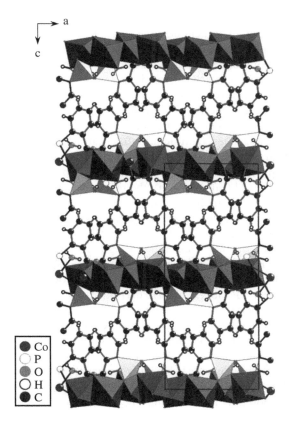

$Co(OOC–C_6H_4–PO_3H)(H_2O)$, orthorhombic, $Pbca$, $a = 8.999(5)$ Å,
$b = 8.614(5)$ Å, $c = 22.506(13)$ Å.

FIGURE 9.31 Crystal structure of the cobalt layered compound obtained from 2-phospho-nobenzoic acid. Adapted from Ref. [95].

indicative of a strong ferromagnetic exchange coupling between the Ni centres, and unlike the Co derivatives, the Ni-based material displays a hysteresis in the field-dependent magnetization data below the ordering temperature indicative of remnant magnetization.

A Co derivative of the 1,2-regioisomer of phosphonobenzoic acid, $Co_3(2\text{-}OOCC_6H_4PO_3)_2(H_2O)_3 \cdot H_2O$ (Figure 9.17a), was reported by Wang et al. [94] along with an isostructural Zn-based material. Like the previous examples, this material has a layered structure, but in this instance, there are three crystallographically independent Co atoms, which are linked through corner sharing of their respective $\{CoO_6\}$ octa-hedra to form a triangular $\{Co_3O_3\}$ motif; these units form columns that are connected to form the inorganic layer. The room temperature effective magnetic moment is 8.08 μ_B/Co_3 unit that is larger than the expected spin-only value of 6.70 8.08 μ_B/Co_3

unit and accounted for by an appreciable orbital contribution. The connectivity of the trimeric units through corner and edge sharing gives rise to antiferromagnetic coupling through the single-atom bridges (Co–O–Co). However, the odd number of Co atoms in the asymmetric structural unit must therefore give rise to a net ferro-magnetic magnetic moment for the inorganic layer as a whole, and hence, interactions between the layers can subsequently be either antiferromagnetic or ferromagnetic depending on the distance between them; for this material, the distance is greater than 9 Å, which favours a ferromagnetic interaction.

Derivatives of lanthanides yield quite different results with respect to magnetic properties, as illustrated by an isostructural series of layered Ln phosphonates based on 1,4-phenylbis(phosphonate), $Ln[O_3P(C_6H_4)PO_3H]$ (Ln = La, Ce, Pr, Nd, Sm, Eu, Gd, Tb, Dy and Ho) [61] (Figure 9.33a); the related series of Y, Er, Tm, Yb and Lu derivatives differed slightly in that they were hydrated and their structure has not yet been determined. The magnetic properties of the Gd, Dy, Ho, Er, Tm and Yb were investigated. All of the materials, hydrated or anhydrous, behaved paramagnetically across the entire experimental temperature range (2–300 K), following the Curie–Weiss law at $T > 20$ K, with effective magnetic moments expected of lanthanide ions with a significant orbital contribution and no evidence of magnetic ordering. The structure of the anhydrous Gd, Dy and Ho derivative materials is layered with the 8-coordinate Ln^{3+} ions being linked by a number of single-atom (O) bridges. The lack of ordering shows that there is no coupling between metal ions despite the relatively short distances between them and the presence of multiple connections, largely due to the core-like nature of the valence f-electrons.

It is clear then that the magnetic behaviour in metal phosphonates can be explained on the basis of the structure and is dependent on the identity of metal and their local environment as imposed by the structure. The role of the phosphonate anion there-fore is arguably a minor one, providing the connectivity between the metal centres, either through single-atom bridges or, less favourably for coupling interactions, through three-atom (O–P–O) bridges. Contrarily, it might also be argued that the role of the phosphonate is the dominant driving force in determining the formation of dimers and trimers that can give rise to ferromagnetic behaviour.

9.7.2 Fluorescence

Luminescence (fluorescence or phosphorescence) in inorganic–organic hybrid mate-rials can arise from either the metal ions or the organic moiety. In discussing mate-rials based on rigid phosphonates in which the rigidity arises from unsaturated (conjugated) systems, there are an appreciable number of photoluminescent mate-rials reported in the literature that might be included. The organophosphonate anion can act in one of two ways with regard to luminescence; it can be the source of the luminescent response, or it can act as an antenna or activator, passing the gathered energy to the inorganic ions, thereby creating an excited state from which energy is released as light as the system relaxes to the ground state. Lanthanide ions are well known for their photoluminescent properties and can be activated (and hence give rise to a response) either directly or be absorbing radiation re-emitted from antenna anions. Arguably, less well known is transition metal- or main group element-based luminescence. There have been several extensive reviews of the luminescent behaviour of metal phosphonates in general [4, 161]; comments herein will be restricted to materials representative examples containing rigid phosphonate anions.

Luminescent behaviour in transition metal-containing materials is associated with transitions involving the valence d electrons. d–d transitions are forbidden under the rules governing electronic transitions, but unlike the lanthanides, vibronic coupling usually reduces the local symmetry and the transitions become partially allowed. Many of the active species are associated with d^5 or d^{10} configurations, and the transitions that give rise to luminescence are perhaps best described as charge transfers, where electrons move between d- and p-orbitals (e.g. $nd^{10} \leftrightarrow nd^9$ $(n+1)p^1$). Compounds of metals such as Ru^{2+} and Zn^{2+} are well known for exhibiting luminescent behaviour, but other species such as Ag^+ and Mn^{2+} are also active. In a lot of cases, a metal phosphonate will display luminescence very similar to other compounds containing the active metal ion, for example, zinc phenylphosphonate, $Zn(O_3PC_6H_5) \cdot H_2O$, has a green luminescence (385 nm) very much like that of ZnO [162]. Similarly, the zinc phosphonobenzoate materials $Zn_3(pbc)_2(bpy)$ $(H_2O) \cdot H_2O$ and $Zn_2(pbc)_2 \cdot Zn(bpy)(H_2O)_4 \cdot 2H_2O$ [84] (bpc = 1,4-phosphonobenzoate anion and bpy = 2,2-bipyridyl) exhibit luminescent responses at 352 nm and 340 nm, respectively, which differs only minimally from that of $[Zn(bpy)_2]^{2+}$ at 360 nm. In other cases, the observed luminescence arises directly from the organic components, for example, $[Zn_3(4-O_3PC_6H_4CO_2)_2(1,2\text{-bis(imidazol-1-ylmethyl)}$ benzene)] with a violet emission at 433 nm ($\lambda_{ex} = 335$ nm) [163] arising from the imidazole moiety.

Of more interest are the cases where the luminescent response varies from that expected. Upon changing the regioisomerism of the phosphonate anion in the last example to a 1,3-substitution, luminescence is observed for the layered Zn-based derivative, $Zn_3(m\text{-}O_3PC_6H_4CO_2)_2$, in the absence of an ancillary organic group [161]. The interest in this material arises from the fact that when it is illuminated with UV light ($\lambda = 265$ nm), a yellow emission, with a single broad (FWHM ~75 nm) line in the emission spectrum centred at 578 nm, is observed that is not directly related to either the metal or the phosphonate anion. More interestingly, the silver-based analogue, $Ag_3(m\text{-}O_3PC_6H_4CO_2)$, displays a bright green luminescence under the same conditions. There is clearly a complex interplay between the absorption and emission characteristics of the metal ions and the organic moieties that is poorly understood at this point in time.

Fluorescent emission from lanthanides arises from f–f transitions, which are also forbidden under the selection rules covering electronic transitions, and hence, observed fluorescence is relatively weak. As such, the lanthanide emission can often be overwhelmed by fluorescent emission from any organic species that are present. Given the electronic structure of the lanthanides and the fact that the valence 4f orbitals are essentially core-like, there is little variance in the wavelengths of their emission lines. Europium and terbium give the most intense emissions in the visible spectrum being red and green, respectively. In both of these examples, the emission is made up of a number of peaks: $^5D_0 \rightarrow {}^7F_J$ ($J = 0$–6) for Eu^{3+} and $^5D_4 \rightarrow {}^7F_J$ ($J = 0$–6) for Tb^{3+}. Both of these ions, along with other lanthanides, can be reacted with 1,4-phenylenebis(phosphonate) to yield an isostructural series of materials, $Ln[O_3P(C_6H_4)POH_3]$ (Ln=La, Ce, Pr, Nd, Sm, Eu, Gd, Tb, Dy, Ho) [61], which possess a 3D pillared structure. The Ln^{3+} ions have bicapped trigonal prismatic $\{LnO_8\}$ coordination geometry. Both the Eu and Tb derivatives demonstrate luminescence when illuminated at $\lambda_{ex}=464$ nm and $\lambda_{ex}=376$ nm, respectively, with it being noted that the lanthanide luminescence dominates that of the organophosphonate anion.

It is noticeable in the emission spectrum of the Eu derivative that the $^5D_0 \rightarrow {}^7F_2$ line at 610 nm is more intense than the $^5D_0 \rightarrow {}^7F_1$ line at 595 nm, and a red emission is observed. A similar observation is made from the 3D pillared 1,4-phosphonocarboxylate $Eu(O_3PC_6H_4CO_2)$ [89]. This observation is attributed to the fact that the Eu occupies a non-symmetric crystallographic site; conversely, where Eu occupies a site on a mirror plane, the $^5D_0 \rightarrow {}^7F_1$ line is more intense than the $^5D_0 \rightarrow {}^7F_2$ line, for example, for layered $EuH(O_3PC_6H_5)$, in which case the emission is visibly orange in colour. The emission spectra of the Tb derivatives show little difference regardless of the identity of the phosphonate anion present.

Uranium has been known as a fluorophore for many years [164]. A large number of uranyl compounds (but certainly not all) display a characteristic fluorescence around 520 nm upon illumination with a UV source, which contains five peaks arising from coupling to the vibrational modes of the uranyl ion. There have been a number of materials reported that are based on uranium and containing diphosphonate anions and occasionally additional transition metal species [62, 64, 67]. In the **Ubbp-n** (n=1–3) series of materials reported by Adelani et al. [62], the fluorescent response appears to be directly linked to the presence of fluorine in the material. The structures of the three materials are quite varied, ranging from 3D pillared (**Ubbp-1**) to 1D chains (**Ubbp-2**) and to a 3D open framework (**Ubbp-3**). The first of these, $[H_3O]_2\{(UO_2)_6[C_6H_4(PO_3)(PO_2OH)]_2[C_6H_4(PO_2OH)_2]_2[C_6H_4(PO_3)_2]\}(H_2O)_2$ (**Ubbp-1**), contains no fluorine and exhibits the strongest fluorescent response upon illumination with a mercury lamp ($\lambda_{ex}=365$ nm) (Figure 9.32). The second, $[H_3O]_4\{(UO_2)_4[C_6H_4(PO_3)_2]_2F_4\}$ (**Ubpp-2**), contains fluorine as an ancillary anion, and the response is significantly reduced, especially the peaks in the lower wavelength region at 500 and 520 nm. The third, $\{(UO_2)[C_6H_2F_2(PO_2OH)_2]_2(H_2O)\}_2 \cdot H_2O$ (**Ubbp-3**), displays no luminescent response at all.

These differences are poorly understood and are attributed not only to the presence of the fluorine but also to the different topologies of the materials [62]. The local environment of the U centre in **Ubbp-1** and **Ubbp-2** is pentagonal bipyramidal, but the composition is UO_7 for Bbbp-1 and UO_5F_2 for **Ubbp-2**. It is likely the disruption of the local electric field has made some of the electronic transitions less favourable. What is clear is that it is the metal ion giving rise to the fluorescent response rather than the phosphonate anion. It may be that the phosphonate anion is acting as an antenna and in the case of **Ubbp-3** fluorine is present on the phenyl ring, which, due to its electron-withdrawing nature, is likely to raise the energy gap between the π (HOMO) and π^* (LUMO). Thus, the lack of fluorescent response might simply then arise from the fact that the 365 nm light source is not sufficiently energetic to excite the phosphonate anion.

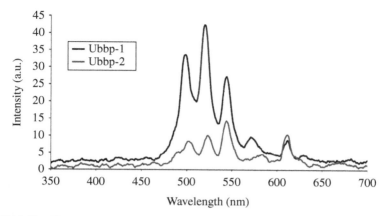

FIGURE 9.32 Fluorescence spectra for $[H_3O]_2\{(UO_2)_6[C_6H_4(PO_3)(PO_2OH)]_2[C_6H_4(PO_2OH)_2]_2$ $[C_6H_4(PO_3)_2]\}(H_2O)_2$ (**Ubbp-1**) and $[H_3O]_4\{(UO_2)_4[C_6H_4(PO_3)_2]_2F_4\}$ (**Ubpp-2**) ($\lambda_{ex} = 365$ nm). Reprinted with permission from Ref. [62]. © Elsevier.

Replacing the 1,4-diphosphonate anion with a 1,4-phosphonocarboxylate anion [90] yields $UO_2(PO_3HC_6H_4CO_2H)_2 \cdot 2H_2O$, which has a 3D pillared structure. The local environment of the U centres is tetragonal bipyramidal UO_6, and the intense fluorescence spectrum closely resembles that of Ubbp-1; the phosphonate anion also fluoresces with an emission centred around 430 nm, but the fluorescence spectrum is dominated by the U emission. One might thus draw the conclusion that the symmetry of the local environment is less important than the composition. In the same paper, Adelani et al. report the structure and properties of a Th derivative of the same anion [90], $ThF_2(PO_3C_6H_4CO_2H)$. Again, this has a 3D pillared structure with both functional groups involved in coordinating the metal ions. The Th centre has eightfold $[ThO_4F_4]$ dodecahedral (distorted hexagonal bipyramid) coordination. The fluorescent response of the material is now solely that of the phosphonocarboxylate ion, though much reduced in intensity, which is attributed by quenching by the Th centre.

Metal phosphonates provide a wealth of opportunity in the production and study of luminescent materials. Varying the metals from transition metals, main group elements and lanthanides allows a large part of the visible colour gamut to be covered, as well as the non-visible parts of the spectrum. Another factor that can be brought into play in the design of new materials is that the relatively similar sizes and chemistries of some ions, in particular the lanthanides, can allow mixtures of ions to be incorporated into the same material, exponentially increasing the number of potentially luminescent materials that may be synthesized. In addition to varying the metal ions, some creativity in terms of functionality in the phosphonate anion, particularly in terms of extended conjugated systems, can allow the adsorption characteristic of the materials to be tuned. Although much research remains to be done in identifying the properties of the various components, combining these two facets should allow the chemist to be able to select the metal and phosphonate anion in order to synthesize a luminescent material with properties specific to a particular requirement or application.

9.7.3 Thermal Stability

Research concerning new functional materials with high performances cannot be considered without these materials exhibiting, in addition to their intrinsic properties, an important thermal stability, which is a key parameter for any application in everyday life. This thermal stability is encountered in various inorganic materials, polymers or organic molecules. Concerning hybrid materials, the use of rigid polyfunctional ligands as building block precursors is one way to confer these materials with a high thermal stability. In order to illustrate this point, it is interesting to consider and to compare the structure of the two anhydrous compounds $GdH(O_3P–C_6H_4–PO_3)$ [61] and $GdH(O_3P–(CH_2)_3–PO_3)$ [165] (Figure 9.33) obtained, respectively, from 1,4-phenylbisphosphonic acid and propylenediphosphonic acid. These compounds present a similar inorganic layer made of gadolinium ions surrounded by eight oxygen atoms forming polyhedra sharing one edge to form chains. These chains are linked by the phosphonate functions leading to the creation of inorganic layer. Two successive inorganic layers are connected together via the organic molecule acting as pillar, and both compounds present an interlayer distance close to 9.5Å. Despite the common nature of their inorganic network, these materials exhibit different thermal behaviour and stability, which could only be explained here by the difference of flexibility of their organic part. By considering the characteristics defined earlier in this work about the flexibility of ligands (Figure 9.2), it is obvious that 1,4-phenylbisphosphonic acid is more rigid than propylenediphosphonic acid. Thus, the results of TGA studies conducted under O_2 atmosphere have shown that the compound $GdH(O_3P–C_6H_4–PO_3)$ was stable up to 500 °C, whereas the compound $GdH(O_3P–(CH_2)_3–PO_3)$ started to decompose around 200 °C.

As already mentioned, the thermal stability of these materials is due to the lack of water in the coordination sphere of the cation present in the inorganic subnetwork of the material and is also reinforced when two neighbouring inorganic subnetworks are connected together by the homo- or heterofunctional reactive groups grafted on the rigid building block. Coupling this thermal stability with another property, such as magnetism or luminescence, could lead to a multifunctional material exhibiting coupled properties. This strategy has been used to synthesize the compound $Eu(p-O_3P–C_6H_4–COO)$ obtained from 1,4-phosphonobenzoic acid and Eu^{3+} cations [89]. This material presents a 3D network made of an inorganic layer connected together via the phosphonic and the carboxylic functions. Each layer is constituted of chains of EuO_7 pentagonal bipyramids connected by the PO_3C tetrahedra and COO functional groups. The combination of TGA and thermodiffractogram demonstrated the remarkable thermal stability of this compound up to 500 °C. The luminescent properties of $Eu(p-O_3PC_6H_4COO)$ have been compared to those of $Eu(O_3P–C_6H_5)(HO_3P–C_6H_5)$ [166], and the differences observed, orange or red during excitation, were explained by the structural resolution as described in the previous section (Figure 9.34).

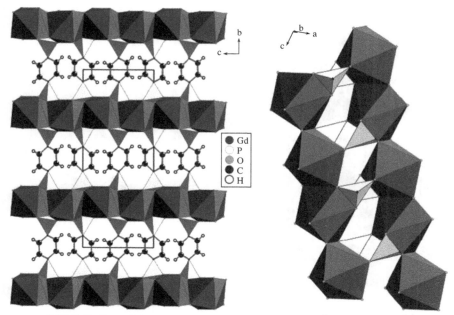

(a)

GdH(O$_3$P–C$_6$H$_4$–PO$_3$), monoclinic, $C2/c$, $a = 5.4805(12)$ Å, $b = 19.943(4)$ Å, $c = 8.1317(15)$ Å, $\beta = 108.367(8)°$

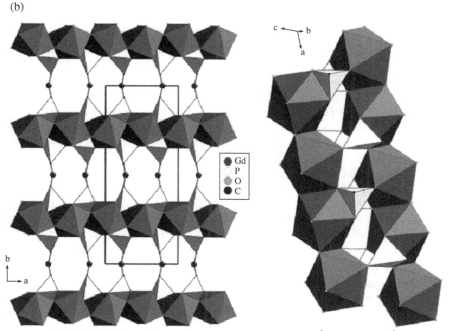

(b)

GdH(O$_3$P–(CH$_2$)$_3$–PO$_3$), monoclinic, $C2/m$, $a = 8.2141$ Å, $b = 18.9644$ Å, $c = 5.2622$ Å, $\beta = 111.9990°$.

FIGURE 9.33 Structure of [GdH(O$_3$P–C$_6$H$_4$–PO$_3$) (a) and GdH[O$_3$P–(CH$_2$)$_3$–PO$_3$] (b). Adapted from Refs (a) [61] and (b) [165].

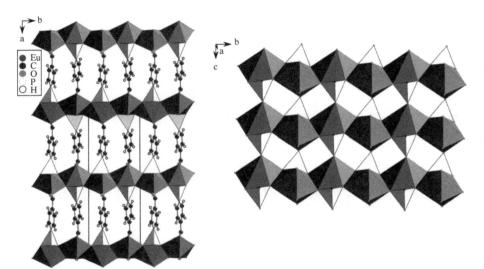

Eu(p-O$_3$P–C$_6$H$_4$–COO), orthorhombic, $Pn2_1a$, $a = 20.399(2)$Å,
$b = 7.0657(3)$Å, $c = 5.4558(4)$Å.

FIGURE 9.34 Structure of Eu(p-O$_3$P–C$_6$H$_4$–COO) compound obtained from 1,4-phosphonobenzoic acid and Eu^{3+} cations. Adapted from Ref. [89].

9.7.4 Drug Release

The synthesis of hybrid materials possessing the capacity to release bioactive compounds has been explored over the last 10 years. One strategy to reach this goal is based on the use of porous materials in which an active drug can be included in the pore. The size of the pore and the hydrophilic/hydrophobic balance of these pores are some of the parameters that impact directly upon the kinetics of release of the bioactive drug. According to this strategy, ibuprofen [167], zoledronate [168], anticancer drugs [169] and biologically active gas [170] such as NO have been stored within hybrid structures. A second strategy was based on the use of organic compounds that possess a biological effect and simultaneously possessed the suitable functional groups to form hybrid material (e.g. carboxylate functional group). Accordingly, the bioactive organic compounds interact with a metallic salt (selected for its biological harmlessness) to produce hybrid structures. As an example, nicotinic acid was used to produce Fe$_2^{III}$Fe$_{1-x}^{III}$Fe$_x^{II}$O$_{1-y}$(OH)$_y$-[O$_2$C–C$_5$H$_4$N]$_5$[O$_2$CCH$_3$] ($x \sim 0.15$) [171]. When this material was placed in water, nicotinic acid was released over time. A similar strategy has also been reported for the release of peptides [172]. A parallel strategy employs a bioactive metallic salt that can be included as metallic partner in the structure of hybrid materials. Silver, which is known for its bactericidal action, was found to be suitable for this strategy. The first results were obtained with coordination polymers formed by isonicotinic-functionalized ethylene glycol [173]. Besides these results that made use of nitrogen atom from pyridyl moiety as coordination site, the use of phosphonic acids was investigated for this purpose. The idea was to use a hard

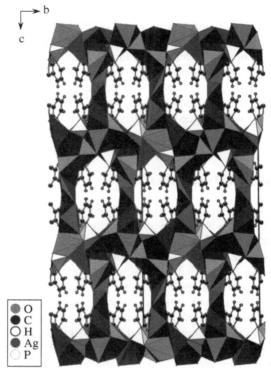

Ag$_6$(m-PO$_3$–C$_6$H$_4$–COO)$_2$, *Pbca*, orthorhombic
$a = 7.9377(7)$ Å, $b = 16.1575(10)$ Å, $c = 29.873(7)$ Å.

FIGURE 9.35 Structure of Ag$_6$(m-PO$_3$–C$_6$H$_4$–COO)$_2$. Adapted from Ref. [162].

base (deprotonated phosphonic acid), as defined by the Pearson theory of hard and soft acid and base (HSAB), in association with a soft acid (Ag$^+$) in order to produce a material possessing a limited stability; this instability favours the release of silver salts when the material is placed in a hydrated environment. Accordingly, Ag$_3$(m-O$_3$P–C$_6$H$_4$–CO$_2$) (Figure 9.35), which was synthesized from the rigid hetero-difunctional phosphonobenzoic acid **3** [162], was able to inhibit the development of Gram-negative (*Escherichia coli, Pseudomonas aeruginosa*) and, to a lesser extent, Gram-positive (*Staphylococcus aureus*) bacterial strains [174] (both types included clinically relevant bacterial strains). The quantification of silver release indicated that the silver concentration of the supernatant water solution was equal to 0.6 mM. However, it was shown that this silver release represents an equilibrium since the replacement of the supernatant water solution by freshwater solution leads to a new equilibrium. The repetition of this procedure naturally led to a full dissolution of this material in water. In addition, it was shown that the bactericidal effect was due to the silver salts and not from the organic compound **3** that was simultaneously release in solution.

This study demonstrated that phosphonic acid-based hybrid was able to act as a reservoir of silver ions with progressive release of this bactericidal salt in water solution in function of time. However, the next issue, which is worth to be addressed, involves the kinetics of silver release. Indeed, it would be of a great interest to produce hybrid materials exhibiting diverse silver ion release capacities in order to select one material for one specific application (e.g. topical treatment, dressing). The design of specific organic precursors that would produce silver-based hybrid material with a controllable stability when placed in water or in wet media is worthy of further investigation.

9.8   CONCLUSION AND PERSPECTIVES

The use of rigid organic precursors functionalized with phosphonic acid functional groups likely favours the formation of crystalline hybrids as it was previously observed with MOF synthesized from polycarboxylic acid. Interestingly, many experimental parameters can be tuned including the temperature, the stoichiometry, the pressure, the pH, etc., thus leading to the production of a wide range of topologies that are so far very difficult to predict. Moreover, the addition of auxiliary ligands (phenanthroline, bipyridine) or structure-directing agent that can be neutral (e.g. amine, crown ether), cationic (e.g. ammonium salt or inorganic cation) or negative (e.g. fluorine) increases the number of synthetic possibilities and thus multiplies the number of materials. Moreover, the structure of the rigid platform (e.g. benzene ring, fused aromatic ring, heteroaromatic, etc.) constitutes another parameter of material diversity. However, from all the structures published from rigid polyphosphonic acid derivatives, it is clearly evident that the geometry of these rigid precursors directly influences the topology of the hybrids as reported for materials based on 1,4-diphosphonic acid **2** and 1,2-diphosphonic acid **4**. Moreover, these materials, especially when they are dehydrated, possessed a high thermal stability that can be partly attributed to the stability of the aromatic ring present in the organic precursor. The use of rigid heterofunctional precursors extends the synthetic possibilities and produces additional original structures. Indeed, when two different functional groups were bonded to a rigid platform, the pH of the reaction media directly influenced the protonation state of these functional groups and therefore can influence their connection with the inorganic network. Phosphonobenzoic acid (3- or 4-regiosisomer) provides an excellent illustration. At low pH, only the phosphonic acid was connected to the inorganic framework, while at higher pH, both functional groups were connected, thus producing a 3D framework, and interestingly, in a few cases, a homochiral structure was produced. In terms of properties, the thermal stability is an important feature of this type of material. As discussed previously and depending on the metal involved in the hybrids, original luminescent and magnetic properties were reported. More recently, the use of hybrid materials as reservoir of bioactive metallic salts constitutes another interesting field of development.

The question is now, 'What could be the next development of phosphonic acid-based hybrids within the next decades?' The use of rigid precursors that possess a limited number of conformations should render the type of hybrid structure produced more predictable. Accordingly, it can be anticipated that on the basis of computational developments, prediction should be possible allowing a reduction in the number of syntheses and also the ability to predict the properties of these hybrids (e.g. luminescence, magnetism). For this purpose, all the materials already reported and all those that will be reported in the future will form a database that will be the basis of these computational approaches. The synthesis of new rigid platforms, and particularly the synthesis of precursors with increased dimensionality or exhibiting different geometries (e.g. planar, tetrahedral), will certainly yield hybrids with singular properties. The post-functionalization of hybrid by reaction on the aromatic ring or the use of functional groups that are not connected to the inorganic framework is worthy of further study. Finally, we can also expect further developments towards applications in the field of biology.

REFERENCES

1. K. Biradha, A. Ramanan, J.J. Vittal, *Cryst. Growth Des.*, **2009**, *9*, 2969–2970.
2. M. Bujoli-Doeuff, M. Evain, P.A. Jaffres, V. Caignaert, B. Bujoli, *Int. J. Inorg. Mater.*, **2000**, *2*, 557–560.
3. A. Clearfield, *J. Alloys Compd.*, **2006**, *418*, 128–138.
4. J.-G. Mao, *Coord. Chem. Rev.*, **2007**, *251*, 1493–1520.
5. K. Maeda, *Microporous Mesoporous Mater.*, **2004**, *73*, 47–55.
6. M. Zheng, Y. Liu, C. Wang, S. Liu, W. Lin, *Chem. Sci.*, **2012**, *3*, 2623–2627.
7. F. Siméon, P.A. Jaffrès, D. Villemin, *Tetrahedron*, **1998**, *54*, 10111–10118.
8. R. Engel, J.L.I. Cohen, *Synthesis of carbon-phosphorus bonds*, 2nd edition, **2004**, CRC Press.
9. P. Tavs, *Chem. Ber.*, **1970**, *103*, 2428–2436.
10. G.B. Hix, V. Caignaert, J.M. Rueff, L. Le Pluart, J.E. Warren, P.A. Jaffrès, *Cryst. Growth Des.*, **2007**, *7*, 208–211.
11. D. Villemin, P.-A. Jaffrès, F. Siméon, *Phosphorus Sulf. Silicon*, **1997**, *130*, 59–63.
12. (a) T. Hirao, T. Masunaga, Y. Ohshiro, T. Agawa, *Tetrahedron Lett.*, **1980**, *21*, 3595; (b) T. Hirao, T. Masunaga, N. Yamada, *Bull. Chem. Soc. Jpn.*, **1982**, *55*, 909–913.
13. Y. Belabassi, S. Alzghari, J.L. Montchamp, *J. Org. Chem.*, **2008**, *693*, 3171–3178.
14. M. Kaleck, A. Aiadl, J. Stawinski, *Org. Lett.*, **2008**, *10*, 4637–4640.
15. D. Gelman, L. Jiang, S.L. Buchwald, *Org. Lett.*, **2003**, *5*, 2315–2318.
16. N.B. Karlstedt, I.P. Beletskaya, *Russ. J. Org. Chem.*, **2011**, *47*, 1011–1014.
17. R. Zhuang, J. Xu, Z. Cai, G. Tang, M. Fang, Y. Zhao, *Org. Lett.*, **2011**, *13*, 2110–2113.
18. J.-M. Rueff, V. Caignaert, S. Chausson, A. Leclaire, C. Simon, O. Perez, L. le Pluart, P.-A. Jaffrès, *Eur. J. Inorg. Chem.*, **2008**, *26*, 4117–4125.
19. L. Delain-Bioton, J.F. Lohier, D. Villemin, J. Sopkova, G.B. Hix, P.A. Jaffrès, *Acta cryst.*, **2008**, *C64*, o47–o49.
20. P.-A. Jaffrès, N. Bar, D. Villemin, *J. Chem. Soc. Perkin Trans. 1*, **1998**, *13*, 2083–2089.
21. C.E. McKenna, M.T. Higa, N.H. Cheung, M.C. McKenna, *Tetrahedron Lett.*, **1977**, *18*, 155–158.

22. N. Bischofberger, E. Waldmann, O. Saito, E.S. Simon, A. Lees, M.D. Bednarski, G.M. Whitesides, *J. Org. Chem.*, **1988**, *53*, 3457–3465.

23. P.-A. Jaffrès, V. Caignaert, D. Villemin, *Chem. Commun.*, **1999**, *19*, 1997–1998.

24. A. Holy, *Synthesis*, **1998**, *4*, 381–385.

25. S.S. Iremonger, J. Liang, R. Vaidhyanathana, G.K.H. Shimizu, *Chem. Comm.*, **2011**, *47*, 4430–4432.

26. C. Sanchez, B. Julián, P. Belleville, M. Popall, *J. Mater. Chem.*, **2005**, *15*, 3559–3592.

27. C. Sanchez, F. Ribot, *New J. Chem.*, **1994**, *18*, 1007.

28. C.S. Cundy, P.A. Cox, *Chem. Rev.*, **2003**, *103*, 663–701.

29. A.K. Cheetham, G. Férey, T. Loiseau, *Angew. Chem. Int. Ed.*, **1999**, *38*, 3268–3292.

30. O.M. Yaghi, H. Li, *J. Am. Chem. Soc.*, **1995**, *117*, 10401–10402.

31. G. Ferey, *Chem. Soc. Rev.*, **2008**, *37*, 191–214.

32. (a) C. Mellot Draznieks, J.M. Newsam, A.M. Gorman, C.M. Freeman, G. Férey, *Angew. Chem. Int. Ed.*, **2000**, *39*, 2270–2275; (b) C. Mellot Draznieks, *J. Mater. Chem.*, **2007**, *17*, 4348–4358.

33. (a) M. Haouas, C. Gérardin, F. Taulelle, C. Estournes, T. Loiseau, G. Ferey, *J. Chim. Phys.*, **1998**, *95*, 302–309; (b) L. Allouche, C. Gérardin, T. Loiseau, G. Ferey, F. Taulelle, *Angew. Chem. Int. Ed.*, **2000**, *39*, 511–514; (c) F. Taulelle, M. Pruski, J.-P. Amoureux, D. Lang, A. Bailly, C. Huguenard, M. Haouas, C. Gérardin, T. Loiseau, G. Férey, *J. Am. Chem. Soc.*, **1999**, *121*, 12148–12153.

34. (a) T.B. Liao, Y. Ling, Z.-X. Chen, Y.-M. Zhou, L.-H. Weng, *Chem. Commun.*, **2010**, *46*, 1100–1102; (b) Z. Chen, Y. Zhou, L. Weng, D. Zhao, *Cryst. Growth Des.*, **2008**, *8*, 4045–4053; (c) Z. Chen, Y. Zhou, L. Weng, C. Yuan, D. Zhao, *Chem. Asian J.*, **2007**, *2*, 1549–1554.

35. A. Clearfield and K. Demadis, ed., *Metal Phosphonate Chemistry: From Synthesis to Applications*, RSC, Cambridge, **2012**.

36. G. Alberti, U. Costantino, S. Allulli, N. Tomassini, *J. Inorg. Nucl. Chem.*, **1978**, *40*, 1113–1117.

37. M.D. Poojary, H.-L. Hu, F.L. Campbell, A. Clearfield, *Acta Cryst.*, **1993**, *B49*, 996–1001.

38. K.J. Gagnon, H.P. Perry, A. Clearfield, *Chem. Rev.*, **2012**, *112*, 1034–1054.

39. T.J.R. Weakley, *Acta Cryst.*, **1976**, *B32*, 2889–2890.

40. A. Clearfield, *J. Mol. Catal.*, **1984**, *27*, 251–262.

41. (a) G. Alberti, S. Murcia-Mascaros, R. Vivani, *Mater. Chem. Phys.*, **1993**, *35*, 187–192; (b) G. Alberti, E. Giontella, S. Murcia-Mascaros, R. Vivani, *Inorg. Chem.*, **1998**, *37*, 4672–4676.

42. J.D. Wang, A. Clearfield, P. Guang-Zhi, *Mater. Chem. Phys.*, **1993**, *35*, 208–216.

43. G. Alberti, F. Marmottini, R. Vivani, P. Zappelli, *J. Porous Mater.*, **1998**, *5*, 221–226.

44. G. Alberti, F. Marmottini, S. Murcia-Mascaros, R. Vivani, *Angew. Chem. Int. Ed.*, **1994**, *33*, 1594–1597.

45. G. Alberti, U. Costantino, F. Marmottini, R. Vivani, P. Zappelli, *Angew. Chem. Int. Ed.*, **1993**, *32*, 1357–1359.

46. A. Clearfield, Z. Wang, P. Bellinghausen, *J. Solid State Chem.*, **2002**, *167*, 376–385.

47. D.M. Poojary, B. Zhang, P. Bellinghausen, A. Clearfield, *Inorg. Chem.*, **1996**, *35*, 4942–4949.

48. D.M. Poojary, B. Zhang, P. Bellinghausen, A. Clearfield, *Inorg. Chem.*, **1996**, *35*, 5254–5263.

49. K.J. Martin, P.J. Squattrito, A. Clearfield, *Inorg. Chim. Acta*, **1989**, *155*, 7–9.

50. Y. Zhang, A. Clearfield, *Inorg. Chem.*, **1992**, *31*, 2821–2826.

51. B. Zhang, D.M. Poojary, A. Clearfield, *Inorg. Chem.*, **1998**, *37*, 1844–1852.

52. Z. Wang, J.M. Heising, A. Clearfield, *J. Am. Chem. Soc.*, **2003**, *125*, 10375–10383.

53. (a) E. Brunet, H.M.H. Alhendawi, C. Cerro, M. José de la Mata, O. Juanes, J.C. Rodríguez-Ubis, *Angew. Chem. Int. Ed.*, **2006**, *45*, 6918–6920; (b) E. Brunet, H.M.H. Alhendawi, C. Cerro, M. José de la Mata, O. Juanes, J.C. Rodríguez-Ubis, *Chem. Eng. J.*, **2010**, *158*, 333–344.

54. A. Subbiah, N. Bhuvanesh, A. Clearfield, *J. Solid State Chem.*, **2005**, *178*, 1321–1325.

55. S. Kirumakki, J. Huang, A. Subbiah, J. Yao, A. Rowland, B. Smith, A. Mukherjee, S. Samarajeewa, A. Clearfield, *J. Mater. Chem.*, **2009**, *19*, 2593–2603.

56. W. Ouellette, G. Wang, H. Liu, G.T. Yee, C.J. O'Connor, J. Zubieta, *Inorg. Chem.*, **2009**, *48*, 953–963.

57. W. Ouellette, M.H. Yu, C.J. O'Connor, J. Zubieta, *Inorg. Chem.*, **2006**, *45*, 3224–3239.

58. P. De Burgomaster, W. Ouellette, H. Liu, C.J. O'Connor, J. Zubieta, *CrystEngComm*, **2010**, *12*, 446–469.

59. P. De Burgomaster, H. Liu, W. Ouellette, C.J. O'Connor, J. Zubieta, *Inorg. Chim. Acta*, **2010**, *363*, 4065–4073.

60. (a) J.M. Breen, W. Schmitt, *Angew. Chem. Int. Ed.*, **2008**, *47*, 6904–6908; (b) J.M. Breen, R. Clerac, L. Zhang, S.M. Cloonan, E. Kennedy, M. Feeney, T. McCabe, D.C. Williams, W. Schmitt, *Dalton Trans.*, **2012**, *41*, 2918–2926.

61. (a) Z. Amghouz, J.R. Garcia, S. Garcia-Granda, A. Clearfield, J. Rodriguez Fernandez, I. de Pedro, J.A. Blanco, *J. Alloys Compd.*, **2012**, *536*, S499–S503; (b) Z. Amghouz, S. Garcia-Granda, J.R. Garcia, A. Clearfield, R. Valiente, *Cryst. Growth Des.*, **2011**, *11*, 5289–5297.

62. P.O. Adelani, T.E. Albrecht-Schmitt, *J. Solid State Chem.*, **2011**, *184*, 2368–2373.

63. P.O. Adelani, T.E. Albrecht-Schmitt, *Inorg. Chem.*, **2009**, *48*, 2732–2734.

64. P.O. Adelani, A.G. Oliver, T.E. Albrecht-Schmitt, *Cryst. Growth Des.*, **2011**, *11*, 1966–1973.

65. P.O. Adelani, T.E. Albrecht-Schmitt, *Angew. Chem. Int. Ed.*, **2010**, *49*, 8909–8911.

66. P.O. Adelani, T.E. Albrecht-Schmitt, *Inorg. Chem.*, **2011**, *50*, 12184–12191.

67. P.O. Adelani, T.E. Albrecht-Schmitt, *Cryst. Growth Des.*, **2011**, *11*, 4227–4237.

68. J. Diwu, S. Wang, J.J. Good, V.H. DiStefano, T.E. Albrecht-Schmitt, *Inorg. Chem.*, **2011**, *50*, 4842–4850.

69. H. Hall, J.C. Sullivan, P.G. Rickert, K.L. Nash, *Dalton Trans.*, **2005**, *11*, 2011–2016.

70. D. Kong, J. Zoń, J. McBee, A. Clearfield, *Inorg. Chem.*, **2006**, *45*, 977–986.

71. J.M. Taylor, R.K. Mah, I.L. Moudrakovski, C.I. Ratcliffe, R. Vaidhyanathan, G.K.H. Shimizu, *J. Am. Chem. Soc.*, **2010**, *132*, 14055–14057.

72. S.R. Kim, K.W. Dawson, B.S. Gelfand, J.M. Taylor, G.K.H. Shimizu, *J. Am. Chem. Soc.*, **2013**, *135*, 963–966.

73. T. Araki, A. Kondo, K. Maeda, *Chem. Commun.*, **2013**, *49*, 552–554.

74. M. Pramanick, A.K. Patra, A. Bhaumik, *Dalton Trans.*, **2013**, *42*, 5140–5149.

75. J.M. Taylor, R. Vaidhyanathan, S.S. Iremonger, G.K.H. Shimizu, *J. Am. Chem. Soc.*, **2012**, *134*, 14338–14340.

76. J. Svoboda, V. Zima, L. Benes, K. Melanova, M. Vlcek, *Inorg. Chem.*, **2005**, *44*, 9968–9976.

77. V. Zima, J. Svoboda, L. Benes, K. Melanova, M. Trchova, J. Dybal, *J. Solid State Chem.*, **2007**, *180*, 929–939.

78. J. Svoboda, V. Zima, L. Benes, K. Melanova, M. Trchova, M. Vlcek, *Solid State Sci.*, **2008**, *10*, 1533–1542.

79. D.M. Zang, D.K. Cao, L.M. Zheng, *Inorg. Chem. Comm.*, **2011**, *14*, 1920–1923.

80. V. Zima, J. Svoboda, L. Benes, K. Melanova, M. Trchova, A. Ruzicka, *J. Solid State Chem.*, **2009**, *182*, 3155–3161.

81. V. Zima, J. Svoboda, Y.C. Yang, S.L. Wang, *CrystEngComm*, **2012**, *14*, 3469–3477.

82. J.-T. Li, D.K. Cao, B. Liu, Y.Z. Li, L.M. Zheng, *Cryst. Growth Des.*, **2008**, *8*, 2950–2953.

83. Z. Chen, Y. Ling, H. Yang, Y. Guo, L. Weng, Y. Zhou, *CrystEngComm*, **2011**, *13*, 3378–3382.

84. Z. Chen, H. Yang, M. Deng, Y. Ling, L. Weng, Y. Zhou, *Dalton Trans.*, **2012**, *41*, 4079–4083.

85. P.F. Wang, Y. Duan, D.K. Cao, Y.Z. Li, L.M. Zheng, *Dalton Trans.*, **2010**, *39*, 4559–4565.

86. P. De Burgomaster, A. Aldous, H. Liu, C.J. O'Connor, J. Zubieta, *Cryst. Growth Des.*, **2010**, *10*, 2209–2218.

87. J.T. Li, L.R. Guo, Y. Shen, L.M. Zheng, *CrystEngComm*, **2009**, *11*, 1674–1678.

88. J.T. Li, D.K. Cao, T. Akutagawa, L.M. Zheng, *Dalton Trans.*, **2010**, *39*, 8606–8608.

89. J.-M. Rueff, N. Barrier, S. Boudin, V. Dorcet, V. Caignaert, P. Boullay, G. Hix, P.-A. Jaffrès, *Dalton Trans.*, **2009**, *47*, 10614–10620.

90. P.O. Adelani, T.E. Albrecht-Schmitt, *Inorg. Chem.*, **2010**, *49*, 5701–5705.

91. P.O. Adelani, A.G. Oliver, T.E. Albrecht-Schmitt, *Cryst. Growth Des.*, **2011**, *11*, 3072–3080.

92. J.-M. Rueff, O. Perez, A. Leclaire, H. Couthon-Gouvrès, P.-A. Jaffres, *Eur. J. Inorg. Chem.*, **2009**, *32*, 4870–4876

93. C. Fang, Z. Chen, X. Liu, Y. Yang, M. Deng, L. Weng, Y. Jia, Y. Zhou, *Inorg. Chim. Acta*, **2009**, *362*, 2101–2107.

94. P.F. Wang, D.K. Cao, S.S. Bao, H.J. Jin, Y.Z. Li, T.W. Wang, L.M. Zheng, *Dalton Trans.*, **2011**, *40*, 1307–1312.

95. J.T. Li, T.D. Keene, D.K. Cao, S. Decurtins, L.M. Zheng, *CrystEngComm*, **2009**, *11*, 1255–1260.

96. X.J. Yang, S.S. Bao, T. Zheng, L.M. Zheng, *Chem. Commun.*, **2012**, *48*, 6565–6567.

97. P.O. Adelani, T.E. Albrecht-Schmitt, *Cryst. Growth Des.*, **2011**, *11*, 4676–4683.

98. (a) J.P. Mercier, P. Morin, M. Dreux, A. Tambute, *Chromatographia*, **1998**, *48*, 529–534; (b) W. El Malti, D. Laurencin, G. Guerrero, M.E. Smith, P.H. Mutin, *J. Mater. Chem.*, **2012**, *22*, 1212–1218.

99. J.-M. Rueff, V. Caignaert, A. Leclaire, C. Simon, J.-P. Haelters, P.-A. Jaffrès, *CrystEngComm*, **2009**, *11*, 556–559.

100. S. Bauer, N. Stock, *J. Solid State Chem.*, **2007**, *180*, 3111–3120.

101. S. Bauer, N. Stock, *Z. Anorg. Allg. Chem.*, **2008**, *634*, 131–136.

102. Y. Ling, T.B. Liao, Z.X. Chen, Y.M. Zhou, L.H. Weng, *Dalton Trans.*, **2010**, *39*, 10712–10718.

103. Y. Ling, M. Deng, Z. Chen, B. Xia, X. Liu, Y. Yang, Y. Zhou, L. Weng, *Chem. Commun.*, **2013**, *49*, 78–80.

104. M. Deng, Y. Ling, B. Xia, Z. Chen, Y. Zhou, X. Liu, B. Yue, H. He, *Chem. Eur. J.*, **2011**, *17*, 10323–10328.

105. T. Yamada, H. Kitagawa, *CrystEngComm*, **2012**, *14*, 4148–4152.

106. P.F. Wang, Y. Duan, T.W. Wang, Y.Z. Li, L.M. Zheng, *Dalton Trans.*, **2010**, *39*, 10631–10636.

107. H.J. Jin, P.F. Wang, C. Yao, L.M. Zheng, *Inorg. Chem. Commun.*, **2011**, *14*, 1677–1680.

108. H.J. Jin, P.F. Wang, S.S. Bao, L.M. Zheng, C. Yao, *Sci. China Chem.*, **2012**, *55*, 1047–1054.

109. J.-M. Rueff, O. Perez, C. Simon, C. Lorilleux, H. Couthon-Gouvrès, P.-A. Jaffres, *Cryst. Growth Des.*, **2009**, *9*, 4262–4268.

110. L.H. Kullberg, A. Clearfield, *Solv. Extr. Ion Exch.*, **1990**, *8*, 187–197.

111. M. Curini, O. Rosati, E. Pisani, *Synlett*, **1996**, *4*, 333–334.

112. G. Alberti, M. Casciola, R. Palombari, A. Peraio, *Solid State Ionics*, **1992**, *58*, 339–344.

113. G. Alberti, M. Casciola, U. Costantino, A. Peraio, E. Montoneri, *Solid State Ionics*, **1992**, *50*, 315–322.

114. R. Ferreira, P. Pires, B. de Castro, R.A. Sa Ferreira, L.D. Carlos, U. Pischel, *New J. Chem.*, **2004**, *28*, 1506–1513.

115. G. Alberti, M. Casciola, A. Donnadio, P. Piaggio, M. Pica, M. Sisani, *Solid State Ionics*, **2005**, *176*, 2893–2898.

116. J.L. Colon, D.S. Thakur, C.Y. Yang, A. Clearfield, C.R. Martin, *J. Catal.*, **1990**, *124*, 148–159.

117. M. Curini, F. Epifano, M.C. Marcotullio, O. Rosati, U. Costantino, *Tetrahedron Lett.*, **1998**, *39*, 8159–8162.

118. M. Curini, F. Epifano, M.C. Marcotullio, O. Rosati, U. Costantino, *Synthetic Comm.*, **1999**, *29*, 541–546.

119. O. Rosati, M. Curini, F. Montanari, M. Nocchetti, S. Genovese, *Catal. Lett.*, **2011**, *141*, 850–853.

120. O. Rosati, M. Curini, F. Messina, M.C. Marcotullio, G. Cravotto, *Catal. Lett.*, **2013**, *143*, 169–175.

121. B. Bonnet, D.J. Jones, J. Roziere, L. Tchicaya, G. Alberti, M. Casciola, L. Massinelli, B. Bauer, A. Peraio, E. Ramunni, *J. New. Mater. Electrochem. Syst.*, **2000**, *3*, 87–92.

122. G. Alberti, M. Casciola, E. D'Alessandro, M. Pica, *J. Mater. Chem.*, **2004**, *14*, 1910–1914.

123. M. Casciola, G. Alberti, A. Ciarletta, A. Cruccolini, P. Piaggio, M. Pica, *Solid State Ionics*, **2005**, *176*, 2985–2989.

124. (a) M. Casciola, D. Capitani, A. Comite, A. Donnadio, V. Frittella, M. Pica, M. Sganappa, A. Varzi, *Fuel Cells*, **2008**, *8*, 217–224; (b) M. Casciola, D. Capitani, A. Donnadio, V. Frittella, M. Pica, M. Sganappa, *Fuel Cells*, **2009**, *9*, 381–386.

125. V. Zima, J. Svoboda, K. Melanova, L. Benes, M. Casciola, M. Sganappa, J. Brus, M. Trchova, *Solid State Ionics*, **2010**, *181*, 705–713.

126. (a) G. Alberti, U. Costantino, M. Casciola, S. Ferroni, L. Massinelli, P. Staiti, *Solid State Ionics*, **2001**, *145*, 249–255; (b) E. W. Stein, A. Clearfield, M.A. Subramanian, *Solid State Ionics*, **1996**, *83*, 113–124.

127. A. Clearfiled, *Dalton Trans.*, **2008**, 6089–6102.

128. Z.Y. Du, H.B. Xu, J.G. Mao, *Inorg. Chem.*, **2006**, *45*, 6424–6430.

129. Z.Y. Du, H.B. Xu, J.G. Mao, *Inorg. Chem.*, **2006**, *45*, 9780–9788.

130. Z.Y. Du, X.L. Li, Q.Y. Liu, J.G. Mao, *Cryst. Growth Des.*, **2007**, *7*, 1501–1507.

131. Z.Y. Du, A.V. Prosvirin, J.G. Mao, *Inorg. Chem.*, **2007**, *46*, 9884–9894.

132. Z.Y. Du, H.B. Xu, X.L. Li, J.G. Mao, *Eur. J. Inorg. Chem.*, **2007**, *28*, 4520–4529.

133. Z.Y. Du, J.J. Huang, Y.R. Xie, H.R. Wen, *J. Mol. Struct.*, **2009**, *919*, 112–116.

134. P. Maniam, N. Stock, *Acta Cryst. Section C Cryst. Struct. Comm.*, **2011**, *67*, m73–m76.

135. P. Maniam, N. Stock, *Zeit. Anorg. All. Chem.*, **2011**, *637*, 1145–1151.

136. P. Maniam, C. Naether, N. Stock, *Eur. J. Inorg. Chem.*, **2010**, *24*, 3866–3874.

137. O.R. Evans, H.L. Ngo, W. Lin, *J. Am. Chem. Soc.*, **2001**, *123*, 10395–10396.

138. H.L. Ngo, W. Lin, *J. Am. Chem. Soc.*, **2002**, *124*, 14298–14299.

139. H.L. Ngo, A. Hu, W. Lin, *J. Mol. Catal. A Chem.*, **2004**, *215*, 177–186.

140. A. Hu, H.L. Ngo, W. Lin, *Angew. Chem. Int. Ed.*, **2003**, *42*, 6000–6003.

141. A. Hu, H.L. Ngo, W. Lin, *J. Am. Chem. Soc.*, **2003**, *125*, 11490–11491.

142. P. Ayyappan, O.R. Evans, B.M. Foxman, K.A. Wheeler, T.H. Warren, W. Lin, *Inorg. Chem.*, **2001**, *40*, 5954–5961.

143. H. Perry, J. Zon, J. Law, A. Clearfield, *J. Solid State Chem.*, **2010**, *183*, 1165–1173.

144. S. Konar, J. Zon, A.V. Prosvirin, K.R. Dunbar, A. Clearfield, *Inorg. Chem.*, **2007**, *46*, 5229–5236.

145. J.L. Zhou, X.Q. He, B. Liu, Y. Huo, S.C. Zhang, X.G. Chun, *Trans. Metal Chem.*, **2010**, *35*, 795–800.

146. C.R. Samanamu, M.M. Olmstead, J.L. Montchamp, A.F. Richards, *Inorg. Chem.*, **2008**, *47*, 3879–3887.

147. J.A. Fry, C.R. Samanamu, J.L. Montchamp, A.F. Richards, *Eur. J. Inorg. Chem.*, **2008**, *3*, 463–470.

148. F. Odobel, B. Bujoli, D. Massiot, *Chem. Mater.*, **2001**, *13*, 163–173.

149. P.F. Wang, S.S. Bao, S.M. Zhang, D.K. Cao, X.G. Liu, L.M. Zheng, *Eur. J. Inorg. Chem.*, **2010**, *6*, 895–901.

150. P.F. Wang, Y. Duan, L.M. Zheng, *Sci. China Chem.*, **2010**, *53*, 2112–2117.

151. P.F. Wang, Y. Duan, J.M. Clemente-Juan, Y. Song, K. Qian, S. Gao, L.M. Zheng, *Chem. Eur. J.*, **2011**, *17*, 3579–3583.

152. Y.F. Yang, Y.S. Ma, L.R. Guo, L.M. Zheng, *Cryst. Growth Des.*, **2008**, *8*, 1213–1217.

153. Y.F. Yang, Y.S. Ma, S.S. Bao, L.M. Zheng, *Dalton Trans.*, **2007**, *37*, 4222–4226.

154. D. Kong, A. Clearfield, *Cryst. Crowth Des.*, **2005**, *5*, 1263–1270.

155. P. Gerbier, C. Guérin, B. Henner, J.-R. Unal, *J. Mater. Chem.*, **1999**, *9*, 2559–2565.

156. M. Gulea, A.C. Gaumont, in *Targets in Heterocyclic Systems*, eds. O.A. Attanasi, D. Spinelli, Italian Society of Chemistry, **2008**, vol. 12, pp. 328–348.

157. J.-M. Rueff, O. Perez, A. Pautrat, N. Barrier, G.B. Hix, S. Hernot, H. Couthon-Gourvès, P.-A. Jaffrès, *Inorg. Chem.*, **2012**, *51*, 10251–10261.

158. (a) J. Salta, J. Zubieta, *J. Cluster Sci.*, **1996**, *7*, 531–551; (b) G. Yucesan, W. Ouellette, V. Golub, C.J. O'Connor, J. Zubieta, *Solid State Sci.*, **2005**, *7*, 445–458; (c) Y.-D. Chang, J. Zubieta, *Inorg. Chim. Acta*, **1996**, *245*, 177–198.

159. L.-R. Guo, J.-W. Tong, X. Liang, J. Köhler, J. Nuss, Y.-Z. Li, L.-M. Zheng, *Dalton Trans.*, **2011**, *40*, 6392–6400.

160. L.-R. Guo, S.-S. Bao, Y.-Z. Li, L.-M. Zheng, *Chem. Commun.*, **2009**, *20*, 2893–2895.

161. G. Hix, in *Metal Phosphonate Chemistry: From Synthesis to Applications*, eds. A. Clearfield, K. Demadis, RSC, Cambridge, **2012**, pp. 525–550.

162. R. Singleton, J. Bye, J. Dyson, G. Baker, R.M. Ranson, G.B. Hix, *Dalton Trans.*, **2010**, *39*, 6024–6030.

163. P.F. Wang, Y. Duan, D.K. Cao, Y.Z. Li and L.M. Zheng, *Dalton Trans.*, **2010**, *39*, 4559–4565.

164. (a) G. Liu, J.V. Beitz, in *The Chemistry of the Actinide and Transactinide Elements*, eds. L.R. Morss, N.M. Edelstein, J. Fuger, Springer, Berlin, **2006**, p. 2088; (b) R.G. Denning, J.O.W. Norris, I.G. Short, T.R. Snellgrove, D.R. Woodwark, in *Lanthanide and Actinide Chemistry and Spectroscopy* (ACS Symposium Series No. 131), ed. N.M. Edelstein, American Chemical Society, Washington, DC, **1980**, 472 pages.

165. F. Serpaggi, G. Férey, *J. Mater. Chem.*, **1998**, *8*, 2749–2755.

166. G. Cao, V.M. Lynch, J.S. Swinnea, T.E. Mallouk, *Inorg. Chem.*, **1990**, *29*, 2112–2117.

167. (a) P. Horcajada, C. Serre, M. Vallet-Regi, M. Sebban, F. Taulelle, G. Férey, *Angew. Chem. Int. Ed.*, **2006**, *45*, 5974–5978; (b) P. Horcajada, C. Serre, G. Maurin, N.A. Ramsahye, F. Balas, M. Vallet-Regi, M. Seban, F. Taulelle, G. Férey, *J. Am. Chem. Soc.*, **2008**, *130*, 6774–6780.

168. S. Josse, C. Faucheux, A. Soueidan, G. Grimandi, D. Massiot, B. Alonso, P. Janvier, S. Laïb, P. Pilet, O. Gauthier, G. Daculsi, J. Guicheux, B. Bujoli, J.M. Bouler, *Biomaterials*, **2005**, *26*, 2073–2080.

169. P. Horcajada, T. Chalati, C. Serre, B. Gillet, C. Sebrie, T. Baati, J.F. Eubank, D. Heurtaux, P. Clayette, C. Kreuz, et al. *Nat. Mater.*, **2010**, *9*, 172–178.

170. (a) P.S. Wheatley, A.R. Butler, M.S. Crane, S. Fox, B. Xiao, A.G. Rossi, I.L. Megson, R.E. Morris, *J. Am. Chem. Soc.*, **2006**, *128*, 502–509; (b) A.C. McKinlay, B. Xiao, D.S. Wragg, P.S. Wheatley, I.L. Megson, R.E. Morris, *J. Am. Chem. Soc.*, **2008**, *130*, 10440–10444.

171. S.R. Miller, D. Heurtaux, T. Baati, P. Horcajada, J.M. Grenèche, C. Serre, *Chem. Commun.*, **2010**, *46*, 4526–4528.

172. J. Rabone, Y.F. Yue, S.Y. Chong, K.C. Stylianou, J. Bacsa, D. Bradshaw, G.R. Darling, N.G. Berry, Y.Z. Khimyak, A.Y. Ganin, et al. *Science*, **2010**, *329*, 1053–1057.

173. T.V. Slenters, J.L. Sagué, P.S. Brunetto, S. Zuber, A. Fleury, L. Mirolo, A.Y. Robin, M. Meuwly, O. Gordon, R. Landmann, et al., *Materials*, **2010**, *3*, 3407–3429.

174. M. Berchel, T. Le Gall, C. Denis, S. Le Hir, F. Quentel, C. Elléouet, T. Montier, J.-M. Rueff, J.-Y. Salaün, J.-P. Haelters, et al. *New J. Chem.*, **2011**, *35*, 1000–1003.

10

DRUG CARRIERS BASED ON ZIRCONIUM PHOSPHATE NANOPARTICLES

JORGE L. COLÓN AND BARBARA CASAÑAS

Department of Chemistry, University of Puerto Rico-Río Piedras Campus, San Juan, Puerto Rico

10.1 INTRODUCTION

Since the crystalline form of the tetravalent metal phosphate α-zirconium phosphate (α-ZrP) was reported 50 years ago by Clearfield and Stynes, along with its ion-exchange behaviour, a wide range of applications have been developed based on this acidic layered inorganic compound and its derivatives [1]. The applications for this material range from being used as catalyst [2–12], energy and electron transfer system [13–30], and drug carrier [31–33] to modified electrodes [30, 34–40]. In addition, this inorganic material can be modified by manipulating certain variables in its synthesis to obtain a material in the nanoscale, which provides additional possible applications in the area of nanotechnology [41]. Due to this diversity, ZrP has been extensively studied [1, 42–46].

Dedicated to Professor Abraham Clearfield, for introducing us to the fascinating chemistry of zirconium phosphates and phosphonates and for being a true mentor and better friend.

Tailored Organic–Inorganic Materials, First Edition. Edited by Ernesto Brunet, Jorge L. Colón and Abraham Clearfield.
© 2015 John Wiley & Sons, Inc. Published 2015 by John Wiley & Sons, Inc.

10.1.1 Zirconium Phosphates

Zirconium phosphates (ZrP) are members of the class of water-insoluble phosphates of tetravalent metals with a layered structure [5]. There are different phases of ZrP that vary in their interlayer structural arrangement and distance. Zirconium bis(monohydrogen orthophosphate) monohydrate ($Zr(HPO_4)_2 \cdot H_2O$, α-ZrP) has an interlayer distance of 7.6Å (Figure 10.1) with a layer thickness of 6.6Å and is the best characterized ZrP. This layered phosphate is an acidic ion exchanger that has been used for the immobilization of several photo- [1, 26–29, 47–50], bio- [31–33] and redox-active compounds [30, 39, 40, 51]. Figure 10.1 shows the structure and unit cell of this layered ZrP [52].

α-ZrP crystallizes in a monoclinic system, space group $P_{21/n}$ with $a = 9.060$Å, $b = 5.297$Å, $c = 15.414$Å and $\beta = 101.71$Å [53]. In α-ZrP, the zirconium atoms in each layer align nearly to a plane with bridging phosphate groups located alternately above and below the metal atom plane. Six oxygen atoms from different phosphate groups octahedrally coordinate the Zr atoms. Three oxygen atoms of each phosphate group are bonded to different Zr atoms, while the fourth one has a proton pointing into the interlayer space that can be exchanged [54]. This arrangement creates a zeolitic cavity per formula unit containing a water molecule each in the interlayer region (Figure 10.2) [55]. These cavities interconnect to each other by entrances with a diameter of 2.61Å [53]. Just a few cations can diffuse through this narrow space, limiting the direct ion-exchange capacity of this material [56, 57].

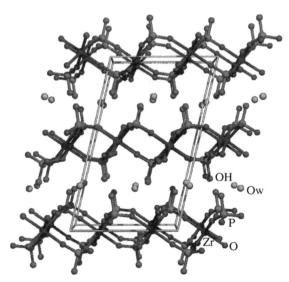

FIGURE 10.1 Structure and unit cell of α-$Zr(HPO_4)_2 \cdot H_2O$ viewed down the b-axis. Taken from Ref. [52].

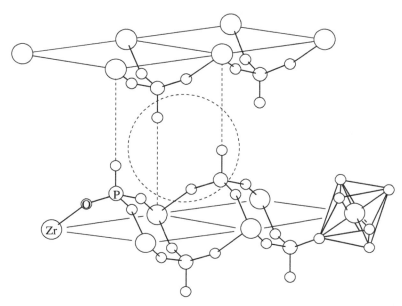

FIGURE 10.2 Schematic illustration of α-ZrP with the cavity indicated by the dashed circle. Reprinted from *J. Inorg. Nucl. Chem., 30*, Clearfield, A.; Blessing, R. H.; Stynes, J. A. , New Crystalline Phases of Zirconium Phosphate Possessing Ion-Exchange Properties., 2249–2258, Copyright 1968, with permission from Elsevier.

Other methods of intercalation that have been used in order to achieve intercalation of larger cations and molecules will be discussed next followed by explanation of the direct ion-exchange method and its modifications.

10.1.2 Pre-intercalation and the Exfoliation (Layer-by-Layer) Method

Direct exchange of large cations into the α-phase of ZrP cannot be performed due to the small interlayer distance [53, 57]. Previous researchers have overcome this limitation by using a pre-intercalated phase of α-ZrP with an expanded inter-layer distance produced either by ion exchange with sodium ions or by adding *n*-butylammonium or tetrabutylammonium (TBA) ions to α-ZrP [44, 58]. When an aqueous solution of TBA hydroxide $((C_4H_9)_4N^+OH^-)$ is used in this pre-intercalation method, α-ZrP can be expanded to 16.8 Å in interlayer distance (Figure 10.3) [59] This expanded phase allows the intercalation of larger species such as metal complexes through ion exchange of the alkylammonium ions. However, one of the problems of using this type of pre-intercalated phase to prepare exchange phases with large cations is that the pre-intercalated phase may not be completely ion exchanged with the desired large cation, depending on the amount of metal complex used in the ion-exchange reaction, and the presence of the pre-intercalant may interfere with the reactions under study.

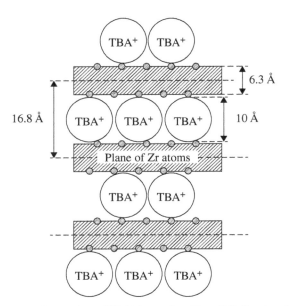

FIGURE 10.3 Schematic drawing of TBA⁺ intercalated in α-ZrP. Reprinted with permission from Kim, H.-N.; Keller, S. W.; Mallouk, T. E.; Schmitt, J.; Decher, G., Characterization of Zirconium Phosphate/Polycation Thin Films Grown by Sequential Adsorption Reactions. *Chem. Mater.* **1997**, *9*, 1414–1421, Copyright 1997 American Chemical Society.

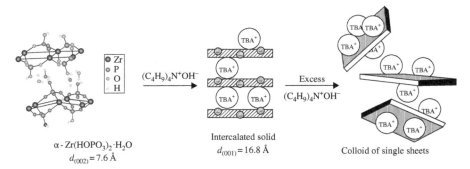

FIGURE 10.4 Exfoliation process of layered α-ZrP with an excess of TBA⁺. Reprinted with permission from Kim, H.-N.; Keller, S. W.; Mallouk, T. E.; Schmitt, J.; Decher, G., Characterization of Zirconium Phosphate/Polycation Thin Films Grown by Sequential Adsorption Reactions. *Chem. Mater.* **1997**, *9*, 1414–1421, Copyright 1997 American Chemical Society.

Adding an excess of TBA will cause the complete exfoliation of the α-ZrP layers obtaining separate layers suspended in the aqueous media (Figure 10.4). When a cationic modified surface is put in contact with the suspension, it will displace the loosely held TBA⁺ on one of the sides of the ZrP layer and anchor the layer to the surface due to electrostatic forces. Then the process is repeated but with the cationic species of interest and repeated until the desired multilayer thin film is achieved in what is called the layer-by-layer method [60].

10.1.3 Direct Ion Exchange of ZrP

The second main method of intercalation is the direct ion exchange of ZrP. Crystalline α-ZrP has Brönsted acid groups (P–OH), which, driven by an acid–base reaction, can intercalate cationic species such as amines. Other molecules can be inserted in the ZrP layers with the help of external driving forces, oxidation–reduction and esterification reactions as, for example, alkanols, ketones and amides, among others. As the ion-exchange intercalation process proceeds, the molecule goes through the interlayer space from the edge towards the inside. This process can be affected by the intercalant/host molar ratio, medium of the intercalation, crystallinity of ZrP and not least the structure and dimensions of the molecule.

Recently, Clearfield et al. reported the effect of the intercalation energy barrier on intercalation using hexylamine (planar) and cyclohexylamine (non-planar) cations in the intercalation process [61]. In addition, these authors studied the effect of the crystallinity of ZrP in the mechanism of the intercalation reaction using samples with either low or high crystallinity. Clearfield et al. found that the intercalation process takes place by two major steps. In the first step, the molecules enter through the edges of the α-ZrP layers, and in the second step, which is believed to be the slower process, the molecules diffuse from the edge towards the centre of the layers. When the intercalation energy barrier was low, these authors observed only a single-phase intercalation where the interlayer expansion is continuous as demonstrated with X-ray powder diffractometry (Figure 10.5a). This process occurred with the low

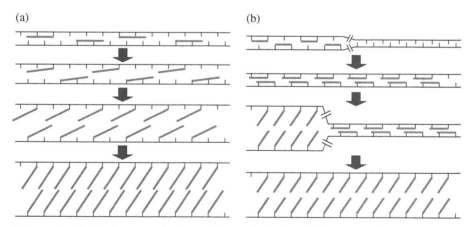

FIGURE 10.5 Different mechanisms of intercalation in ZrP depending if the reaction has a (a) low intercalation energy barrier or (b) high intercalation energy barrier. Reprinted from *J. Colloid Interf. Sci.*, 333, Sun, L.; O'Reilly, J. Y.; Kong, D.; Su, J. Y.; Boo, W. J.; Sue, H. J.; Clearfield, A., The effect of guest molecular architecture and host crystallinity upon the mechanism of the intercalation reaction, 503–509, Copyright 2009, with permission from Elsevier.

crystallinity ZrP and the planar guest molecule. When the intercalation energy barrier was high, two phases coexisted during the intercalation having a stepwise expansion of the layers (Figure 10.5b). This was observed when the high crystallinity ZrP was used with the hexylamine. This research demonstrated that the intercalation energy barrier has a significant effect in the intercalation process. It is important to clarify that apart from the structure and crystallinity of the host and guest, other factors can affect this energy barrier.

10.1.4 Direct Ion Exchange Using θ-ZrP

To overcome the high energy barrier of the direct intercalation of large metal complexes in α-ZrP, we used instead a highly hydrated phase of ZrP (containing six molecules of water per formula unit), known as the θ-ZrP phase, which permits the intercalation of larger inorganic complexes without the need of the pre-intercalation process [26, 28–32, 39, 40, 47–51, 62–64]. Although first synthesized by Clearfield [57], we have used the synthesis reported by Kijima et al. where $ZrOCl_2$ is mixed with phosphoric acid with constant stirring at 94 °C for 48 h [65]. This robust layered structure has a 10.4 Å separation distance (10.3 Å in our hands) between the layers but maintains the α-ZrP-type layers (Figure 10.6) [56].

Using X-ray powder diffraction (XRPD), the interlayer distance is determined by the first peak of the pattern (the peak at the lowest angle), which is the 002-plane reflection of ZrP corresponding to the interlayer distance. Apart from the interlayer

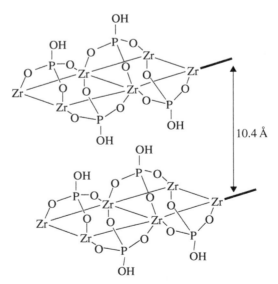

FIGURE 10.6 Structure of $Zr(HPO_4)_2 \cdot 6H_2O$, θ-ZrP (10.3 Å ZrP).

distance, the other difference between the θ-phase and the α-phase is that θ-ZrP has six molecules of water per formula unit and α-ZrP has only one, which explains the expanded interlayer distance of θ-ZrP. If the θ-phase is dehydrated, it converts to the α-phase of ZrP [57]. This known conversion is the basis of the method we use to monitor our intercalation reactions using θ-ZrP. In a typical intercalation reaction, we mix θ-ZrP in a solution of our intercalant, and after a few days, we take an aliquot of the solution, filter it, rinse the wet solid and then dry it to obtain a powder suitable for XRPD analysis. If the lowest angle peak observed in the XRPD pattern of the dry product of the intercalation reaction corresponds to an interlayer distance larger than 7.6Å, then the intercalation was successful. If instead the lowest angle peak corresponds to an interlayer distance of 7.6Å, then the intercalation was unsuccessful, since we would have only obtained the α-ZrP phase formed by the conversion of θ-ZrP to α-ZrP by dehydration of θ-ZrP.

Since we reported in 2003 our first successful intercalation via direct ion exchange of a large luminescent metal complex ([Ru(bpy)$_3$]$^{2+}$) [26], we have reported numerous successful intercalations of different metal complexes using θ-ZrP [28–30, 39, 40, 47–50, 63, 64]. The development of the θ-ZrP phase as a host for inorganic and organometallic compounds gives us the possibility to construct many new materials that otherwise we would not be able to construct using α-ZrP. In addition, ZrP can be modified by manipulating certain variables in its synthesis to obtain a material in the nanoscale, which provides additional applications in the area of nanotechnology [41].

For these reasons, we have focused our research in recent years to directly intercalate various bioactive species without pre-intercalating butylammonium ions, using this highly hydrated form of ZrP for possible applications in nanobiotechnology, specifically in the area of drug carriers (Figure 10.7) [31–33, 64]. Intercalation of this type of compounds using either of the two methods of intercalation mentioned previously has not been reported in the literature.

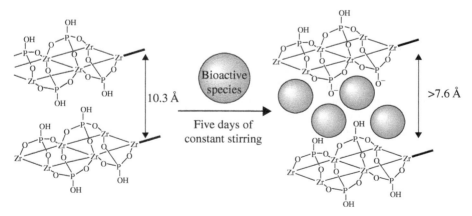

FIGURE 10.7 Direct intercalation process of bioactive species into zirconium phosphate.

10.2   DRUG NANOCARRIERS BASED ON θ-ZrP

Nanotechnology is defined as the manipulation and understanding of matter at the nanoscale of 1–100 nm. At this scale, unique properties of materials emerge that can be applied to different and novel applications [66]. When these applications are used in medicine, the field is usually known as nanomedicine. Nanomedicine has introduced a number of nanomaterials to develop new diagnostics, therapeutics and drug delivery systems, among other applications [66, 67].

The process of releasing a bioactive agent at a specific rate and at a specific site is called drug delivery. Drug delivery is an application in the field of nanomedicine that has been gaining the attention of researchers over the years and of pharma- and biotechnological industries as well. This is because one of the main limitations that these industries encounter is the selective delivery of medicines to their sites of therapeutic action. The field of drug delivery can offer a tool to expand current drug markets and new delivery technologies that will offer them a competitive edge, increase patient compliance and reduce healthcare cost, among other advantages [68]. This is why the development of targeted drug delivery systems using nanoparticles is an important area of research. Nanoparticles are envisioned as the future of the drug delivery technology due to their many potential applications in medicine, from therapeutic to diagnostic tools.

Among the materials that are being suggested and studied for drug delivery applications are polymers, dendrimers, liposomes and inorganic nanoparticles such as those based on gold, iron oxide and layered materials [31–33, 67]. The use of inorganic layered nanomaterials (ILN) as potential drug carriers has been growing since the discovery of their ability to encapsulate and control release bioactive compounds via a chemical switch [31–33, 69]. An extensively studied ILN is ZrP. ZrP have already been used as host materials in biomedical applications such as in dialysis systems and most recently as chemical delivery materials [69, 70]. We present now examples of new drug delivery nanomaterials that we have developed in our laboratory by intercalating various drug species into θ-ZrP via direct ion exchange.

10.2.1   Insulin

Insulin is one of the biomolecules that have currently attracted much attention by researchers interested in stabilizing biomolecules for medicinal applications through their immobilization. This is due to the high number of patients that need the hormone and the many disadvantages of insulin delivery by injection [31].

Insulin is a peptidic hormone composed of two peptide chains, designated as A and B, which are linked together by two disulphide bonds. In most species, the structure of insulin is extremely well preserved with 21 amino acids in the A chain and 30 amino acids in the B chain. In a pH ~ 6 solution, insulin tends to form aggregates that form dimers, which in turn form hexamers in the presence of Zn^{2+} or Ca^{2+} ions (Figure 10.8) [51].

The principal role of insulin in the body is the regulation of the glucose levels in the bloodstream. Insulin is secreted to the blood serum by endocytosis in its hexamer form when glucose levels increase in the bloodstream. Once in the bloodstream, it stores the excess glucose in the liver and muscles and stops the use of fats as an energy source. Each time that a meal is consumed, glucose concentration increases in the body, which induces the secretion of insulin into the bloodstream. If insulin secretion fails, the glucose levels increase excessively in the blood, causing the body to depend on fats as its energy source. This lack of control in the secretion of insulin when the glucose levels increase is a common chronic disease called diabetes mellitus [51]. According to the National Diabetes Statistics (2011), diabetes is the seventh leading cause of death in the United States and affects 25.8 million people of all ages, which represents 8.3% of the US population. Of those patients, 26% need insulin as part of their daily treatment [51, 71].

Recently, new insulin administration methods are being developed, and more alternatives have become available instead of the principal insulin intake treatment based on daily injections, which is a painful and tortuous regimen that patients have to suffer for the rest of their life [51]. The search of a way to avoid the use of a needle in the treatment of diabetes has led to the use of nanomedicine in the research of synthesis of nanomaterials that can provide an alternative route for insulin delivery [72]. One of the most convenient ways for the uptake of insulin can be the oral way, where the patient would take a tablet with every meal. New nanoparticles with the capacity of sequestering insulin, surviving the stomach environment and releasing the insulin in the intestine directly into the bloodstream are under constant investigation [51].

Kumar et al. reported that the ideal support matrix should prevent protein aggregation or spontaneous denaturation and leave the native properties of the immobilized protein intact [73]. In addition, the matrix must be biocompatible and nontoxic and have the potential to protect the peptidic hormone of possible biodegradation by natural biological processes [51]. Furthermore, as Díaz et al. reported, this ideal matrix should release the peptidic hormone once it is needed, and then the matrix

FIGURE 10.8 From left to right are the ribbon representations of insulin in its monomer, dimer and hexamer form (the Zn^{2+} central metal in the hexamer form is omitted). Image taken from the Cambridge protein data bank (code 3I3Z).

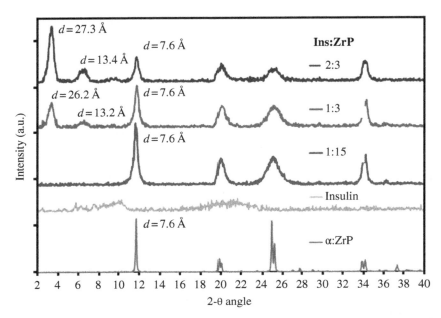

FIGURE 10.9   XRPD patterns of α-ZrP, insulin and insulin/ZrP at different molar ratios. Adapted from Ref. [51].

should be completely decomposed to be excreted, in the case that it cannot be excreted in their native state, by natural processes [31].

In 2010, Díaz et al. published the direct intercalation of insulin in ZrP using θ-ZrP as a precursor without any pre-intercalation procedure that would contaminate the product [31]. Figure 10.9 shows the XRPD patterns obtained from the dry products of the intercalation reaction of insulin with ZrP at various insulin/ZrP molar ratios. The diffraction peak at the lowest 2θ angle corresponds to the interlayer distance. The XRPD patterns of insulin-exchanged ZrP at insulin/ZrP (Ins:ZrP) molar ratios of 2 : 3 and 1 : 3 show that a new mixed phase has formed whose first-order diffraction peak corresponds to an interlayer distance of about 26–27Å. In addition, those two XRPD patterns show a diffraction peak corresponding to a distance of 13.2–13.4Å (second-order peak), another peak for the 2 : 3 Ins : ZrP material corresponding to a distance of about 9.4Å (third-order peak) and a peak at 7.6Å, the characteristic interlayer distance of α-ZrP, indicating the presence of α-ZrP; hence, those two materials are mixed phases [51].

The XRPD pattern obtained for the Ins : ZrP material at the low 1 : 15 molar ratio (Figure 10.9) is typical of un-intercalated α-ZrP with an interlayer distance of 7.6Å. The α-ZrP phase is formed upon dehydration of θ-ZrP without any intercalated species present in the layers [57]. This result indicates that at that low insulin loading level, a new phase is not formed in the materials and any exchanged insulin in this sample is mainly bound on the surface of the agglomerated ZrP nanoparticles. In addition, these results indicated that the energy barrier for the intercalation reaction of insulin with ZrP is high, which explains why a mixed phase was obtained and two

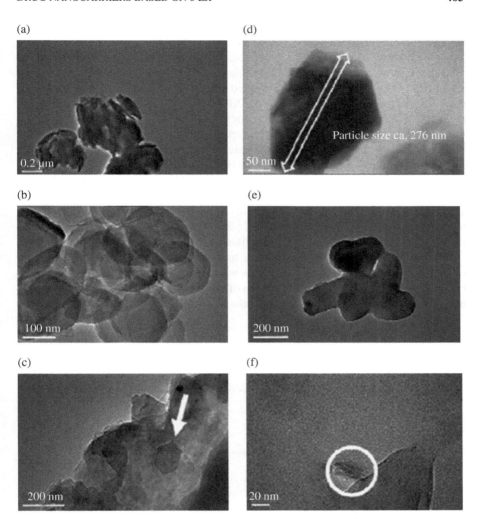

FIGURE 10.10 TEM images of Ins : ZrP at 1 : 3 (a, b, c) and 2 : 3 (d, e, f) molar ratios. In image (d), the double head arrow represents the length of the particle that is ca. 275 nm; in image (c), the white arrow shows the a hexagonal shape crystal, characteristic of ZrP. Reprinted with permission from Díaz, A.; David, A.; Pérez, R.; González, M. L.; Báez, A.; Wark, S. E.; Clearfield, A.; Colón, J. L., Nanoencapsulation of insulin into zirconium phosphate for oral delivery applications. *Biomacromolecules* **2010**, *11*, 2465–2470. Copyright 2010 American Chemical Society.

interlayer distances are observed for the 2 : 3 and 1 : 3 molar ratio samples, one for the intercalated materials and one for the un-intercalated layers [51].

The TEM images of the insulin intercalation products correlate remarkably well with the structure that the XRPD patterns suggest (Figure 10.10). TEM images of the 2 : 3 and 1 : 3 Ins : ZrP materials show marked differences in morphology, homogeneity and agglomeration between them (Figure 10.10) [31].

The TEM image obtained of the 1 : 3 Ins : ZrP material at lower magnification (100–200 nm, Figure 10.10a and b) shows the heterogeneity of the sample and a considerable amount of particles agglomerated. Hexagonal nanocrystals, characteristic of α-ZrP, can be identified in the TEM images of the 1 : 1 sample at intermediate magnification (Figure 10.10b and c marked with a white arrow) in agreement with the 7.6 Å diffraction peak observed in the XRPD patterns for mixed-phased samples (Figure 10.9). The highly transparent particles shown in the TEM images are most likely the Ins : ZrP nanoparticles; the transparency is due to the presence of a large amount of organic molecules in the materials, which usually produce transparent-like images in the TEM. Likewise, these transparent nanoparticles shown in the TEM images should be less crystalline if they are produced by Ins : ZrP nanoparticles and are consistent with the broad peaks shown in their XRPD patterns (Figure 10.9). The TEM image indeed shows less crystallinity in those nanoparticles suggesting that these nanoparticles correspond to the Ins : ZrP intercalation product, in addition to the shape of the nanoparticles that looks like bloated hexagonal nano-platelets [31].

To determine if any structural change had occurred to the hormone during the intercalation process, FTIR measurements were performed on the intercalated materials. Figure 10.11 shows the FTIR spectra of insulin intercalated into ZrP at different molar ratios. The IR spectra of the intercalation products with Ins : ZrP molar ratios

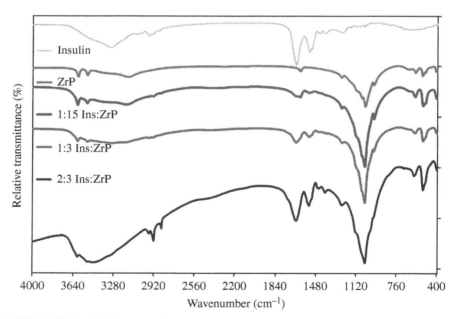

FIGURE 10.11 FTIR spectra of (upper to lower) insulin, α-ZrP and insulin/ZrP intercalated into the θ-ZrP framework at room temperature and different molar ratios. Reprinted with permission from Díaz, A.; David, A.; Pérez, R.; González, M. L.; Báez, A.; Wark, S. E.; Clearfield, A.; Colón, J. L., Nanoencapsulation of insulin into zirconium phosphate for oral delivery applications. *Biomacromolecules* **2010**, *11*, 2465–2470. Copyright 2010 American Chemical Society.

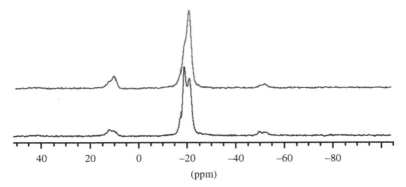

FIGURE 10.12   ^{31}P MAS NMR spectra of insulin-intercalated ZrP at 2 : 3 (top) and 1 : 3 (bottom) Ins : ZrP molar ratios [51].

of 2 : 3 and 1 : 3 show the characteristic amide bands of insulin at the approximately 1500 cm^{-1} region. On the other hand, the IR spectrum of the intercalation product with an Ins : ZrP molar ratio of 1 : 15 shows very weak amide region bands, as expected if just a small amount of insulin is adsorbed on the surface of the aggregated nanoparticles, but the overall concentration of the hormone is low, which is in agreement with the XRPD results shown in Figure 10.9. The orthogonal phosphate bands at approximately 1050 cm^{-1} have disappeared in the Ins : ZrP intercalation products with molar ratios of 2 : 3 and 1 : 3, indicative of a successful intercalation [74]. Since there are no significant shifts of the IR bands of amide I to higher frequencies or amide II to lower frequencies, these results suggest that there is no significant denaturation of the hormone after intercalation [51. 73].

Figure 10.12 shows the ^{31}P MAS NMR spectra of Ins : ZrP intercalation products at various loading levels [51]. Insulin-exchanged ZrP has just two kinds of interactions with the phosphate groups of the layers, ionic interactions and H-bonding with the phosphates.

The ^{31}P MAS NMR spectrum for the Ins : ZrP with a molar ratio of 1 : 3 shows two peaks that are part of a broad band at ca. 19.2 and 21.3 ppm. In the case of Ins : ZrP with a molar ratio of 2 : 3, just one broad band is observed (at 21.3 ppm) with a shoulder at 19.2. The two resonances of the Ins : ZrP intercalation product with a molar ratio of 1 : 3 are assigned to two different phosphates in the layers that are interacting in different ways with the insulin. At the 1 : 3 molar ratio, there is not enough insulin to completely saturate the galleries of ZrP. In that case, there are going to be deprotonated phosphate groups present interacting directly with the basic residues of the insulin, producing a chemical shift to 21.3 ppm, and also protonated phosphate groups are going to be present that are too far for a direct interaction but that form H bonds with cointercalated water molecules, resulting in a chemical shift of 19.1 ppm. In the spectrum of Ins : ZrP with a molar ratio of 2 : 3, the 19.1 ppm peak is now just a shoulder embedded in a more intense peak at 21.3 ppm corresponding to phosphates in direct interaction with the insulin. The cointercalated water was displaced by the insulin causing an increased intensity of the 21.3 ppm signal and a weakening of the 19.2 ppm peak. These results are in agreement with the XRPD, UV–vis and FTIR results that show that the Ins : ZrP material with a molar ratio of 2 : 3 contains more insulin than the Ins : ZrP intercalation product with a molar ratio of 1 : 3 [51].

To obtain the actual loading of the intercalated solids, thermogravimetric analysis (TGA) was performed. The TGA results indicated that the experimental Ins : ZrP molar ratio is 1 mol of insulin for every 42 formula units of ZrP, for the 1 : 3 intercalation product (24% loading) and 1 mol of insulin for every 32 formula units of ZrP for the 2 : 3 intercalation product (28% loading). Taking into account the dimensions of insulin and the interlayer distance, obtained by XRPD, we can determine the cross-sectional area of the insulin that will be parallel to the layer ($31.0 \times 33.4 Å^2$). Because the area of a single ZrP formula unit is $24 Å^2$, the product of the ratio between the insulin area and the ZrP unit area should produce the theoretical full loading of Ins : ZrP. The value of this calculation is 1–43; in other words, theoretically, at full loading, there should be 1 insulin molecule for every 43 ZrP formula units. This result is in agreement with the TGA results; the 1 : 3 loading level is a full loaded material, and the 2 : 3 loading level is fully loaded with the excess of insulin adsorbed on the surface [31].

The controlled-release experiments of Ins : ZrP suspensions were carried out between pH 8.2 and 7.4 to study the release of the hormone from the layers using a pH stimulus. The release was monitored with UV–vis spectrophotometry by observing the change in absorbance of the characteristic band of insulin at 280 nm of the

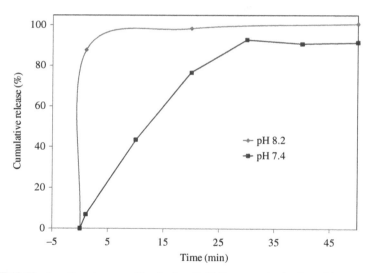

FIGURE 10.13   Insulin release profiles for insulin/ZrP upon agitation in NaPi pH 8.2 (◊) and 7.4 (□); μ = 0.1 buffer at room temperature. Reprinted with permission from Díaz, A.; David, A.; Pérez, R.; González, M. L.; Báez, A.; Wark, S. E.; Clearfield, A.; Colón, J. L., Nanoencapsulation of insulin into zirconium phosphate for oral delivery applications. *Biomacromolecules* **2010**, *11*, 2465–2470. Copyright 2010 American Chemical Society.

centrifuge aliquots of the suspensions. Figure 10.13 shows the release profile of Ins : ZrP at pH 8.2 and 7.4.

At pH 8.2, insulin is released from the layers at a fast pace (~5 min) until it reaches a plateau; at pH 7.4, the release is slower (~30 min). Because the isoelectric point of insulin is 5.4, at pH between 8.2 and 7.4, the six carboxylic acid groups of the hormone are deprotonated. The produced carboxylate groups disrupt the hydrogen bond interactions between the insulin and the phosphate groups of the layers producing a permanent negatively charged insulin. The overall negative charge of insulin at pH 8.2–7.4 will be repelled by the negatively charged phosphate groups of the ZrP layers, resulting in the rapid release of insulin to the solution and an uptake of the Na^+ ions from the buffer. This release mechanism is very similar to the biological release of insulin, where the insulin is in the hexamer conformation bonded by a Zn^{2+} cation in the storage vesicle; once the insulin hexamer is released to the serum, it experiences a jump in pH from approximately 5.5 to 7.4. The pH jump causes the carboxylic acid groups to deprotonate, and the repulsions lead to a rapid dissociation from the metal complex hexamer; the insulin ends up as the monomer in the bloodstream [31]. Díaz et al. reported that the rapid release of the hormone from the ZrP galleries at pH 7.4 is comparable to the time it takes for the insulin level to rise in the body between meals (30 min) [31]. The acidic nature of the ZrP material and the relatively high pH of the buffer are the keys for the controlled-release mechanism of the nanoparticle, making possible a new therapy for patients with diabetes [31].

To determine the potential toxicity of the ZrP nanoparticles to human cells, the cell viability of a MCF-7 human breast carcinoma cell line grown for 24 h was measured in the presence of different concentrations of nanoparticles [31]. The MTT cell viability assay revealed an absence of overt toxicity to MCF-7 cells following exposure to the formulations containing ZrP for up to 24 h [31]. Díaz et al. proposed that the release of insulin into the duodenal duct will cause the absorption of the hormone into the bloodstream. The activity of the insulin is going to be less compromised via this route because the initial steps of digestion are going to be avoided. Based on previous methods of gastrointestinal insulin delivery, the insulin uptake should be around 5%, which would lead to an oral administration of 162 and 139 mg for the 1 : 3 and 2 : 3 Ins : ZrP intercalation product, respectively, per meal. This amount should be equivalent to the usual insulin dose per meal when injected by the conventional injection route [31].

Díaz concluded that this type of material represented a strong candidate to developing a non-invasive insulin carrier for the treatment of diabetes mellitus for the following reasons: (i) for the first time, the cytotoxicity of ZrP to human cells *in vitro* has been investigated and no toxicity was demonstrated and (ii) a controlled-release study of nanoencapsulated insulin using pH changes as stimuli has been proposed and investigated where the hormone remains stable during the entire process and does not appear to have any significant chemical or structural changes over a period of time (up to 6 months). Although more detailed studies are warranted, this system may pose an alternative to the painful and tedious administration of insulin through injection that diabetic patients suffer every day [31].

10.2.2 Anticancer Agents

10.2.2.1 *Nanoparticles and the Enhanced Permeability and Retention Effect* Nanoparticles are being designed to deliver drugs to specific targets such as cancer cells. They can help to tackle certain disadvantages that nowadays currently used drugs present, such as non-selectivity, high drug toxicity, poor solubility, instability and non-localized release. These disadvantages are often the reason for the manifestation of side effects that makes the patient sick before it makes him/her better [75].

Currently, many of the drugs designed, discovered or already in use for therapeutics are limited by their poor solubility, high toxicity, high dosage, aggregation, non-specific delivery, *in vivo* degradation and short circulating half-lives. This is why the development of targeted drug delivery systems is currently one of the most important areas of drug research. Targeted drug delivery via nanoparticles can be achieved either by two types of targeting: passive or active. Passive targeting can occur through the enhanced permeability and retention (EPR) effect based on anatomical differences between normal and diseased tissue specifically because of the leaky vasculature and poor lymphatic drainage in tumours where nanoparticles can accumulate more than in normal tissue [67, 68, 76]. On the other hand, active targeting requires the conjugation of receptor-specific ligands that promote site-specific targeting [68].

The nanoparticles of the inorganic layered ZrP can be prepared with the appropriate size to be specifically absorbed by the tumour cells through the EPR effect displayed by cancerous cells [77, 78]. The drug, while being excluded from the external medium when entrapped in ZrP, will remain biologically inactive, but once the drug-loaded ZrP is inside the cancer cells, it will be released selectively in the tumour cell. The drug release can occur through dissolution of ZrP by the acidic microenvironment characteristic of cancer cells or by cell lysosomes, by delamination and/or by direct ion exchange of the intercalated drug with ions in the cells' interior [51].

10.2.2.2 *Cisplatin* Cisplatin (*cis*-diamminedichloroplatinum(II)) is one of the most potent anticancer agents approved for use in humans. Cisplatin has been shown to be active against ovarian and testicular cancer, Hodgkin's lymphoma and certain other malignancies [51]. The biological activity of cisplatin stems from the coordination of the drug to DNA, which blocks DNA replication and transcription resulting in apoptosis [79]. However, the clinical effectiveness of cisplatin is limited by significant side effects (due to lack of specificity) and the emergence of drug resistance [79]. Delivery of anticancer drugs such as cisplatin to the tumour cells without damaging healthy organs or tissues is highly challenging.

We have investigated the intercalation of cisplatin into α-ZrP, using θ-ZrP as precursor, for possible use in drug delivery applications [33]. The intercalation of cisplatin within the layers of θ-ZrP at different loading levels (molar ratios) produced new materials with unique structure and morphology. Figure 10.14 shows the XRPD patterns of α-ZrP and cisplatin-intercalated ZrP using θ-ZrP at different cisplatin–ZrP molar ratios (loading levels) [33]. The XRPD patterns show the formation of a new phase with an interlayer distance of ca. 9.3 Å for all the intercalation products. In this new phase, since the ZrP layer thickness is 6.6 Å, cisplatin has increased the interlayer distance by 2.7 Å, confirming that intercalation has taken place [53, 54].

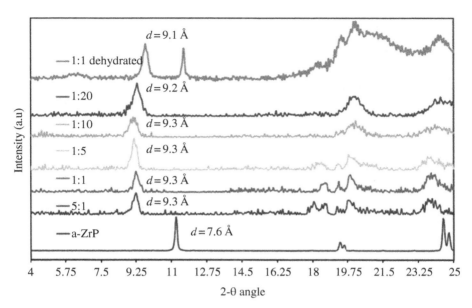

FIGURE 10.14   X-ray powder diffractograms of α-ZrP and of the intercalation products of the reaction of θ-ZrP and cisplatin at several cisplatin–ZrP molar ratios. Díaz, A.; González, M. L.; Pérez, R. J.; David, A.; Mukherjee, A.; Báez, A.; Clearfield, A.; Colón, J. L., Direct intercalation of cisplatin into zirconium phosphate nanoplatelets for potential cancer nanotherapy. *Nanoscale* **2013**, *5*, 11456–11463. Reproduced with permission from the Royal Society of Chemistry.

For the intercalation product with low concentration of cisplatin (1 : 20), the expected interlayer distance, based on cisplatin dimensions, is 8.3Å, because the square planar molecule has enough room to lie flat, parallel to the layers. We had expected that as the concentration of cisplatin increases (such as in the 1 : 5 loading level), the interlayer distance would increase, because there would be more cisplatin in the interlayer space and the molecules will tend to orient themselves with a progressive increase in the inclination angle to accommodate more molecules until they reach the maximum level of intercalation where the cisplatin molecules would be oriented perpendicular to the layers. Contrary to our expectations, every intercalation product produced the same interlayer distance, which suggests that the 9.3Å interlayer distances observed are not from materials with cisplatin oriented at 45° with respect to the ZrP planes, but rather that the intercalation process is not a simple ion-exchange mechanism [33]. In order to explore this possibility, we performed a structure modelling analysis to elucidate how cisplatin is accommodated in the interlayer galleries [33].

The labile nature of the chloride ligands in the cisplatin complex makes highly likely that in the new phase the chloride ions are displaced by the phosphate groups from the layer [80]. Taking into account the experimental interlayer distance (obtained by XRPD), two different kinds of structures were predicted: one where the Pt is bonded to two phosphates in the same layer and the other with the Pt bonded to phosphates in adjacent layers in a cross-linking fashion. Molecular modelling was performed to elucidate which of these two structures is more likely [51]. The results of the molecular modelling suggest that the cisplatin is covalently bonded to the ZrP layer in a cross-linking fashion where a theoretical interlayer distance of 9.1 Å is obtained (Figure 10.15); the experimental interlayer distance obtained was 9.3 Å (Figure 10.14). However, the molecular modelling did not consider the presence of water molecules in the interlayer space of the sample materials submitted to XRPD analysis. Therefore, the sample was heated at 150 °C for 2 h to dehydrate it and compare the interlayer distance of the dehydrated intercalated layered compound with the distance obtained in the molecular modelling simulation. The XRPD pattern of the dehydrated sample (Figure 10.14) shows an interlayer distance of 9.1 Å, in exact agreement with the molecular modelling result, and it is a mixed phase since some un-intercalated ZrP with interlayer distance of 7.6 Å is also observed [51]. The un-intercalated ZrP should correspond to the central part of the ZrP nanoparticles where the platinum complexes could not reach in the intercalation process due to the steric hindrance of the initially intercalated platinum complexes that react in the outer parts of the interlayer galleries. The agreement between the experimental interlayer distance and the theoretical one expected for a cross-linked structure indicated a successful intercalation of cisplatin into ZrP and suggested our hypothesis that the cisplatin chloride ligands are substituted with the phosphate groups of the ZrP layers [33].

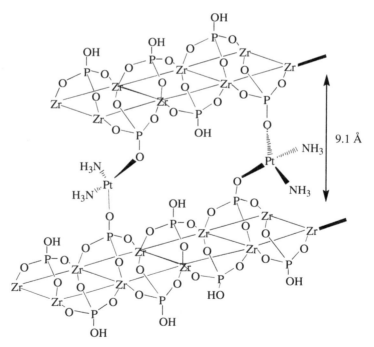

FIGURE 10.15 Idealized representation of cisplatin in the galleries of zirconium phosphate, bound to the phosphates of the layers in a cross-linked fashion [51].

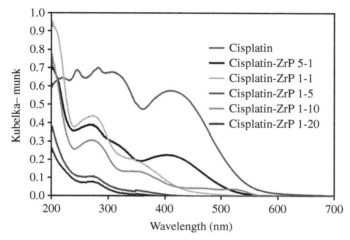

FIGURE 10.16 Diffuse reflectance for cisplatin and the intercalation products of the reaction of ZrP and cisplatin at 5 : 1, 1 : 1, 1 : 5, 1 : 10 and 1 : 20 cisplatin–ZrP molar ratios. Díaz, A.; González, M. L.; Pérez, R. J.; David, A.; Mukherjee, A.; Báez, A.; Clearfield, A.; Colón, J. L., Direct intercalation of cisplatin into zirconium phosphate nanoplatelets for potential cancer nanotherapy. *Nanoscale* **2013**, *5*, 11456–11463. Reproduced with permission from the Royal Society of Chemistry.

The chloride ligand substitution in cisplatin-intercalated ZrP was corroborated by spectroscopic methods. Diffuse reflectance was performed to elucidate the amount of cisplatin that was intercalated at different molar ratios and to corroborate the chloride substitution hypothesis. Figure 10.16 shows the diffuse reflectance spectra for cisplatin and for the cisplatin-intercalated ZrP material at different molar ratios. The spectrum of cisplatin showed its characteristic band at 417 nm, which is also present in the spectrum for the intercalation product with a 5 : 1 molar ratio material; however, the spectra of intercalation products with 1 : 20, 1 : 10, 1 : 5 and 1 : 1 cisplatin–ZrP molar ratios do not exhibit this characteristic band. The spectra of all the intercalated products, including the 5 : 1 molar ratio, show two new UV–vis absorption bands at ca. 278 and 360 nm, which suggested a change in the chemical composition of cisplatin in the intercalated samples [51]. Bose and co-workers reported that the UV–vis absorption spectrum of a 0.1 M cis-Pt(NH$_3$)$_2$H$_2$PO$_7$ solution in NaClO$_4$ at pH = 2.8 showed two UV bands at 252 and 330 nm with a peak intensity ratio of 4 : 1 [81]. These authors also observed that those bands were red shifted with changes in pH. We assign the observed bands at 278 and 360 nm to the bands previously observed by Bose and co-workers in the phosphate-bound platinum complex but red shifted by ca. 30 nm due to the difference in pH and electronic environment [51]. The similarity of the phosphate-bound cis-diammine platinum complex spectrum for the cisplatin-intercalated ZrP to that of cis-Pt(NH$_3$)$_2$H$_2$PO$_7$ is consistent with our hypothesis that the loss of the chloride ligands upon intercalation into ZrP is due to the nucleophilic displacement by the phosphate groups of the ZrP layer [51].

In addition to the diffuse reflectance measurements, we performed ^{31}P MAS NMR measurements. Figure 10.17 shows the ^{31}P MAS NMR of cisplatin–ZrP at various loading levels. Two distinctive chemical shift signals were detected, one upfield and the other downfield. The characteristic chemical shift of α-ZrP at −18.7 ppm is absent in all the ^{31}P MAS NMR spectra for the cisplatin–ZrP intercalation product. This is typical in a successful intercalation reaction, which is in agreement with the XRPD and FTIR results obtained. Although ^{31}P has a very high natural abundance (100%), the resolution of the ^{31}P MAS NMR spectra is poor for this kind of material, making it very difficult for the complete and concrete identification of the chemical shifts in the materials. On the other hand, there are at least three signals that can be identified in the spectra and can be assigned based on the literature. The first signal at ca. −21.2 ppm is assigned to the interaction, via H bond, between the amino groups in the platinum complex and the deprotonated phosphate of the layer. In all the intercalation products, this peak shows minimal positional variation, but the full width at half maximum (FWHM) of this peak is considerably larger (ca. 2.6 ppm) especially at lower loading levels (1 : 20). The widening of this peak can be explained if we take into account the number of different interactions that are taking place in a very heterogeneous environment, particularly for lower loading levels where there are no steric hindrance and hydrophobic and hydrophilic effects that are causing a structured arrangement of the intercalated molecules in the intergalleries space. From the spectra, it is noticeable that the peak is more defined and intense at higher loading levels (such

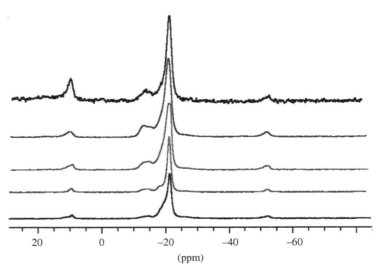

FIGURE 10.17   ^{31}P MAS NMR spectra of cisplatin-intercalated ZrP at (top to bottom) 5 : 1, 1 : 1, 1 : 5, 1 : 10 and 1 : 20 cisplatin–ZrP molar ratios. Díaz, A.; González, M. L.; Pérez, R. J.; David, A.; Mukherjee, A.; Báez, A.; Clearfield, A.; Colón, J. L., Direct intercalation of cisplatin into zirconium phosphate nanoplatelets for potential cancer nanotherapy. *Nanoscale* **2013**, *5*, 11456–11463. Reproduced with permission from the Royal Society of Chemistry.

as in the 5 : 1 molar ratio sample) because the packing is more tight, minimizing the free displacement and rotation of the amino groups in the cisplatin [33].

The second resonance at ca. -13.5 ppm is attributed to the phosphate bonded to the platinum(II) of the cisplatin molecule, causing a significant deshielding of the phosphorus. There is an appreciable little peak at -18.3 ppm for the 1 : 10 loading level that could be also present in the 1 : 20 loading level spectra (the shoulder at ca. -19 ppm). That peak appears to be moving downfield and overlapping the peak at ca. -13 ppm. That little peak that varies from spectrum to spectrum could be produced by the phosphorus interaction between the amino groups of cisplatin, the water molecules in the interlayer galleries and the protonated phosphate of the layer [33]. In summary, the ^{31}P MAS NMR spectroscopy results are also consistent with our hypothesis that upon intercalation the cisplatin chloride ligands are substituted by the phosphate groups of the ZrP layers.

The chloride ligand substitution by phosphate groups of the ZrP layer was further confirmed by microprobe quantitative compositional analyses (Table 10.1). Table 10.1 shows that the amount of chloride in the intercalation products is far lower than expected for free cisplatin, which has a Pt : Cl ratio of 1 : 2. For instance, the sample made with a 1 : 1 cisplatin–ZrP molar ratio has a Pt : Cl ratio of 31 : 1. In addition, only in the sample with the highest loading level, the 5 : 1 intercalation product, did the Pt : Cl ratio reached 4 : 1, which we assign to the presence of cisplatin molecules that retain their chloride ligands on the surface of the ZrP particles at that loading level. This cisplatin with chlorides ligands in the 5 : 1 intercalation product results from the excess amount of cisplatin used in the synthesis procedure for that loading level, causing agglomerations on the surface of the ZrP nanoparticles. Furthermore, the chemical formula obtained indicates that a maximum loading of cisplatin in the interior of the ZrP layers was achieved when the intercalation reaction was performed with a cisplatin–ZrP solution molar ratio of 1 : 1. In contrast, when an excess amount of cisplatin was used in the intercalation reaction (i.e. cisplatin–ZrP ratio = 5 : 1), no further uptake of the platinum complex into the interior of the layers occurred [51].

X-ray elemental distribution 'mappings' obtained for the 5 : 1, 1 : 1 and 1 : 5 intercalation products showed the atomic distribution of Pt, P and Cl in a $62 \times 62 \, \mu m^2$ section of the surface of the intercalation products [51]. A very uniform distribution of Pt and P was observed in the 1 : 1 intercalation product with practically no Cl in the material. On the other hand, for the 5 : 1 intercalation product, there is clearly an

TABLE 10.1 Molecular formula determination for cisplatin–ZrP at different loading levels based on microprobe and ICP–MS analysis (molar ratio based on the initial reaction ratio used between Pt and Zr) [51]

Molar ratio	Chemical formula	Pt : Cl ratios
1 : 20	$Zr(H_{0.9935}PO_4)_2(Pt(NH_3)_2)_{0.008}Cl_{0.003} \cdot 0.8H_2O$	17 : 1
1 : 10	$Zr(H_{0.9920}PO_4)_2(Pt(NH_3)_2)_{0.012}Cl_{0.008} \cdot 0.3H_2O$	23 : 1
1 : 5	$Zr(H_{0.8635}PO_4)_2(Pt(NH_3)_2)_{0.140}Cl_{0.007} \cdot 0.3H_2O$	27 : 1
1 : 1	$Zr(H_{0.6870}PO_4)_2(Pt(NH_3)_2)_{0.318}Cl_{0.010} \cdot 0.2H_2O$	31 : 1
5 : 1	$Zr(H_{0.9935}PO_4)_2(Pt(NH_3)_2)_{0.286}Cl_{0.077} \cdot 0.5H_2O$	4 : 1

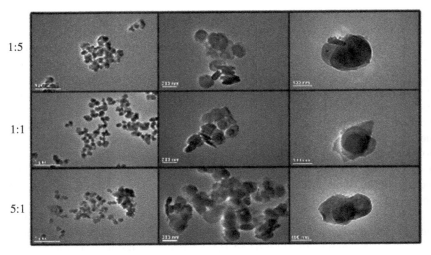

FIGURE 10.18   TEM images of cisplatin–ZrP intercalation products for the 1 : 5, 1 : 1 and 5 : 1 molar ratios at different magnifications. Díaz, A.; González, M. L.; Pérez, R. J.; David, A.; Mukherjee, A.; Báez, A.; Clearfield, A.; Colón, J. L., Direct intercalation of cisplatin into zirconium phosphate nanoplatelets for potential cancer nanotherapy. *Nanoscale* **2013**, *5*, 11456–11463. Adapted with permission from the Royal Society of Chemistry.

agglomeration of $Pt(NH_3)_2Cl_2$ on the surface where there are parts that are rich in chloride, while others are not. In addition, for the 1 : 5 intercalation product, there is a lack of Pt atoms, which was expected since in the 1 : 5 intercalation product, there are 1/3 of the available intergallery spaces unfilled in the layered material as indicated by the chemical formula obtained from the microprobe analysis [51].

Figure 10.18 shows the TEM images of the cisplatin–ZrP intercalation products at loading levels of 5 : 1, 1 : 1 and 1 : 5. The images show that the hexagonal-like shape of the ZrP crystallites is partially retained on the intercalation product. The particle size for the entire intercalation product is practically the same, ca. 180 nm, and also the shape is virtually retained. The fact that the three different intercalation products preserve the morphological structure of ZrP is indicative of the robustness and stability of the ZrP nano-platelets. The nano-platelet shape is the same in all the intercalation products with the same thickness of ca. 40 nm; this is in agreement with the XRPD results where all these three intercalation products present the same interlayer distance [51]. Nanoparticle size should be optimized in the range of 20–200 nm to avoid renal filtration and to keep the particles in the bloodstream until they reach their targets, while particle sizes of 100–200 nm are desirable to take advantage of the enhanced permeation and retention effect.

The controlled-release experiments were carried out at pH 7.4 and 4.5 in simulated biological fluids to study the release of the Pt complex from the layers using a pH stimulus under the possible biological environments. Simulated body fluid (SBF) simulates the suspension of the nanoparticles within human plasma (pH = 7.4), whereas artificial lysosomal fluid (ALF) simulates the nanoparticles in the cell lyso-somes (pH = 4.5). Figure 10.19 shows the release profiles for cisplatin-intercalated ZrP at different loading levels in both SBF and ALF. The release was monitored with ICP–MS by following the platinum (Pt195) signal over time. Figure 10.19 shows that at pH = 7.4 in SBF platinum is released within the first 12 h, but only 2–3% of the total platinum is released.

We assign the platinum released under these conditions to platinum on the surface and edges to the ZrP nanoparticles, but not intercalated within ZrP layers. A slow release of cisplatin at pH = 7.4 is desirable to avoid the general administration of the drug and its many side effects. In contrast, rapid release of the platinum complex at the low pH of 4.5 is desirable since it approaches the typical pH of the acidic envi-ronment of the tumour endosomes and lysosomes.

Figure 10.19 shows that in ALF with pH = 4.5 the release profile is much faster, compared to the release in SBF, with over 60% release achieved after the first 12 h for the 1 : 5 molar ratio sample [51]. The release of the 1 : 10 and 1 : 1 molar ratio samples is slower, probably due to the differences in the structure of the cisplatin-intercalated ZrP nanoparticles at those molar ratios. In the 1 : 1 molar ratio sample, most of the platinum complexes are covalently bonded to the layers in a cross-linked fashion, making the release of the Pt complex more difficult by the steric hindrance of the system. In contrast, for the 1 : 10 molar ratio sample, the platinum complex is less hindered but is more diluted throughout the whole nanoparticle, making the release slower. Finally, for the 1 : 5 molar ratio sample, there is less hindrance than in the 1 : 1 molar ratio sample but a higher concentration than in the 1 : 10 molar ratio sample, producing a faster release.

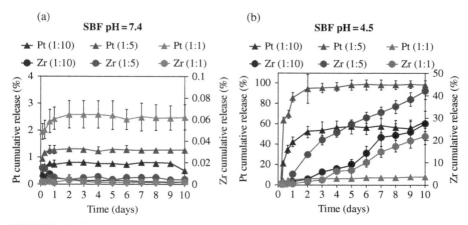

FIGURE 10.19 *In vitro* platinum and zirconium release from cisplatin–ZrP nanoparticles (1 : 1, 1 : 5 and 1 : 10 molar ratio) in SBF (a) and ALF (b), at pH 7.4 and 4.5, respectively. Díaz, A.; González, M. L.; Pérez, R. J.; David, A.; Mukherjee, A.; Báez, A.; Clearfield, A.; Colón, J. L., Direct intercalation of cisplatin into zirconium phosphate nanoplatelets for potential cancer nanotherapy. *Nanoscale* **2013**, *5*, 11456–11463. Reproduced with permission from the Royal Society of Chemistry.

Zr release was also monitored during the Pt release experiments, since the release of Zr atoms is indicative of hydrolysis of the particles, which is important for the clearance of the inorganic nanoparticles from the body. ZrP nanoparticle hydrolysis produces a soluble Zr salt and inorganic phosphate that can be reused in the biological system. Figure 10.19b shows that the release of Zr is slower than the release of Pt from the particle, suggesting that the hydrolysis of the platinum complex is faster than the hydrolysis of the nanoparticles under ALF conditions.

The proposed mechanism for the uptake and release of cisplatin in the cancerous tumour is via endocytosis, where a nanoparticle approaches the cancerous cell and the phosphate of the loaded nanoparticle interacts, via hydrogen bonding, with the phosphate of the phospholipid cell membrane. Once this interaction takes place, the cell takes in the nanoparticle via endocytosis where the phospholipid membrane folds and entraps the nanoparticle in an endosome vessel. The pH of a normal cell endosome is around 6, but cancerous cell endosomes should have a lower pH value than 6 since cancerous cells are acidic [82, 83]. This acidic environment is convenient for cisplatin release from the cisplatin–ZrP nanoparticles, while normal cells should be less affected diminishing the many side effects that cisplatin produces. The rapid release of the cisplatin from the ZrP galleries at low pH is desirable since it approaches the typical pH of the acidic environment of cancer cells. This controlled-release mechanism, together with the controllable particle size, is viable to design this material for drug therapy [51].

Inhibition of MCF-7 cancer cell growth by ZrP nanoparticles and cisplatin-intercalated ZrP nanoparticles was monitored after 24 and 48 h treatment using the MTT assay (Figure 10.20). Figure 10.20 shows that ZrP alone does not affect the viability of MCF-7 cells; however, treatment with cisplatin alone for 48 h reduced the viability of MCF-7 cells by 50% ($IC_{50} = 7 \, \mu M$). A higher concentration of cisplatin-intercalated ZrP (at 1 : 1 molar ratio) was needed to reduce MCF-7 cell growth by 40%. These results are consistent with the cisplatin release profile results (Figure 10.20). Both at 24 and 48 h, the effective cisplatin concentration released from cisplatin-intercalated ZrP nanoparticles is much slower, and therefore, its inhibitory concentration. Burst release profiles of drug-loaded polymeric nanoparticles and liposomes have been shown to be undesirable since they increase local and systemic toxicity, lower drug availability at the tumour site and, consequently, reduce therapeutic effect. Hence, the slow release profile for cisplatin-intercalated ZrP nanoparticles might be of pharmacological advantage by increasing the time of exposure of the tumour cells to the drug. ZrP nano-platelets as carriers will allow cisplatin, through the EPR effect, to be released mainly into the tumour cells, which might result in lower systemic toxicity to the patient.

10.2.2.3 Doxorubicin   The intercalation of the anticancer drug doxorubicin into θ-ZrP has also been achieved by direct ion exchange [32]. Figure 10.21 shows the XRPD patterns of α-ZrP (dried θ-ZrP) and of θ-ZrP before and after the intercalation reaction with doxorubicin. The XRPD pattern of the doxorubicin-intercalated ZrP (DOX : ZrP) shows an intense first peak corresponding to an interlayer distance of 20.3 Å, followed by the second-, third- and fourth-order diffraction peaks at 10.2, 6.8

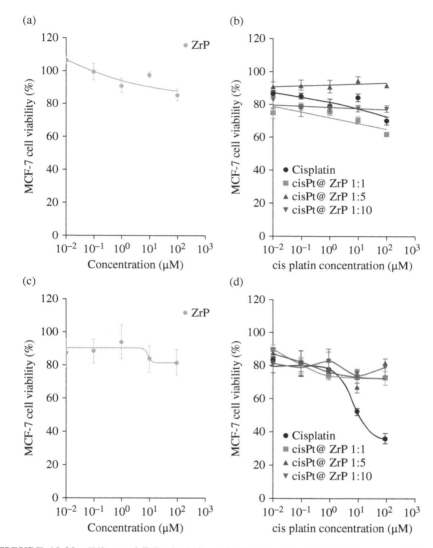

FIGURE 10.20 Effects of ZrP, cisplatin and cisplatin-intercalated ZrP nanoparticles on MCF-7 cell growth viability. Cells were treated with varying concentrations of ZrP, cisplatin or cisplatin-intercalated ZrP nanoparticles for 24 (a and b) and 48 h (c and d). The cisplatin concentration in micromolar units was calculated from cisplatin-intercalated ZrP based on the molecular formula at different loading levels as shown in Table 1. Díaz, A.; González, M. L.; Pérez, R. J.; David, A.; Mukherjee, A.; Báez, A.; Clearfield, A.; Colón, J. L., Direct intercalation of cisplatin into zirconium phosphate nanoplatelets for potential cancer nanotherapy. *Nanoscale* **2013**, *5*, 11456–11463. Adapted with permission from the Royal Society of Chemistry.

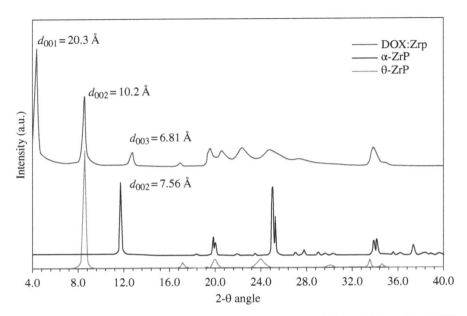

FIGURE 10.21 X-Ray powder diffraction (XRPD) of α-ZrP, θ-ZrP (highly hydrated ZrP) and doxorubicin-intercalated ZrP. Díaz, A.; Saxena, V.; González, J.; David, A.; Casañas, B.; Batteas, J.; Colón, J. L.; Clearfield, A.; Hussain, M., Zirconium phosphate nano-platelets: A platform for drug delivery in cancer therapy. *Chem. Comm.* **2012**, *48*, 1754–1756. Adapted with permission from the Royal Society of Chemistry.

and 5.1 Å, respectively. In addition, the peaks corresponding to the 020 and 312 planes are also present at 33.81° and 34.21° (2θ) showing that the individual α-ZrP-type layers remain intact.

Figure 10.22 shows scanning electron microscopy (SEM) and transmission microscopy (TEM) images of the DOX : ZrP intercalation product. The images show the electron characteristic platelet-like hexagonal shape of the ZrP nanoparticles. The average particle size of DOX : ZrP nano-platelets is 175 nm in diameter (Figure 10.22). The interlayer distance corresponds to the dimension of the doxorubicin intercalated perpendicularly relative to the plane of the layer. In addition, doxorubicin has a permanent positive charge on the amine group in one of the sides of the molecule. We expect this positive charge to strongly interact with the negatively charged phosphate layer causing this orientation. Moreover, in order to maximize the intermolecular forces and minimize the electrostatic repulsions between doxorubicin molecules within the layers, the molecules should be accommodated in a specific fashion. The doxorubicin molecules would be parallel to each other in a π-π stacking fashion with the protonated amino groups alternating on a tail-to-head conformation to maximize the opposite charge interactions, which is in complete agreement with the interlayer distance obtained (Figure 10.21) [32]. The high-resolution TGA (Hi-Res TGA) for the DOX : ZrP intercalation product showed the amount attributed to doxorubicin is 0.329 mol per ZrP formula unit or 34.9% of DOX per weight ($Zr(H_{0.84}PO_4)_2(DOX)_{0.33}$ 2.8 H_2O) [32].

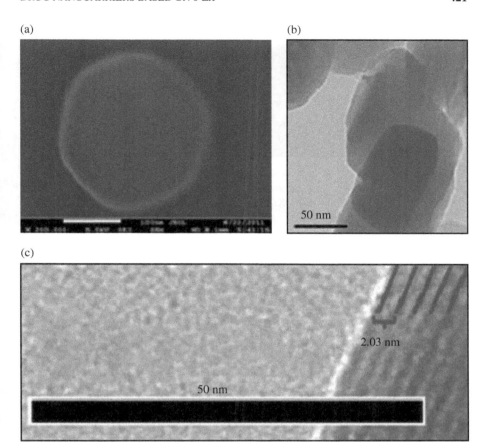

FIGURE 10.22 SEM/TEM images of ZrP; SEM image of an α-ZrP nano-platelet (a), TEM images of DOX : ZrP nano-platelets tilted (b), zoom-in of the TEM image (b) showing the layers and the interlayer distance (c). Díaz, A.; Saxena, V.; González, J.; David, A.; Casañas, B.; Batteas, J.; Colón, J. L.; Clearfield, A.; Hussain, M., Zirconium phosphate nano-platelets: A platform for drug delivery in cancer therapy. *Chem. Comm.* **2012**, *48*, 1754–1756. Adapted with permission from the Royal Society of Chemistry.

Confocal laser scanning microscopy (CLSM) images showed higher uptake and retention of the DOX from DOX : ZrP nano-platelets in the human breast cancer cells, MCF-7, as compared to free doxorubicin within the first 4 h of incubation (Figure 10.23) [32]. Cell viability assays for the DOX : ZrP nano-platelets were performed on MCF-7 cells, using the MTT assay. The cell viability assay reveals that the DOX : ZrP nano-platelets exhibited higher cytotoxicity than free doxorubicin (Figure 10.23) at lower concentrations. The overall IC_{50} value for the DOX : ZrP nano-platelets (52 ± 3 nM) was found to be lower than that of free doxorubicin (66 ± 3 nM) at 24 h of exposure. After 48 h of exposure, the overall IC_{50} value of DOX : ZrP nanoparticles (0.18 ± 0.05 nM) was found to be approximately 50 times lower than that of free doxorubicin

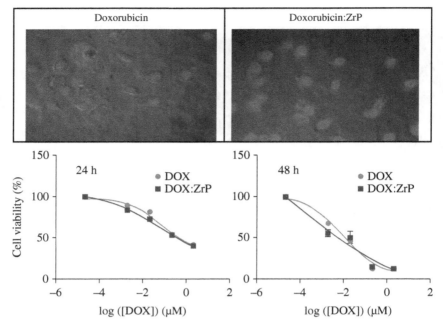

FIGURE 10.23   CLSM images and MTT results in breast cancer cells; top panels show CLSM images of MCF-7 cells treated with DOX (left) and DOX : ZrP (right) nano-platelets showing the higher uptake of the DOX : ZrP nano-platelets into the cells. The bottom panel shows MTT assay results in MCF-7 cell lines at 24 and 48 h of exposure of DOX : ZrP, with their respective controls ($n = 3$, $p < 0.05$ for both 24 and 48 h of exposure). Díaz, A.; Saxena, V.; González, J.; David, A.; Casañas, B.; Batteas, J.; Colón, J. L.; Clearfield, A.; Hussain, M., Zirconium phosphate nano-platelets: A platform for drug delivery in cancer therapy. *Chem. Comm.* **2012**, *48*, 1754–1756. Reproduced with permission from the Royal Society of Chemistry.

(9 ± 2 nM) (Figure 10.23). Cell viability assays for the ZrP nano-platelets alone on MCF-7 cells show no cytotoxicity at 24 and 48 h of exposure [32].

These results show that successful intercalation of doxorubicin into α-ZrP was achieved by direct ion exchange producing a new phase with an interlayer distance of 20.3 Å. The obtained DOX : ZrP nano-platelets had an impressive 34.9% (w/w) drug loading. Cellular uptake studies showed higher uptake of DOX : ZrP in breast cancer cells. The cell viability assay indicated a significant improvement in the cytotoxicity. The Clearfield group is currently modifying the surface of ZrP nano-platelets for increasing circulation half-life and specifically target cancer cells. Further studies are needed for determining the efficacy of the targeted ZrP nano-platelets in animal models of human cancer [32].

10.2.2.4   *Metallocenes*   After titanocene dichloride's (Cp$_2$TiCl$_2$, Figure 10.24a) antitumour properties were discovered in 1979 [84], investigations of other related metal-based potential drugs emerged due to the lower toxicity exhibited in comparison

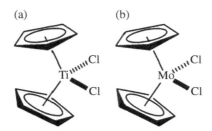

(a)   (b)

FIGURE 10.24   (a) Structure of titanocene dichloride and (b) molybdocene dichloride.

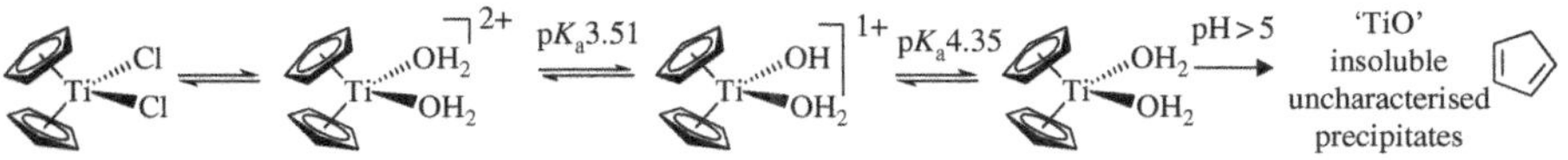

FIGURE 10.25   Hydrolysis chemistry of titanocene dichloride showing the major species present at physiological pH. Buck, D. P.; Abeysinghe, P. M.; Culliname, C.; Day, A. I.; Collins, J. G.; Harding, M. M. Inclusion complexes of the antitumor metallocenes Cp_2MCl_2 (M=Mo, Ti) with cucurbit[n]urils. *Dalton Trans*. **2008**, 2328–2334. Reproduced with permission from the Royal Society of Chemistry.

with the known anticancer drug cisplatin (*cis*-Pt(NH$_3$)Cl$_2$). The metallocenes of general formula Cp_2MX_2 (M=Ti, V, Nb, Mo; X=halides and pseudohalides) showed antitumour activity against a wide variety of tumour cells [85–87].

From all the metallocenes studied, titanocene dichloride showed to be the most active complex [85, 88]. Two phase II clinical trials have been reported, but the results were not promising enough in comparison with other treatments [87, 89, 90]. Furthermore, Cp_2TiCl_2 is unstable in water and at physiological pH hydrolyses extensively forming insoluble species (Figure 10.25). The hydrolysis of Cp_2TiCl_2 presents major difficulties in the preparation of a stable formulation suitable for administration [85, 87, 91, 92]. Therefore, researchers are looking for ways to stabilize and deliver this drug to its intended target.

Given the drawbacks of titanocene dichloride, we have also been exploring the intercalation of molybdocene dichloride (Cp_2MoCl_2) (Figure 10.24b). The mechanism of action of Cp_2MoCl_2 is not yet clear in terms of the cellular uptake, distribution and damage in the tumour cells. Molybdocene dichloride, having a softer metal than titanocene, prefers coordination to thiols over phosphates, amino and carboxylate groups suggesting that this compound would be more attracted to thiol-containing biomolecules as a cellular target compared to the titanocene dichloride, which has been reported to interact with cellular DNA [86, 87].

In addition, Cp_2MoCl_2 has a poor aqueous solubility, which limits its utility as an antitumour drug. In contrast to titanocene dichloride, the aqueous chemistry of Cp_2MoCl_2 has been well characterized (Figure 10.26). The Cp ligands are reported to be very stable, but the hydrolysis of the chloride ligands is very fast, with more than half the chloride atoms being displaced upon dissolution, and at pH 7.4, the chloride hydrolysis was too rapid to measure [87]. Potentiometric titrations of

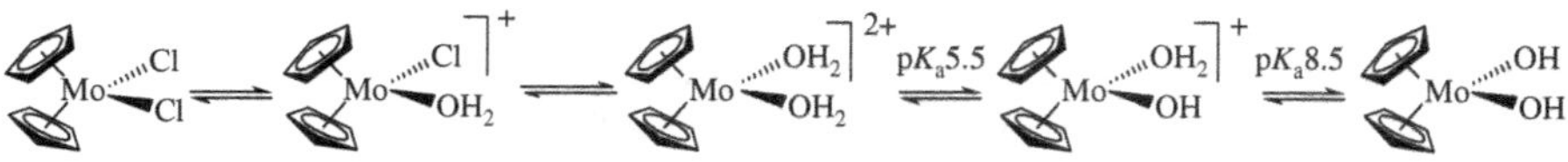

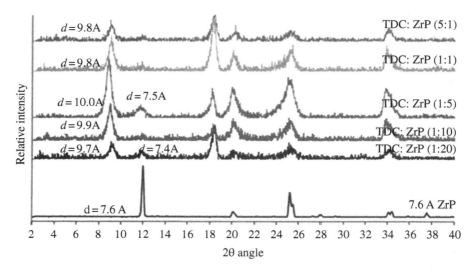

FIGURE 10.26 Hydrolysis chemistry of molybdocene dichloride. Reprinted from *J. Organomet. Chem.*, *689*, Waern, J. B.; Harding, M. M., Bioorganometallic chemistry of molybdocene dichloride, 4655–4668. Copyright 2004, with permission from Elsevier.

FIGURE 10.27 X-ray powder diffractograms of titanocene dichloride-exchanged zirconium phosphate at various loading levels and α-ZrP [64].

aqueous solutions of Cp_2MoCl_2 showed two deprotonations with pKa (1)=5.5 and pKa (2)=8.5 (Figure 10.26), and therefore, it was concluded that at physiological conditions the monocation 1a is the predominant species present [93].

The drawbacks of low solubility in water, instability and non-localized release of the drug are the reasons we intended to directly intercalate these two metallo-cenes, without pre-intercalating butylammonium ions, using the highly hydrated form of ZrP, which is the θ-ZrP phase. Intercalation of this type of drugs proposed here in θ-ZrP has not been reported previously in the literature.

Titanocene Dichloride   For the intercalation of titanocene dichloride, the XRPD patterns show the formation of a new phase with an interlayer distance of ca. 9.7–10.0Å (Figure 10.27). The 1 : 5 and 1 : 20 TDC : ZrP molar ratio materials appear to be mixed phases. As the metallocene derivative is entering through the edges of the layers of ZrP (fast step), the diffusion towards the interior of the layer becomes slower because of the high energy barrier leading to the mixed phases seen in the diffractograms [61].

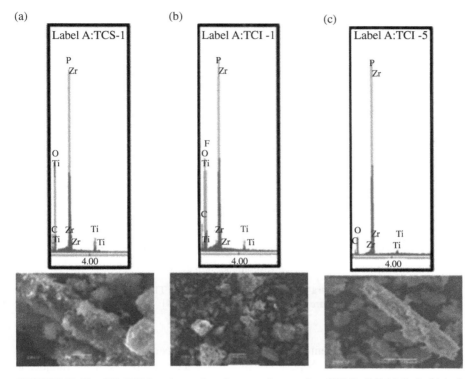

FIGURE 10.28   SEM-EDS and scanning electron micrographs of TDC : ZrP (a) 5 : 1, (b) 1 : 1 and (c) 1 : 5 loading level [64].

From the literature, we know that titanocene dichloride is rapidly hydrolyzed; the loss of the first chloride ligand is too fast to be measured, and the second aquation step occurs much slower [85, 94]. For this reason, and our previous results with cisplatin, we expected substitution of the chloride ligands in the metallocene derivative by aquo ligands that are subsequently substituted by the phosphate groups of the ZrP layers. This is why we performed qualitative elemental analysis by energy-dispersive X-ray spectroscopy (EDS) by means of SEM of the titanocene-intercalated ZrP (TDC : ZrP) materials at the 5 : 1, 1 : 1, 1 : 5, 1 : 10 and 1 : 20 loading levels. Figure 10.28 shows the EDS analyses for the TDC : ZrP 5 : 1, 1 : 1 and 1 : 5 loading levels. These analyses show the characteristic peaks indicating the presence of Ti, Zr, P, O and C, except in the 1 : 20 TDC : ZrP sample where the Ti signal does not appear probably due to the low concentration of the metal in this sample. The presence of F in the 1 : 1 TDC : ZrP intercalated material could be from contamination with material from the teflon-coated stirring bars used in the intercalation reaction. The absence of the chloride peak in the EDS analyses suggests that the chloride ligands might have been substituted in the TDC : ZrP materials [64].

The IR spectra of the TDC : ZrP intercalated materials (Figure 10.29) showed the decrease in intensity of the water vibrational peaks (at ~3600 and 3550 cm^{-1})

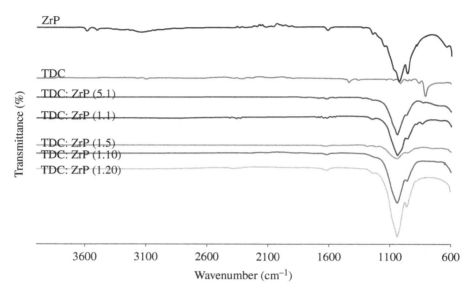

FIGURE 10.29 FTIR spectra of (upper to lower) α-ZrP and TDC intercalated into the θ-ZrP framework at room temperature and different molar ratios [64].

indicating the displacement of water by the metallocene derivative, as expected [39, 40, 50]. However, the appearance of the characteristic peaks of the titanocene dichloride (expected at 3100 and 1018 cm^{-1} for the v_{C-H} bands in the Cp rings, 1440 cm^{-1} for the $v_{C=C}$ bands and 1370 cm^{-1} for the v_{C-C} bands) in the spectra of the intercalated compounds was not observed. Only the peak at 820 cm^{-1}, characteristic of the vibrational band of the C–H bond in the Cp rings, is observed in the 1 : 1 TDC : ZrP sample. The absence of the other titanocene vibrational bands in the spectra is probably due to the overlap of the ZrP vibrational bands with those of the metallocene derivative. The loading level and the intensity of the titanocene bands is too low for the other TDC bands to be observed [64].

The successful intercalation of this new material has potential as a drug carrier for cancer treatment. *In vitro* release of this Ti(IV) complex after the intercalation and cell toxicity are being evaluated.

Molybdocene Dichloride Intercalation at various molybdocene dichloride–ZrP loading levels (5 : 1, 1 : 1, 1 : 5, 1 : 10 and 1 : 20) was performed [63].

Figure 10.30 shows the X-ray powder diffractogram of the dried intercalation products, which showed formation of a new phase with an interlayer distance of ca. 10.3–11.0 Å. The absence of the 7.6 Å diffraction peak (except in the 1 : 20 loading level) confirms that the metallocene derivative was successfully intercalated into the ZrP layers. A mixed phase is observed for the lowest loading level (1 : 20), which is characteristic of the presence of intercalated parts of the layers and parts of the layers with no intercalation, therefore resulting in the peak at ca. 7.6 Å.

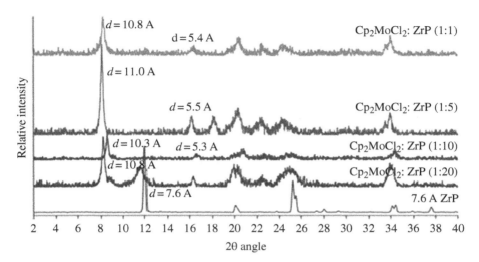

FIGURE 10.30   X-ray powder diffractograms of Cp_2MoCl_2 : ZrP at various loading levels [64].

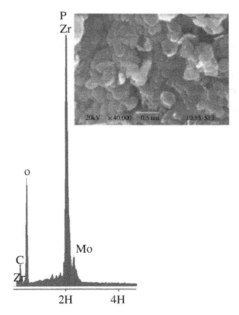

FIGURE 10.31   SEM-EDS of $Cp_2Mo(H_2O)(OH)^+$: ZrP (1 : 1).

At pH 7, Cp_2MoCl_2 hydrolyses to $[Cp_2Mo(H_2O)(OH)]^+$, which we expected would intercalate in the ZrP layers by ion exchange [87]. This is why we performed the qualitative elemental analysis by EDS by means of SEM of $Cp_2Mo(H_2O)(OH)^+$: ZrP (1 : 1). Figure 10.31 shows the characteristic peaks indicating the presence of Mo, Zr, P, O and C. The important result was the absence of a chloride peak in this

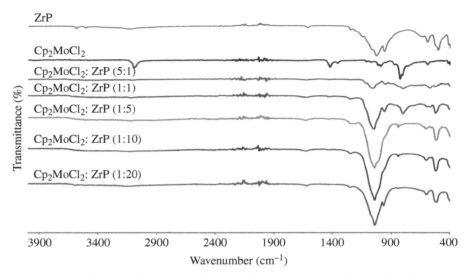

FIGURE 10.32   FTIR spectra of $Cp_2Mo(H_2O)(OH)^+$: ZrP at various loading levels.

elemental analysis, which suggests that the species intercalated is the $[Cp_2Mo(H_2O)(OH)]^+$ cation.

In addition, we performed IR studies of both un-intercalated and intercalated molybdocene dichloride to determine any structural change that might have occurred after the intercalation reaction. Figure 10.32 shows the disappearance of the ZrP lattice water bands (from water molecules held in the crystal lattice) that appear in the region $3580-3200\,cm^{-1}$ and $1630-1600\,cm^{-1}$ for ZrP in the various loading levels of $Cp_2Mo(H_2O)(OH)^+$: ZrP IR spectra [74]. This is consistent with the displacement of interlayer water by the metallocene derivative upon intercalation. However, the appearance of the characteristic peaks of the molybdocene dichloride (expected at 3100 and $1018\,cm^{-1}$ for the v_{C-H} bands in the Cp rings, $1440\,cm^{-1}$ for the $v_{C=C}$ bands and $1370\,cm^{-1}$ for the v_{C-C} bands) in the spectra of the intercalated compounds was not observed. Only the peak at $820\,cm^{-1}$ characteristic of the vibrational band of the C–H bond in the Cp rings is observed in the various loading levels of Cp_2MoCl_2 : ZrP. This could be due to the low concentration of the metallocene derivative inside the layers and the overlapping of the bands of the ZrP material and the complex. These results suggest that the intercalation of the metallocene derivative was successfully achieved without any significant chemical change in the intercalated molecules. *In vitro* drug release studies are being pursued before cell studies are performed.

10.2.3   Neurological Agents

The drawbacks of low solubility in water, instability and non-localized release of the drug are not limited to cancer therapy but are also encountered in drugs for the treatment of brain disorders. A complicating matter is the presence of the blood–brain barrier

(a) (b)

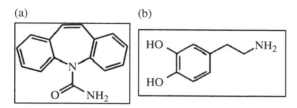

FIGURE 10.33 (a) Structure of carbamazepine and (b) structure of dopamine.

(BBB), a structure that limits the access of drugs to their sites of action in the central nervous system [95].

One drug used for the treatment of epilepsy is carbamazepine (CBZ) (Figure 10.33a). CBZ is usually given by oral administration, but due to its poor water solubility, it is characterized by slow and irregular gastrointestinal absorption. Furthermore, CBZ is characterized by a considerable hepatic first-pass effect owing to the enzymatic auto-induction of its metabolism [96].

Another neurochemical drug is dopamine (DA), a neurotransmitter associated with several diseases of the nervous system that depend on its activity (Figure 10.33b). One example is Parkinson's disease, which has been associated with a lack of DA in the brain. DA is characterized by low or insufficient water solubility and oxidation and decarboxylation reactivity with intestinal and blood enzymes [97].

For these reasons, we have been studying the direct intercalation of these neurochemical drugs, CBZ and DA, without pre-intercalating butylammonium ions, using the highly hydrated theta phase of ZrP. Hayashi et al. have previously reported the intercalation of DA in α-ZrP [98]. We will compare our results with those of Hayashi et al.

10.2.3.1 CBZ Our preliminary results show that we have obtained an XRPD pattern of the intercalation product of CBZ in ZrP (Figure 10.34) in which the first peak has a corresponding distance of 10.3 Å. Further studies are being conducted to completely interpret this distance.

We also performed FTIR measurements of the 5 : 1 CBZ : ZrP material using KBr pellets (data not shown) where the spectra show the decrease of the ZrP lattice water bands (from water molecules held in the crystal lattice) that normally appear in the region 3580–3200 cm⁻¹ for ZrP, in the CBZ : ZrP spectra in the 5 : 1 loading level [74]. This could suggest partial displacement of interlayer water by the neurochemical drug upon intercalation. Still, the characteristic peaks of CBZ (expected at 3460 cm⁻¹ for the N–H stretch, 3160 cm⁻¹ for the aromatic C–H stretch band and 1680 cm⁻¹ for the amide C=O band) [99] in the spectra of the intercalated compound were not observed. Only the peaks at 3050 and 3020 cm⁻¹ characteristic of the stretching band of the C–H bond in the alkenes are observed displaced to 2925 and 2854 cm⁻¹, respectively, in the 5 : 1 loading level of CBZ : ZrP. This could be due to low concentration of the drug inside the layers and the overlapping of the bands of the ZrP material and the complex.

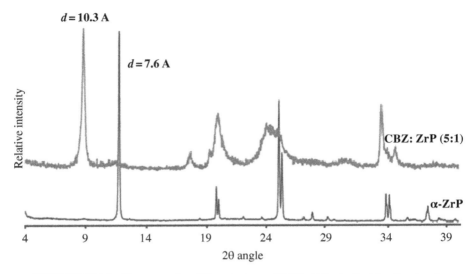

FIGURE 10.34   X-ray powder diffractograms of CBZ : ZrP at a 5 : 1 loading level.

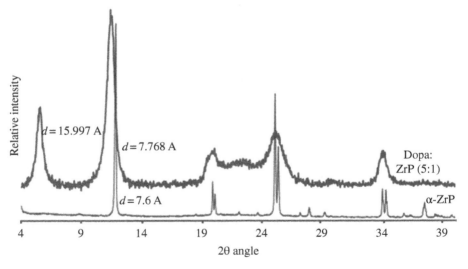

FIGURE 10.35   X-ray powder diffractograms of α-ZrP and DA : ZrP at a 5 : 1 loading level.

10.2.3.2   DA   For DA, we also have obtained preliminary results. XRPD of the 5 : 1 loading level DA : ZrP intercalated product was obtained (Figure 10.35). The first-order peak of the samples has a corresponding distance of 16.0Å, and a second-order peak is also observed with a corresponding distance of 7.77Å. This distance is similar to what Hayashi et al. obtained when intercalating directly DA at pH 7–8 with α-ZrP [98]. Further studies are needed to verify if the DA has been oxidized and which is the orientation of the neurotransmitter inside the layers. Although we are early in our endeavours into intercalation of neurochemicals into ZrP, our preliminary results are stimulating our investigations into this area.

10.3 CONCLUSION

The development of the intercalation chemistry of the theta phase of ZrP in our laboratories has yielded successful intercalations of a variety of molecules that could not have been directly intercalated into α-ZrP. Our success using θ-ZrP prompted our recent excursion into the intercalation of biological, anticancer and neurochemical drugs. Initial results are very promising; future animal model studies should demonstrate the viability of this new approach into nanotherapeutic treatment using ZrP nanoparticles. The applications of ZrP, after more than 50 years of studies, continue to expand.

REFERENCES

1. Clearfield, A.; Stynes, J. A., The preparation of crystalline zirconium phosphate and some observations on its ion exchange behaviour. *J. Inorg. Nucl. Chem.* **1964**, *26*, 117–129.

2. Constantino, U.; Marmottini, F., Metal exchanged layered zirconium hydrogen phosphate as base catalyst of the Michael reaction. *Catal. Lett.* **1993**, *22*, 333–336.

3. Hu, H.; Martin, J. C.; Zhang, M.; Southworth, C. S.; Xiao, M.; Meng, Y.; Sun, L., Immobilization of ionic liquids in θ-zirconium phosphate for catalyzing the coupling of CO_2 and epoxides. *RSC Adv.* **2012**, *2*, 3810–3815.

4. Ginestra, A. L.; Patrono, P.; Berardelli, M. L.; Galli, P.; Ferragina, C.; Massucci, M. A., Catalytic activity of zirconium phosphate and some derived phases in the dehydration of alcohols and isomerization of butenes. *J. Catal.* **1987**, *103*, 346–356.

5. Curini, M.; Montanari, F.; Rosati, O.; Lioy, E.; Margarita, R., Layered zirconium phosphate and phosphonate as heterogeneous catalyst in the preparation of pyrroles. *Tetrahedron Lett.* **2003**, *44*, 3923–3925.

6. Clearfield, A.; Thakur, D. S., Zirconium and titanium phosphates as catalysts: A review. *Appl. Catal.* **1986**, *26*, 1–26.

7. Segawa, K.; Nakajima, Y.; Nakata, S.; Asaoka, S.; Takahashi, H., P-31-MASNMR spectroscopic studies with zirconium phosphate catalysts. *J. Catal.* **1986**, *101*, 81–89.

8. Colón, J. L.; Thakur, D. S.; Yang, C. Y.; Clearfield, A.; Martin, C. R., X-ray photoelectron spectroscopy and catalytic activity of α-zirconium phosphate and zirconium phosphate sulfophenylphosphonate. *J. Catal.* **1990**, *124*, 148–159.

9. Niño, M. E.; Giraldo, S. A.; Páez-Mozo, E. A., Olefin oxidation with dioxygen catalyzed by porphyrins and phthalocyanines intercalated in α-zirconium phosphate. *J. Mol. Catal. A Chem.* **2001**, *175*, 139–151.

10. Álvaro, V. F. D.; Johnstone, R. A. W., High surface area Pd, Pt and Ni ion-exchanged Zr, Ti and Sn(IV) phosphates and their application to selective heterogeneous catalytic hydrogenation of alkenes. *J. Mol. Catal. A Chem.* **2008**, *280*, 131–141.

11. Khare, S.; Chokhare, R., Synthesis, characterization and catalytic activity of Fe(Salen) intercalated α-zirconium phosphate for the oxidation of cyclohexene. *J. Mol. Catal. A Chem.* **2011**, *344*, 83–92.

12. Khare, S.; Chokhare, R., Oxidation of cyclohexene catalyzed by Cu(Salen) intercalated α-zirconium phosphate using dry tert-butylhydroperoxide. *J. Mol. Catal. A Chem.* **2012**, *353*, 138–147.

13. Yeates, R. C.; Kuznicki, S. M.; Lloyd, L. B.; Eyring, E. M., Tris(2,2'-Bipyridine) Ruthenium(II) electron-transfer in a layered zirconium phosphate lattice. *J. Inorg. Nucl. Chem.* **1981**, *43*, 2355–2358.

14. Vliers, D. P.; Schoonheydt, R. A.; Deschrijver, F. C., Synthesis and luminescence of ruthenium tris(2,2'-Bipyridine) zirconium phosphates. *J. Chem. Soc. Faraday Trans. 1* **1985**, *81*, 2009–2019.

15. Vliers, D. P.; Collin, D.; Schoonheydt, R. A.; Deschryver, F. C., Synthesis and characterization of aqueous tris(2,2'-Bipyridine)ruthenium(II)-zirconium phosphate suspensions. *Langmuir* **1986**, *2*, 165–169.

16. Rosenthal, G. L.; Caruso, J., Photochemical behavior of metal complexes intercalated in zirconium phosphate. *J. Solid State Chem.* **1991**, *93*, 128–133.

17. Kumar, C. V.; Chaudhari, A.; Rosenthal, G. L., Enhanced energy transfer between aromatic chromophores bound to hydrophobically modified layered zirconium phosphate suspensions. *J. Am. Chem. Soc.* **1994**, *116*, 403–404.

18. Kumar, C. V.; Williams, Z. J., Supramolecular assemblies of tris(2,2'-bipyridine) ruthenium(II) bound to hydrophobically modified α-zirconium phosphate: Photophysical studies. *J. Phys. Chem.* **1995**, *99*, 17632–17639.

19. Kumar, C. V.; Williams, Z. J.; Turner, R. S., Supramolecular assemblies of metal complexes: Light-Induced electron transfer in the galleries of α-zirconium phosphate. *J. Phys. Chem. A* **1998**, *102*, 5562–5568.

20. Krishna, R. M.; Kevan, L., Photoinduced electron transfer from *N, N, N', N'*-tetramethyl-benzinide incorporated into layered zirconium phosphate studied by ESR and diffuse reflectance spectroscopies. *Microporous Mesoporous Mater.* **1999**, *32*, 169–174.

21. Odobel, F.; Massiot, D.; Harrison, B. S.; Schanze, K. S., Photoinduced energy transfer between ruthenium and osmium tris-bipyridine complexes covalently pillared into γ-ZrP. *Langmuir* **2003**, *19*, 30–39.

22. Brunet, E.; Alonso, M.; Cerro, C.; Juanes, O.; Rodríguez-Ubis, J. C.; Kaifer, A. E., A luminescence and electrochemical study of photoinduced electron transfer within the layers of zirconium phosphate. *Adv. Funct. Mater.* **2007**, *17*, 1603–1610.

23. Wang, Q. F.; Yu, D. Y.; Wang, Y.; Sun, J. Q.; Shen, J. C., Incorporation of water-soluble and water-insoluble ruthenium complexes into zirconium phosphate films fabricated by the layer-by-layer adsorption and reaction method. *Langmuir* **2008**, *24*, 11684–11690.

24. Brunet, E.; Alonso, M.; Quintana, M. C.; Atienzar, P.; Juanes, O.; Rodríguez-Ubis, J. C.; García, H., Laser flash-photolysis study of organic-inorganic materials derived from zirconium phosphates/phosphonates of Ru(bpy)(3) and C60 as electron donor-acceptor pairs. *J. Phys. Chem. C* **2008**, *112*, 5699–5702.

25. Shi, S. K.; Peng, Y. L.; Zhou, J., Enhanced stable luminescence and photochemical properties of ruthenium complex/zirconium phosphate hybrid assemblies. *J. Nanosci. Nanotechnol.* **2009**, *9*, 2746–2752.

26. Martí, A. A.; Colón, J. L., Direct ion exchange of tris(2,2'-bipyridine)ruthenium(II) into an α-zirconium phosphate framework. *Inorg. Chem.* **2003**, *42*, 2830–2832.

27. Martí, A. A.; Mezei, G.; Maldonado, L.; Paralitici, G.; Raptis, R. G.; Colón, J. L., Structural and photophysical characterisation of fac-[Tricarbonyl(chloro)-(5,6-epoxy-1,10-phenanthroline)rhenium(I)]. *Eur. J. Inorg. Chem.* **2005**, *118–124*.

28. Martí, A. A.; Paralitici, G.; Maldonado, L.; Colón, J. L., Photophysical characterization of methyl viologen ion-exchanged within α-zirconium phosphate framework. *Inorg. Chim. Acta.* **2007**, *360*, 1535–1542.

29. Martí, A. A.; Rivera, N.; Soto, K.; Maldonado, L.; Colón, J. L., Intercalation of Re(phen) (CO)$_3$Cl into zirconium phosphate: A water insoluble inorganic complex immobilized in a highly polar rigid matrix. *Dalton Trans.* **2007**, *1713–1718*.

30. Santiago, M.; Declet-Flores, C.; Díaz, A.; Velez, M. M.; Bosques, M. Z.; Sanakis, Y.; Colón, J. L., Layered inorganic materials as redox agents: Ferrocenium-intercalated zirconium phosphate. *Langmuir* **2007**, *23*, 7810–7817.

31. Díaz, A.; David, A.; Pérez, R.; González, M. L.; Báez, A.; Wark, S. E.; Clearfield, A.; Colón, J. L., Nanoencapsulation of insulin into zirconium phosphate for oral delivery applications. *Biomacromolecules* **2010**, *11*, 2465–2470.

32. Díaz, A.; Saxena, V.; González, J.; David, A.; Casañas, B.; Batteas, J.; Colón, J. L.; Clearfield, A.; Hussain, M., Zirconium phosphate nano-platelets: A platform for drug delivery in cancer therapy. *Chem. Comm.* **2012**, *48*, 1754–1756.

33. Díaz, A.; González, M. L.; Pérez, R. J.; David, A.; Mukherjee, A.; Báez, A.; Clearfield, A.; Colón, J. L., Direct intercalation of cisplatin into zirconium phosphate nanoplatelets for potential cancer nanotherapy. *Nanoscale* **2013**, *5*, 11456–11463.

34. Ruan, C. M.; Yang, F.; Xu, J. S.; Lei, C. H.; Deng, J. Q., Immobilization of methylene blue using α-zirconium phosphate and Its application within a reagentless amperometric hydrogen peroxide biosensor. *Electroanalysis* **1997**, *9*, 1180–1184.

35. Yuan, Y.; Wang, P.; Zhu, G. Y., Sol-gel derived carbon ceramic electrode containing methylene blue-intercalated α-zirconium phosphate micro particles. *Anal. Bioanal. Chem.* **2002**, *372*, 712–717.

36. Liu, Y.; Lu, C. L.; Hou, W. H.; Zhu, J. J., Direct electron transfer of hemoglobin in layered α–zirconium phosphate with a high thermal stability. *Anal. Biochem.* **2008**, *375*, 27–34.

37. Yang, X. S.; Chen, X.; Yang, L.; Yang, W. S., Direct electrochemistry and electrocatalysis of horseradish peroxidase in α-zirconium phosphate nanosheet film. *Bioelectrochemistry* **2008**, *74*, 90–95.

38. Yang, X. S.; Chen, X.; Zhang, J. H.; Yang, W. S., Fabrication of electroactive layer-by-layer films with myoglobin and zirconium phosphate nanosheets. *Chem. Lett.* **2008**, *37*, 240–241.

39. Santiago, M. B.; Velez, M. M.; Borrero, S.; Diaz, A.; Casillas, C. A.; Hoffman, C.; Guadalupe, A. R.; Colón, J. L., NADH electrooxidation using bis(1,10-phenanthroline-5,6-dione)(2,2′-bipyridine) ruthenium(II)-exchanged zirconium phosphate modified carbon paste electrodes. *Electroanalysis* **2006**, *18*, 559–572.

40. Santiago, M.; Daniel, G.; David, A.; Casañas, B.; Hernández, G.; Guadalupe, A.; Colón, J. L., Effect of enzyme and cofactor immobilization on the response of ethanol amperometric biosensors modified with layered zirconium phosphate. *Electroanalysis* **2010**, *22*, 1097–1105.

41. Sun, L.; Boo, W. J.; Sue, H.-J.; Clearfield, A., Preparation of α-zirconium phosphate nanoplatelets with wide variations in aspect ratios. *New J. Chem.* **2007**, *31*, 39–43.

42. Clearfield, A., Metal-phosphonate chemistry. *Prog. Inorg. Chem.* **1998**, *47*, 371–510.

43. Alberti, G., Synthesis, crystalline structure, and ion-exchange properties of insoluble acid salts of tetravalent metals and their salt forms. *Acc. Chem. Res.* **1978**, *11*, 163–170.

44. Alberti, G.; Constantino, U.; Gill, J. S., Recent progress in the intercalation chemistry of layered α-zirconium phosphate and its derivatives, and future perspectives for their use in catalysis. *J. Mol. Catal.* **1984**, *27*, 235–250.

45. Kumar, C. V.; Bhambhani, A.; Hnatiuk, N., In *Handbook of Layered Materials*, Auerbach, S. M.; Carrado, K. A.; Dutta, P. K., Eds. Marcel Dekker, Inc.: New York, **2004**; pp. 313–372.

46. Clearfield, A., Role of ion exchange in solid-state chemistry. *Chem. Rev.* **1988**, *88*, 125–148.

47. Rivera, E. J.; Barbosa, C.; Torres, R.; Grove, L.; Taylor, S.; Connick, W. B.; Clearfield, A.; Colón, J. L., Vapochromic and vapoluminescent response of materials based on platinum(II) complexes intercalated into layered zirconium phosphate. *J. Mater. Chem.* **2011**, *21*, 15899–15902.

48. Bermúdez, R. A.; Colón, Y.; Tejada, G. A.; Colón, J. L., Intercalation and photophysical characterization of 1-Pyrenemethylamine into zirconium phosphate layered materials. *Langmuir* **2005**, *21*, 890–895.

49. Bermúdez, R. A.; Arce, R.; Colón, J. L., Photolysis of 1-pyrenemethylamine ion-exchanged into a zirconium phosphate framework. *J. Photochem. Photobiol. A* **2005**, *175* (2–3), 201–206.

50. Rivera, E. J.; Figueroa, C.; Colón, J. L.; Grove, L.; Connick, W. B., Room-temperature emission from platinum(II) complexes intercalated into zirconium phosphate-layered materials. *Inorg. Chem.* **2007**, *46*, 8569–8576.

51. Díaz, A., Structural characterization of bioactive species intercalated into α-zirconium phosphate for drug delivery applications. University of Puerto Rico, San Juan, **2010**.

52. Alberti, G., Laboratorio di Chimica Inorganica Homepage. http://www.chm.unipg.it/chimino/1/fig/gstick.html (accessed 12 April 2010).

53. Troup, J. M.; Clearfield, A., Mechanism of ion exchange in zirconium phosphates. 20. Refinement of the crystal structure of α-zirconium phosphate. *Inorg. Chem.* **1977**, *16*, 3311–3314.

54. Yang, C. Y.; Clearfield, A., The preparation and ion-exchange properties of zirconium sulphophosphonates. *React. Polym.* **1987**, *5*, 13–21.

55. Clearfield, A.; Blessing, R. H.; Stynes, J. A., New crystalline phases of zirconium phosphate possessing ion-exchange properties. *J. Inorg. Nucl. Chem.* **1968**, *30*, 2249–2258.

56. Alberti, G.; Constantino, U.; Gill, J. S., Crystalline insoluble acid salts of tetravalent metals – XXIII: Preparation and main ion exchange properties of highly hydrated zirconium bis monohydrogen orthophosphates. *J. Inorg. Nucl. Chem.* **1976**, *38*, 1733–1738.

57. Clearfield, A.; Duax, W. L.; Medina, A. S.; Smith, G. D.; Thomas, J. R., On the mechanism of ion exchange in crystalline zirconium phosphates. Sodium ion exchange of α-zirconium phosphate. *J. Phys. Chem.* **1969**, *73*, 3424–3430.

58. Clearfield, A., Group IV phosphates as catalysts and catalyst supports. *J. Mol. Catal.* **1984**, *27*, 251–262.

59. Kim, H. N.; Keller, S. W.; Mallouk, T. E.; Schmitt, J.; Decher, G., Characterization of zirconium phosphate polycation thin films grown by sequential adsorption reactions. *Chem. Mater.* **1997**, *9*, 1414–1421.

60. Kim, H. N.; Keller, S. W.; Mallouk, T. E., Layer-by-layer assembly of intercalation compounds and heterostructures on surfaces: Toward molecular 'Beaker' epitaxy. *J. Am. Chem. Soc.* **1994**, *116*, 8817–8818.

61. Sun, L.; O'Reilly, J. Y.; Kong, D.; Su, J. Y.; Boo, W. J.; Sue, H. J.; Clearfield, A., The effect of guest molecular architecture and host crystallinity upon the mechanism of the intercalation reaction. *J. Colloid Interf. Sci.* **2009**, *333*, 503–509.

62. Díaz, A., *Structural characterization of bioactive species intercalated into a-zirconium phosphate for drug delivery applications.* M.S. Thesis, University of Puerto Rico, San Juan, **2010**.

63. Rivera, E. J.; Barbosa, C.; Torres, R.; Rivera, H.; Fachini, E. R.; Green, T. W.; Connick, W. B.; Colon, J. L., Luminescence rigidochromism and redox chemistry of pyrazolate-bridged binuclear platinum(II) diimine complex intercalated into zirconium phosphate layers. *Inorg. Chem.* **2012**, *51* (5), 2777–2784.

64. Casañas-Montes, B., Chemical and electrochemical characterization of metallocene derivatives in zirconium phosphate layers. M.S. Thesis, University of Puerto Rico, San Juan, **2010**.

65. Kijima, T., Direct Preparation of θ-zirconium phosphate. *Bull. Chem. Soc. Jpn.* **1982**, *55*, 3031–3032.

66. Clearfield, A.; Duax, W. L.; Medina, A. S.; Smith, G. D.; Thomas, J. R.. On the mechanism of ion exchange in crystalline zirconium phosphates. Sodium ion exchange of alfa-zirconium phosphate. *J. Phys. Chem.* **1969**, *73*, 3424–3430.

67. NIH Nanotechnology at NIH. http://www.nih.gov/science/nanotechnology/ (accessed 15 October 2012).

68. Wang, M.; Thanou, M., Targeting nanoparticles to cancer. *Pharmacol. Res.* **2010**, *62*, 90–99.

69. Parveen, S.; Misra, R.; Sahoo, S. K., Nanoparticles: A boon to drug delivery, therapeutics, diagnostics and imaging. *Nanomed Nanotechnol.* **2012**, *8*, 147–166.

70. Bringley, J. F.; Liebert, N. B., Controlled chemical and drug delivery via the internal and external surfaces of layered compounds. *J. Disper. Sci. Tech.* **2003**, *24*, 589–605.

71. Zhu, Z. Y.; Zhang, F. Q.; Xie, Y. T.; Chen, Y. K.; Zheng, X. B., In vitro assessment of anti-bacterial activity and cytotoxicity of silver contained antibacterial HA coating material. *Mater. Sci. Forum* **2009**, *620–622*, 307–310.

72. National Institute of Diabetes and Digestive and Kidney Diseases (NIDDK), National Institutes of Health (NIH) National Diabetes Statistics, **2011**. http://diabetes.niddk.nih.gov/dm/pubs/statistics/-fast/ (accessed 11 September 2014).

73. Khafagy, E.-S.; Morishita, M.; Onuki, Y.; Takayama, K., Current challenges in non-invasive insulin delivery systems: A comparative review. *Adv. Drug Deliv. Rev.* **2007**, *59*, 1521–1546.

74. Kumar, C. V.; Chaudhari, A., Proteins immobilized at the galleries of layered α-zirconium phosphate: Structure and activities studies. *J. Am. Chem. Soc.* **2000**, *122*, 830–837.

75. Horsley, S. E.; Nowel, D. V.; Stewart, D. T., The infrared and raman spectra of α-zirconium phosphate. *Spectrochim. Acta* **1974**, *30A*, 535–541.

76. Halliday, A. J.; Moulton, S. E.; Wallace, G. G.; Cook, M. J., Novel methods of antiepileptic drug delivery-polymer based implants. *Adv. Drug Deliver. Rev.* **2012**, *64*, 953–964.

77. Maeda, H., The enhanced permeability and retention (EPR) effect in tumor vasculature: The key role of tumor-selective macromolecular drug targeting. *Adv. Enzyme Regul.* **2001**, *41*, 189–207.

78. Cao, Z.; Tong, R.; Mishra, A.; Xu, W.; Wong, G. C. L.; Cheng, J.; Lu, Y., Reversible cell-specific drug delivery with aptamer-functionalized liposomes. *Angew. Chem. Int. Ed.* **2009**, *48*, 6494–6498.

79. Matsumura, Y.; Maeda, H., A new concept for macromolecular therapeutics in cancer che-motherapy: Mechanism of tumoritropic accumulation of proteins and the antitumor agent Smancs. *React. Polym.* **1986**, *5*, 13.

80. Bertini, I.; Gray, H. B.; Stiefel, E. L.; Valentine, J. S., Biological inorganic chemistry: Structure and reactivity. Books, U. S., Ed. Sausalito, CA, **2007**; pp. 95–104.

81. Heudi, O.; Cailleux, A.; Allain, P., Interactions between cisplatin derivatives and mobile phase during chromatographic separation. *Chromatographia* **1997**, *44*, 19–24.

82. Bose, R. N.; Cornelius, R. D.; Viola, R. E., Kinetics and mechanisms of platinum(II)-promoted hydrolysis of inorganic polyphosphates. *Inorg. Chem.* **1985**, *24*, 3989–3996.

83. Dhar, S.; Daniel, W. L.; Giljohann, D. A.; Mirkin, C. A.; Lippard, S. J., Polyvalent oligo-nucleotide gold nanoparticle conjugates as delivery vehicles for platinum (IV) warheads. *J. Am. Chem. Soc.* **2009**, *131*, 14652–14653.

84. Brewer, A. K., The high pH therapy for cancer test on mice and humans. *Pharmacol. Biochem. Behav.* **1984**, *21*, 1–5.

85. Kopf, H.; Kopf-Maier, P., Titanocene dichloride-the first metallocene with cancerostatic activity. *Angew. Chem. Int. Ed.* **1979**, *18*, 477–478.

86. Meléndez, E., Titanium complexes in cancer treatment. *Crit. Rev. Oncol. Hemat.* **2002**, *42*, 309–315.

87. Waern, J. B.; Harding, M. M., Coordination chemistry of the antitumor metallocene molybdocene dichloride with biological ligands. *Inorg. Chem.* **2004**, *43*, 206–213.

88. Manohari, P.; Harding, M. M., Antitumour bis(cyclopentadienyl) metal complexes: Titanocene and molybdocene dichloride and derivatives. *Dalton Trans.* **2007**, 3474–3482.

89. Meléndez, E., Metallocenes as target specific drugs for cancer treatment. *Inorg. Chim. Acta* **2012**, *393*, 36–52.

90. Lummen, G.; Sperling, S.; Luboldt, H.; Otto, T.; Rubben, H., Phase II trial of titanocene dichloride in advanced renal-cell carcinoma. *Cancer Chemoth. Pharm.* **1998**, *42* (5), 415–417.

91. Kroger, N.; Kleeberg, U. R.; Mross, K.; Edler, L.; SaB, G.; Hossfeld, D. K., Phase II clinical trial of titanocene dichloride in patients with metastatic breast cancer. *Onkologie* **2000**, *23*, 60–62.

92. Gasser, G.; Ott, I.; Metzler-Nolte, N., Organometallic anticancer compounds. *J. Med. Chem.* **2011**, *54*, 3–25.

93. Buck, D. P.; Abeysinghe, P. M.; Culliname, C.; Day, A. I.; Collins, J. G.; Harding, M. M., Inclusion complexes of the antitumour metallocenes Cp_2MCl_2 (M= Mo, Ti) with cucurbit[*n*]urils. *Dalton Trans.* **2008**, 2328–2334.

94. Waern, J. B.; Harding, M. M., Bioorganometallic chemistry of molybdocene dichloride. *J. Organomet. Chem.* **2004**, *689*, 4655–4668.

95. Ravera, M.; Gabano, E.; Baracco, S.; Osella, D., Electrochemical evaluation of the inter-action between antitumoral titanocene dichloride and biomolecules. *Inorg. Chim. Acta* **2009**, *362*, 1303–1306.

96. Malakoutikah, M.; Teixidó, M.; Giralt, E., Shuttle-mediated drug delivery to the brain. *Angew. Chem.* **2011**, *50*, 7998–8014.

97. Gavini, E.; Hegge, A. B.; Rassau, G.; Sanna, V.; Testa, C.; Pirisino, G.; Karlsen, J.; Giunchedi, P., Nasal administration of carbamazepine using chitosan microspheres: In vitro/in vivo studies. *Int. J. Pharm.* **2006**, *307*, 9–15.

98. Shityakov, S.; Broscheit, J.; Forster, C., α-Cyclodextrin dimer complexes of dopamine and levodopa derivatives to asses drug delivery to the central nervous system: ADME and molecular docking studies. *Int. J. Nanomed.* **2012**, *7*, 3211–3219.

99. Hayashi, A.; Yoshikawa, Y.; Ryu, N.; Nakayama, H.; Tsuhako, M.; Eguchi, T., Intercalation of biogenic amines into layered zirconium phosphates. *Phosphorus Res. Bull.* **2008**, *22*, 48–53.

100. Kumar, T. S.; Umamaheswari, S., FTIR, FTR and UV-Vis Analysis of Carbamazepine. *Res. J. Pharm. Biol. Chem. Sci.* **2011**, *2* (4), 685–693.

INDEX

Note: Page numbers in *italics* refer to Figures; those in **bold** to Tables.

acid–base chemistry, phosphonic acids
  acidity, 86, *86*
  bis-phosphonic acids, 87
  deprotonation processes, 84, *84*
  (poly)phosphonic acids, structures, 84, *85*
  p*K*a values, 84, **87**
  protonation constants, AMP, HEDP and
    DTPMP, 86, **86**
  stepwise protonation, 87, **87**
acrylamide (AAm), 267, 269
allotropic forms, ZrP, 47, *47*
3-amino-5-(dihydroxyphosphoryl)benzoic acid,
    90, *91*
aminomethylenephosphonic acids (AMP)
  complexation constants with metal cations, 88
  distribution, 86, *86*
  equilibrium constant values and titration curves, 88
  protonation constants, 86, **86**
  stability constants and carboxy analogues, 87, **88**
  structure, 84, *85*
amino-tris-(methylenephosphonates), 167
anticancer agents
  cisplatin *see* cisplatin
  doxorubicin *see* doxorubicin

enhanced permeability and retention, 410
  metallocenes *see* metallocenes
2-(-arylamino phosphonate)-chitosan (2-AAPCS),
    synthetic pathway, 102, *103*
atom transfer radical polymerization (ATRP), 324
auxiliary ligands
  classes, 194
  N-donor co-ligands
    coordination modes, with metal ions, 196, 198
    directionality and chelating properties, 196
    molecular structure, 196, 197
    non-covalent interactions, instauration, 196
    phosphonate-based materials, 196
  O-donor co-ligands
    carboxylic ligands, 198
    CPs formation, 198
    molecular structure, 198, *199*
    phosphonates and phosphinates, 198
aza-heterocyclic derivatives
  2,2′-bipyridine-5,5′-bisphosphonic acid, 370
  bis-imidazole derivative, 372
  Cd(4-O$_3$P–C$_5$H$_5$N)$_2$, 370
  channel structure, 370, *371*
  Co(4-O$_3$P–C$_5$H$_5$N)(H$_2$O), 370

Tailored Organic–Inorganic Materials, First Edition. Edited by Ernesto Brunet, Jorge L. Colón
and Abraham Clearfield.
© 2015 John Wiley & Sons, Inc. Published 2015 by John Wiley & Sons, Inc.